AF389090

50th Anniversary of the Metaphorical Butterfly Effect since Lorenz (1972): Multistability, Multiscale Predictability, and Sensitivity in Numerical Models

50th Anniversary of the Metaphorical Butterfly Effect since Lorenz (1972): Multistability, Multiscale Predictability, and Sensitivity in Numerical Models

Editors

Bo-Wen Shen
Roger A. Pielke Sr.
Xubin Zeng

Basel • Beijing • Wuhan • Barcelona • Belgrade • Novi Sad • Cluj • Manchester

Editors

Bo-Wen Shen
Department of Mathematics
and Statistics
San Diego State University
San Diego
United States

Roger A. Pielke Sr.
Cooperative Institute
for Research in
Environmental Sciences
University of Colorado
Boulder
United States

Xubin Zeng
Department of Hydrology
and Atmospheric Science
The University of Arizona
Tucson
United States

Editorial Office
MDPI
St. Alban-Anlage 66
4052 Basel, Switzerland

This is a reprint of articles from the Special Issue published online in the open access journal *Atmosphere* (ISSN 2073-4433) (available at: https://www.mdpi.com/journal/atmosphere/special_issues/predictability_numerical_models).

For citation purposes, cite each article independently as indicated on the article page online and as indicated below:

Lastname, A.A.; Lastname, B.B. Article Title. *Journal Name* **Year**, *Volume Number*, Page Range.

ISBN 978-3-0365-8910-7 (Hbk)
ISBN 978-3-0365-8911-4 (PDF)
doi.org/10.3390/books978-3-0365-8911-4

Contents

About the Editors

Bo-Wen Shen

Dr. Bo-Wen Shen holds the position of Associate Professor in the Department of Mathematics and Statistics at San Diego State University (SDSU). His academic journey includes earning a Ph.D. from North Carolina State University in 1998. Over his career, spanning more than 25 years, he has accumulated expertise in various areas such as mesoscale/global modeling, high-performance computing, the application of models in numerical weather prediction, and nonlinear dynamics. He has published more than 50 refereed journal articles and more than 50 technical or conference reports. Dr. Shen joined the modeling team at the NASA Goddard Space Flight Center (GSFC) in 1999, where he contributed to the development of a unified weather and climate model. During his earlier career at GSFC, he and his colleagues successfully utilized a global mesoscale model for real-time hurricane prediction, receiving recognition in notable publications like the American Geophysical Union, Science, and other media outlets. Between 2009 and 2015, Dr. Shen assumed the role of principal investigator for NASA's CAMVis (Coupled Advanced Global Modeling and Concurrent Visualization Systems) projects. During this time, he led initiatives to enhance the computational efficiency of the Goddard Multiscale Modeling Framework (MMF) and introduced three-level parallelism to the ensemble Empirical Mode Decomposition (EMD). In 2011, Dr. Shen shifted his focus to nonlinear dynamics and chaos theory, aiming to gain insights into the predictability of high-resolution global model simulations. His journey led him to join the faculty at SDSU in 2014. Since then, he has authored significant publications, including the development of a generalized Lorenz model. He and his co-authors presented a revised perspective on the dual nature of weather. This fresh perspective challenges the conventional notion that "weather is chaotic".

Roger A. Pielke Sr.

Over half a century, Professor Roger A. Pielke Sr. has played different roles, as a researcher, educator, and leader in atmospheric science and related fields. As one of the most accomplished atmospheric scientists in the world with an H-index of 100, he is the authority on the understanding and modeling of the land surface impact (e.g., land cover and land use change, surface heterogeneities) on weather and climate, the co-developer of the Regional Atmospheric Modeling System (RAMS), the author of the most popular textbook on mesoscale meteorological modeling, and a pioneer in interdisciplinary science. He has also provided exceptional service to our field, including three Chief Editor positions that require substantial time commitments (including the AMS' *Monthly Weather Review* and *Journal of the Atmospheric Sciences*). The Pielke research group focuses on land–atmosphere interactions on the local, meso-, regional, and global scales. These interactions include biophysical, biogeochemical, and biogeographic effects. The RAMS model is a major tool used in this research. RAMS has been coupled to two different ecosystem dynamics models (CENTURY and GEMTM) as part of these studies. Also applied is the CCM3 atmospheric global model from the National Center for Atmospheric Research (NCAR) in Boulder, Colorado. His studies range from the tropics into the high Arctic regions. His group has also applied RAMS to atmosphere–ocean interactions, including Arctic sea-ice feedbacks. He and his group members have investigated these nonlinear interactions within the Earth's climate system using the coupled RAMS model, as well as simplified nonlinear mathematical models.

Xubin Zeng

Dr. Xubin Zeng is the Agnese N. Haury Chair in Environment, Professor of Atmospheric Sciences, and Director of the Climate Dynamics and Hydrometeorolgy Collaborative at the University of Arizona. He is an affiliated professor of the Applied Mathematics, Global Change, and Remote Sensing and Spatial Analysis Interdisciplinary Programs. He also directs the Land–Atmosphere–Ocean Interaction (LAOI) Group. Through over 250 peer-reviewed papers, Dr. Zeng's research has focused on land–atmosphere–ocean interface processes, weather and climate modeling, hydrometeorology, remote sensing, nonlinear dynamics, and big data analytics. Dr. Zeng is a fellow of the American Meteorological Society (AMS) and served on its Council and Executive Committee. He received the AMS Charles Franklin Brooks Award for Outstanding Service to the Society in 2021. He is also a fellow of the American Geophysical Union (AGU) and American Association for the Advancement of Science (AAAS). He received the Special Creativity Award from the National Science Foundation and the Outstanding Faculty Award from the University's Asian American Faculty, Staff and Alumni Association. He received the Colorado State University Atmospheric Science Outstanding Alum Award. He was named a Distinguished Visiting Scientist at the NASA/Caltech Jet Propulsion Laboratory (JPL) in 2022. Currently, Dr. Zeng co-chairs the Scientific Steering Group of Global Energy and Water Exchange (GEWEX) with 25–30 active international projects, chairs the Science Team of Global Precipitation Experiment (GPEX)—a new World Climate Research Programme Lighthouse Activity, and chairs the Advisory Committee on Earth and Biological Sciences (with 650 people and USD 200M/year budget) at DOE PNNL.

Preface

The 50th anniversary of the metaphorical butterfly effect, celebrated in 2022, traces its origin back to the pioneering work of Lorenz, who, in 1963, rekindled the concept of sensitive dependence on initial conditions. In 1972, Lorenz introduced the term "butterfly effect" as a metaphor to discuss whether a tiny perturbation could (or could not) eventually create a tornado with a three-dimensional organized coherent structure. These seminal studies set the stage for chaos theory, sparking interdisciplinary explorations. While linked to the original butterfly effect based on sensitive dependence on initial conditions, its metaphorical variant carries distinct nuances, prompting debates. To mark this milestone, we launched a Special Issue, inviting contributions that deepen our grasp of both butterfly effects.

Lorenz's elegant yet simplified toy weather model, while not an exact representation of atmospheric dynamics, unveiled the inherent chaos within these systems, ultimately contributing to the Nobel Physics Prize awarded in 2021. This Special Issue was designed to nurture fundamental research through the application of theoretical and real-world models, as well as contemporary machine learning techniques, all of which aimed to address critical topics such as multistability, multiscale predictability, and sensitivity. The call for papers preceded this Issue by 15 months, yielding ten noteworthy contributions.

Among the intriguing subjects explored in these papers, a central focus is predictability and unique trajectory prediction in phase spaces, along with high-impact vortex-type weather phenomena in physical space. The image displayed below depicts an infrared METEOSAT 7 satellite snapshot from May 9, 2002, revealing two pairs of twin tropical cyclones in the Indian Ocean. These two cyclones in either hemisphere appear as the infinity symbol or the horizontal number eight (8). In non-linear dynamics, the non-disssipative Lorenz model produces homoclinic orbits resembling infinity symbols, while the classical or generalized Lorenz model exhibits chaotic attractors reminiscent of tilted number eights.

The selected papers are categorized into the following themes: butterfly effects and sensitivities, atmospheric dynamics and the applications with theoretical models, predictability and prediction, and computational and machine learning methods. In this Special Issue, we first provide a concise overview of Lorenz's models spanning from 1960 to 2008 and introduce the published papers. Just before the deadline, Professor W.-Y. Sun authored a comprehensive review on computational geophysical fluid dynamics. We aim to equip readers with a comprehensive understanding of advancements in the butterfly effect, multistability, almost intransitivity, multiscale predictability, and numerical sensitivity. The concept of time-varying multistability mirrors "changes in weather regimes" and "the edge of chaos". Finally, we express our gratitude to all of the authors and reviewers.

Bo-Wen Shen, Roger A. Pielke Sr., and Xubin Zeng
Editors

Image courtesy of The European Organisation for the Exploitation of Meteorological Satellites (EUMETSAT).
In METEOSAT 7 satellite imagery captured on May 9, 2002, infrared data reveals the presence of two pairs of twin tropical cyclones (TCs) accompanied by well-organized convection patterns in the Indian Ocean.
Documented by Shen et al. (2012): Genesis of Twin Tropical Cyclones as Revealed by a Global Mesoscale Model: The Role of Mixed Rossby Gravity Waves. J. Geophys. Res., 117, D13114, doi:10.1029/2012JD017450.

Editorial

The 50th Anniversary of the Metaphorical Butterfly Effect since Lorenz (1972): Multistability, Multiscale Predictability, and Sensitivity in Numerical Models

Bo-Wen Shen [1,*], Roger A. Pielke, Sr. [2] and Xubin Zeng [3]

1   Department of Mathematics and Statistics, San Diego State University, San Diego, CA 92182, USA
2   Cooperative Institute for Research in Environmental Sciences, University of Colorado Boulder, Boulder, CO 80203, USA; pielkesr@gmail.com
3   Department of Hydrology and Atmospheric Science, The University of Arizona, Tucson, AZ 85721, USA; xubin@arizona.edu
*   Correspondence: bshen@sdsu.edu

Abstract: Lorenz rediscovered the butterfly effect, which is defined as the sensitive dependence on initial conditions (SDIC), in 1963. In 1972, he used the term "butterfly" as a metaphor to illustrate how a small perturbation can lead to a tornado with a complex structure. The metaphorical butterfly effect, which celebrated its 50th anniversary in 2022, is not precisely the same as the original butterfly effect with SDIC. To commemorate the 50th anniversary, a Special Issue was launched and invited the submission of research and review articles that can help to enhance our understanding of both the original and metaphorical butterfly effects. The Special Issue also sought recent developments in idealized Lorenz models and real-world models that address multistability, multiscale predictability, and sensitivity. The call for papers was opened 15 months prior to the completion of the Special Issue and features nine selected papers. This editorial provides a brief review of Lorenz models, introduces the published papers, and summarizes each one of them.

Keywords: dual nature; chaos; generalized Lorenz model; predictability; multistability

Citation: Shen, B.-W.; Pielke, R.A., Sr.; Zeng, X. The 50th Anniversary of the Metaphorical Butterfly Effect since Lorenz (1972): Multistability, Multiscale Predictability, and Sensitivity in Numerical Models. *Atmosphere* **2023**, *14*, 1279. https://doi.org/10.3390/atmos14081279

Academic Editor: Anthony R. Lupo

Received: 22 May 2023
Accepted: 28 July 2023
Published: 12 August 2023

1. Introduction

A well-accepted definition of the butterfly effect is the sensitive dependence on initial conditions (SDIC), which was rediscovered by Lorenz in 1963 (Lorenz 1963a) [1]. Usage of the term "butterfly" appeared in 1972 when Lorenz applied a metaphor to discuss the possibility of whether a tiny perturbation may eventually create a tornado with a three-dimensional, organized coherent structure (Lorenz 1972 [2]). Since 1972, the metaphorical butterfly effect has received increasing attention and has become associated with the SDIC. Lorenz's work in 1963 and 1972 established the foundation for chaos theory, inspiring many studies in different fields. The metaphorical butterfly effect celebrated its 50th anniversary in 2022 (Glick 1987 [3]; Lorenz 1993 [4]).

Although Lorenz's 1963 theoretical model is considered a toy weather model that cannot accurately depict weather and climate, the model's simplicity allowed for the identification of chaotic weather and climate, leading to the widely accepted view that weather is chaotic. The announcement of the 2019 Nobel Physics Prize cited Lorenz's pioneering chaos study as a foundation for the awarded studies, which proposed physics-based mathematical models for improving our understanding of predictability in complex systems [5]. The proposed physics-based mathematical models are capable of revealing crucial, fundamental physical processes, including co-existing rapidly and slowly varying systems and the metastability of spin glass [6]. By extending Lorenz's 1963 model in order to improve our understanding of predictability in complex systems such as weather and climate, which possess co-existing attractors [7], Shen et al. (2021a, b) [8,9] proposed a revised view in that "weather possesses chaos and order; it includes emerging organized systems (such as tornadoes) and recurrent seasons", in contrast to the conventional view of "weather is chaotic".

To celebrate the 50th anniversary of the metaphorical butterfly effect and to promote fundamental research through the use of both theoretical and real-world modeling, as well as modern machine learning technology, a Special Issue was established, seeking investigations into various topics related to multistability, multiscale predictability, and sensitivity. These topics included:

1. The validity of existing analogies and metaphors for butterfly effects.
2. The insightful analysis of various Lorenz models reveals the role of monostability with single types of solutions and multistability with attractor co-existence in contributing to the multiscale predictability of weather and climate.
3. The development of conceptual, theoretical, and real-world models to reveal fundamental physical processes and multiscale interactions that contribute to our understanding of the butterfly effect on the predictability of weather and climate.
4. Innovative machine learning methods that (1) classify chaotic and non-chaotic processes and identify weather and climate systems at various spatial and temporal scales (e.g., sub-seasonal to seasonal time scales) and (2) detect computational chaos and saturation dependence on various types of solutions.
5. The impact of tiny perturbations on emergent pattern formation with self-organization (e.g., stripes and rolls), the formation of high-impact weather (e.g., tornados and hurricanes), etc.

Nine papers, covering topics related to the first four categories, were published during the 15-month period of the Special Issue. The published papers are summarized herein based on the four categories: (A) butterfly effects and sensitivities, (B) atmospheric dynamics and the application of theoretical models, (C) predictability and prediction, and (D) computational and machine learning methods.

To complement the contributions of the published papers to the goals of the Special Issue, in Section 2, we first provide a brief review of Lorenz's models between 1960 and 2008. Further mathematical analysis of the Lorenz models can be found in a companion article by Shen (2023) [10]. Thereafter, Section 3 provides a summary of the papers. Appendix A briefly compares the mathematical similarities of the Lorenz 12-variable model [4] and the many-mode model, and Appendix B documents the relationship between the logistic map and the logistic differential equation used by Lorenz in the 1960s. For detailed definitions of the selected concepts, please refer to Table A1 in Appendix C.

2. A Brief Review of Lorenz Models and Butterfly Effects

2.1. A Review of Lorenz Models

In this paper, we provide an overview of Prof. Lorenz's mathematical models that illustrate chaotic and unstable features in the atmosphere. The primary emphasis is on models that present two key concepts, namely chaos and almost intransitivity (for further details, please refer to Table A1 in Appendix C), both of which are related to predictability. Lorenz published 61 papers during his lifetime, with 58 of them being single authored (Chen 2020 [11]). Tables 1 and 2 summarize his mathematical models in the 1960s and between 1970 and 2008, respectively.

Table 1. A list of Lorenz models in the 1960s. * By a strict definition, the Lorenz 1969 model cannot be categorized as a turbulence model, but it does illustrate the multiscale turbulent features of the Atmosphere.

Year	1960	1960/62	1963	1964	1965	1969	1969
Equations	3 ODEs,	12 ODEs	3 ODEs,	Logistic map	28 variables	Logistic ODE	21 2nd-order ODEs,
Origins	PDE; vorticity Eq	PDEs; a 2-layer, QG model	PDEs; convection		PDEs; a 2-layer, QG model		PDE; vorticity Eq
Features of Solutions	oscillatory solutions with elliptic functions	irregular fluctuations	chaos	steady, periodic, non-periodic	irregular solutions	error growth and saturation	'turbulence' *

Table 2. A list of Lorenz models between 1970 and 2008.

Year	1972	1976	1980	1984	1986	1996/2006	2005	2008
Equations	turbulence models	cubic map	low-order PE or QG (9 or 3) ODEs	3 ODEs	5 or 3 ODEs	N ODEs	N ODEs	Henon map
Origins	PDE based		PDE based	Not PDE based	PDE based, (the 1980 model)	Not PDE based	Not PDE based	
Features of Solutions	turbulence	transitivity	(modified) shallow water Eqs.	general circulation, transitivity	slow manifolds (w elliptic functions)	chaos	chaos	chaos

2.1.1. Lorenz Models in the 1960s

In 1960, Lorenz proposed a low-order system of ordinary differential equations (ODEs) based on a single partial differential equation (PDE) for the conservation of vorticity (Lorenz 1960 [12]). This system, called Maximum Simplification of the Dynamic Equations, produced oscillatory solutions as elliptic functions ([12,13]). As discussed below, Lorenz's 1960 model also appeared as a simplified version of the Lorenz 1986 system for studying slow variables. In the same year, Lorenz reported "irregular solutions" based on a system of twelve ODEs, derived from the geostrophic form of the two-layer baroclinic model, which is discussed below. The system includes two variables that represent the average and the difference of the stream functions for the two layers. The second variable was identified as the temperature field through the thermal wind relationship. To express each variable, six Fourier modes with six coefficients were used, resulting in a system of twelve ODEs with twelve mathematical variables but only two physical fields (i.e., a stream function and temperature). The mathematical equations are presented in Equation (A1) in Appendix A. In 1960, Lorenz presented the "irregular solutions", which were later published in a conference proceeding in 1962 [14], at the International Symposium on Numerical Weather Prediction in Tokyo.

Motivated by the irregular solutions observed within the Lorenz (1960/1962) model, Lorenz simplified the Saltzman seven-variable model [15,16] into a three-variable model, which is known as the Lorenz 1963 model with three ODEs [1,17]. Since then, thanks to its rich array of features, the model has been valuable in explaining the nonlinear and chaotic dynamics of diverse fields [1,3,4].

Within a continuous dynamical system, three ODEs are required for the appearance of chaos. While the Lorenz 1963 model represents one such simple system for producing chaos, a simple logistic map (which is a difference equation) has also been used to illustrate chaos (e.g., May 1976 [18]; Li and Yorke 1975 [19]). Although cited in [18], Lorenz (1964) [20], indeed, applied the logistic map in order to illustrate irregular solutions earlier than May (1976). Lorenz (1964) specifically reported the transition from steady-state to periodic to non-periodic behavior as the control parameter increases. Simply based on the findings of Lorenz (1962, 1963a, and 1964), we may conclude that Lorenz was fully aware of the potential of Saltzman's model in illustrating chaos. As compared to the logistic map in a discrete form, the continuous form of the logistic equation was applied in order to illustrate the monotonic increase and the saturation of errors (e.g., Lorenz 1969a [21]). The relationship between the logistic map and the logistic ODE is briefly documented in Appendix B.

The aforementioned low-order models were constructed in order to illustrate various atmospheric phenomena which are intrinsically nonlinear (Lorenz 1982a [22]). During the early 1960s, numerous multi-layer models existed. One of the two-layer models was developed by Lorenz in 1960 (Lorenz 1960 [23]). The geostrophic form of the two-layer model was applied in order to derive the Lorenz 1962 8-variable model [24] and the Lorenz 1960/1962 12-variable model and it was subsequently changed to become a 14-variable model for investigating vacillation in a dishpan, as described in Lorenz's work in 1963 (Lorenz 1963b [25]). Within the eight-variable model, Lorenz was able to obtain analytical solutions to describe essential features of the transition between the Hadley and Rossby regimes. Building upon these endeavors, Lorenz proposed a 28-variable model in 1965 to explore non-periodic solutions (Lorenz 1965 [26]). The four quasi-geostrophic (QG) models are also referred to as the Lorenz 1962-8v, 1960/1962-12v, 1963-14v, and 1965-28v models. Whilst the 1960/1962-12v model and the 1965-28v model are listed in Table 1, the Lorenz 1960 two-layer model, the 1962-8v model, and the 1963-14v model are not listed in Table 1.

The Lorenz 1965 28-variable model is a highly simplified QG model used for capturing some of the gross features of atmospheric behavior and contains two physical variables, the stream function (i.e., vorticity) and potential temperature. This model yielded a significant finding of flow-dependent predictability: *"The time required for errors comparable to observational errors in the atmosphere to grow to intolerable errors is strongly dependent upon the current circulation pattern, and varies from a few days to a few weeks."* This finding implies that the predictability limit depends on a synoptic situation (Lorenz 1982b [27]). In fact, some results obtained using the 28-variable model were also reported in Lorenz (1963c) [28].

While this review primarily focuses on the features of Lorenz models, it is worth mentioning the following important "analysis method." In Lorenz (1965), the growth of small errors, such as during the linear stage, was analyzed using the singular value (SV) method (Kalnay 2002 [29]; Lewis 2005 [30]). This method required the calculation of the system's Jacobian matrix (for a linear tangent model) and its singular values (for growth rates). Such a method was applied to determine the local or finite-time Lyapunov exponent over a finite time interval (Nese 1989 [31]; Abarbanel et al., 1992 [32]; Eckhardt and Yao 1993 [33]; Krishnamurthy 1993 [34]; Szunyogh et al. 1997 [35]; Yoden 2007 [36]). On the other hand, Oseledec (1968) [37] independently proposed a general approach for calculating the (global) Lyapunov exponent over an infinite time interval. The singular value method was also employed in the systems of the European Centre for Medium-Range Weather Forecasts (ECMWF) (Molteni et al., 1996 [38]; Buizza et al., 2008 [39]). In a recent study conducted by Cui and Shen (2021) [40], the relationship between the SV decomposition (SVD), principal component analysis (PCA), and kernel PCA methods was documented. Furthermore, they deployed kernel PCA to identify co-existing attractors within the phase space. Empirical orthogonal function (EOF), which was initially introduced into the field of meteorology by Lorenz in 1956 [41], is another term commonly used to denote PCA.

A two-layer, quasi-geostrophic model was also applied by Pedlosky in the 1970s, referred to as the Pedlosky model (Pedlosky 1971, 1972, 1987 [42–44]), in order to derive a low-order system to study nonlinear baroclinic waves. Although the Pedlosky and Lorenz 1963 models were derived from very different PDEs with different physical processes, research indicates that two systems of ODEs can be mathematically identical when time-independent parameters are properly selected (Pedlosky and Frenzen, 1980 [45]; Shen 2021 [46]). Similar to the Lorenz 1963 model, the Pedlosky model also produced three types of solutions, including steady-state, chaotic, and limit cycle solutions.

2.1.2. Lorenz Models between 1970 and 2008

While Lorenz 1963 has been highly cited within the nonlinear dynamics community, the Lorenz 1969 many-mode model (Lorenz 1969b [47]) is better known by individuals within the meteorology community. Similar to the Lorenz 1960 model, the Lorenz 1969 model was also derived based on the conservation of vorticity. Since the conservative PDE does not include friction, as compared to the turbulence models proposed by Leith 1971 [48], Leith and Kraichnan (1972) [49], and Lorenz himself in 1972 [50,51], the Lorenz 1969 model is not a turbulence model. Thus, while the Lorenz 1963 model was simplified from the Saltzman model in order to qualitatively illustrate the chaotic nature of the atmosphere, the Lorenz 1969 model was proposed to qualitatively illustrate multiscale interactions of the "turbulent" atmosphere. As a result of its importance, as summarized in Section 3, detailed features of the Lorenz 1969 model have recently been documented. To illustrate the concept of "intransitivity" highlighted in Section 2.1.3, Lorenz (1976) [52] employed a cubic map. This cubic map can be obtained by substituting a cubic term for a quadratic term in Equation (B4), as found in Appendix B.

Based on a shallow water equation, Lorenz 1980 [53] obtained a low-order system of nine ODEs, referred to as the nine-variable barotropic primitive-equation (PE) model (Lorenz 1982a [22]), to describe the evolution of the velocity potential, stream function, and the height of a free surface. By applying a quasi-geostrophic assumption, Lorenz showed that a simplified version of the low-order system can be mathematically identical to the Lorenz 1963 model. In 1986, by applying the Lorenz 1980 low-order system, Lorenz (1986) [54] proposed a system of five ODEs in order to illustrate the features of the slow manifold. Using different initial conditions, Lorenz discussed two types of solutions: (1) a linear gravity wave solution for two fast variables and (2) a nonlinear Rossby wave with a dependence of oscillatory frequency on the amplitude of the wave for three slow variables, U, V, and W. As a result, the term "slow" indicates a state of the system where time scales are sufficiently separated and fast gravity modes are effectively separated. Both solutions are non-chaotic. Lorenz (1986) did not present the concept of co-existing chaotic and non-chaotic solutions. For a detailed discussion on the absence of universal slow-manifold evolution in rotating stratified fluids, the readers may find the studies by Lorenz and Krishnamurthy (1987) [55], Lorenz (1992) [56], and McWilliams (2019) [57] of interest. These studies delve into the subject matter and provide further insights.

Table 3 compares the "subsystem" of a Lorenz 1986 model for the three slow variables and two simplified versions of the Lorenz 1963 model [58–60]. The first two models, listed in the second and third columns, have closed-form solutions with elliptic functions, showing the modulation of frequency within nonlinear periodic solutions. Although the models can produce periodic solutions, they cannot generate limit cycle solutions because there are no dissipative terms. A limit cycle is defined as an isolated closed orbit within the phase space [8]. A system with U, V, and W variables (e.g., Table 3) is comparable to Equations (15)–(17) in Lorenz (1960)l, whose solutions are also the elliptic functions dn, sn, and cn (see Table 3).

Table 3. A comparison of the Lorenz 1986 model for slow variables with two simplified versions of the Lorenz 1963 model (Sparrow 1982 [58]; Shen 2018 [59]).

	The Lorenz (1986) Model	The Limiting Equations	The Non-Dissipative Lorenz Model
References	Lorenz (1986)	Sparrow(1982)	Shen (2018)
Equations	$\frac{dU}{dt} = -VW,$ $\frac{dV}{dt} = UW,$ $\frac{dW}{dt} = -UV.$	$\frac{dX}{d\tau} = \sigma Y,$ $\frac{dY}{d\tau} = -XZ,$ $\frac{dZ}{d\tau} = XY.$	$\frac{dX}{d\tau} = \sigma Y,$ $\frac{dY}{d\tau} = -XZ + rX,$ $\frac{dZ}{d\tau} = XY.$
Solutions	$U = U^* cn(W^*T)$ $V = V^* sn(W^*T)$ $W = W^* dn(W^*T)$	$X = \sqrt{2}\sigma cn(\sqrt{\sigma}\tau + 3K)$ nonlinear periodic orbits	three types of solutions, including two types of periodic solutions and homoclinic orbits
Remarks	$T = T(t)$	See Shen (2018) for details	See Shen (2018) for details

To depict atmospheric circulation, Lorenz (1984) [61] proposed a different system of three ODEs, viewed as the simplest general circulation model (GCM). In the past, the model has been applied in order to qualitatively illustrate features of the atmosphere (e.g., chaotic winters and non-chaotic summers (Lorenz 1990 [62]), the long-term variability of the climate (Pielke and Zeng (1994) [63]), etc.). To help readers understand the validity of the model and related findings, the following comments are provided. Lorenz (1990) [62] documented that (1) the system was constructed in a somewhat ad hoc manner (i.e., a lack of physical foundations) and (2) there is no assurance that the system produces bounded solutions. Although a recent study has provided a method to derive the Lorenz 1984 model [64,65], transformations of variables made it difficult to interpret the physical meanings of variables within the 1984 system. As compared to typical, fully dissipative systems where the time change rate of the volume of the solutions is negative, the volume of the solution within the 1984 model does not necessarily shrink to zero. Specifically, Lorenz (1991) [66] stated that: (1) *We do not claim or intend that it (i.e., the Lorenz 1984 model) should reproduce atmospheric behavior in any quantitative sense.* (2) *The point is not that the model is a good representation of the atmosphere.*

Until 1996/2006, as indicated by the title of Lorenz (1996, 2006) [67,68], Lorenz acknowledged that the predictability problem has not been fully solved. While the Lorenz 1963 model is a forced, dissipative, but limited-scale model, the Lorenz 1969 model is a linear, multiscale system with no dissipations. In 1996, to discuss growth rates at different scales and their saturations, Lorenz [67] proposed a many-mode "nonlinear" model that included dissipations and forcing terms. Lorenz (1996) was a technical report and was later published as a book chapter in 2006 (Lorenz 2006 [68]). As shown in Equation (A2) in Appendix A, the system contains nonlinear quadratic terms for advection and linear terms for dissipative and forcing terms. The system produces chaotic responses. However, without providing detailed justifications for how rescaling was performed, the coefficients for all terms at all scales were one (e.g., Equation (1) in Lorenz 1996). As compared to the Lorenz 1960/1962-12v model (e.g., Equation (A1) in Appendix A) and the Lorenz 1969 model, as well as the generalized Lorenz model (Shen 2019 [7]), the coupling terms (or the nonlinear quadratic terms), which originate from mode–mode interactions, should not have the same coefficients. Similarly, coefficients for the dissipative terms should not be the same.

Although a new chaotic many-mode system was proposed in order to address the predictability problem in 1996, Lorenz rescaled the variables to make the time unit (i.e., a time scale) equal to 5 days and chose a time step of 0.05 units that is equivalent to 6 h (page 5 of Lorenz 1996). The time scale of 5 days is different from the time scale of 6 days used by Lorenz (1969b), who determined the time scale by assuming a velocity scale of 17.2 m/s. Thus, the impact of a selected time scale on the estimate of predictability horizons within idealized models (that may or may not be physics based) should be examined. While two copies of the Lorenz 1996 model (i.e., Equation (1) in Lorenz 1996 or Equation (A2) in Appendix A) were applied in order to construct a two-layer system (i.e., Equations (2) and (3) in Lorenz 1996), additional linear coupling terms were included, calling for additional justifications for the choice of linear coupling terms that should be consistent with existing nonlinear quadratic terms. Other chaotic models in Lorenz (2005) [69] shared similar pros and cons with the 1996 model. Lorenz (2006) [70] utilized the Lorenz 1996/2005 model to demonstrate the existence of regimes that occur when the external forcing parameter (e.g., F in Equation (A2)) exceeds its critical value for the onset of chaos. In the last year of his life, Lorenz (2008) [71] published a paper based on the Hénon map which is a second-order difference equation with a quadratic term. A PDE-based classification of Lorenz models is provided in Table 4.

Table 4. The PDE-based classification of Lorenz models.

Quasi-geostrophic (QG) System	1962-8v, 1960/1962-12v, 1963-14v, 1965-28v
Conservative Vorticity Equation	1960, 1969
Rayleigh Benard Convection Equations	1963 (& Generalized Lorenz Models)
Shallow Water Equations	1980, 1986
No PDEs	1984, 1996, 2005
(Discrete) Maps	1964 (Logistic), 1976 (Cubic), 2008 (Henon)

2.1.3. Transitivity, Intransitivity, and Almost Intransitivity

While chaos with SDIC and linearly unstable solutions for multiscale predictability have been well explored using Lorenz's original models, the concept of "intransitivity" has been suggested and explored using different models since the 1960s (Lorenz, 1968 [72], 1975 [73], 1976 [52], 1982c [74], 1990 [62], and 1997 [75]). The idea of "intransitivity" addresses whether or not a specific type of weather or climate regime may last forever. In mathematical terms, the relation is classified as transitive if, whenever it establishes a link between A and B, as well as a separate link between B and C, a connection is implied between A and C. Thus, a relation is called intransitive if it is not transitive. When examining a system with two distinct states or regimes (such as warm and cold weather characterized by positive and negative temperature anomalies, respectively), if either warm or cold weather indefinitely persists, the system demonstrates "intransitivity". In other words, transitions from one state to another state are impossible. On the other hand, if the two weather regimes are alternate, the system exhibits "transitivity". The third scenario, known as "almost intransitivity", occurs when a regime change is possible but happens infrequently, falling between the two aforementioned scenarios.

Therefore, drawing from the preceding information, we can observe that "transitivity", "almost intransitivity", and "(complete) intransitivity" are demonstrated by the occurrence of "frequent", "rare", and "no" changes in the signs of solutions (i.e., oscillations between two distinct regimes), correspondingly. As shown in Figure 1, these features were effectively illustrated by Lorenz (1976), who applied a cubic map, $X_{n+1} = \alpha \left(3X_n - 4X_n^3\right)$ (i.e., a cubic finite difference equation, e.g., Holmes (1979) [76] and May (1984) [77]). The parameter α symbolizes a combined effect of linear forcing and dissipations, referred to as effective linear forcing. Additionally, the ratio between the coefficients of linear and nonlinear components determines the level of nonlinearity.

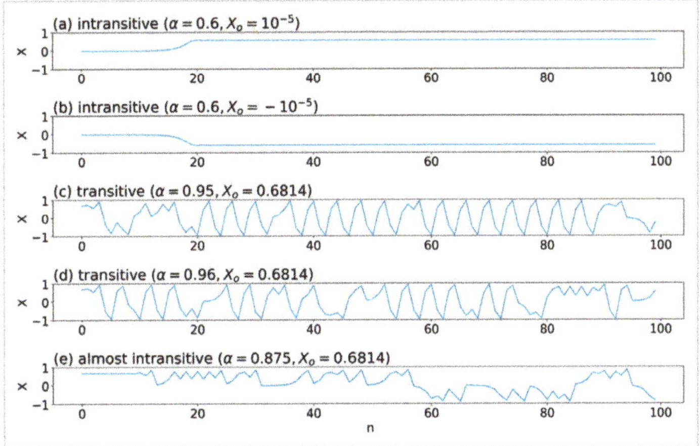

Figure 1. Results obtained using the Lorenz (1976) cubic map, $X_{n+1} = \alpha \left(3X_n - 4X_n^3\right)$, to illustrate intransitivity (**a**,**b**), transitivity (**c**,**d**), and almost intransitivity (**e**). X_o indicates an initial condition. Each panel may have a different value of α and/or X_o.

As shown in Figure 2 for the bifurcation diagram of the cubic map, when the parameter α is greater than $1/3$, the model displays a pitchfork bifurcation that introduces two non-trivial equilibrium points. Such non-trivial equilibrium points become unstable when α is greater than $2/3$. Then, each critical point experiences period doubling bifurcations, yielding a chaotic attractor (De Oliveira et al., 2013 [78]). Transitivity arises from the merger of two chaotic attractors [79,80] when α exceeds the critical threshold $\alpha_c = \sqrt{3}/2$. Lorenz (1976) suggested that while transitivity arises when α is greater than α_c, almost intransitivity occurs when α slightly surpasses α_c.

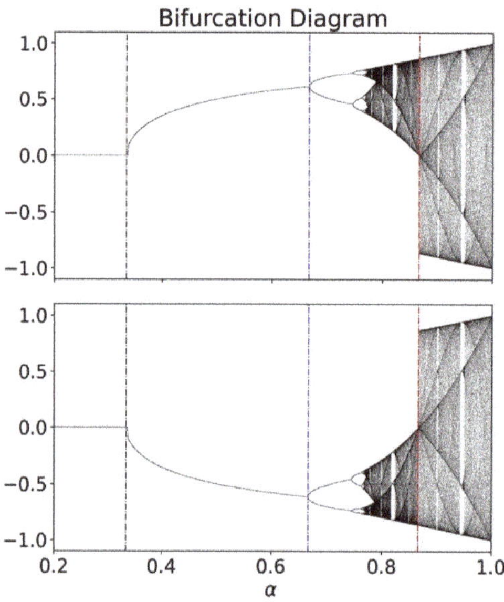

Figure 2. Bifurcation diagrams for the Lorenz 1976 cubic map. The black, blue, and red vertical dashed lines indicate the values of $\alpha = 1/3$, $\alpha = 2/3$, and $\alpha = \sqrt{3}/2$, displaying critical values for the pitchfork bifurcation, the appearance of unstable non-trivial equilibrium points, and the attractor merging crisis, respectively. The top and bottom panels have initial conditions of $X_0 = 10^{-5}$ and $X_0 = -10^{-5}$, respectively.

In the chaotic regime of the Lorenz 1963 model, which consists of one saddle point and two unstable equilibrium points, as evidenced by the presence of a persistent chaotic solution, the model demonstrates "transitivity" in the phase space. According to Devaney's definition of chaos [81], topological transitivity, SDIC (sensitive dependence on initial conditions), and dense periodic orbits are the three conditions that define chaos. In comparison, the concept of "intransitivity" and "almost intransitivity" emerges as an intriguing topic within systems that allow for the co-existence of two chaotic attractors associated with two non-trivial equilibrium points (such as the Lorenz 1976 cubic map), the co-existence of two oscillatory solutions (like the Lorenz 1984 model), and/or various types of attractor co-existence (such as the generalized Lorenz model [7]).

For example, given time-independent parameters, co-existing attractors may occupy different portions of the phase space. In such cases, both transitivity and "almost intransitivity" become possible when one or more parameters change over time. According to the following description of an intransitive system by Lorenz (1968) [72]:

> *"There are two or more sets of long-term statistics, each of which has a greater-than-zero probability of resulting from randomly chosen initial conditions"*

the intransitivity and the final state sensitivity (Grebogi et al. (1983) [82]; Table A1) are related. On the other hand, considering the presence of chaotic solutions within a transitive regime, such as those observed in the Lorenz 1963 model, whether or not "almost intransitivity" could manifest during a specific time period is an intriguing question.

When the concepts of "intransitivity" and "almost intransitivity" were initially introduced to explore the potential longevity of different climate regimes, if they exist, investigating regime changes on weather or short-term climate scales may also be valuable. For instance, Charney and DeVore (1979) [83] proposed that the presence of multiple, stable steady-state solutions in a low-order QG system could lead to the emergence of atmospheric blocking in the form of a stable steady-state solution (e.g., Chen and Xiong 2016 [84]). This phenomenon arises from the interplay of nonlinearity and mechanical and thermal forcing. Understanding the factors that govern the transitions between zonal and blocked atmospheric circulations, as well as the persistence of these regimes, remains an active area of research (Faranda et al., 2016 [85]; Dorrington and Palmer 2023 [86]). At longer time scales, transitions between El Niño and La Niña events exhibit varying durations for each regime ([87], Wallace et al. [88]), which raises scientific inquiries regarding the factors influencing the frequency and duration of these transitions. This prompts the exploration of whether the notions of "almost intransitivity" or "transitivity" can be applicable in describing such transitions and understanding the conditions (i.e., nonlinear and/or chaotic dynamics) that govern them.

2.1.4. Analogues and Recurrence

Lorenz's models revealed the potential existence of multiple attractors, such as co-existing chaotic attractors in Lorenz (1976) and co-existing periodic oscillations in Lorenz (1984). However, despite these findings, the prevailing assumption regarding the presence of a single dominant chaotic attractor has been implicitly applied in order to explain why identical weather systems are rare (e.g., Van Den Dool 1989 and 1994 [89,90]). Such an explanation gained widespread acceptance due to the fact that neighboring trajectories in chaotic systems, such as the Lorenz 1963 model, do not interact within the phase space.

However, it is important to note that nearby oscillatory trajectories, which eventually reach a steady state, do not intersect prior to reaching their final destination. Additionally, as illustrated in Figure 3 in Faghih-Naini and Shen (2018) [91], who applied the non-dissipative five-dimensional Lorenz model, a quasi-periodic solution with multiple incommensurate frequencies does not form a closed orbit but rather a torus. Therefore, when considering the temporal and spatial variations of forcing and dissipation processes, which are often neglected in autonomous systems, it becomes essential to consider "recurrence" instead of "periodicity". As defined in Thompson and Stewart (2002) [92] and studied using high-dimensional Lorenz models by Reyes and Shen (2019, 2020) [93,94], recurrence refers to a trajectory returning to the vicinity of its previous location. Recurrence can be seen as a more inclusive concept that encompasses quasi-periodicity with multiple frequencies and chaos.

In fact, the idea of recurrence can be compared to the quasi-periodicity introduced in Lorenz's earlier studies during the 1960s (e.g., Lorenz 1963a and 1963c [28]). Lorenz defined a "quasi-periodic flow" as:

> *"A flow with no transient component eventually comes arbitrarily close to assuming a state which it has assumed before, and the history following the latter occurrence remains arbitrarily close to the history following the former."*

Additionally, the notion of "recurrence" aligns well with Lorenz's concept of "analogues" (Lorenz 1969a [21], 1973 [95]), defined as follows:

> *"Analogues are two states of the atmosphere that exhibit resemblance to each other. Either state in a pair of analogues can be considered equivalent to the other state plus a small superposed 'error'."*

While weather may not repeat itself exactly, the idea of "analogues" or "recurrence" is also consistent with the classification of various weather systems such as quasi-biennial oscillations (QBOs), atmospheric blocking, African Easterly Waves (AEWs), and others

(Shen et al., 2021a,b and references therein). The natural occurrence of analogues, i.e., similar weather situations, has been applied as an empirical approach for estimating growth rates (Lorenz 1969c [96]). When two states are comparable, their differences are viewed as an error, and the growth of the error can be estimated by observing the evolution of the two states. Lorenz (1969a) reported both quasi-exponentially growing and decaying systems. While Lorenz's analogue approach in a follow-up study (Lorenz 1973) suggested a predictability horizon of at least 12 days for instantaneous weather patterns, recent advances in nonlinear dynamics, including attractor co-existence and almost intransitivity, can serve as a basis for understanding the potential for studying predictability at sub-seasonal, seasonal, or climate scales.

2.1.5. Simplifications and Generalizations of Lorenz Models

The three primary categories of chaotic systems include Hamiltonian systems (such as the three-body problem of Poincare [97,98]), forced and dissipative differential ODEs, and difference equations (like May 1976 and Li and Yorke 1975). Lorenz models can be categorized as belonging to the second and third categories. As such, numerous studies have been conducted over the past five decades, drawing inspiration from Lorenz's previously mentioned models. These studies have explored mathematical and physical simplifications and the generalizations of Lorenz models (e.g., Curry 1978 [99]; Curry et al., 1984 [100]; Howard and Krishnamurti 1986 [101]; Hermiz et al., 1995 [102]; Thiffeault and Horton 1996 [103]; Musielak et al., 2005 [104]; Roy and Musielak 2007a, b, c [105–107]; Moon et al., 2017 [108]), resulting in various noteworthy findings. Some of these include mathematical equivalence amongst the non-dissipative Lorenz 1963 model, the Korteweg–de Vries (KdV) equation in the traveling wave coordinate, the nonlinear Schrodinger equation for the amplitude of a traveling wave, the SIR model with weakly nonlinear assumptions, and the inviscid Pedlosky model (Shen 2020, 2021 [46,60]; Paxson and Shen 2022 [109]). SIR represents the three categories of individuals: susceptible (S), infected (I), and recovered (R) in that order. Consequently, such simplified systems have proven effective in revealing the dynamics of solitary waves, homoclinic orbits, epidemic waves, and nonlinear baroclinic waves. As a result of orbital instability and the unique sensitivity to initial conditions exhibited by homoclinic orbits, one could argue that the conservative, non-dissipative Lorenz 1963 model generates limited chaos.

Recently, Saiki et al. (2017) [110] proposed mathematical unification of the Lorenz 1963, 1984, and 1996 models. These systems, whether of dimension 3 or N, share quadratic characteristics and possess a generalized Lyapunov function that ensures bounded trajectories, with all trajectories entering a bounded trapping region. Furthermore, in terms of generalizing the Lorenz 1963 model, Shen (2019a, b) [7,111] proposed an expanded version that incorporates the following features: (1) any odd number of state variables greater than three, (2) the presence of three common types of solutions (steady-state, chaos, and limit cycle solutions), (3) two kinds of attractor co-existence, (4) aggregated negative feedback, (5) hierarchical spatial scale dependence, and (6) energy conservation under the dissipationless condition.

The presence of two types of attractor co-existence makes the generalized model suitable for investigating transitivity and intra-transitivity. Revealing the predictability of different attractors can be achieved through the effective detection of attractors, which involves determining their boundary (referred to as the basin boundary; Cui and Shen 2021 [40]). The concept of "edge of chaos" suggests that the interplay of chaos and order could lead to complexity as well as the emergence of self-organized phenomena [112,113]. The simplest system for illustrating the edge of chaos is the self-adjusting logistic map [114], i.e., a logistic map with a time-varying control parameter. Thus, the generalized Lorenz model with time-varying parameters can be applied to illustrate the transition between regular and chaotic solutions.

The Lorenz 1963 model revealed not only SDIC (i.e., chaos) but also the intriguing discovery of fractal geometry within its solution (Lorenz 1963a, 1993; Gleick 1987; Palmer

2009 [115]; Emanuel 2011 [116]; Feldman 2012 [117]). Fractals represent geometric objects that exhibit self-similarity. Various techniques exist for calculating fractal dimensions, employing diverse mathematical definitions and types of fractal dimensions (Grassberger and Procaccia 1983 [118]; Nese et al., 1987 [119]; Ruelle 1989 [120]; Zeng et al., 1992 [121]). One such measure is the Kaplan–Yorke dimension (Kaplan and Yorke 1987 [122]), which requires the computation of Lyapunov exponents (Wolf et al., 1985 [123]; Shen 2014 [124]). Consequently, these dimensions are influenced by system parameters, such as the Rayleigh parameter. By considering typical parameter values (e.g., Shen 2014 and 2015 [125]), the fractal dimension of the Lorenz 1963 model was determined to be 2.06, indicating a dependence amongst the three selected Fourier models. The five-, six-, and seven-variable Lorenz models were derived by extending the nonlinear feedback loop, incorporating five, six, and seven Fourier modes, respectively (commonly referred to as the five-, six-, and seven-dimensional Lorenz models, Shen 2014, 2015, and 2016 [124–126]). These higher-dimensional Lorenz models exhibit fractal dimensions, and the 7D or higher-dimensional Lorenz models clearly demonstrate hierarchical scale dependence (Shen 2016 and 2017 [126,127]).

2.1.6. Error Growth Analysis Using the First- and Second-Order ODEs

Throughout his entire career, Lorenz employed the logistic differential equation (e.g., Equation (A3) in Appendix B) in order to examine errors, which became known as the Lorenz error growth model (Lorenz 1969a [21], 1982 [27]; Nicolis 1992 [128], Zhang et al. (2019) [129]). This equation is a first-order ODE containing a quadratic nonlinear term (e.g., X^2, with X representing a variable). Lorenz (1969a) provided explanations for this nonlinear term and referred to his assumption as the quadratic hypothesis. While his analogue approach resulted in a doubling time of 8 days (Lorenz 1969a), the logistic ODE yielded an estimated doubling time of 2.5 days. To account for the disparities, he substituted the quadratic term (X^2) with a cubic term (X^3), which produced a different doubling time of 5 days. However, Lorenz remained cautious and acknowledged that the choice of the cubic term had less justification. In fact, later, Lorenz (1982a) suggested that results with the quadratic hypothesis are "reasonable but not readily verifiable". Apart from discrepancies arising from different nonlinear terms (such as quadratic and cubic terms), variations can also emerge when applying different types of ODEs (with the same initial growth rate), including first and second-order ODEs.

For instance, whilst incorporating nonlinearity within a single first-order ODE can generate new equilibrium points, first-order ODEs can only exhibit a "monotonic" evolution between successive equilibrium points (Alligood et al., 1996 [130]; Meiss 2007 [131]; Strogatz 2015 [132]). The classical logistic ODE cannot reveal both quasi-exponentially growing and decaying errors reported using the analogue approach in Lorenz (1969b). In contrast, introducing nonlinearity within a single second-order ODE can yield different outcomes, leading to homoclinic orbits and/or oscillatory solutions. The X (or Z) component of the non-dissipative Lorenz 1963 model is, indeed, a second-order ODE with a cubic (or quadratic) nonlinear term.

Recent studies have documented differences in error representation between first-order and second-order ODEs using the linear system (such as the logistic equation without the quadratic term), the logistic equation (i.e., the Lorenz error growth model), and the KdV-SIR equation, which is a second-order ODE with a cubic nonlinear term (Paxson and Shen 2022 and 2023 [109,133]). Mathematically, the KdV-SIR equation is identical to the Z component of the non-dissipative Lorenz 1963 model and the Kortewegde Vries (KdV) equation in the traveling wave coordinate; it represents the epidemic SIR model under a weakly nonlinear assumption. The classical SIR model, proposed by Kermack and McKendrick (1927) [134], describes the evolution of susceptible (S), infected (I), and recovered (R) individuals. Using real-world models, the presence of oscillatory root-mean-square averaged forecast errors in ensemble runs was previously reported when oscillatory solutions prevailed over specific regions or time periods (Liu et al., 2009 [135]). Shen et al. (2010) [136] documented oscillatory correlation coefficients in 30-day simulations of multiple African easterly waves using a global mesoscale model (Shen et al., 2006a, b, [137,138]).

2.2. A Review of Butterfly Effects within Lorenz Models

Over the past five decades, various types of butterfly effects (BEs) have impacted our lives in direct or indirect ways. However, misunderstandings and misinterpretations of these effects are prevalent. In order to provide clarity and support the goals of the Special Issue, Shen et al. (2022b) [139] define and describe the three major kinds of BEs: the sensitive dependence on initial conditions (SDIC), the ability of small disturbances to create large-scale circulation, and the hypothetical role of small-scale processes and/or instability in contributing to finite predictability [140]. These three kinds of BEs are referred to as BE1, BE2, and BE3, respectively (see Table A1 in Appendix C). In order to prevent inaccurate generalizations, understanding the physical source of energy through a brief exploration can aid in the appropriate application of related discoveries. One of the major limitations of a linear model is that its instability remains unchanged over time.

The following folklore has been a popular analogy for SDIC (i.e., the BE1) (e.g., [3,141]):

For want of a nail, the shoe was lost.
For want of a shoe, the horse was lost.
For want of a horse, the rider was lost.
For want of a rider, the battle was lost.
For want of a battle, the kingdom was lost.
And all for the want of a horseshoe nail.

As illustrated by Shen et al., 2022b [139], such an analogy is not accurate (e.g., Lorenz 2008 [142].) because the verse does not indicate boundedness.

2.3. A Review of Lorenz's Perspective on the Predictability Limit

In relation to the origin of the inherent limit to predictability in the atmosphere (e.g., Charney et al. [143], GARP 1969 [144], Lorenz 1969a, b, c, d, e, 1982a, b, 1984b, 1985, 1993, and 1996 [145–148]; Smagorinsky (1969) [149]; Lighthill 1986 [150]; Lewis 2005 [30]; Rotunno and Synder 2008 [151]; Durran and Gingrich 2014 [152]; Reeve 2014 [153]; and Zhang et al., 2019 [129]), a recent literature review was conducted by Shen et al., 2023 [154] to present the following perspective as proposed by Lorenz:

A.   The Lorenz 1963 model qualitatively revealed the essence of finite predictability within a chaotic system, such as the atmosphere. However, the Lorenz 1963 model did not determine a precise limit for atmospheric predictability.

B.   In the 1960s, using real-world models, the two-week predictability limit was originally estimated based on a doubling time of 5 days. Since then, this finding has been documented in Charney et al. (1966) [143] and has become a consensus.

Further examination of predictability within the Lorenz 1963 and 1969 models was carried out by Shen et al. (2022a), with a summary provided in Section 3.

3. Overview of the Published Papers

Our summary of the papers published in the Special Issue covers four topics: (A) butterfly effects and sensitivities, (B) atmospheric dynamics and the application of theoretical models, (C) predictability and prediction, and (D) computational and machine learning methods.

3.1. (A) Butterfly Effects and Sensitivities

Saiki and Yorke (2023) [155] challenged the notion that a positive Lyapunov exponent is a necessary condition for the occurrence of a butterfly effect, which is defined as temporary exponential growth resulting from a small perturbation. A positive Lyapunov exponent refers to the rate at which two infinitesimally close trajectories diverge from each other (Wolf et al., 1985 [123]; Shen 2014 [124]). Using simple linear map models, the paper demonstrated that a butterfly effect can occur even without chaos or a positive Lyapunov exponent. Specifically, the authors examined a 24-dimensional version of the map, which exhibits a significant butterfly effect despite having a Lyapunov exponent of 0.

The paper introduced a linear "infected mosquito" model to illustrate how off-diagonal matrix entries can contribute to a finite-time growth rate. The authors suggested that the degree of instability in a system can be better characterized by its finite-time growth rate, which takes into account the impact of off-diagonal matrix entries. The study's findings indicated that in higher-dimensional systems, off-diagonal matrix entries may have a more significant impact on a system's behavior than Lyapunov exponents. The paper concluded that understanding the dynamics of simpler, non-meteorological models can shed light on additional aspects of the butterfly effect. By focusing on the finite-time growth rate, valuable insights into a system's behavior can be gained, even in linear systems that lack chaos or positive Lyapunov exponents.

The study by Chou and Wang (2023) [156] investigated the sensitivity of the ventilation coefficient parameterization to the life span of a simulated storm. The authors modified the parameterization of the ventilation coefficient for different precipitation particles, such as rain, snow, and hail, and compared the results to their previous study where the ventilation effect of precipitation particles was either halved or doubled as a whole. The results indicated that changing the ventilation coefficient of rain, snow, or hail categories leads to different storm evolution paths. It was found that reducing the ventilation effect of rain leads to quick dissipation, whereas enhancing the ventilation effect of rain or snow/hail leads to the development of multicellular storms. The study additionally demonstrated the impact of changes in cloud microphysical parameterization on larger-scale dynamical processes.

3.2. (B) Atmospheric Dynamics and the Application of Theoretical Models

The study of Lewis and Lakshmivarahan (2022) [17] focused on the challenge of data assimilation in chaotic and non-chaotic regimes, specifically the placement of observations to induce convexity of the cost function in the space of control. The authors used Saltzman's spectral model of convection, which has both chaotic and non-chaotic regimes controlled by two parameters, to examine this problem. They simplified the problem by stripping the seven-variable constraint to a three-variable constraint and used the forecast sensitivity method (FSM) to deliver sensitivities within the placement. The authors used the observability Gramian to locate observations at places that force the Gramian positive definite, and the locations were chosen such that the condition number of $V^T V$ was small and guaranteed convexity in the vicinity of the cost function minimum. Four numerical experiments were executed, and the results were compared with the structure of the cost function independently determined through arduous computation over a wide range of the two non-dimensional numbers. The authors reported good results based on a reduction in the cost function value and a comparison with the cost function structure.

Shen et al. (2022a) [157] compared similarities and differences between the Lorenz 1963 and 1969 models, both of which suggested finite predictability. The two models represent different physical systems—one for convection and the other for barotropic vorticity—and have different mathematical complexities. The Lorenz 1963 model is limited scale and nonlinear, while the Lorenz 1969 model is closure based, physically multiscale, mathematically linear, and numerically ill-conditioned.

The authors highlighted that the existence of a saddle point at the origin is a common feature that produces instability in both systems. Within the chaotic regime of the 1963 nonlinear model, unstable growth is constrained by nonlinearity and dissipation, leading to time-varying growth rates along an orbit, and a dependence of finite predictability on initial conditions. In comparison, within the 1969 linear model, multiple unstable modes at various growth rates appear, and the growth of a specific unstable mode is constrained by imposing a saturation assumption, also leading to a time-varying system growth rate.

Shen et al. (2022a) concluded that while both the Lorenz 1963 model with SDIC and the Lorenz 1969 model with ill-conditioning have been collectively or separately applied to qualitatively reveal the nature of finite predictability in the weather and climate, only single-type solutions were examined. Moreover, the 1969 system easily captures numerical instability. Thus, an estimate of the predictability limit using either of the models, with or without additional assumptions (e.g., saturation), should be interpreted with caution and should not be generalized as an upper limit for atmospheric predictability.

Zeng (2023) [158] investigated the interaction between tropical clouds and radiation and its connection to atmospheric instability and oscillations. He used a very low-order system to derive criteria for instability and explained distinct timescales observed in atmospheric oscillations and suggested that the instability of boundary layer quasi-equilibrium leads to the quasi-two-day oscillation, that the instability of radiative-convective equilibrium leads to the Madden–Julian oscillation (MJO), and that the instability of radiative-convective flux equilibrium leads to the El Niño–southern oscillation. The paper introduced a novel cloud parameterization for weather and climate models and suggested that models can predict oscillations if they properly represent cirrus clouds and convective downdrafts within the tropics.

His low-order system revealed that the atmosphere exhibits predictability through limit cycles when its parameters are understood. Conversely, when parameters slowly vary with time, bifurcations associated with the limit cycle may introduce a new level of uncertainty in atmospheric predictability. The incorporation of a conceptual model featuring limit cycles and bifurcations, as a particular instance, aligns with the notion of co-existing determinism and chaos within the atmosphere (Shen et al., 2021a; Shen et al., 2022c).

3.3. (C) Predictability and Predictions

By extending the studies of Shen (2021a, b) [8,9], Shen et al. (2022c) [159] further explored the dual nature of chaos and order in weather and climate using the Lorenz 1963 and 1969 models. While a conventional view in previous research has shown that the weather and climate are chaotic, a revised view was previously proposed to suggest that the atmosphere possesses both chaos and order with distinct predictability. The revised view is supported by attractor co-existence, which suggests limited predictability for chaotic solutions and unlimited predictability for non-chaotic solutions. Real-world examples of non-chaotic weather systems were also provided to support the revised view. The concept of attractor co-existence (to be specific, the first kind of attractor co-existence) was first documented using the Lorenz 1963 model by Yorke and Yorke (1979) [160]. Using kayaking and skiing as analogies, the paper discussed multistability and monostability, respectively, for attractor co-existence and single-type solutions. The paper also emphasized the predictable nature of recurrence for slowly varying solutions and the less predictable or unpredictable nature of emerging solutions. The paper concluded, with a refined view, that the atmosphere possesses chaos and order, including emerging organized systems from favorite environmental conditions (such as tornadoes) and time-varying forcing from recurrent seasons.

In "Predictability and Predictions", Anthes (2022) [161] described his experience with predictability theory and weather prediction. The essay discussed various historical figures who attempted to predict the future through occult methods, as well as more modern attempts to predict complex systems based on mathematics and physics. The author described his early experiences as a meteorologist working at the then U.S. Weather Bureau, where he noticed some predictability in thunderstorms initiated by daytime heating over the Blue Ridge Mountains in Virginia, propagating eastward toward the coast. He then became interested in numerical simulations and developed a simple nonlinear one-dimensional gravity wave model while studying at the University of Wisconsin. Later, he worked at the National Hurricane Research Laboratory in Miami, where he developed the first nonlinear, baroclinic, three-dimensional model of tropical cyclones.

The essay also discussed Lorenz's classical 1963 paper, which laid the theoretical foundation for modern predictability theory for nonlinear fluid dynamics. Lorenz's work has been applied conceptually to numerical prediction of the atmosphere and oceans, as well as to other complex systems, including human societies. The author noted that the lack of infinite predictability of chaotic nonlinear systems has come to be known as the "butterfly effect".

Finally, the essay touched on the author's friendship with Fred Sanders, a leading synoptic meteorologist from MIT. They sailed together from Miami to Marblehead, Massachusetts, discussing weather predictability, including Laplace's demon which argued for the possibility of unlimited prediction.

The essay concluded by noting that while predictability theory and predictions have made great strides in recent decades, the author believed that there will always be an element of uncertainty in weather prediction and other complex systems.

Wang et al. (2022) [162] presented a study that used the Cloud-Resolving Storm Simulator (CReSS) to forecast the rainfall of three landfalling typhoons in the Philippines: Mangkhut (2018), Koppu (2015), and Melor (2015). By applying a time-lagged strategy for ensemble simulations, the study verified the track and quantitative precipitation forecasts (QPFs) against observations at 56 rain-gauge sites at seven thresholds up to 500 mm. The predictability of rainfall was found to be highest for Koppu, followed by Melor and the lowest for Mangkhut. The study found that at lead times of 3–7 days before typhoon landfall in the Philippines, there is a fair chance to produce decent QPFs with useful information on rainfall scenarios for early preparation. The QPF results were found to be encouraging and comparable to the skill level for typhoon rainfall in Taiwan.

3.4. (D) Computational and Machine Learning Methods

Tseng (2022) [163] presented a study that compared traditional principal component analysis (PCA) and isometric feature mapping (ISOMAP) for the classification and detection of El Niño and La Niña events based on sea surface temperature (SST) data. ISOMAP is a nonlinear dimensionality reduction method that measures the distance between two data points based on the geodesic distance, which more accurately reflects the actual distance in a nonlinear space. The results indicated that the classification based on ISOMAP-reconstructed SST data points outperformed traditional PCA-reconstructed data points in terms of accurately differentiating between points in different events and measuring differences amongst points in the same event. The study additionally determined that the trajectories of the leading three temporal eigen components of the SST ISOMAP were similar to the Lorenz 63 model within a phase space, which could have implications for tracing NWP perturbations. The paper also included a comparison of the spatial eigenmodes between traditional PCA and ISOMAP.

Author Contributions: B.-W.S. designed and performed research; B.-W.S. wrote the paper and R.A.P.S. and X.Z. provided input and suggested edits. All authors have read and agreed to the published version of the manuscript.

Funding: This research received no external funding.

Institutional Review Board Statement: Not applicable.

Informed Consent Statement: Not applicable.

Data Availability Statement: Not applicable.

Acknowledgments: The Guest Editors of this Special Issue of the journal *Atmosphere* express their gratitude to all of the authors and reviewers. They anticipate that the papers compiled in this Special Issue will inspire further advancements in enhancing the present comprehension of butterfly effects, multistability, multiscale predictability, and numerical sensitivities. The first author would like to acknowledge and show their appreciation for the invaluable inspiration provided by several studies (Pielke 2008, Anthes 2021, and Zeng et al., 1993 [164–166]), which played a pivotal role in shaping a series of related studies that ultimately resulted in the completion of this research.

Conflicts of Interest: The authors declare no conflict of interest.

Appendix A. The Lorenz (1960/1962) Model and the Lorenz (1996/2006) Model

For the readers' convenience, the original mathematical symbols are used to display the Lorenz (1960/1962) model with 12 ODEs for 12 variables, written as:

$$\frac{dX_i}{dt} = \sum_{m,n} a_{imn} X_m X_n + \sum_m b_{im} X_m + c_i. \tag{A1}$$

In the above equation, the nonlinear terms $a_{imn} X_m X_n$ represent nonlinear advection. The linear terms $b_{im} X_m$ represent the effects of friction and part of the effects of heating. The constant terms c_i are only non-zero in the last six equations. For the 12 variables, 6 variables represent amplitudes of the averaged stream function, and 6 variables indicates the amplitudes of temperature.

In comparison, the Lorenz (1996) model is written as follows:

$$\frac{dX_k}{dt} = -X_{k-2}X_{k-1} + X_{k-1}X_{k+1} - X_k + F, \tag{A2}$$

which contains K ODEs for K variables, and F is a constant. On the right-hand side, the nonlinear quadric terms represent advection, while the linear terms $(-X_k)$ and F represent thermal or mechanical damping (or dissipation) and external forcing, respectively. Additional conditions used in Lorenz (1996) and Lorenz (2005, p 1575) are provided as follows: for $n < 1$ (or $n > N$) X_n indicate X_{n+N} (or X_{n-N}).

Appendix B. The Logistic Map and Logistic ODE

The relationship between the logistic map and the logistic ODE is illustrated below.

$$\frac{dX}{d\tau} = rX(1 - X). \tag{A3}$$

Here, τ, X, and r represent the time variable, time-dependent variable, and a parameter, respectively. Applying the Euler method, we obtain:

$$\frac{X_{n+1} - X_n}{\Delta\tau} = rX_n(1 - X_n), \tag{A4}$$

where $\Delta\tau$ represents the time step and $X_n = X(t = n\Delta\tau)$, and n indicates the number of time integrations. After defining a new variable Y_n and a new parameter ρ, the following applies:

$$Y_n = \frac{r\Delta\tau X_n}{1 + r\Delta\tau}, \tag{A5a}$$

and

$$\rho = 1 + r\Delta\tau. \tag{A5b}$$

Equation (A4) yields the following equation:

$$Y_{n+1} = \rho Y_n(1 - Y_n). \tag{A6}$$

Equation (A6) is called the logistic map, which is a discrete version of the logistic differential equation in Equation (A3). The logistic map is a type of quadratic map, defined by the recurrence relation:

$$Z_{n+1} = aZ_n^2 + bZ_n + c, \tag{A7}$$

where a, b, and c are constant coefficients. While Lorenz (1964) applied the quadratic map in Equation (A6), Lorenz (1969d) [145] utilized the following quadratic map:

$$Z_{n+1} = 1.64 - Z_n^2 \tag{A8}$$

to illustrate the dependence of solutions on initial conditions. The cubic map that was used in Lorenz 1976 can be obtained by substituting a cubic term for a quadratic term in Equation (A6).

Appendix C. Definitions of Selected Concepts in the Special Issue

Table A1. Definitions of the selected concepts included in the Special Issue.

Name	Definitions	Recommendations
First kind of attractor co-existence	The co-existence of chaotic and steady-state solutions.	[7–9,160]
Second kind of attractor co-existence	The co-existence of nonlinear oscillatory and steady-state solutions.	[7–9]
Analogues	*Analogues are two states of the atmosphere that exhibit resemblance to each other.*	[21]
Attractor	The smallest attracting point set that, itself, cannot be decomposed into two or more subsets with distinct basins of attraction.	[8]
Butterfly effect (BE), general	*The phenomenon in that a small alteration in the state of a dynamical system will cause subsequent states to differ greatly from the states that would have followed without the alteration.*	[4]
BE of the first kind (BE1)	The sensitive dependence on initial conditions (SDICs).	[1,4,139]
BE of the second kind (BE2)	*The capability of a small disturbance to create an organized circulation at large distances.*	[2,4,139]
BE of the third kind (BE3)	*(1) The hypothetical role of small-scale processes in contributing to finite predictability; (2) Ill-conditioning and instability.*	[139,140,157]
BE in Saiki and Yorke (2023)	*Instability in high-dimensional linear systems.*	[155]
Chaos	Bounded aperiodic orbits exhibit a sensitive dependence on ICs.	[4,159]
Final state sensitivity	Nearby orbits settle to one of multiple attractors for a finite but arbitrarily long time.	[82]
Intransitivity	A specific type of solution lasts forever.	[52,72]
Monostability	The appearance of single-type solutions.	[159]
Multistability	A system with multistability contains more than one bounded attractor that only depends on ICs.	[159]
Quasi-periodicity	A quasi-periodic solution consists of two or more incommensurate frequencies, the ratios of which are irrational.	[91]
Recurrence	This refers to a trajectory returning to the vicinity of its previous location.	[92,93]
Sensitive dependence	*The property characterizing an orbit if most other orbits that pass close to it at some point do not remain close to it as time advances.*	[4]
Ventilation coefficient parameterization	A parameter used in numerical models of atmospheric convection to represent the effect of environmental wind on convective updrafts.	[156]

References

1. Lorenz, E.N. Deterministic nonperiodic flow. *J. Atmos. Sci.* **1963**, *20*, 130–141. [CrossRef]
2. Lorenz, E.N. Predictability: Does the flap of a butterfly's wings in Brazil set off a tornado in Texas? In Proceedings of the 139th Meeting of AAAS Section on Environmental Sciences, New Approaches to Global Weather, GARP, AAAS, Cambridge, MA, USA, 29 December 1972.
3. Gleick, J. *Chaos: Making a New Science*; Penguin: New York, NY, USA, 1987; 360p.
4. Lorenz, E.N. *The Essence of Chaos*; University of Washington Press: Seattle, WA, USA, 1993; 227p.
5. The Nobel Committee for Physics. *Scientific Background on the Nobel Prize in Physics 2021 "For Groundbreaking Contributions to Our Understanding of Complex Physical Systems"*; The Royal Swedish Academy of Sciences: Stockholm, Sweden, 2021.
6. Fischer, K.H.; Hertz, J.A. *Spin Glasses*; Cambridge University Press: Cambridge, UK, 1993; ISBN 9780521447775.
7. Shen, B.-W. Aggregated Negative Feedback in a Generalized Lorenz Model. *Int. J. Bifurc. Chaos* **2019**, *29*, 1950037. [CrossRef]
8. Shen, B.-W.; Pielke, R.A., Sr.; Zeng, X.; Baik, J.-J.; Faghih-Naini, S.; Cui, J.; Atlas, R. Is weather chaotic? Coexistence of chaos and order within a generalized Lorenz model. *Bull. Am. Meteorol. Soc.* **2021**, *2*, E148–E158. [CrossRef]
9. Shen, B.-W.; Pielke, R.A., Sr.; Zeng, X.; Baik, J.-J.; Faghih-Naini, S.; Cui, J.; Atlas, R.; Reyes, T.A. Is Weather Chaotic? Coexisting Chaotic and Non-Chaotic Attractors within Lorenz Models. In *The 13th Chaos International Conference CHAOS 2020; Springer Proceedings in Complexity*; Skiadas, C.H., Dimotikalis, Y., Eds.; Springer: Cham, Switzerland, 2021.
10. Shen, B.-W. Attractor Coexistence, Butterfly Effects, and Chaos Theory (ABC): A Review of Lorenz Models and a Generalized Lorenz Model. In Proceedings of the 16th Chaos International Conference CHAOS 2023, Heraklion, Greece, 13–16 June 2023; submitted.
11. Chen, G.-R. Butterfly Effect and Chaos. 2020. Available online: https://www.ee.cityu.edu.hk/~gchen/pdf/Lorenz_T.pdf (accessed on 1 July 2023). (In Chinese)
12. Lorenz, E.N. Maximum simplification of the dynamic equations. *Tellus* **1960**, *12*, 243–254. [CrossRef]
13. Lewis, J.; Lakshmivarahan, S.; Dhall, S. *Dynamic Data Assimilation: A Least Squares Approach*; Cambridge University Press: Cambridge, UK, 2006; 654p.
14. Lorenz, E.N. The statistical prediction of solutions of dynamic equations. In Proceedings of the International Symposium on Numerical Weather Prediction, Tokyo, Japan, 7–13 November 1962; pp. 629–635.
15. Saltzman, B. Finite Amplitude Free Convection as an Initial Value Problem-I. *J. Atmos. Sci.* **1962**, *19*, 329–341. [CrossRef]
16. Lakshmivarahan, S.; Lewis, J.M.; Hu, J. Saltzman's Model: Complete Characterization of Solution Properties. *J. Atmos. Sci.* **2019**, *76*, 1587–1608. [CrossRef]
17. Lewis, J.M.; Lakshmivarahan, S. Role of the Observability Gramian in Parameter Estimation: Application to Nonchaotic and Chaotic Systems via the Forward Sensitivity Method. *Atmosphere* **2022**, *13*, 1647. [CrossRef]
18. May, R.M. Simple mathematical models with very complicated dynamics. *Nature* **1976**, *261*, 459–467. [CrossRef]
19. Li, T.-Y.; Yorke, J.A. Period Three Implies Chaos. *Am. Math. Mon.* **1975**, *82*, 985–992. [CrossRef]
20. Lorenz, E.N. The problem of deducing the climate from the governing equations. *Tellus* **1964**, *16*, 1–11. [CrossRef]
21. Lorenz, E.N. Atmospheric predictability as revealed by naturally occurring analogues. *J. Atmos. Sci.* **1969**, *26*, 636–646. [CrossRef]
22. Lorenz, E.N. Low-order models of atmospheric circulations. *J. Meteor. Soc. Jpn.* **1982**, *60*, 255–267. [CrossRef]
23. Lorenz, E.N. Energy and numerical weather prediction. *Tellus* **1960**, *12*, 364–373. [CrossRef]
24. Lorenz, E.N. Simplified dynamic equations applied to the rotating-basin experiments. *J. Atmos. Sci.* **1962**, *19*, 39–51. [CrossRef]
25. Lorenz, E.N. The mechanics of vacillation. *J. Atmos. Sci.* **1963**, *20*, 448–464. [CrossRef]
26. Lorenz, E.N. A study of the predictability of a 28-variable atmospheric model. *Tellus* **1965**, *17*, 321–333. [CrossRef]
27. Lorenz, E.N. Atmospheric predictability experiments with a large numerical model. *Tellus* **1982**, *34*, 505–513. [CrossRef]
28. Lorenz, E. The predictability of hydrodynamic flow. *Trans. N. Y. Acad. Sci.* **1963**, *25*, 409–432. [CrossRef]
29. Kalnay, E. *Atmospheric Modeling, Data Assimilation and Predictability*; Cambridge University Press: Cambridge, UK, 2002. [CrossRef]
30. Lewis, J. Roots of ensemble forecasting. *Mon. Weather Rev.* **2005**, *133*, 1865–1885. [CrossRef]
31. Nese, J.M. Quantifying local predictability in phase space. *Phys. D Nonlinear Phenom.* **1989**, *35*, 237–250. [CrossRef]
32. Abarbanel, H.D.I.; Brown, R.; Kennel, M.B. Local Lyapunov exponents computed from observed data. *J. Nonlinear Sci.* **1992**, *2*, 343–365. [CrossRef]
33. Eckhardt, B.; Yao, D. Local Lyapunov exponents in chaotic systems. *Phys. D* **1993**, *65*, 100–108. [CrossRef]
34. Krishnamurthy, V. A predictability study of Lorenz's 28-variable model as a dynamical system. *J. Atmos. Sci.* **1993**, *50*, 2215–2229. [CrossRef]
35. Szunyogh, I.; Kalnay, E.; Toth, Z. A comparison of Lyapunov and optimal vectors in a low-resolution GCM. *Tellus* **1997**, *49A*, 200–227. [CrossRef]
36. Yoden, S. Atmospheric Predictability. *J. Meteorol. Soc. Jpn.* **2007**, *85B*, 77–102. [CrossRef]
37. Oseledec, V.I. A multiplicative ergodic theorem. Ljapunov characteristic numbers for dynamical systems. *Trans. Mosc. Math. Sci.* **1968**, *19*, 197–231.
38. Molteni, F.; Buizza, R.; Palmer, T.N.; Petroliagis, T. The ECMWF ensemble prediction system: Methodology and validation. *Quart. J. Roy. Meteor. Soc.* **1996**, *122*, 73–119. [CrossRef]
39. Buizza, R.; Leutbecher, M.; Isaksen, L. Potential use of an ensemble of analyses in the ECMWF Ensemble Prediction System. *Quart. J. Roy. Meteor. Soc.* **2008**, *134*, 2051–2066. [CrossRef]

40. Cui, J.; Shen, B.-W. A Kernel Principal Component Analysis of Coexisting Attractors within a Generalized Lorenz Model. *Chaos Solitons Fractals* **2021**, *146*, 110865. [CrossRef]
41. Lorenz, E.N. *Empirical Orthogonal Functions and Statistical Weather Prediction*; Scientific Report No. 1, Statistical Forecasting Project; Air Force Research Laboratories, Office of Aerospace Research, USAF: Bedford, MA, USA, 1956.
42. Pedlosky, J. Finite-amplitude baroclinic waves with small dissipation. *J. Atmos. Sci.* **1971**, *28*, 587–597. [CrossRef]
43. Pedlosky, J. Limit cycles and unstable baroclinic waves. *J. Atmos. Sci.* **1972**, *29*, 53–63. [CrossRef]
44. Pedlosky, J. *Geophysical Fluid Dynamics*, 2nd ed.; Springer: New York, NY, USA, 1987; p. 710.
45. Pedlosky, J.; Frenzen, C. Chaotic and periodic behavior of finite-amplitude baroclinic waves. *J. Atmos. Sci.* **1980**, *37*, 1177–1196. [CrossRef]
46. Shen, B.-W. Solitary Waves, Homoclinic Orbits, and Nonlinear Oscillations within the non-dissipative Lorenz Model, the inviscid Pedlosky Model, and the KdV Equation. In *The 13th Chaos International Conference CHAOS 2020; Springer Proceedings in Complexity*; Skiadas, C.H., Dimotikalis, Y., Eds.; Springer: Cham, Switzerland, 2021.
47. Lorenz, E.N. The predictability of a flow which possesses many scales of motion. *Tellus* **1969**, *21*, 289–307. [CrossRef]
48. Leith, C.E. Atmospheric predictability and two-dimensional turbulence. *J. Atmos. Sci.* **1971**, *28*, 145–161. [CrossRef]
49. Leith, C.E.; Kraichnan, R.H. Predictability of turbulent flows. *J. Atmos. Sci.* **1972**, *29*, 1041–1058. [CrossRef]
50. Lorenz, E.N. Investigating the predictability of turbulent motion. Statistical Models and Turbulence. In Proceedings of the Symposium Held at the University of California, San Diego, CA, USA, 15–21 July 1971; Springer: Berlin/Heidelberg, Germany, 1972; pp. 195–204.
51. Lorenz, E.N. Low-order models representing realizations of turbulence. *J. Fluid Mech.* **1972**, *55*, 545–563. [CrossRef]
52. Lorenz, E.N. Nondeterministic theories of climatic change. *Quat. Res.* **1976**, *6*, 495–506. [CrossRef]
53. Lorenz, E.N. Attractor sets and quasi-geostrophic equilibrium. *J. Atmos. Sci.* **1980**, *37*, 1685–1699. [CrossRef]
54. Lorenz, E.N. On the existence of a slow manifold. *J. Atmos. Sci.* **1986**, *43*, 1547–1557. [CrossRef]
55. Lorenz, E.N.; Krishnamurthy, V. On the nonexistence of a slow manifold. *J. Atmos. Sci.* **1987**, *44*, 29402950. [CrossRef]
56. Lorenz, E.N. The slow manifold. What is it? *J. Atmos. Sci.* **1992**, *49*, 24492451. [CrossRef]
57. McWilliams, J.C. A perspective on the legacy of Edward Lorenz. *Earth Space Sci.* **2019**, *6*, 336–350. [CrossRef]
58. Sparrow, C. *The Lorenz Equations: Bifurcations, Chaos, and Strange Attractors*; Applied Mathematical Sciences; Springer: New York, NY, USA, 1982; 269p.
59. Shen, B.-W. On periodic solutions in the non-dissipative Lorenz model: The role of the nonlinear feedback loop. *Tellus A* **2018**, *70*, 1471912. [CrossRef]
60. Shen, B.-W. Homoclinic Orbits and Solitary Waves within the non-dissipative Lorenz Model and KdV Equation. *Int. J. Bifurc. Chaos* **2020**, *30*, 15. [CrossRef]
61. Lorenz, E.N. Irregularity: A fundamental property of the atmosphere. Crafoord Prize Lecture, presented at the Royal Swedish Academy of Sciences, Stockholm, September 28, 1983. *Tellus* **1984**, *36A*, 98–110. [CrossRef]
62. Lorenz, E.N. Can chaos and intransitivity lead to interannual variability? *Tellus* **1990**, *42A*, 378–389. [CrossRef]
63. Pielke, R.A.; Zeng, X. Long-Term Variability of Climate. *J. Atmos. Sci.* **1994**, *51*, 155–159. [CrossRef]
64. Van Veen, L.; Opsteegh, T.; Verhulst, F. Active and passive ocean regimes in a low-order climate model. *Tellus A* **2001**, *53*, 616–627. [CrossRef]
65. Van Veen, L. Baroclinic Flow and the Lorenz-84 Model. *Int. J. Bifurc. Chaos* **2003**, *13*, 2117–2139. [CrossRef]
66. Lorenz, E.N. Chaos, spontaneous climatic variations and detection of the greenhouse effect. In *Greenhouse-Gas-Induced Climatic Change: A Critical Appraisal of Simulations and Observations*; Schlesinger, M.E., Ed.; Elsevier Science Publishers B. V.: Amsterdam, The Netherlands, 1991; pp. 445–453.
67. Lorenz, E.N. Predictability—A problem partly solved. In Proceedings of the Seminar on Predictability, Shinfield Park, Reading, UK, 4–8 September 1995; ECMWF: Reading, UK, 1996; Volume I.
68. Lorenz, E.N. Predictabilitya problem partly solved. In *Predictability of Weather and Climate*; Palmer, T., Hagedorn, R., Eds.; Cambridge University Press: Cambridge, UK, 2006; pp. 40–58.
69. Lorenz, E.N. Designing Chaotic Models. *J. Atmos. Sci.* **2005**, *62*, 1574–1587. [CrossRef]
70. Lorenz, E.N. Regimes in simple systems. *J. Atmos. Sci.* **2006**, *63*, 2056–2073. [CrossRef]
71. Lorenz, E.N. Compound windows of the Hénon map. *Phys. D* **2008**, *237*, 1689–1704. [CrossRef]
72. Lorenz, E.N. Climatic determinism. Meteor. Monographs, Amer. *Meteor. Soc.* **1968**, *8*, 1–3.
73. Lorenz, E.N. *Climatic Predictability*; GARP Publications Series; GARP: Jersey City, NJ, USA, 1975; pp. 132–136.
74. Lorenz, E.N. Some aspects of atmospheric predictability. European Centre for Medium Range Weather Forecasts, Seminar 1981. In Proceedings of the Problems and Prospects in Long and Medium Range Weather Forecasting, Reading, UK, 14–18 September 1982; pp. 1–20.
75. Lorenz, E.N. *Climate Is What You Expect*; NCAR: Boulder, CO, USA, 1997; unpublished work. Available online: https://eapsweb.mit.edu/sites/default/files/Climate_expect.pdf (accessed on 1 July 2023).
76. Holmes, P. A Nonlinear Oscillator with a Strange Attractor. *Phil. Trans. R. Soc.* **1979**, *A191*, 419.
77. May, R. The cubic map in theory and practice. *Nature* **1984**, *311*, 13–14. [CrossRef]
78. De Oliveira, J.A.; Papesso, E.R.; Leonel, E.D. Relaxation to Fixed Points in the Logistic and Cubic Maps: Analytical and Numerical Investigation. *Entropy* **2013**, *15*, 4310–4318. [CrossRef]

79. Grebogi, C.; Ott, E.; Yorke, J.A. Chaotic attractors in crisis. *Phys. Rev. Lett.* **1982**, *48*, 1507–1510. [CrossRef]
80. Grebogi, C.; Ott, E.; Yorke, J.A. Crises, sudden changes in chaotic attractors, and transient chaos. *Phys. D* **1983**, *7*, 181–200. [CrossRef]
81. Hirsch, M.; Smale, S.; Devaney, R.L. *Differential Equations, Dynamical Systems, and an Introduction to Chaos*, 3rd ed.; Academic Press: Waltham, MA, USA, 2013; 432p.
82. Grebogi, C.; McDonald, S.W.; Ott, E.; Yorke, J.A. Final state sensitivity: An obstruction to predictability. *Phys. Lett. A* **1983**, *99*, 415–418. [CrossRef]
83. Charney, J.G.; DeVore, J.G. Multiple flow equilibria in the atmosphere and blocking. *J. Atmos. Sci.* **1979**, *36*, 1205–1216. [CrossRef]
84. Chen, Z.-M.; Xiong, X. Equilibrium states of the Charney-DeVore quasi-geostrophic equation in mid-latitude atmosphere. *J. Math. Anal. Appl.* **2016**, *444*, 1403–1416. [CrossRef]
85. Faranda, D.; Masato, G.; Moloney, N.; Sato, Y.; Daviaud, F.; Dubrulle, B.; Yiou, P. The switching between zonal and blocked mid-latitude atmospheric circulation: A dynamical system perspective. *Clim. Dyn.* **2016**, *47*, 1587–1599. [CrossRef]
86. Dorrington, J.; Palmer, T. On the interaction of stochastic forcing and regime dynamics. *Nonlinear Process. Geophys.* **2023**, *30*, 49–62. [CrossRef]
87. Wikipedia. El Niño–Southern Oscillation—Wikipedia, The Free Encyclopedia. Available online: https://en.wikipedia.org/wiki/El_Ni%C3%B1o%E2%80%93Southern_Oscillation (accessed on 1 July 2023).
88. Wallace, J.M.; Battisti, D.S.; Thompson, D.W.J.; Hartmann, D.L. *The Atmospheric General Circulation*, 1st ed.; Cambridge University Press: Cambridge, UK, 2023; 456p.
89. Van den Dool, H.M. A new look at weather forecasting trough analogues. *Mon. Wea. Rev.* **1989**, *117*, 2230–2247. [CrossRef]
90. Van den Dool, H.M. Searching for analogues, how long must we wait? *Tellus A* **1994**, *46*, 314–324. [CrossRef]
91. Faghih-Naini, S.; Shen, B.-W. Quasi-periodic orbits in the five-dimensional non-dissipative Lorenz model: The role of the extended nonlinear feedback loop. *Int. J. Bifurc. Chaos* **2018**, *28*, 1850072. [CrossRef]
92. Thompson, J.M.T.; Stewart, H.B. *Nonlinear Dynamics and Chaos*, 2nd ed.; John Wiley & Sons, Ltd.: Hoboken, NJ, USA, 2002; p. 437.
93. Reyes, T.; Shen, B.-W. A Recurrence Analysis of Chaotic and Non-Chaotic Solutions within a Generalized Nine-Dimensional Lorenz Model. *Chaos Solitons Fractals* **2019**, *125*, 1–12. [CrossRef]
94. Reyes, T.; Shen, B.-W. A Recurrence Analysis of Multiple African Easterly Waves during Summer 2006. In *Current Topics in Tropical Cyclone Research*; IntechOpen: London, UK, 2020. [CrossRef]
95. Lorenz, E.N. On the existence of extended range predictability. *J. Appl. Meteor.* **1973**, *12*, 543–546. [CrossRef]
96. Lorenz, E.N. Three approaches to atmospheric predictability. *Bull. Am. Meteor. Soc.* **1969**, *50*, 345–351.
97. Poincaré, H. Sur le problème des trois corps et les équations de la dynamique. *Acta Math.* **1890**, *13*, 1–270.
98. Poincaré, H. *Science et Méthode, Flammarion*, 1908 ed.; English Translated; Maitland, F., Ed.; Thomas Nelson and Sons: London, UK, 1914.
99. Curry, J.H. Generalized Lorenz systems. *Commun. Math. Phys.* **1978**, *60*, 193–204. [CrossRef]
100. Curry, J.H.; Herring, J.R.; Loncaric, J.; Orszag, S.A. Order and disorder in two- and three-dimensional Benard convection. *J. Fluid Mech.* **1984**, *147*, 1–38. [CrossRef]
101. Howard, L.N.; Krishnamurti, R.K. Large-scale flow in turbulent convection: A mathematical model. *J. Fluid Mech.* **1986**, *170*, 385–410. [CrossRef]
102. Hermiz, K.B.; Guzdar, P.N.; Finn, J.M. Improved low-order model for shear flow driven by Rayleigh–Benard convection. *Phys. Rev. E* **1995**, *51*, 325–331. [CrossRef] [PubMed]
103. Thiffeault, J.-L.; Horton, W. Energy-conserving truncations for convection with shear flow. *Phys. Fluids* **1996**, *8*, 1715–1719. [CrossRef]
104. Musielak, Z.E.; Musielak, D.E.; Kennamer, K.S. The onset of chaos in nonlinear dynamical systems determined with a new fractal technique. *Fractals* **2005**, *13*, 19–31. [CrossRef]
105. Roy, D.; Musielak, Z.E. Generalized Lorenz models and their routes to chaos. I. Energy-conserving vertical mode truncations. *Chaos Solit. Fract.* **2007**, *32*, 1038–1052. [CrossRef]
106. Roy, D.; Musielak, Z.E. Generalized Lorenz models and their routes to chaos. II. Energyconserving horizontal mode truncations. *Chaos Solit. Fract.* **2007**, *31*, 747–756. [CrossRef]
107. Roy, D.; Musielak, Z.E. Generalized Lorenz models and their routes to chaos. III. Energyconserving horizontal and vertical mode truncations. *Chaos Solit. Fract.* **2007**, *33*, 1064–1070. [CrossRef]
108. Moon, S.; Han, B.-S.; Park, J.; Seo, J.M.; Baik, J.-J. Periodicity and chaos of high-order Lorenz systems. *Int. J. Bifurc. Chaos* **2017**, *27*, 1750176. [CrossRef]
109. Paxson, W.; Shen, B.-W. A KdV-SIR Equation and Its Analytical Solutions for Solitary Epidemic Waves. *Int. J. Bifurc. Chaos* **2022**, *32*, 2250199. [CrossRef]
110. Saiki, Y.; Sander, E.; Yorke, J. Generalized Lorenz equations on a three-sphere. *Eur. Phys. J. Spec. Top.* **2017**, *226*, 1751–1764. [CrossRef]
111. Shen, B.-W. On the Predictability of 30-Day Global Mesoscale Simulations of African Easterly Waves during Summer 2006: A View with the Generalized Lorenz Model. *Geosciences* **2019**, *9*, 281. [CrossRef]
112. Lawler, E.; Thye, S.; Yoon, J. *Order on the Edge of Chaos Social Psychology and the Problem of Social Order*; Cambridge University Press: Cambridge, UK, 2015; ISBN 9781107433977.

113. Crutchfield, J.P. Between order and chaos. *Nat. Phys.* **2011**, *8*, 17–24. [CrossRef]
114. Melby, P.; Kaidel, J.; Weber, N.; Hübler, A. Adaptation to the edge of chaos in the self-adjusting logistic map. *Phys. Rev. Lett.* **2000**, *84*, 5991–5993. [CrossRef]
115. Palmer, T.N. Edward Norton Lorenz. 23 May 1917–16 April 2008. *Biogr. Mem. Fellows R. Soc.* **2009**, *55*, 139–155. [CrossRef]
116. Emanuel, K. *Edward Norton Lorenz (1917–2008)*; National Academy of Sciences: Washington, DC, USA, 2011; p. 4.
117. Feldman, D. *Chaos and Fractals: An Elementary Introduction*; Oxford University Press: Oxford, UK, 2012; 408p.
118. Grassberger, P.; Procaccia, I. Characterization of strange attractors. *Phys. Rev. Lett.* **1983**, *5*, 346–349. [CrossRef]
119. Nese, J.M.; Dutton, J.A.; Wells, R. Calculated attractor dimensions for low-order spectral models. *J. Atmos. Sci.* **1987**, *44*, 1950–1972. [CrossRef]
120. Ruelle, D. Chaotic Evolution and Strange Attractors. In *Lezioni Lincee*; Cambridge University Press: Cambridge, UK, 1989. [CrossRef]
121. Zeng, X.; Pielke, R.A.; Eykholt, R. Estimate of the fractal dimension and predictability of the atmosphere. *J. Atmos. Sci.* **1992**, *49*, 649–659. [CrossRef]
122. Kaplan, J.L.; Yorke, J.A. Chaotic behavior of multidimensional difference equations. In *Functional Differential Equations and the Approximations of Fixed Points*; Lecture Notes in Mathematics; Peitgen, H.O., Walther, H.O., Eds.; Springer: New York, NY, USA, 1979; Volume 730, pp. 228–237.
123. Wolf, A.; Swift, J.B.; Swinney, H.L.; Vastano, J.A. Determining Lyapunov exponents from a time series. *Phys. D* **1985**, *16*, 285–317. [CrossRef]
124. Shen, B.-W. Nonlinear feedback in a five-dimensional Lorenz model. *J. Atmos. Sci.* **2014**, *71*, 1701–1723. [CrossRef]
125. Shen, B.-W. Nonlinear feedback in a six-dimensional Lorenz Model. Impact of an additional heating term. *Nonlin. Process. Geophys.* **2015**, *22*, 749–764. [CrossRef]
126. Shen, B.-W. Hierarchical scale dependence associated with the extension of the nonlinear feedback loop in a seven-dimensional Lorenz model. *Nonlin. Process. Geophys.* **2016**, *23*, 189–203. [CrossRef]
127. Shen, B.-W. On an extension of the nonlinear feedback loop in a nine-dimensional Lorenz model. *Chaotic Model. Simul. (CMSIM)* **2017**, *2*, 147–157.
128. Nicolis, C. Probabilistic aspects of error growth in atmospheric dynamics. *Quart. J. Roy. Meteorol. Soc.* **1992**, *118*, 553–568. [CrossRef]
129. Zhang, F.; Sun, Y.Q.; Magnusson, L.; Buizza, R.; Lin, S.-J.; Chen, J.-H.; Emanuel, K. What is the predictability limit of midlatitude weather? *J. Atmos. Sci.* **2019**, *76*, 1077–1091. [CrossRef]
130. Alligood, K.; Saucer, T.; Yorke, J. *Chaos An Introduction to Dynamical Systems*; Springer: New York, NY, USA, 1996; 603p.
131. Meiss, J.D. *Differential Dynamical Systems*; Society for Industrial and Applied Mathematics: Philadelphia, PA, USA, 2007; 412p.
132. Strogatz, S.H. *Nonlinear Dynamics and Chaos: With Applications to Physics, Biology, Chemistry, and Engineering*; Westpress View: Boulder, CO, USA, 2015; 513p.
133. Paxson, W.; Shen, B.-W. A KdV-SIR Equation and Its Analytical Solutions: An Application for COVID-19 Data Analysis. *Chaos Solitons Fractals* **2023**, *173*, 113610. [CrossRef] [PubMed]
134. Kermack, W.; McKendrink, A. A contribution to the mathematical theory of epidemics. *Proc. R. Soc. London. Ser. A Math. Phys. Sci.* **1927**, *115*, 700–721. [CrossRef]
135. Liu, H.-L.; Sassi, F.; Garcia, R.R. Error growth in a whole atmosphere climate model. *J. Atmos. Sci.* **2009**, *66*, 173–186. [CrossRef]
136. Shen, B.-W.; Tao, W.-K.; Wu, M.-L. African Easterly Waves in 30-day High-resolution Global Simulations: A Case Study during the 2006 NAMMA Period. *Geophys. Res. Lett.* **2010**, *37*, L18803. [CrossRef]
137. Shen, B.-W.; Atlas, R.; Oreale, O.; Lin, S.-J.; Chern, J.-D.; Chang, J.; Henze, C.; Li, J.-L. Hurricane Forecasts with a Global Mesoscale-Resolving Model: Preliminary Results with Hurricane Katrina (2005). *Geophys. Res. Lett.* **2006**, *33*, L13813. [CrossRef]
138. Shen, B.-W.; Tao, W.-K.; Atlas, R.; Lee, T.; Reale, O.; Chern, J.-D.; Lin, S.-J.; Chang, J.; Henze, C.; Li, J.-L. Hurricane Forecasts with a Global Mesoscale-resolving Model on the NASA Columbia Supercomputer. In Proceedings of the AGU 2006 Fall Meeting, San Francisco, CA, USA, 11–16 December 2006.
139. Shen, B.-W.; Pielke, R.A., Sr.; Zeng, X.; Cui, J.; Faghih-Naini, S.; Paxson, W.; Atlas, R. Three Kinds of Butterfly Effects within Lorenz Models. *Encyclopedia* **2022**, *2*, 1250–1259. [CrossRef]
140. Palmer, T.N.; Doring, A.; Seregin, G. The real butterfly effect. *Nonlinearity* **2014**, *27*, R123–R141. [CrossRef]
141. Drazin, P.G. *Nonlinear Systems*; Cambridge University Press: Cambridge, UK, 1992; p. 333.
142. Lorenz, E.N. The butterfly effect. In *Premio Felice Pietro Chisesi E Caterina Tomassoni Award Lecture*; University of Rome: Rome, Italy, 2008.
143. Charney, J.G.; Fleagle, R.G.; Lally, V.E.; Riehl, H.; Wark, D.Q. The feasibility of a global observation and analysis experiment. *Bull. Am. Meteor. Soc.* **1966**, *47*, 200–220.
144. GARP. GARP topics. *Bull. Am. Meteor. Soc.* **1969**, *50*, 136–141.
145. Lorenz, E.N. How much better can weather prediction become? *MIT Technol. Rev.* **1969**, 39–49. Available online: https://eapsweb.mit.edu/sites/default/files/How_Much_Better_Can_Weather_Prediction_1969.pdf (accessed on 1 July 2023).
146. Lorenz, E.N. Studies of atmospheric predictability. In *[Part 1] [Part 2] [Part 3] [Part 4] Final Report, February, Statistical Forecasting Project*; Air Force Research Laboratories, Office of Aerospace Research, USAF: Bedford, MA, USA, 1969; 145p. Available online: https://eapsweb.mit.edu/about/history/publications/lorenz (accessed on 1 July 2023).

147. Lorenz, E.N. Estimates of atmospheric predictability at medium range. In *Predictability of Fluid Motions*; Holloway, G., West, B., Eds.; American Institute of Physics: New York, NY, USA, 1984; pp. 133–139.
148. Lorenz, E.N. The growth of errors in prediction. In *Turbulence and Predictability in Geophysical Fluid Dynamics and Climate Dynamics*; Società Italiana di Fisica: Bologna, Italy, 1985; pp. 243–265.
149. Smagorinsky, J. Problems and promises of deterministic extended range forecasting. *Bull. Amer. Meteor. Soc.* **1969**, *50*, 286–312. [CrossRef]
150. Lighthill, J. The recently recognized failure of predictability in Newtonian dynamics. *Proc. R. Soc. Lond. A* **1986**, *407*, 35–50.
151. Rotunno, R.; Snyder, C. A generalization of Lorenz's model for the predictability of flows with many scales of motion. *J. Atmos. Sci.* **2008**, *65*, 1063–1076. [CrossRef]
152. Durran, D.; Gingrich, M. Tmospheric predictability: Why atmospheric butterflies are not of practical importance. *J. Atmos. Sci.* **2014**, *71*, 2476–2478. [CrossRef]
153. Reeves, R.W. Edward Lorenz Revisiting the Limits of Predictability and Their Implications: An Interview from 2007. *BAMS* **2014**, *95*, 681–687. [CrossRef]
154. Shen, B.-W.; Pielke, R.A., Sr.; Zeng, X.; Zeng, X. Lorenz's View on the Predictability Limit. *Encyclopedia* **2023**, *3*, 887–899. [CrossRef]
155. Saiki, Y.; Yorke, J.A. Can the Flap of a Butterfly's Wings Shift a Tornado into Texas—Without Chaos? *Atmosphere* **2023**, *14*, 821. [CrossRef]
156. Chou, Y.-L.; Wang, P.-K. An Expanded Sensitivity Study of Simulated Storm Life Span to Ventilation Parameterization in a Cloud Resolving Model. *Atmosphere* **2023**, *14*, 720. [CrossRef]
157. Shen, B.-W.; Pielke, R.A., Sr.; Zeng, X. One Saddle Point and Two Types of Sensitivities Within the Lorenz 1963 and 1969 Models. *Atmosphere* **2022**, *13*, 753. [CrossRef]
158. Zeng, X. Atmospheric Instability and Its Associated Oscillations in the Tropics. *Atmosphere* **2023**, *14*, 433. [CrossRef]
159. Shen, B.-W.; Pielke, R.A., Sr.; Zeng, X.; Cui, J.; Faghih-Naini, S.; Paxson, W.; Kesarkar, A.; Zeng, X.; Atlas, R. The Dual Nature of Chaos and Order in the Atmosphere. *Atmosphere* **2022**, *13*, 1892. [CrossRef]
160. Yorke, J.; Yorke, E. Metastable chaos: The transition to sustained chaotic behavior in the Lorenz model. *J. Stat. Phys.* **1979**, *21*, 263–277. [CrossRef]
161. Anthes, R.A. Predictability and Predictions. *Atmosphere* **2022**, *13*, 1292. [CrossRef]
162. Wang, C.-C.; Tsai, C.-H.; Jou, B.J.-D.; David, S.J. Time-Lagged Ensemble Quantitative Precipitation Forecasts for Three Landfalling Typhoons in the Philippines Using the CReSS Model, Part I: Description and Verification against Rain-Gauge Observations. *Atmosphere* **2022**, *13*, 1193. [CrossRef]
163. Tseng, J.C.-H. An ISOMAP Analysis of Sea Surface Temperature for the Classification and Detection of El Niño & La Niña Events. *Atmosphere* **2022**, *13*, 919. [CrossRef]
164. Pielke, R., Sr. The Real Butterfly Effect. 2008. Available online: https://pielkeclimatesci.wordpress.com/2008/04/29/the-real-butterfly-effect/ (accessed on 9 July 2023).
165. Anthes, R. Turning the Tables on Chaos: Is the Atmosphere More Predictable than We Assume? UCAR Magazine, Spring/Summer, 6 May 2011. Available online: https://news.ucar.edu/4505/turning-tables-chaos-atmosphere-more-predictable-we-assume (accessed on 9 July 2023).
166. Zeng, X.; Pielke, R.A., Sr.; Eykholt, R. Chaos theory and its applications to the atmosphere. *Bull. Am. Meteorol. Soc.* **1993**, *74*, 631–644. [CrossRef]

 *atmosphere*

Article

Can the Flap of a Butterfly's Wings Shift a Tornado into Texas—Without Chaos?

Yoshitaka Saiki [1,*] **and James A. Yorke** [2,3,4]

1   Graduate School of Business Administration, Hitotsubashi University, Tokyo 186-8601, Japan
2   Institute for Physical Science and Technology, University of Maryland, College Park, MD 20742, USA; yorke@umd.edu
3   Department of Mathematics, University of Maryland, College Park, MD 20742, USA
4   Department of Physics, University of Maryland, College Park, MD 20742, USA
*   Correspondence: yoshi.saiki@r.hit-u.ac.jp

Abstract: In our title, "chaos" means there is a positive Lyapunov exponent that causes the tornado to move. We are asserting that a positive Lyapunov exponent is not always needed to have a butterfly effect. Lorenz's butterfly effect initially appeared in meteorology and has captured the imaginations of people for applications to all kinds of fields. We feel it is important to understand simpler non-meteorological models to understand the additional aspects of the butterfly effect. This paper presents simple linear map models that lack "chaos" but exhibit a butterfly effect: our simplest model does not have any positive Lyapunov exponents but still exhibits a butterfly effect, that is, temporary exponential growth from a tiny perturbation such as one infected mosquito setting off an epidemic outbreak. We focus on a 24-dimensional version of the map where a significant butterfly effect is observed even though the only Lyapunov exponent is 0. We introduce a linear "infected mosquito" model that shows how off-diagonal matrix entries can cause a finite-time growth rate. We argue that the degree of instability in our systems can be better measured by its finite-time growth rate. Our findings suggest that even in linear systems, off-diagonal matrix entries can significantly impact the system's behavior and be more important than the Lyapunov exponents in higher-dimensional systems. A focus on finite-time growth rates can yield valuable insights into the system's dynamics.

Keywords: butterfly effect; sensitive dependence on initial conditions; instability; Zika toy model

Citation: Saiki, Y.; Yorke, J.A. Can the Flap of a Butterfly's Wings Shift a Tornado into Texas—Without Chaos? *Atmosphere* **2023**, *14*, 821. https://doi.org/10.3390/atmos14050821

Academic Editors: Bo-Wen Shen, Roger A. Pielke Sr. and Xubin Zeng

Received: 24 March 2023
Revised: 21 April 2023
Accepted: 28 April 2023
Published: 2 May 2023

1. Introduction

The butterfly effect described by Edward Lorenz [1,2] has become a central concept in the study of nonlinear systems. This effect is typically associated with "chaos" and characterized by positive Lyapunov exponents. Of course, the Lyapunov exponents of a trajectory are a partial description of what happens near the trajectory based on a linearization of the dynamics near the trajectory. That analysis is greatly simplified if the system is linear since the linearization of all trajectories is the same. By studying linear systems, we gain clarity about the relationship between Lyapunov exponents and the butterfly effect, which can be quantified by what we call the "butterfly number" in Section 2. By presenting a butterfly effect in simple linear systems, we hope to contribute to a deeper understanding of the role of dimensionality and off-diagonal terms in the linearization of the dynamics of physical systems. We are not creating a new butterfly effect here, but we are simply pointing out how it can occur in systems for which there is no positive exponent, and we use only elementary mathematical techniques.

Sensitive dependence on initial conditions. There are several ways of characterizing sensitive dependence on initial conditions [3–5]. Sensitive dependence on initial conditions is defined by J. Guckenheimer [5] in any dimension as follows: there exists a positive ε such that, for all x in the phase space and all $\delta > 0$, there is some y that is within a distance δ of x and there is some n such that $d(T^n x, T^n y) > \varepsilon$. The phrase "sensitive dependence on initial conditions" was used by D. Ruelle [6] to indicate some exponential rate of divergence of the orbits of nearby points, which has often been used to characterize chaos in the literature. Guckenheimer's definition does not require there to be a positive Lyapunov exponent, but it is challenging to find a natural system that satisfies their definition but has no trajectories with a positive Lyapunov exponent.

The ancient history of sensitivity to initial conditions. The following seems to describe how a tiny perturbation can lead to large changes through a pattern of increasingly big effects over a finite amount of time. Benjamin Franklin included a version of the tale about horseshoes and battles in their Poor Richard's Almanack. Franklin's was far from the first version. (Benjamin Franklin, Poor Richards Almanack, June 1758, The Complete Poor Richard Almanacks, facsimile ed., vol. 2, pp. 375–377) [7].

The horseshoe nail cascade effect: "For Want of a Nail", an example of "finite-time sensitive dependence"

For want of a nail the shoe was lost;
For want of a shoe the horse was lost;
For want of a horse the rider was lost;
For want of a rider the message was lost;
For want of a message the battle was lost;
For want of a battle the kingdom was lost;
Furthermore, all for the want of a horseshoe nail.

We believe that scientists and mathematicians were among the last people to learn about sensitivity to initial conditions. The above nail tale has been discussed by other authors, including Lorenz [8] and Shen [9]. We mention it here to contrast it with our "parable about angles" in Section 3.

Dynamics of the Lorenz's 1969 paper [2]. The Lorenz paper from 1969 analyzed a high dimensional spatial multi-scale model, describing how energy at the model's smallest scales could propagate quickly to the largest scales, provided there was lots of energy in the smallest scales. Perhaps huge numbers of energetic butterflies could provide that energy, although the article did not mention butterflies. Wikipedia's "butterfly effect" [10] says: "According to Lorenz, when he failed to provide a title for a talk he was to present at the 139th meeting of the American Association for the Advancement of Science in 1972, Philip Merilees concocted 'Does the flap of a butterfly's wings in Brazil set off a tornado in Texas?' as a title." A single flapping of a butterfly causes chaos in the Lorenz paper from 1963 [1], whereas the butterfly effect in the Lorenz paper from 1969 [2] is caused by the propagation of huge numbers of flapping butterflies. The analysis of the Lorenz's 1969 model's linearly unstable solutions was first documented in Shen et al. [11].

Outline of the paper. In Section 2, we begin with a mosquito model instead of a butterfly. We introduce a simple linear one-space-dimension map. We show that the map shows a butterfly effect, "the sensitive dependence on initial conditions" for some finite times, although the map is not chaotic in the sense that it does not have positive Lyapunov exponents. In Section 3, we describe another well-known multi-dimensional torus map where the coordinates are angles. We investigate the numerics of the map in Section 4. In Section 5, we explain the known mathematical results and introduce a conjecture together with some numerical evidence. We discuss the related works and summarize our results in Section 6.

2. Linear K-Dimensional System Showing the Butterfly Effect

Models for the spread of Zika-infected mosquitoes. Jeffery Demers et al. [12] investigated a heterogeneous, two-space-dimensional model of infected Aedes mosquitoes. Some species of mosquitoes, such as Aedes, can become infected and can transmit diseases such as Zika, and they have little mobility during their lifetimes. This means it can be effective to focus the killing of mosquitoes in the small areas where the disease exists. See our discussion section (Section 6.2) for New York Times articles that discuss the Zika outbreak in Miami, Florida, and elsewhere. The World Health Organization recently declared mosquito-borne Zika an "international public health emergency". Our focus is on how a new exponentially growing outbreak is local early in the outbreak. Hence, a new outbreak can sometimes be interrupted using only local strategies for killing mosquitoes, as opposed to expensive strategies that would try to (temporarily) eradicate mosquitoes in a large region. Of course, "exponential growth" is temporary. Our model is spatially one-dimensional, but it is easy to convert it into a two-space dimension model, provided the winds generally blow mosquitoes in one direction.

A one-space-dimensional infected-mosquito model. We assume that there is a line of land tracts numbered $k = 1, 2, 3, \ldots$, perhaps along a road. We imagine that these regions all have a similar area, perhaps one square hectare. Our model's infected mosquitoes live for one time period, and in one time period, each infected mosquito in tract k has $a_k + b_k$ progeny, which we imagine is approximately 2. We assume $a_k \approx 1$ mosquitoes stay in tract k and $b_k \approx 1$ are blown by a constant wind into tract $k + 1$. Writing x_k for the number of infected mosquitoes in tract k yields the following dynamical model, which we call our "(infected-) mosquito model".

$$(x_1, x_2, \ldots, x_K) \mapsto (a_1 x_1, a_2 x_2 + b_1 x_1, \ldots, a_K x_K + b_{K-1} x_{K-1}), \tag{1}$$

This equation can be viewed as a linearization about $y_j = 0$ for all coordinates j—of a spatial logistic-based map. One possibility uses $g(y) := y(1 - y)$ where $y = y_k$ is a density at each tract k with a maximum sustainable value of 1. The nonlinear equation is the following:

$$(y_1, y_2, \ldots, y_K) \mapsto \big(g(a_1 y_1), g(a_2 y_2 + b_1 y_1), \ldots, g(a_K y_K + b_{K-1} y_{K-1})\big). \tag{2}$$

Recall that y_k is a density while x_k is the number of mosquitoes. Huge values of x_k correspond to moderate values of y_k.

We can write (1) in vector format, writing $X = (x_1, x_2, \ldots, x_K)$. For time $n \geq 0$, let $X^n = (x_1^n, x_2^n, \ldots, x_K^n)$ be a trajectory of (1) determined by

$$X^{n+1} = M_{gen} X^n, \tag{3}$$

where we define M_{gen} as follows:

$$M_{gen} = \begin{pmatrix}
a_1 & 0 & 0 & 0 & 0 & 0 & \cdots & 0 \\
b_1 & a_2 & 0 & 0 & 0 & 0 & \cdots & 0 \\
0 & b_2 & a_3 & 0 & 0 & 0 & \cdots & 0 \\
0 & 0 & b_3 & a_4 & 0 & 0 & \cdots & 0 \\
0 & 0 & 0 & b_4 & a_5 & 0 & \cdots & 0 \\
0 & 0 & 0 & 0 & b_5 & a_6 & \cdots & 0 \\
 & & & \vdots & & & & \\
0 & 0 & 0 & & \cdots & & b_{K-1} & a_K
\end{pmatrix}. \tag{4}$$

There can be rapid growth over time for large K when $a_k + b_k > 1$ for all k. We invite the reader to try various choices of coefficients. This model gives many cases to consider.

Notation for "spatially homogeneous" matrices. Tri-diagonal matrix. In this part, we comment on the case where a map is determined by the tri-diagonal matrix M_{abc}. Assume that the $K \times K$ matrix M_{abc} is a tri-diagonal matrix with the following form:

$$M_{abc} = \begin{pmatrix} a & b & 0 & 0 & & 0 & 0 & \cdots & 0 \\ c & a & b & 0 & & 0 & 0 & \cdots & 0 \\ 0 & c & a & b & & 0 & 0 & \cdots & 0 \\ 0 & 0 & c & a & & b & 0 & \cdots & 0 \\ 0 & 0 & 0 & c & & a & b & \cdots & 0 \\ 0 & 0 & 0 & 0 & & c & a & \cdots & 0 \\ & & & \vdots & & & & & \\ 0 & 0 & 0 & & & \cdots & & c & a \end{pmatrix}, \tag{5}$$

where a, b, and c are non-negative real numbers. This type of matrix is considered to commonly appear in a linearized system in various nonlinear phenomena. An example with a linearization that is somewhat similar is the Gledzer–Okhitani–Yamada (GOY) shell model of fluid turbulence that mimics the Galerkin spectral equations of the Navier–Stokes equations (A detailed analysis of this matter will be reported elsewhere as a joint work with Miki U. Kobayashi.). It is a system of N-dimensional complex-valued ordinary differential equations with the following form [13]:

$$\left(\frac{d}{dt} + \nu k_n^2 \right) u_n = i[c_n^{(1)} u_{n+1}^* u_{n+2}^* + c_n^{(2)} u_{n-1}^* u_{n+1}^* + c_n^{(3)} u_{n-1}^* u_{n-2}^*] + f\delta_{n,1},$$

where f is a forcing parameter, $\delta_{n,1}$ is a Kronecker delta, and $c_1^{(2)} = c_1^{(3)} = c_2^{(3)} = c_{N-1}^{(1)} = c_N^{(1)} = c_N^{(2)} = 0$, and $c_n^{(1)} = k_n, c_n^{(2)} = -k_{n-2}, c_n^{(3)} = -k_{n-3}$ with $k_n = 2^{n-4}$ for other n.

The eigenvalues the matrix M_{abc} are

$$\lambda_k = a + 2\sqrt{bc} \, \cos\left[\frac{k\pi}{K+1} \right], \quad k = 1, \ldots, K; \tag{6}$$

the **Lyapunov number** is $\max_k |\lambda_k|$, and the Lyapunov exponent is the log of the Lyapunov number. From this formula, we obtain an inequality:

$$\lambda_k \leq a + 2\sqrt{bc}, \quad k = 1, \ldots, K. \tag{7}$$

See [14] for details. For large K, $\max_k \lambda_k \simeq a + 2\sqrt{bc}$. Of course if $b = c$, the right-hand side is $a + b + c$.

Define the **butterfly number** (or **the local Lyapunov number**) of the tri-diagonal matrix to be the column sum of the typical column of the matrix,

$$\beta(M_{abc}) := a + b + c. \tag{8}$$

For positive a, b, c,

$$\beta(M_{abc}) > \max |\lambda_k|,$$

which is the largest Lyapunov number. Hence, the temporary growth rate is $\beta(M_{abc})$. If $a = b = 0$ and $c = 2$, the infected mosquitoes double each generation (until the open boundary at tract K is reached) while the Lyapunov number is 0 (and the Lyapunov exponent is $-\infty$. If $X = (\ldots, x_k, \ldots)$ has no negative entries and the first and last entries are 0, then $\sum_k x_k' = \beta(M_{abc}) \sum_k x_k$ where $x_k' = (M_{abc}X)_k$. Hence, the sum of the entries increases by a factor of $a + b + c$ from one application of M_{abc}.

To focus on a truly elementary case, we focus on the case where all $a = c = 1$ and $b = 0$ or, even more simply $a = 0, c = 1$, and $b = 0$.

Our model still has a one-space dimension and is spatially homogeneous (except at entries 1 and K).

$$(x_1, x_2, \ldots, x_K) \mapsto (x_1, x_2 + x_1, \ldots, x_K + x_{K-1}), \tag{9}$$

the one-mosquito-generation map with K land tracts. Assume that, at time 0, there is one infected mosquito, and it is in tract 1.

As we describe later, tract k at time n for $k \geq n \geq 1$ will have $\binom{k}{n}$ infected mosquitoes. Recall that $\binom{k}{n} = \frac{k!}{n!(k-n)!}$, which is 0 if $n > k$. Figure 1 displays the distribution. It seems to show a pattern analogous to the pipe flow [15], where the center of the distribution at time n is tract $\frac{n}{2}$. Of course, pipe flow is a nonlinear process, while our model emphasizes only the linear growth phase. See our remarks in Section 6 on the work of Kaneko and Crutchfield on pipe flow. Later, we discuss how the center of the infected mosquito distribution moves, as seen in Figure 1. Clearly, in reality, the population of infected mosquitoes exists against a background of a large uninfected population. The number of infected mosquitoes would eventually saturate, a fact not reflected in our model.

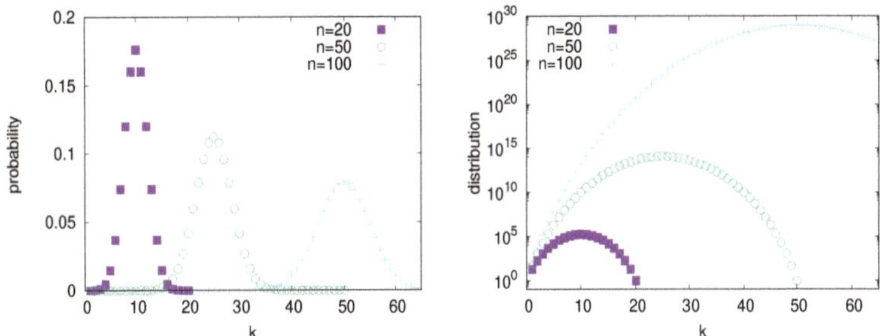

Figure 1. The mosquito model solution $\binom{n}{k}$ where k is the number of the tract (or location), and n is time. Normalized binomial probability distributions $\binom{n}{k}2^{-n}$ **(left)** and log plot of $\binom{n}{k}$ **(right)** are shown for $n = 20, 50, 100$. Both have a mean $\frac{n}{2}$ and standard deviation $\frac{\sqrt{n}}{2}$. The center of the distribution, which is the mean, moves as n changes, and the distribution on the left has sum $= 1$, and on the right, $\binom{n}{k}$ has sum $= 2^n$. See Figure 2 for the movement of the peak as time n increases.

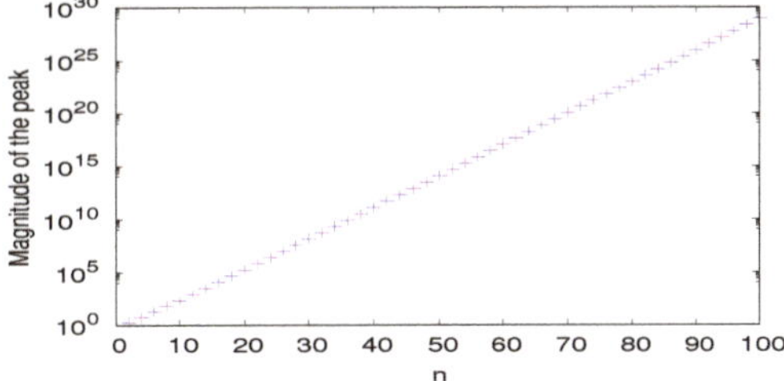

Figure 2. Peak value $\binom{n}{n/2}$ of the distribution $\binom{n}{k}$. The binomial is the vector $X^n = M^n(X)$ for the initial value $X = e_1 = (1,0,0,\ldots)$—with respect to n for even n, where n can be thought of as time measured in mosquito generations.

One infected Miami mosquito. We examine a finite line of K tracts where all $x_k \geq 0$, so eventually, if n is large, the largest value of X^n occurs at the end, i.e., at x_K. In Miami, the winds tend to blow from west to east, and since Miami is on the western edge of the Atlantic Ocean, its mosquitoes in the last tract can be blown into the ocean and cease to be a problem. We assume an outbreak starts with one mosquito infected with the Zika virus. Here, we aim at simplicity. A more detailed model such as Demers' would be spatially heterogeneous and two-dimensional, which would include people infecting mosquitoes and mosquitoes infecting people. The resulting behavior could be similar to the explosive growth we observe here, depending on the details, possibly growing even faster.

Again, we can write (9) in vector format, writing $X = (x_1, x_2, \ldots, x_K)$. For time $n \geq 0$, let $X^n = (x_1^n, x_2^n, \ldots, x_K^n)$ be a trajectory of (9) determined by

$$X^{n+1} = MX^n, \tag{10}$$

where M is the two-diagonal matrix $K \times K$ matrix M_{101}, i.e., $a = c = 1, b = 0$.

Let e_k denote the vector whose k^{th} entry is 1 and its other entries are 0. The finite-dimensional matrix M (with dimension K) has only one eigenvector, $e_K := (0, \ldots, 0, 1)$, and its eigenvalue is 1, so all of the map's Lyapunov exponents are 0. The multiplicity of the eigenvector is K. The matrix M is a Jordan block matrix.

The time n mosquito distribution $M^n e_1$ for Equation (10). The natural initial condition for the mosquito model is $e_1 = (1, 0, \ldots, 0)$. For this initial state e_1, i.e., one mosquito in tract 1 at time 0, the number of infected mosquitoes at any later time $n \geq 0$ on tract k where $k \leq n, K$ is $\binom{n}{k}$, and the total number of infected mosquitoes is 2^n provided $n \leq K$.

The vector $M^n e_1$ at time n is the left-most column of M^n, and its last coordinate x_K is $\binom{n}{K-1}$, which, of course, is 0 for $n < K - 1$. Since K is fixed, x_K^n grows slower than exponentially as n increases.

If the last coordinate x_K is 0 at time n, then the total number of infected mosquitoes at time $n + 1$ is twice the total at time n, i.e., for time n,

$$\sum_{k=1}^{K} x_k^{n+1} = 2 \sum_{k=1}^{K} x_k^n. \tag{11}$$

In this case, generally, when time n is less than K, the mosquitoes have not yet reached the last tract, K, and the application of M doubles the sum. See Figure 1 (right). As mentioned above, there is only one eigenvector of M, and that is $(0, \ldots, 0, 1)$ i.e., all entries except the last are 0, and the last is one. This is an eigenvector with an eigenvalue of 1 with a multiplicity of K. Hence, the rapid growth of the sum of the coordinates seen in Equation (11) is not because of the eigenvalue but is **a result of the off-diagonal terms that do not contribute to the eigenvalues**. While Lyapunov exponents come from linearizations, they do not reflect all of the information in the linearization. Since our map is linear, this point is easier to see.

3. Converting Each x_k into a Direction on a Circle

Skew-product map of a torus. The special mosquito map can be converted into a well-known "skew-product map" on a torus. Each x_k becomes an angle $\theta_k = x_k \bmod 1$ in a circle that we represent by $[0, 1)$, and we identify the ends, 0 and 1, by computing each $\theta_k, \bmod 1$. In a physics context, angles are often referred to as phases.

It is traditional to include an irrational α, in the rotation map on the first circle $\mathbb{T}^1$,

$$\theta_1 \mapsto \theta_1 + \alpha \ \bmod 1, \tag{12}$$

and having α irrational is required for the results in Section 5. We also obtain maps of any finite dimension, including the long-studied two-dimensional map on the two-torus $\mathbb{T}^2$ [16–19],

$$(\theta_1, \theta_2) \mapsto (\theta_1 + \alpha, \theta_2 + \theta_1) \ \bmod 1, \tag{13}$$

where, in this paper, "$v \bmod 1$" signifies that $\bmod 1$ is always applied to each coordinate of a vector v.

We investigate the map of the K-dimensional torus $\mathbb{T}^K$ because it exhibits a butterfly effect (via the perturbation of angles):

$$(\theta_1, \theta_2, \ldots, \theta_K) \overset{F}{\to} (\theta_1 + \alpha, \theta_2 + \theta_1, \ldots, \theta_K + \theta_{K-1}) \ \bmod 1, \tag{14}$$

where K is fairly large, perhaps $K \simeq 24$. We choose 24 for illustrations because the width of Texas is approximately 2^{24} times the size of a 3 cm. butterfly.

Each coordinate value, θ_k, is a point on a circle, so we can refer to it as a "direction" or an angle. Each perturbation of direction θ_k on iterate n perturbs the direction θ_{k+1} on iterate $n+1$ and later. We make a conjecture about the behavior of this equation in Section 5. Write $\Theta = (\theta_1, \theta_2, \ldots, \theta_K)$, and for $n \geq 0$, let Θ^n be a trajectory of (14).

Notice that the kth coordinate at time n, namely Θ_k^n, only depends on Θ_i^{n-1} for $i \leq k$, so that Θ^n is a trajectory of (14), and for $K = 2$, it is a vector of length 2, so then it is a trajectory of (13). Writing (14) in vector format

$$\Theta^{n+1} = M\Theta^n \bmod 1. \tag{15}$$

Notice that because all entries of M are integers, the map $[M^n(\Theta)] \bmod 1$ (applying $\bmod\, 1$ only after n applications of M) is identical to $[M(\Theta) \bmod 1]^n$, applying $\bmod\, 1$ at each iterate. Furthermore, if we make numerical studies in which each coordinate remains less than 1, the results are meaningful regardless of whether $\bmod\, 1$ is applied.

A parable about angles. This part is about angles or orientations. We now propose a parable about changing directions, where one orientation of something causes a change in the orientation of something larger. Imagine a lot of activity in a flat field with a butterfly, a bird, a cat, a dog, a person, a bike, a car, and a truck, all traveling in different directions. The direction of each is a point in a circle, collectively a point on an eight-dimensional torus $\mathbb{T}^8$.

A butterfly flaps its wings;
a nearby bird changes direction;
a running cat watching the bird swerves;
a dog swerves at the motion of the cat;
a walking person swerves to avoid hitting the dog;
a passing bicycle swerves a bit;
causing a passing car to change its direction;
a truck changes directions in response.

The butterfly flapping its wings cascades to ever larger scales, perhaps to Texas-sized scales containing tornadoes. All of these changes in directions need not affect the energy of each component. Or we can follow Lorenz, who investigates packets of air of different sizes, small packets perturbing larger packets, in increasing sequence of increasingly large air packets. Each packet is twice the diameter of its predecessor [2]. Tiny initial perturbations throughout the small-scale packets grow quickly and together become a large-scale perturbation. In this sense, one butterfly flapping its wings in the smallest packet under consideration totally changes the direction of the largest-scale tornado environment, shifting the tornado perhaps from Oklahoma to Texas, or vice versa. Oklahoma and Texas are adjacent states in the USA. These are among the places in the world with the most severe tornado weather.

The perturbations keep influencing the larger scales, adding ever-increasing perturbations to the direction of the next larger packet. The time scale for the perturbations to occur is shorter for the small scales than for the larger scales, but we ignore this inconsistency.

4. Numerical Investigations

We use a superscript to denote the iterate number of vectors such as ϕ, θ, and ψ and a subscript for coordinate number. Choose initial vectors $\phi, \theta \in \mathbb{T}^K$ so that only the first coordinates (denoted by subscripts) have a difference: $(\phi - \theta)_1 := \phi_1 - \theta_1 = 10^{-30}$, and for coordinate $k > 1$, $(\phi - \theta)_k := \phi_k - \theta_k = 0$. We find that the 20th coordinate of the 300th iterate of M is approximately $\frac{1}{2}$, i.e.,

$$(\phi - \theta)_{20}^{300} = M^{300}(\phi - \theta)_{20} \bmod 1 \approx \frac{1}{2}.$$

Note that the value $\frac{1}{2}$ is the maximum possible difference since ϕ_k and θ_k are in a circle where the maximum distance between two points is $\frac{1}{2}$.

We can think of different θ_ks as representing behaviors at different scales of different swirls in the atmosphere, consecutive scales representing the size differences of a factor of 2. Perhaps we can say that a typical butterfly has a size of roughly 0.06 meters, while tornado weather patterns have a scale of at least 10^6 meters, which is the approximate width of Texas. These differ by roughly 2^{24} (or more precisely 23.99). Each swirl can change the direction of the next higher swirl. In order to realistically scale the butterfly problem is $K \approx 24$ for our skew-product system Equation (14). Each swirl might be twice the scale of the next smaller scale.

Of course, there is an immense number of air packets that are the size of a butterfly, and perturbations in each might affect some next larger packet. An intermediate-size packet might be perturbed by several smaller packets. However, here, we look only at a single cascade starting with one butterfly, and for each packet, we only look at its influence on one larger packet. Realistically, the time of propagation from the kth packet to the larger $(k+1)^{\text{st}}$ is longer as k increases, a fact that our maps do not reflect.

A moving peak of the distribution. In the following Table 1, define

$$C(n) := \binom{n}{n/2}$$

for $n = 2, 4, 6, \ldots$. It is the magnitude of the largest coordinate of $M^n e_1$ where $e_1 = (1, 0, 0, \ldots)$ for the case of unbounded space. Hence the location of the peak value is at $\frac{n}{2}$. This location moves to larger values as time n increases.

The number $C(n)^{1/n}$ is a finite-time geometric growth rate (which approaches 2 as can be seen in the table). This value approaches 2, but it is showing that as n increases, some circle (namely circle # $n/2$) is strongly affected by the initial tiny perturbation. See also Figure 2.

Table 1. The magnitude of the largest coordinate $\binom{n}{n/2}$ of $M^n e_1$ where $e_1 = (1, 0, 0, \ldots)$. Convergence of the right column to 2 is slow as n increases.

n	$C(n) = \binom{n}{n/2}$	$C(n)^{1/n}$
2	2	1.414214
4	6	1.565086
22	705,432	1.844331
24	2,704,156	1.853537
26	10,400,600	1.861602
100	1.009×10^{29}	1.950018

Sensitivity is about the behavior of the difference between two different trajectories. Since the equations we study are linear, we can write the equations for such a difference. Let $\psi = \phi - \theta$ where ϕ and θ are two trajectories. Then, the difference ψ satisfies

$$\psi = (\psi_1, \psi_2 \ldots, \psi_K) \mapsto (\psi_1, \psi_2 + \psi_1, \ldots, \psi_K + \psi_{K-1}) \bmod 1. \tag{16}$$

This equation without mod1 is the mosquito model (9). In that case, the sum of coordinates of ψ doubles each iterate in the unbounded case where K is infinite. We consider two trajectories whose initial conditions differ only in the first coordinate. In Figure 3, we see the growth in the difference in coordinate k of ψ between two trajectories with respect to n. It asymptotically converges to an exponential growth.

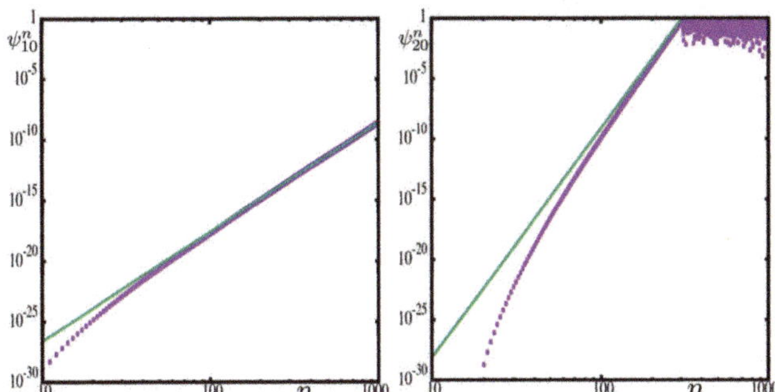

Figure 3. For Equation (16), the growth in a ψ coordinate of the difference between two trajectories. The initial condition for the difference ψ is chosen with the first coordinate $\psi_1 = 10^{-30}$, and all other coordinates are 0. Notice that, for each k, $\psi_k^n=0$ for iterate n when $n < k$. (**Left**), ψ_{10}^n. The dots plot ψ_{10}^n as a function of the number of iterates n of the map, and the straight line (green) is $C_1 \cdot n^9$, where C_1 is chosen so that the line and the curve are tangent. (**Right**), ψ_{20}^n. This panel is similar to the left panel using the straight line $C_2 \cdot n^{19}$. The green lines have "slopes" $s = 9$ and 19 in our "log log" plot, respectively, i.e., $\log(\psi_{10}^n \text{ or } \psi_{20}^n) = s \cdot \log(n) + \text{constant}$. This fact does not depend on the base of the logs since the slope is the ratio of two logs with the same base.

5. Occasional Closest Approach of Two Trajectories

In this section, we discuss the mathematical aspects of the map F in Equation (14) concerning the degree of complexity.

Scrambled pairs. For a map F, we say a pair ϕ, θ of points in the space is **scrambled** [20] if

$$\liminf_{n \to \infty} \text{dist}(F^n(\phi), F^n(\theta)) = 0 \quad \text{and} \quad \limsup_{n \to \infty} \text{dist}(F^n(\phi), F^n(\theta)) > 0.$$

For the map Equation (14), the first coordinate of $F^n(\phi) - F^n(\theta)$ is independent of n. Hence, the infimum of the distance between $F^n(\phi)$ and $F^n(\theta)$ could go to zero only if we choose two points with the same first coordinate.

Let $\mathcal{P}$ be a set of pairs of points. Let $\textbf{diam}(\mathcal{P})$ denote the maximum possible distance between pairs in $\mathcal{P}$. We say a pair ϕ, θ in $\mathcal{P}$ is **totally scrambled** (in $\mathcal{P}$) if the pair is scrambled and satisfies

$$\limsup_{n \to \infty} \text{dist}(F^n(\phi), F^n(\theta)) = \text{diam}(\mathcal{P}).$$

Conjecture 1. *For the map Equation* (14) *on* $\mathbb{T}^K$ *for* $K > 2$, *let* $\mathcal{P}$ *be the set of pairs* ϕ, θ *whose first coordinates are equal. Then, almost every pair* ϕ, θ *in* $\mathcal{P}$ *is totally scrambled.*

We describe the numerical evidence for this conjecture later in this section. Notice that $\mathcal{P}$ is invariant: for each pair ϕ, θ in $\mathcal{P}$, the pair $(F^n(\phi), F^n(\theta))$ is in $\mathcal{P}$ for all $n > 0$.

Results for the map Equation (14) when α is irrational. The following theorems by Furstenberg [19] are fundamental results for the map (14). See also Furstenberg Theorem 4.21 (p. 116) and Corollary 4.22 of the book [18].

Theorem 1. *The map is minimal (every orbit is dense).*

Recall: A Borel set is any set that can be formed from open sets through the operations of countable union, countable intersection, and complements.

Theorem 2. *The map is uniquely ergodic (there is only one invariant Borel probability measure).*

Since the map Equation (14) preserves the Lebesgue measure, its unique ergodicity implies its minimality. In general, minimality does not imply unique ergodicity, for example, interval exchange transformations (IET).

Numerical evidence for our conjecture. Two trajectories might occasionally move close together, only to diverge, repeating this pattern in a non-periodic or apparently stochastic manner. In Figure 4, we can see that the closest approach $y(N)$ between two trajectories during the time interval $[0, N]$ decreases as time N increases and converges to 0. It supports the first condition of the above conjecture. The decay rate of the closest difference at time N seems to obey a power law which can be explained below. Suppose we choose N random points using a uniform distribution on a D-dimensional box $[0, \frac{1}{2}]^D$. For $0 < \varepsilon < \frac{1}{2}$, when choosing N random independent points, the expected number of points in $[0, \varepsilon]^D$ is $(2\varepsilon)^D N$. If we choose ε so that the expected number is 1, we obtain $(2\varepsilon)^D = N^{-1}$, i.e., $2\varepsilon = N^{\frac{-1}{D}}$. In our case, $D = K - 1$. Thus, we plot $\varepsilon = \frac{1}{2}N^{\frac{-1}{K-1}}$ on a log-log scale which yields a straight line $\frac{-1}{K-1}\log(N) - \log 2$ with slope $-\frac{1}{K-1}$. For such an ε, given N, the probability that none of the N points are in the $[0, \varepsilon]^D$ is $(1 - (\frac{2}{\varepsilon})^D)^N = (1 - \frac{1}{N})^N \approx \frac{1}{e}$. Figure 4 shows that the difference between two trajectories behaves in this manner. Note that, if two values of N are chosen close together, whether the plotted points on curves are above or below the line is correlated.

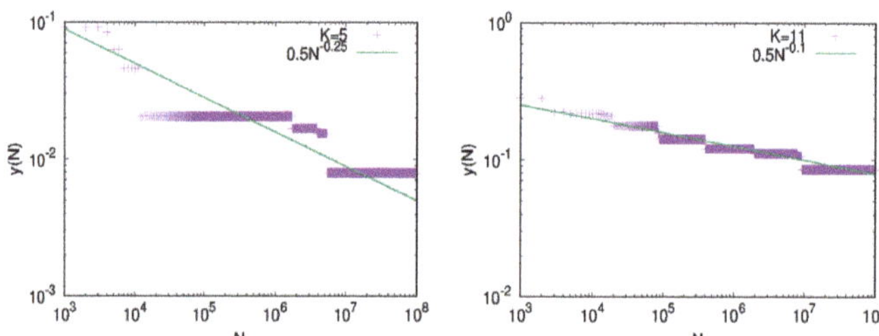

Figure 4. Closest approach between two trajectories θ^n and ϕ^n up to time N. The initial points θ_1^0, ϕ_1^0 were chosen at random. For coordinates $k = 2, \cdots, K$, at time n, define $\delta^n := \max_{k=2,\cdots,K} |\theta_k^n - \phi_k^n|$. Notice coordinate $k = 1$ is excluded since for all $n > 0$, $\theta_1^n - \phi_1^n = \theta_1^0 - \phi_1^0$ is constant. The **(left)** panel is for dimension $K = 5$ and the **(right)** is for $K = 11$. For both, we plot $y(N) := \min_{n \leq N} \delta^n$, the closest approach between two trajectories θ and ϕ on $[1, N]$. This is non-increasing as N increases. For comparison, we also plot $y = \frac{1}{2}N^{-\frac{1}{K-1}}$, which in this log-log plot is a straight line with slope $\frac{-1}{K-1}$.

6. Discussion

6.1. Summary

We present linear models exhibiting large-scale rapid growth that is not reflected in the Lyapunov exponent of the model since there exist no positive Lyapunov exponents. The growth of solutions is due to off-diagonal terms. The off-diagonal terms correspond, in our case, to spatial transmission. This is an often-discussed topic, but our treatment is different in that, here, our system is linear. Since M is a lower-triangular matrix, the off-diagonal terms do not affect the eigenvalue(s) of M. We observe an initial exponential growth in the total number of infected mosquitoes, but the exponential growth is only transient. Note that this is not a chaotic transient [21], as there is no chaotic invariant set.

These models present an alternative source for the sensitivity to initial conditions present in complex systems. See Sander and Yorke [22] for a discussion of some of the many definitions of chaos that reflect different aspects of chaos.

6.2. Zika-Infected Mosquitoes in Miami, Florida, USA

The problem of containing and controlling Zika-infected mosquitoes is based on an actual outbreak in the Americas and specifically in Miami, Florida, where it was a localized outbreak. The New York Times published articles about Zika in Miami with the titles. (The online titles may differ from the print version's titles).

- (3 February 2016) "Fighting the Zika Virus. Subtitle: The World Health Organization was right to declare the mosquito-borne disease an international public health emergency" (by the Editorial Board)
- (1 April 2016) "In Miami, Facing Risk of Zika With Resolve but Limited Resources. Subtitle: Stopping the virus's spread, health experts say, requires vigorous control of the mosquitoes that carry it."
- (9 August 2016) "Zika Cases Rise in Miami, and Officials Try to Soothe Fears"
- (print edition 10 August 2016, page A3) "Zika Cases Rise in Miami, but Remain Contained"

Our simplified model shows that a local outbreak can be treated locally with insecticides, but there is a risk of a rapidly growing outbreak. In particular, it is not necessary to treat all of Miami with insecticides. We rejected the alternative approach of treating the problem of mosquitoes with partial differential equations with diffusion, thereby creating a model no politician in Miami could understand.

6.3. Related Works

Other than simple temporal chaos, complexity in dynamical systems can be generated by several mechanisms, some of which are similar to those in our simple map in some sense. See [21].

Works on Kaneko and Crutchfield on coupled map lattices. There is of course, a huge literature on pipe flow where there can be a moving localized region of turbulence, outside of which the flow is laminar. This can be seen in the earlier paper by Briggs [23] for a similar phenomenon called convective instability in plasma physics. Kerswell [24] is a more recent discussion of pipe flow with real fluid turbulence.

In Kaneko's study [25] of a coupled map lattice model of pipe flow, the isolated regions of chaos move at a fixed rate. To detect the chaos, he uses a moving frame of reference to compute the co-moving Lyapunov exponent, which is related to the propagation of the disturbance in space.

There is also a study [15] on the following coupled system of the Nagumo map, each of which has stably periodic dynamics, but the total system shows similar behavior to that shown in spatio-temporal chaotic dynamics:

$$x_k^{n+1} = \frac{1}{2r+1} \sum_{j=-r}^{r} f(x_{k+j}^n), \tag{17}$$

where $f(x) = sx + \omega \,(\mathrm{mod}\,1)$ for some real values s and ω, and $x_k^n \in [0,1)$ is the state at the kth site ($k = 0, \ldots, K-1$) at time n, r the radius of the coupling, and K the number of sites. In their work, r is fixed as 1, meaning the nearest neighbor coupling. Even if the local dynamics is periodic under conditions such as $s = 0.91, \omega = 0.1$, when it is coupled into a spatial system, "turbulent" behavior can exist for times that grow faster than exponentially with increasing system volume.

Spatially varying chaos. Nonlocally coupled oscillators can exhibit complex spatio-temporal patterns called chimera states, consisting of coexisting domains of spatially coherent (synchronized) and incoherent dynamics [26]. High-dimensional coupled Kuramoto oscillators can show chaotic behavior, although a single oscillator behaves periodically without mutual interactions [27–29]. It is also well known that the degree of chaos in the Kuramoto–Sivashinsky system becomes higher as the spatial size increases [30,31]. The system can show so-called spatio-temporal chaos.

Strange nonchaotic attractor (SNA) [32–34]. The skew-product map above and SNA models have an irrational rotation as in Equation (12). Both can show rapid growth, but neither has a positive Lyapunov exponent. SNA models are nonlinear and usually two-dimensional, whereas our maps are linear and high-dimensional.

Author Contributions: Investigation, Y.S. and J.A.Y.; writing—review and editing, Y.S. and J.A.Y. All authors have read and agreed to the published version of the manuscript.

Funding: JSPS KAKENHI 19KK0067, 21K18584 and 23H04465 and JSPS JPJSBP 120229913.

Institutional Review Board Statement: Not applicable.

Informed Consent Statement: Not applicable.

Data Availability Statement: Not applicable.

Acknowledgments: The authors would like to thank Joseph Auslander, Dima Dolgopyat, Miki U Kobayashi, Yuzuru Sato, Bo-wen Shen, Hiroki Takahasi, Natsuki Tsutsumi, and Michio Yamada for giving us some important references and for their insightful comments. We also thank the referees for their comments to improve the manuscript. This version greatly benefited from their comments.

Conflicts of Interest: The authors declare no conflict of interest.

References

1. Lorenz, E.N. Deterministic nonperiodic flow. *J. Atmos. Sci.* **1963**, *20*, 130–141. [CrossRef]
2. Lorenz, E.N. The predictability of a flow which possesses many scales of motion. *Tellus* **1969**, *21*, 289–307. [CrossRef]
3. Banks, J.; Brooks, J.; Cairns, G.; Davis, G.; Stacey, P. On Devaney's definition of chaos. *Am. Math. Mon.* **1992**, *99*, 332–334. [CrossRef]
4. Glasner, E.; Weiss, B. Sensitive dependence on initial conditions. *Nonlinearity* **1993**, *6*, 1067. [CrossRef]
5. Guckenheimer, J. Sensitive dependence to initial conditions for one dimensional maps. *Commun. Math. Phys.* **1979**, *70*, 133–160. [CrossRef]
6. Ruelle, D. Small random perturbations of dynamical systems and the definition of attractors. *Commun. Math. Phys.* **1981**, *82*, 137–151. [CrossRef]
7. Franklin, B. *Poor Richard's Almanack*; Barnes & Noble Publishing: New York, NY, USA, 2004.
8. Lorenz, E.N. The butterfly effect. In *Premio Felice Pietro Chisesi e Caterina. Tomassoni award Lecture*; University of Rome: Rome, Italy, 2008.
9. Shen, B.W.; Pielke, R.A.; Zeng, X.; Cui, J.; Faghih-Naini, S.; Paxson, W.; Atlas, R. Three Kinds of Butterfly Effects within Lorenz Models. *Encyclopedia* **2022**, *2*, 1250–1259. [CrossRef]
10. Wikipedia. Butterfly Effect—Wikipedia, The Free Encyclopedia. Available online: http://en.wikipedia.org/w/index.php?title=Butterfly%20effect&oldid=1140833034 (accessed on 24 March 2023).
11. Shen, B.W.; Pielke, R.A.; Zeng, X. One Saddle Point and Two Types of Sensitivities within the Lorenz 1963 and 1969 Models. *Atmosphere* **2022**, *13*, 753. [CrossRef]
12. Demers, J.; Bewick, S.; Agusto, F.; Caillouët, K.A.; Fagan, W.F.; Robertson, S.L. Managing disease outbreaks: The importance of vector mobility and spatially heterogeneous control. *PLoS Comput. Biol.* **2020**, *16*, e1008136. [CrossRef]
13. Yamada, M.; Ohkitani, K. Lyapunov spectrum of a chaotic model of three-dimensional turbulence. *J. Phys. Soc. Jpn.* **1987**, *56*, 4210–4213. [CrossRef]
14. Kulkarni, D.; Schmidt, D.; Tsui, S.K. Eigenvalues of tridiagonal pseudo-Toeplitz matrices. *Linear Algebra Its Appl.* **1999**, *297*, 63–80. [CrossRef]
15. Crutchfield, J.P.; Kaneko, K. Are attractors relevant to turbulence? *Phys. Rev. Lett.* **1988**, *60*, 2715. [CrossRef]
16. Assani, I. Wiener-Wintner Ergodic Theorem, in Brief. *Not. Am. Math. Soc.* **2022**, *69*, 198–209. [CrossRef]
17. Auslander, J. *Minimal Flows and Their Extensions*; Elsevier: Amsterdam, The Netherlands, 1988.
18. Einsiedler, M.; Ward, T. *Ergodic Theory*; Springer: Berlin/Heidelberg, Germany, 2013; Volume 4, pp. 4–5.
19. Furstenberg, H. Strict ergodicity and transformation of the torus. *Am. J. Math.* **1961**, *83*, 573–601. [CrossRef]
20. Li, T.Y.; Yorke, J.A. Period Three Implies Chaos. *Am. Math. Mon.* **1975**, *82*, 985–992. [CrossRef]
21. Lai, Y.C.; Tél, T. *Transient Chaos: Complex Dynamics on Finite Time Scales*; Springer: New York, NY, USA, 2011; Volume 173.
22. Sander, E.; Yorke, J.A. The Many Facets of Chaos. *Int. J. Bifurc. Chaos* **2015**, *25*, 1530011. [CrossRef]
23. Briggs, R.J. *Electron-Stream Interaction with Plasmas*; MIT Press: Cambridge, MA, USA, 1964.
24. Kerswell, R.R. Recent progress in understanding the transition to turbulence in a pipe. *Nonlinearity* **2005**, *18*, R17. [CrossRef]
25. Kaneko, K. Lyapunov analysis and information flow in coupled map lattices. *Phys. D Nonlinear Phenom.* **1986**, *23*, 436–447. [CrossRef]
26. Omelchenko, I.; Omel'chenko, E.; Hövel, P.; Schöll, E. When nonlocal coupling between oscillators becomes stronger: Patched synchrony or multichimera states. *Phys. Rev. Lett.* **2013**, *110*, 224101. [CrossRef]
27. Girnyk, T.; Hasler, M.; Maistrenko, Y. Multistability of twisted states in non-locally coupled Kuramoto-type models. *Chaos Interdiscip. J. Nonlinear Sci.* **2012**, *22*, 013114. [CrossRef]
28. Jensen, M.H.; Bak, P. Spatial chaos. *Phys. Scr.* **1985**, *1985*, 64. [CrossRef]
29. Omelchenko, I.; Maistrenko, Y.; Hövel, P.; Schöll, E. Loss of coherence in dynamical networks: Spatial chaos and chimera states. *Phys. Rev. Lett.* **2011**, *106*, 234102. [CrossRef] [PubMed]

30. Kuramoto, Y. Diffusion-induced chaos in reaction systems. *Prog. Theor. Phys. Suppl.* **1978**, *64*, 346–367. [CrossRef]
31. Toh, S. Statistical Model with Localized Structures Describing the Spatio-Temporal Chaos of Kuramoto-Sivashinsky Equation. *J. Phys. Soc. Jpn.* **1987**, *56*, 949–962. [CrossRef]
32. Glendinning, P.; Jäger, T.; Keller, G. How chaotic are strange non-chaotic attractors? *Nonlinearity* **2006**, *19*, 2005–2022. [CrossRef]
33. Grebogi, C.; Ott, E.; Pelikan, S.; Yorke, J.A. Strange attractors that are not chaotic. *Phys. Nonlinear Phenom.* **1984**, *13*, 261–268. [CrossRef]
34. Pikovsky, A.S.; Feudel, U.; Kuznetsov, S.P. *Strange Nonchaotic Attractors: Dynamics between Order and Chaos in Quasiperiodically Forced Systems*; World Scientific: Toh Tuck Link, Singapore, 2006; Volume 56.

Article

An Expanded Sensitivity Study of Simulated Storm Life Span to Ventilation Parameterization in a Cloud Resolving Model

Yen-Liang Chou [1] and Pao-Kuan Wang [1,2,3,4,*]

[1] Research Center for Environmental Changes, Academia Sinica, Taipei 115, Taiwan
[2] Department of Aeronautics and Astronautics, National Cheng Kung University, Tainan 70101, Taiwan
[3] Department of Atmospheric Sciences, National Taiwan University, Taipei 106319, Taiwan
[4] Department of Atmospheric and Oceanic Sciences, University of Wisconsin-Madison, Madison, WI 53706, USA
*   Correspondence: pkwang@gate.sinica.edu.tw; Tel.: +886-2-2787-5882

Abstract: We performed a sensitivity study on the life span of a numerically simulated storm using the parameterization of the ventilation coefficient. This is an expanded sequel to our previous study, where the ventilation effect of precipitation particles (snow, rain, and hail) was either halved or doubled as a whole. In this study, we tested the sensitivity of the ventilation coefficient for different precipitation particles and compared that with the previous results. In the present study, we changed the ventilation coefficient in two scenarios: (1) only the rain category was changed; (2) only the snow and hail categories were changed. The results show that these different scenarios lead to different evolution paths for the storm. In general, reducing the ventilation effect of rain leads to quick dissipation, whereas enhancing the ventilation of either rain or snow/hail leads to the development of multicellular storms. An analysis of the physical mechanisms leading to such results is provided. This study shows yet another example of how a change in a cloud's microphysical parameterization can lead to a profound change in its larger-scale dynamical process.

Keywords: ventilation effect; cloud resolving model; cloud dynamics

Citation: Chou, Y.-L.; Wang, P.-K. An Expanded Sensitivity Study of Simulated Storm Life Span to Ventilation Parameterization in a Cloud Resolving Model. *Atmosphere* **2023**, *14*, 720. https://doi.org/10.3390/atmos14040720

Academic Editors: Bo-Wen Shen, Roger A. Pielke, Sr. and Xubin Zeng

Received: 7 March 2023
Revised: 11 April 2023
Accepted: 14 April 2023
Published: 15 April 2023

1. Introduction

The current field of atmospheric sciences has its main roots in traditional meteorology, and the mainstream studies during the "formation period" of modern meteorology in the early 20th century (for example, Bjerknes, 1919) [1] were undoubtedly synoptic meteorology, which concerns the large-scale motion of the lower atmosphere. During those days, small-scale motions such as convections were mostly ignored in the analysis as they did not seem to be relevant to the motions of air masses, which are usually larger than a few thousand km in horizontal dimension. It was later (e.g., Riehl and Markus, 1958 [2]; see also Yanai and Johnson, 1993 [3], for a later review) that convective scale motions received more attention and the importance of vertical motions to the formation of clouds and their impact on the thermodynamics of the atmosphere were more widely studied. Earlier studies of convective scale motions tended to focus more on the dynamics side, although it was well-known that various microphysical processes, especially those involving the phase change of water substances that either release or consume latent heat, can have a significant impact on the cloud behavior (Byers, 1965 [4]). Convection scale motions are usually considered "mesoscale" phenomena with dimensions of usually tens to hundreds of km as opposed to 1000 km or more for the synoptic scale. Cloud microphysics operates at a scale even smaller than the convection scale, and the most fundamental cloud microphysical processes such as nucleation and crystallization operate at submicron or smaller scales (Fletcher, 1962 [5]; Pruppacher and Klett, 2010 [6]; Wang, 2013 [7]). These processes are justifiably called microscale.

Because of the wide separation of the scales of these atmospheric phenomena, the earlier studies of each scale tended to develop fairly independently, even though most atmospheric scientists were aware of the existence (and hence the potential impact) of other scales. A common consensus among meteorologists in the early days was that small-scale phenomena, such as individual clouds and precipitation processes, do not affect larger-scale motions, for example, the motion of a frontal system. The conventional thought of energy cascade in atmospheric motions assumes that it is the larger scale motions that cascade their energies to produce the driving force to initiate smaller scale motions but not vice versa.

Although such a view was (and likely is) held by most in the community, it has never been proven from the first principle, although it fits the intuition that large-scale matters have a large impact whereas small-scale matters have a small impact. However, such an intuition is obviously linear in nature, and nature itself does not always behave linearly. It was Edward Lorenz, in his famous paper (Lorenz, 1963) [8], who utilized the nonlinear convection equation system established by Saltzman (1962) [9] and showed that the nonlinear effect can make an initially small "error" amplify greatly, such that the end results become unpredictable. This inaugurated the field of nonlinear system studies in many scientific fields, now generally called chaotic systems.

Cloud and precipitation systems are also highly nonlinear. The interactions among different dynamic, thermodynamic, and microphysical mechanisms are so complicated that the classical parcel method (see, e.g., Houze, 2014) [10] that was used to describe the fundamental cloud formation process cannot be used to predict the evolution of a real cloud system. Instead, cloud models that take into account the many interactions in thermodynamic, dynamic, and microphysical mechanisms are necessary to complete this feat (e.g., Cotton et al., 1989 [11]; Wang, 2013 [7]).

However, it is also well known that cloud models cannot capture the cloud microphysical processes precisely because the most accurate form of these processes is written in partial differential equations, and the current generation of computers simply cannot handle calculations that solve these differential equations directly during the execution of cloud model simulations. Consequently, nearly all these microphysical processes are parameterized by simple algebraic equations or even some direct look-up tables. The parameterizations are often based on a few observations that do not cover a wide range of atmospheric conditions, and many models develop their own parameterizations according to their own preferences (e.g., Cotton et al., 1989 [11]; Straka, 2009 [12]). Different parameterizations often lead to completely different predictions of cloud behavior. Hence, there is a need to study the sensitivity of the model results to a certain parameterization so that the accuracy of the model can be assessed.

In some of these sensitivity studies, it has been shown that the small-scale cloud's microphysical processes can indeed affect the larger-scale cloud's dynamical behavior. In an earlier study, Johnson et al. (1993) [13] tested the sensitivity of the life span of a simulated supercell on the parameterization of cloud microphysics by comparing the two simulated results: one with the presence of ice physics and the other without ice. The result showed that the presence of ice, which distributes the water substance in clouds, greatly extends the life of the supercell by modifying the storm's internal circulation such that the downdraft branch of the circulation keeps a sufficient distance from the updraft branch and hence does not cut the ascending branch off. Recently, we also tested the sensitivity of the storm life span on another microphysical parameter—the ventilation coefficient—and the results also showed that the life span is sensitive to such microphysical factors. In this study, we expanded our previous work on ventilation to perform more variations of the ventilation coefficient to test the sensitivity.

There are many cloud microphysical processes operating in cloud development, and many have been parameterized and incorporated into cloud models. For example, the cloud model WISCDYMM-2 (to be described in Section 3) contains 38 such processes. Most of the processes are based on data that have been well established (or there appear to be no new updates), but the ventilation effect is one process that is poorly represented. However, recently there have been new results of ventilation coefficients available, although they have not been parameterized for cloud models yet. The purpose of our previous sensitivity study and the present one is to show that proper ventilation parameterization is important and that there is a need to form new parameterizations based on new results.

In the following sections, we will first provide a background explanation of the ventilation coefficient, followed by a brief description of the model and the storm simulated. Then, we will describe the experimental design of the present study. Results, discussions, and conclusions will follow.

2. Ventilation Coefficient

When a cloud or precipitation particle (hereafter called a hydrometeor) is falling in the air, it is performing a relative motion with respect to the surrounding air. If the surrounding air is subsaturated, the hydrometeor will evaporate, whereas if the environment is supersaturated, the hydrometeor will grow by adding vapor to its surface. The latter process is called diffusion growth, whereas the former is simply evaporation (also called sublimation if the hydrometeor is an ice particle).

When either diffusion growth or evaporation is occurring, the motion of the hydrometeor has a strong influence on the rate. A stationary hydrometeor sitting in a subsaturated environment will also evaporate, but a hydrometeor in motion will evaporate faster than a stationary one due to an effect called the ventilation effect (see, e.g., Pruppacher and Klett, 2010 [6]; Wang, 2013 [7]). The ratio of the evaporation rate of the moving hydrometeor to that of the stationary one is the enhancement factor due to the motion and is called the ventilation coefficient. Mathematically, the ventilation coefficient is defined as the ratio of the mass growth rate $(dm/dt)_0$ of a moving hydrometeor to that of a stationary hydrometeor: $(dm/dt)_0$:

$$\overline{f}_v = \left(\frac{dm}{dt}\right) / \left(\frac{dm}{dt}\right)_0 \tag{1}$$

where m is the mass of the hydrometeor and $\overline{f}_v$ the mean ventilation coefficient (Wang, 2013 [7]). It is called mean because it is the average of the local ventilation coefficients over the surface of the hydrometeor, and the local ventilation coefficients can be different at different surface points. The term "ventilation coefficient" hereafter will represent the "mean ventilation coefficient". In the above, we used evaporation as an example to explain $\overline{f}_v$. In the case of the same hydrometeor falling in a supersaturated environment, it will grow faster than a stationary one by the same ventilation factor. In addition, the evaporation and diffusion growth both involve the consumption or release of latent heat, respectively, and this will also influence the evaporation and diffusion growth as well (see Pruppacher and Klett, 2010 [6]; Wang, 2013 [7]).

At present, the ventilation coefficients used in all cloud models are parameterized based on earlier experimental results (e.g., Beard and Pruppacher, 1971 [14]; Hall and Pruppacher, 1976 [15]). These earlier results are valid for small particles, but they are used for larger particles by extrapolation. Detailed calculations utilizing high precision computational fluid dynamical methods (e.g., Ji and Wang, 1999 [16]; Cheng et al., 2014 [17]; Nettesheim and Wang, 2018 [18]; Wang and Chueh, 2020 [19]) only became available fairly recently, and these new results have not been included in cloud models yet, but the values for large particles are much larger than those extrapolated from earlier formulas. This means that the current ventilation parameterization expressions may be unrealistic, and it is meaningful to study how sensitive the simulated storm development is to ventilation parameterization.

In a previous study (Chou et al., 2021) [20], we performed the first study on the sensitivity of the storm life span to the ventilation parameterization and obtained the conclusion that the sensitivity exists and can be severe. In the present study, we expand that previous study to include more numerical experiments on the ventilation criteria.

3. The Cloud Model and the Simulated Storm Case

The cloud model we used throughout this study for testing the ventilation sensitivity is one developed in P. K. Wang's research group at the University of Wisconsin-Madison, the Wisconsin Dynamical Microphysical Model double moment version (WISCDYMM-II)

(Wang et al. 2016) [21]. This is the same as what we used in our previous study, and we take the following descriptions from it (Chou et al., 2021) [20]. WISCDYMM-II adopts fully compressible non-hydrostatic moisture equations (Klemp et al., 2007) [22], in contrast to its predecessor WISCDYMM (Straka, 1989 [23]; Johnson et al., 1993 [13], 1994 [24]; Lin and Wang, 1997 [25]; Wang, 2003 [26]), which is quasi-compressible. Furthermore, the model uses a double moment scheme as given in Morrison et al. (2005) [27] (hereafter called the Morrison scheme) to predict both the mixing ratio and the number concentration instead of the single moment scheme used in the previous version. On the other hand, the model also adopts the 1.5 order turbulence closure to account for the sub-grid eddy mixing as the previous version. WISCDYMM-II has been used successfully in a study of the storm top internal gravity wave breaking (Wang et al., 2016) [21].

The initial conditions we adopt for this study are also the same as those we used in Chou et al. (2021) [20], which is the sounding recorded right before the mid-latitude supercell storm took place on 2 August 1981, in southeastern Montana by the cooperative convective precipitation experiment (CCOPE) observational network as detailed by Knight (1982) [28]. This supercell has been widely studied since then (Wade, 1982 [29]; Miller et al., 1988 [30]; Johnson et al., 1993 [13], 1994 [24]; Wang et al., 2007 [31]; Fernández-González et al., 2016 [32]). This supercell had been simulated by Johnson et al. (1993 [13], 1994 [24]), who performed detailed analyses of the simulation results using the older version of WIS-CDYMM. We follow the methodology of these previous studies and initiate the storm by perturbing the temperature field confined by a warm ellipsoidal bubble near the surface level. The instability of the warm bubble would evolve into a supercell. The ventilation effect on hydrometeors was parameterized by Morrison et al. (2005) [27] in their Equations (3) and (4) for cloud droplets, cloud ice, rain, and snow.

The ventilation parametrization they are based on that summarized in Pruppacher and Klett (2010) [6]. Morrison et al. (2009) [33] added the ventilation effect of the graupel/hail category based on the time tendency equation given by Reisner et al. (1998) [34].

The motivation to perform a sensitivity test on the ventilation effect in our first study (Chou et al., 2021) [20] was that the present ventilation parameterization is based on a small set of data that is mostly relevant to small particles from the experimental results of Beard and Pruppacher (1971) [14] and Hall and Pruppacher (1976) [15]. Those ventilation coefficients for large particles are extrapolated, and the values are inaccurate. For example, the ventilation coefficients for hailstones were calculated to be much smaller than the results obtained by Cheng et al. (2014) [17] and Wang and Chueh (2020) [19]. Due to such inaccuracies, it is meaningful to ask whether or not the simulated results are sensitive to different ventilation parameterizations, and the answer in Chou et al. (2021) [20] is yes. In this study, we expand that previous study to include more scenarios, and the details will be described in the following sections.

4. Impact of the Domain Size

In our previous work (Chou et al., 2021) [20], we conducted a sensitivity study of the life span of a simulated convective storm to the ventilation effect of rain, snow, and hail utilizing the cloud model WISCDYMM-II described above. We altered the ventilation coefficients of all three hydrometeors by multiplying a common amplification factor, Z, which led to changes in the storm structure and lifespan. These experiments were idealized and performed on a relatively small domain that follows the center of the storm to economize the use of computational resources. The domain was $160 \times 160 \times 25$ km^3. The resolution was 500 m horizontally and 200 m vertically. However, as the storm evolves, it will move out of the domain at a certain stage, and a numerical process needs to be employed to re-center the storm in the domain. This operation implies that the domain would cover new territories where the values of variables are unknown but are assumed to be equal to the values at the nearest boundaries. There is a possibility that numerical artifacts may become introduced into the results, thus causing errors.

To rule out the potential artificial effect arising from the boundary condition of the moving reference frame, we perform a second set of the sensitivity study on a larger domain. The domain size in this second study is 640×640 grids with a horizontal resolution of 500 m, and hence covers an area of 320×320 km^2. The vertical extension is 25 km with a resolution of 200 m. The new domain is therefore four times that of the small domain simulations. This allows the simulated storm to evolve in the domain in a 5-hour simulation without re-centering. In addition, we expand the sensitivity study of the ventilation effects by altering the ventilation coefficients of rain and snow/hail individually instead of lumping them together, as was conducted in the previous study.

In the following, we will report the results of these two batches of experiments. We conducted two batches of numerical experiments on sensitivity. In the first batch, we ran WISCDYMM-2 with the small domain and varied the Z factor from 0.5 to 2 with an increment of 0.1 that applies to all three precipitation particles equally. In the second batch, we ran WISCDYMM-2 in the large domain with the Z factor values of 0.5, 0.8, 1.0, 1.5, and 2.0, and the change in Z also applies to all three precipitation categories equally. In the following, we will first describe some general features of the results of the first batch of experiments, which would also explain why we would only focus on the two Z values in the second batch of experiments.

4.1. Results of the First Batch Experiments

This batch pertains to the case of small domains. Figure 1 shows the precipitation intensity (kg/s) of rain and hail as a function of Z and time. The disappearance of rain indicates the dissipation of the storm. The precipitation intensity patterns fluctuate as they usually do, but a few general features are evident: (1) the smaller Z, the earlier the heaviest precipitation occurs; (2) the heaviest precipitation in small Z cases is more continuous (less fluctuating) than those of larger Z cases; and (3) the larger Z values, the longer the life span of the storm. Another interesting feature is that, when dissipating, storms with a smaller ventilation effect last longer due to the slower evaporation. The pattern of hail resembling rain might indicate most of the rain is coming from melting hail.

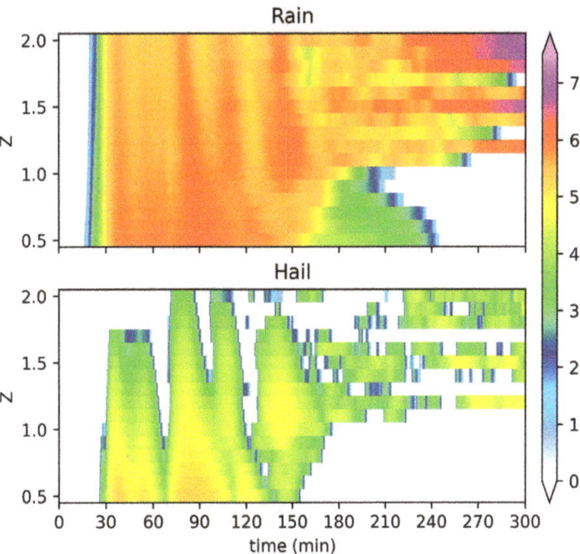

Figure 1. Temporal variation of the domain total precipitation intensity R (kg/s) of the rain and hail categories in the simulated storm with the Z factor for small domain simulations. The color scale is in units of $\log_{10} R$.

4.2. Results of the Second Batch Experiments

As indicated above, the first batch of experiments consist of running WISCDYMM-2 with the Z factor values of 0.5, 0.8, 1.0, 1.5, and 2.0. These changes were applied to all three precipitation categories as a whole.

Figure 2 shows the time evolution of the precipitation intensity categories of rain and hail in the whole domain as a function of Z. We see that the general features are essentially the same as those in Figure 1 and, especially, that the larger Z values usually lead to a longer life span. Differences in details exist as expected; for example, the control case ($Z = 1$) in this batch lasts to the very end of the simulation while that of the small domain dissipates at ~210 min. Thus, the domain size does have an impact on the details; however, the general features of the ventilation sensitivity remain the same.

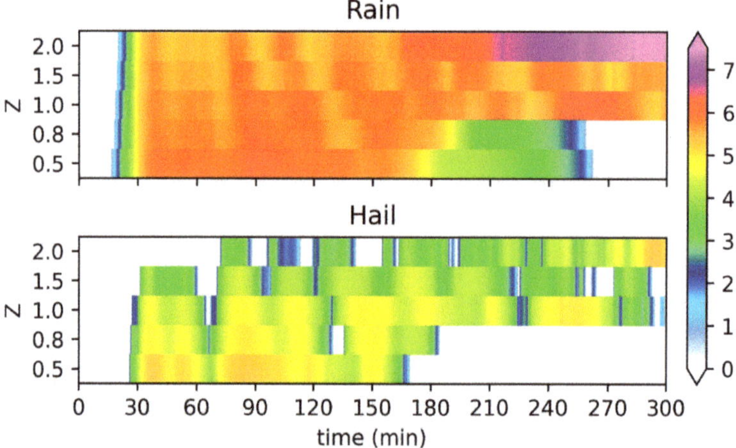

Figure 2. Temporal variation of the domain total precipitation intensity R (kg/s) of the rain and hail categories in the simulated storm with the Z factor for large domain simulations. The color scale is in units of $\log_{10} R$.

We decided to conduct this sensitivity study using the large domain strategy, considering that its boundary conditions are more straightforward (and contain fewer ad hoc assumptions). The simulations were performed similarly to what produced Figure 2, but with each Z factor applying to the liquid (rain) and solid (snow + hail) precipitation categories separately instead of equally to all three categories. We will report the results in the following section. Since the general features of the sensitivity dependence are again similar to those in Figures 1 and 2, we will use the results of the two extreme cases, $Z = 0.5$ and 2.0, to illustrate the sensitivity and make an analysis to explore the possible mechanisms responsible for causing it.

5. Results and Discussion

We conducted numerical experiments to test the sensitivity of changing the ventilation effect by multiplying the original parameterization by 0.5 and 2.0 and applying each change to the following three scenarios: rain, snow/hail, and rain/snow/hail. These 6 runs plus the control run yield 7 case results in total, and their experimental setups are summarized in Table 1.

Figure 3 shows the top view of the simulated storms in all experiments at three suitably selected time steps to represent the early, mature, and later stages. The outermost surface of the storm is represented by the 0.1 g/kg (gray color) isosurface of the mixing ratio of total water condensates. The isosurfaces of the 0.1 g/kg mixing ratios of rain (yellow), snow (purple), and hail (blue) are also plotted.

Table 1. The list of the experiments studied in this manuscript.

Case	Description	Species Affected	Storm Condition at the End
Control	The control run		Multicellular system (early stage)
$Z = 0.5$		Rain, snow, and hail	Dissipated cloud
$Z_r = 0.5$	Reducing the ventilation effect	Rain	Dissipated cloud
$Z_s = 0.5$		Snow and hail	Single cell
$Z = 2$		Rain, snow, and hail	Multicellular system
$Z_r = 2$	Amplifying the ventilation effect	Rain	Multicellular system
$Z_s = 2$		Snow and hail	Multicellular system (early stage)

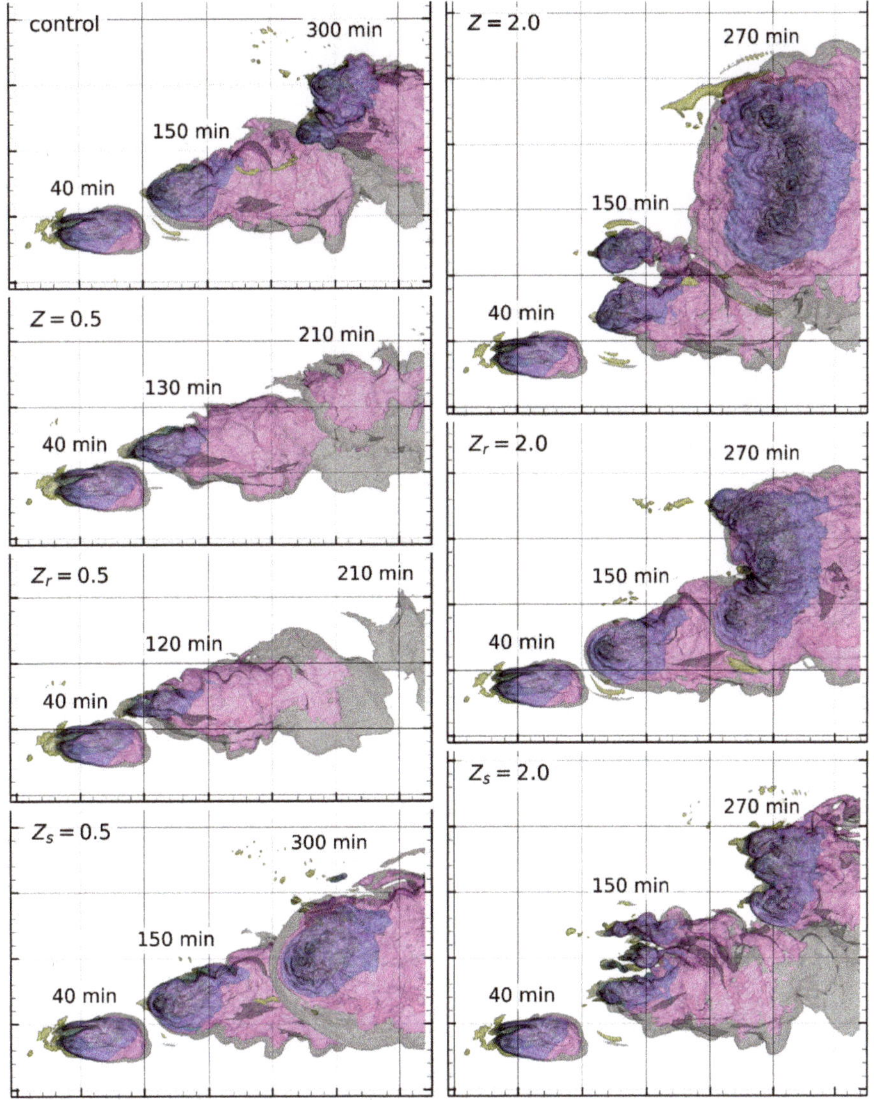

Figure 3. Top view of the storms developed in various experiments. In each panel, the storm at three different time steps is overlapped, with the later one superimposed on the earlier one. The outermost

surface of the storm is represented by the 0.1 g/kg isosurface of the mixing ratio of total water condensates (gray). The 0.1 g/kg isosurfaces of rain (yellow), snow (purple), and hail (blue) are also shown. The full domain is $320 \times 320 \times 25$ km^3 (with part in the north cropped off). The major (minor) grid spacing is 50 (10 km). The case (experiment) names are on the top-left corner of each panel, with timestamps to the north of the storms.

In the control run, the storm lasts to the end of the simulation, when it splits into two cells. Before storm splitting, small cells emerge occasionally, but none of them grows large enough to affect the main cell. This differs from the control case of our previous study, where the storm dissipated at around 180 min. Reducing the ventilation effect of all three species together ($Z = 0.5$) results in early dissipation of the storm, whereas enhancing the ventilation ($Z = 2$) causes the storm to split and eventually evolve into a multicellular system. Despite the difference in the resulting lifespan compared with that in Chou et al.'s (2021) study [20], the response of the storm's development to changing ventilation parameterization agrees well with the earlier study.

Next, we will examine the cases where we reduce or enhance the ventilation separately for liquid and solid precipitation particles. Reducing the ventilation effect of only rain ($Z_r = 0.5$) causes earlier dissipation than reducing the ventilation of all species ($Z = 0.5$). For these two cases, there is neither cell splitting nor new cell development in the simulation. On the contrary, enhancing the ventilation of rain ($Z_r = 2$) causes storm splitting and develops into multicellular structure, but with a smaller scale than in the case of $Z = 2$. These two experiments on varying the ventilation effect of rain show that it could be a major factor affecting the storm's development. The stronger the ventilation effect of rain, the stronger the storm will evolve into a larger system, and vice versa.

Varying the ventilation effect of snow and hail leads to more sophisticated consequences. Both reducing and enhancing the ventilation effect cause the storm to evolve into a larger-scale system than the control case, however, in distinct ways. In the $Z_s = 0.5$ case, although there is a small cell rising to the north of the cell at 210 min (not shown), it dissipates at the end (and its remnants could be seen to the north of the main cell), whereas the main storm remains intact. On the other hand, in the case of $Z_s = 2$, the storm evolves with the characteristic new cell formation, splitting and merging. The appearance of a new cell occurs fairly early, at $t = 80$ min (not shown), and the storm evolves into a triplet at 150 min. These cells evolve independently of each other until $t = 250$ min, when they start to merge.

Examining the history of hydrometeor mass fluctuations provides clues as to possible mechanisms responsible for the storm's evolutionary behavior as depicted in Figure 3. Figure 4 shows that all storm cases have nearly the same hydrometeor mass in each category up to $t \sim 70$ min. In the $Z = 2$ case, the mass of total condensates grows rapidly owing to the fast accumulating hail from 110 min (bottom panel) and of rain (middle panel) later from 150 min. It becomes the largest storm in terms of mass of rain, hail, and total water condensate at 150 min when it forms two firm cells (as shown in Figure 3). Enhancing the ventilation effect of rain ($Z_r = 2$) also causes the storm to grow substantially in mass but at a slower rate than $Z = 2$. Unlike $Z = 2$, it only produces one cell at 150 min, but with a round back-sheared anvil indicating strong divergent flow near the cloud top.

The case of enhancing the ventilation effect of snow and hail ($Z_s = 2$) results in the production of a triplet system, yet its water condensate mass is the smallest during the mature stage among the four cases. The hail category contributes the most to losing the water condensate mass, likely due to the enhanced sublimation rate in the parameterization. The storm has grown rapidly in mass since 250 min ago, when cells started to merge.

On the other hand, in the case of decreasing the ventilation effect of snow and hail ($Z_s = 0.5$), the storm develops nearly linearly in terms of the mass of hail and total water condensates during the period between 120 and 240 min. The rise of the small cell mentioned previously is reflected by the faster increase of rain growth at 210 min, and of hail at 240 min. The dissipation of the small cell is, on the other hand, marked by the slowing down of hail and the total condensate mass growth rate at around 270 min.

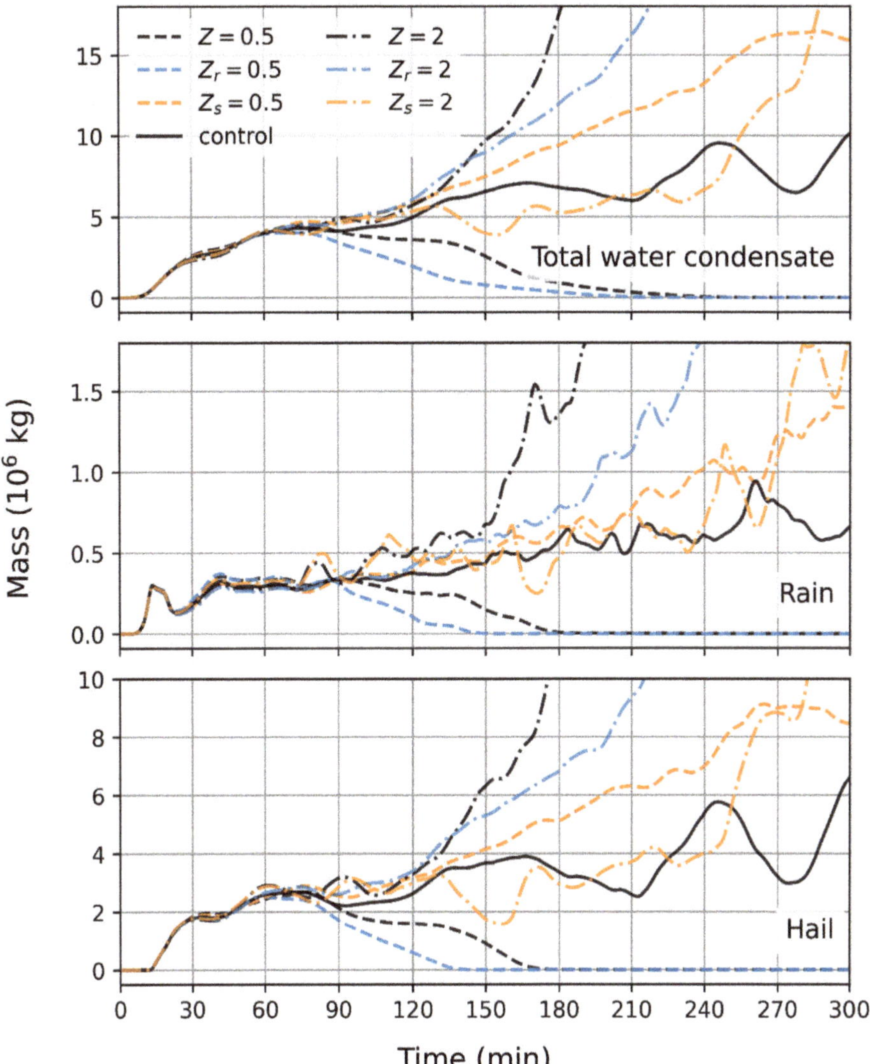

Figure 4. The time evolution of the mass of total water condensate (**upper**), rain (**middle**), and hail (**lower**) in the entire domain. In each panel, the experiments with reduced ventilation are shown in dashed curves, whereas the enhanced cases are in dot-dashed curves. Black is for altering all three species equally, blue for just rain, and orange for just snow and hail. The control case is shown as a black, solid curve. Note that the scales of the vertical axis are different in different panels.

Figure 5 shows the time series of the domain extrema of the vertical wind fields. These are smoothed curves that resulted from convolving with a discrete Gaussian kernel with a standard deviation of 4 min. The large fluctuations of the original curves (sampled every 2 min) make it difficult to identify the trends clearly, whereas it is much easier to do so by examining the smoothed curves.

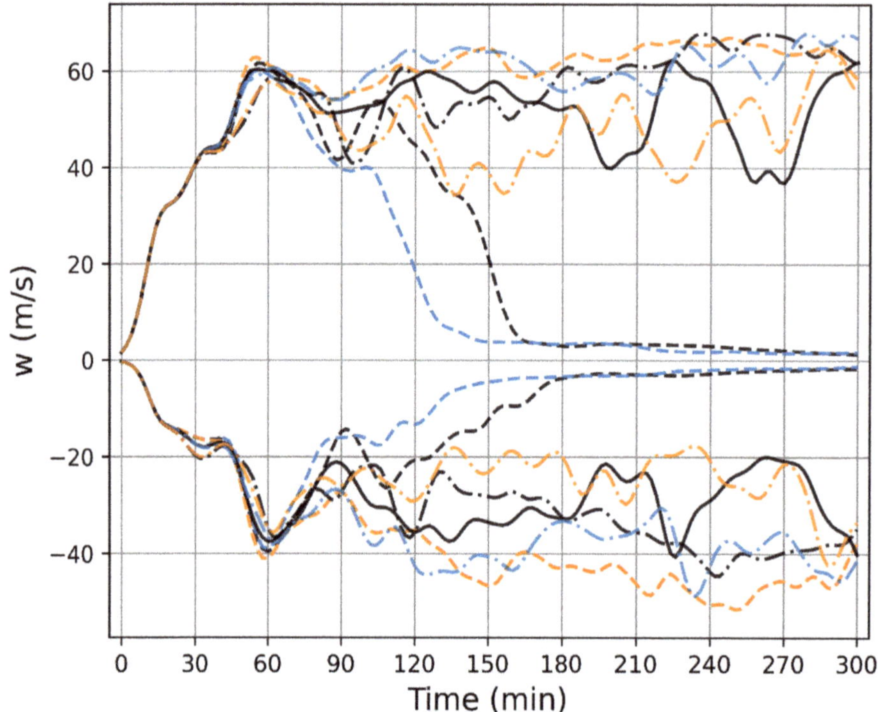

Figure 5. The temporally smoothed domain maximum updrafts (positive values) and downdrafts (negative values). The colors and styles of curves representing different cases are the same as in Figure 4.

A common feature of all the cases is that both the updraft/downdraft reach a peak between 50 and 60 min, followed by a sudden decline in magnitude. The peak occurs at about the same time as the overshooting top of the storm peaks for the first time. Many more fluctuations in maximum updraft/downdraft thereafter may follow, depending on the cases. We find that the large fluctuation occurs when the storm shows active intercellular processes, including frequent formation and dissipation of new cells, storm splitting, and cells merging. These intercellular processes often lead to multicellular complexes.

Inspection of the snapshots in the control run (not shown) reveals that the storm develops steadily before 180 min. Subsequently, new cells emerge vigorously to the north of the storm, followed by a splitting of the main cell at around 260 min. In this period (180–260 min), the maximum updrafts and downdrafts exhibit large fluctuations.

In the $Z = 2$ case, new cell formation and storm splitting occur frequently from 80 min to 150 min when it develops into two large cells. The extrema of vertical winds have larger fluctuations than the control run at this time. A triplet structure emerges at 180 min and a merged multicellular structure at 200 min (not shown). Although the storm changes morphology substantially, it grows steadily in size, and its vertical velocity extrema show less fluctuation than the early period.

In the $Z_s = 2$ case, there are vigorous new cell formation and splitting processes during the 80–250-min period, and the cells merge to form a multicellular system at around 270 min. However, the transformation is unsteady, and consequently, its vertical velocity extrema fluctuate more than the $Z = 2$ case throughout the simulation.

The steadiest case is $Z_s = 0.5$ where the storm evolves gradually into a single cell. Its extrema remain relatively constant at almost the highest values among all the experiments.

In the $Z_r = 2$ case, the storm grows steadily until 170 min, when a cell splits from the north of the main cell (not shown), which initiates a fast development into a multicellular cluster. However, the transformation is smoother than in the cases of the control run and $Z_s = 2$. The extrema after the transformation show larger fluctuations than before but are still smaller than those in the same stage in the control run and $Z_s = 2$ cases.

The domain water condensate masses of all cases already show substantial differences in the 2nd hour, as seen in Figure 4. The differences become even clearer when we examine the vertical distributions of precipitation particles, which suggest different paths in their developments. In Figure 6, the total mass of rain and hail in each layer of the simulation domain is averaged over the period from 30 to 70 min, and the differences with respect to the control run are presented.

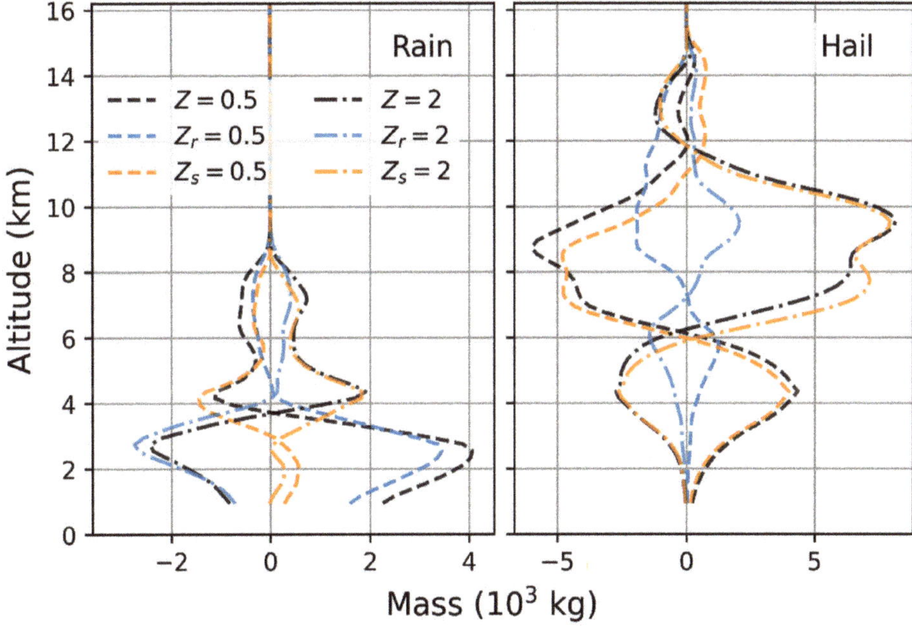

Figure 6. The differences in the averaged mass of rain (**left**) and hail (**right**) from that of the control run in the period from 30 to 70 min as a function of altitude. The colors and styles of curves representing different cases are the same as in Figure 4.

We see in Figure 6 that enhancing the ventilation effect of rain ($Z_r = 2$, left panel) causes the rain mass to be smaller than the control run below the freezing level around 4 km but to be larger above the level. On the other hand, decreasing the ventilation effect of rain ($Z_r = 0.5$) results in the opposite trend. The role that the ventilation effect of rain plays in regulating condensation and evaporation, which have a direct impact on the mass gained and/or lost by rain, is clearly shown by these two cases.

Similarly, enhancing the ventilation effect of snow and hail ($Z_s = 2$, right panel) results in more hail mass at levels between 6 and 12 km, and less hail mass below 6 km than in the control run. As $Z_r = 0.5$ is to $Z_r = 2$, reducing the ventilation effect of snow/hail, $Z_s = 0.5$ (right panel), shows roughly the opposite trend to the case of $Z_s = 2$.

Enhancing the ventilation effect of a specified category helps to enhance the vapor transport between the hydrometeor surface and the environmental air. The growth layer usually locates at higher levels, and the evaporation layer usually locates at lower levels. Hail forms at higher levels than rain, and the sublimation when falling through a subsaturated environment would undergo a longer distance than the evaporation of rain.

Therefore, rain contributes more to gravitational fallout as a sink of water content in the storm than hail does. Reducing the ventilation effect of rain causes the evaporation rate to decrease, and the precipitation to increase as shown in the cases of $Z = 0.5$ and $Z_r = 0.5$.

Reducing the ventilation effect of snow/hail slows down the sublimation, and the hail that falls below the freezing line would melt and form rain. In this case, rain undergoes a normal evaporation process, and the precipitation would be lighter than in the cases of $Z = 0.5$ and $Z_r = 0.5$. This is likely one of the reasons that reducing the ventilation effect of rain would cause the storm to terminate earlier than the control run, but reducing the effect of snow/hail alone would not.

Adjusting the ventilation effects also has a direct impact on the thermal and dynamic processes of the storm because it impacts the rate of latent heat release or consumption. When evaporation/sublimation occur rapidly, the temperature of the surrounding air would become colder quickly, and downdrafts are stronger. Figure 7 shows this effect. Here the updrafts/downdrafts averaged from 50 to 70 min with magnitudes greater than or equal to the 99.9 percentile on each level of domain are shown.

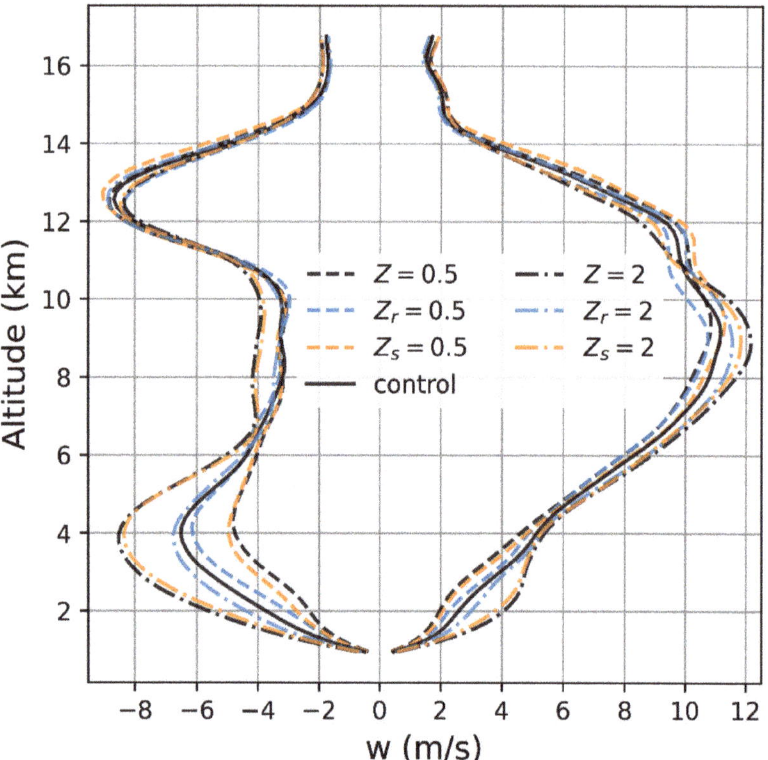

Figure 7. The distribution of the top 99.9 percentile of updrafts/downdrafts averaged in 50–70 min in the entire domain as a function of height. The percentiles of each level are calculated separately at each time step. The data have a temporal spacing of two minutes.

These top updrafts/downdrafts occupy about 10×10 km^2 of area. Below 5 km in height, the cases with enhanced ventilation effects have stronger downdrafts. The sublimation cooling has a stronger effect than the evaporative cooling. This is because (1) enhancing the ventilation effect of snow/hail produces more hail than the increase in rain produced by enhancing the rain ventilation, and (2) the latent heat absorbed is larger in sublimation than in evaporation. As a result, the case of $Z = 2$ produces the strongest cooling and downdrafts, followed by the cases of $Z_s = 2$, $Z_r = 2$, and the control run, in that order, at levels below 5 km. The reduced ventilation cases show the opposite trend that $Z_r = 2$ has stronger effect than $Z_s = 2$, followed by $Z = 2$, which has the smallest effect. Because the air is a continuum, when downdraft hits the ground level updraft will occur somewhere around the main storm cell and, if sufficiently strong, may even initiate new cells. The stronger downdrafts have a larger potential to generate greater updrafts than the weak ones. The updrafts in Figure 7, below 4 km, show that their averaged magnitudes are arranged in the same order as the downdrafts.

The cases $Z = 2$ and $Z_s = 2$ have strongest updrafts on average at the early stage (50–70 min), where the storm becomes mature. In both cases, a strong updraft causes a new cell formation as early as 80 min, which grows to a size large enough to compete and interfere with the mother cell and other cells generated later. In the case of $Z_s = 0.5$, although the cell grows steadily in size and lasts to the end, it has the second-weakest updrafts among all the cases. The weak updraft keeps the storm from remaining as a single cell as it evolves.

6. Conclusions

We performed an expanded set of sensitivity studies utilizing cloud model simulations to investigate how different ventilation parameterizations may impact the life span of the simulated storms. We used the widely adopted ventilation parameterizations as the control case and investigated the sensitivity of the reduced ventilation, normal, and enhanced ventilation effects by multiplying a constant factor Z ranging from 0.5 to 2.0 so as to represent the various degrees of enhancement.

Differing from our previous study (Chou et al., 2022) [16], where the storm develops in a confined storm-following domain, our present study shows that the storm evolving freely in a larger domain lasts longer. This change in the lifespan is likely due to the effect of boundary conditions unrelated to ventilation sensitivity. However, it is worthy of note that the trends in the life span of the simulated storm in response to reducing/increasing the ventilation effects of all three hydrometeors are similar in the storm-following and fixed domain runs.

The impact of the ventilation effect on rain is the major factor affecting the lifespan of the storm. Modulating the ventilation effect of rain alone results in a similar trend as modulating the effect of all three species. Hail, on the other hand, controls the formation of new cells and the following intercellular processes. Increasing the ventilation effect of hail/snow increases cellular activity and the possibility of developing into a multicellular complex.

Despite the similarity in their appearance in the early mature stages, subtle discrepancies in the total hydrometeor masses and vertical wind fields among different cases reveal the effects caused by adjusting the ventilation coefficients, which lead to different paths of their further development.

Enhancing the ventilation effect of rain increases the rain's evaporation rate below the freezing level and decreases the precipitation, and vice versa. On the other hand, enhancing the ventilation effect of snow/hail by increasing the rate of sublimation makes small differences in the total hail masses that fall to the surface in different cases. As an important sink of the total water content, which affects the storm lifespan, the gravitational fallout from rain dominates that from hail/snow. Hence, adjusting the ventilation effect of rain is a key factor affecting the lifespan of the storm.

On the other hand, although the hail/snow category affects less on the storm life span compared to rain, enhanced hail/snow ventilation still contributes to the adjustment of latent heat amount, which seems to be a factor leading to the multicellular structure.

This study thus demonstrates that the storm simulation results are sensitive to the parameterization of ventilation. It also demonstrated that a change in cloud microphysical parameters could lead to a large change in large-scale cloud behavior, such as the life span we investigated here.

Author Contributions: P.-K.W. is the research fund holder who conceptualized the work, provided supervision, and edited the manuscript. Y.-L.C. performed the calculations, analyzed the results, and wrote the first draft of the manuscript. All authors have read and agreed to the published version of the manuscript.

Funding: We thank the support of the Taiwan National Science and Technology Council (NSTC) grant NSTC 111-2111-M-001-008; the Higher Education Sprout Project, Ministry of Education to the Headquarters of University Advancement at National Cheng Kung University (NCKU); and the US NSF grant AGS-1633921. Any opinions, findings, and conclusions for recommendations given in this article are those of the authors and do not necessarily reflect the viewpoints of the National Science Foundation (NSF).

Institutional Review Board Statement: Not applicable.

Informed Consent Statement: Not applicable.

Data Availability Statement: Not applicable.

Conflicts of Interest: The authors declare no conflict of interest.

References

1. Bjerknes, J. On the Structure of Moving Cyclones. *Mon. Weather. Rev.* **1919**, *1*, 95–99. [CrossRef]
2. Riehl, H.; Malkus, J.S. On the Heat Balance in the Equatorial Trough Zone. *Geophysica* **1958**, *6*, 503–538.
3. Yanai, M.; Johnson, R.H. Impacts of Cumulus Convection on Thermodynamic Fields. In *The Representation of Cumulus Convection in Numerical Models*; Emanuel, K.A., Raymond, D.J., Eds.; American Meteorological Society: Boston, MA, USA, 1993; pp. 39–62.
4. Byers, H.R. *Elements of Cloud Physics*; University of Chicago Press: Chicago, IL, USA, 1965; ISBN 978-0226086972.
5. Fletcher, N.H. *The Physics of Rainclouds*; Cambridge University Press: Cambridge, UK, 1962.
6. Pruppacher, H.R.; Klett, J.D. *Microphysics of Clouds and Precipitation*; Atmospheric and Oceanographic Sciences Library; Springer: Dordrecht, The Netherlands, 2010; Volume 18, ISBN 978-0-7923-4211-3.
7. Wang, P.K. *Physics and Dynamics of Clouds and Precipitation*; Cambridge University Press: Cambridge, UK, 2013, ISBN 9780511794285.
8. Lorenz, E.N. Deterministic Nonperiodic Flow. *J. Atmos. Sci.* **1963**, *20*, 130–141. [CrossRef]
9. Saltzman, B. Finite Amplitude Free Convection as an Initial Value Problem—I. *J. Atmos. Sci.* **1962**, *19*, 329–341. [CrossRef]
10. Houze, R. *Cloud Dynamics*; Elsevier: Amsterdam, The Netherlands, 2014.
11. Cotton, W.; Anthes, R. *Storm and Cloud Dynamics*, 1st ed.; Dmowska, R., Ed.; Academic Press: Cambridge, MA, USA, 1989; ISBN 9781483288383.
12. Straka, J.M. *Cloud and Precipitation Microphysics*; Cambridge University Press: Cambridge, UK, 2009; ISBN 9780511581168.
13. Johnson, D.E.; Wang, P.K.; Straka, J.M. Numerical Simulations of the 2 August 1981 CCOPE Supercell Storm with and without Ice Microphysics. *J. Appl. Meteorol.* **1993**, *32*, 745–759. [CrossRef]
14. Beard, K.V.; Pruppacher, H.R. A Wind Tunnel Investigation of the Rate of Evaporation of Small Water Drops Falling at Terminal Velocity in Air. *J. Atmos. Sci.* **1971**, *28*, 1455–1464. [CrossRef]
15. Hall, W.D.; Pruppacher, H.R. The Survival of Ice Particles Falling from Cirrus Clouds in Subsaturated Air. *J. Atmos. Sci.* **1976**, *33*, 1995–2006. [CrossRef]
16. Ji, W.; Wang, P.K. Ventilation Coefficients for Falling Ice Crystals in the Atmosphere at Low-Intermediate Reynolds Numbers. *J. Atmos. Sci.* **1999**, *56*, 829–836. [CrossRef]
17. Cheng, K.-Y.; Wang, P.K.; Wang, C.-K.K. A Numerical Study on the Ventilation Coefficients of Falling Hailstones. *J. Atmos. Sci.* **2014**, *71*, 2625–2634. [CrossRef]
18. Nettesheim, J.J.; Wang, P.K. A Numerical Study on the Aerodynamics of Freely Falling Planar Ice Crystals. *J. Atmos. Sci.* **2018**, *75*, 2849–2865. [CrossRef]
19. Wang, P.K.; Chueh, C.-C. A Numerical Study on the Ventilation Coefficients of Falling Lobed Hailstones. *Atmos. Res.* **2020**, *234*, 104737. [CrossRef]
20. Chou, Y.-L.; Wang, P.K.; Cheng, K.-Y. Sensitivity of Simulated Storm Life Span to Ventilation Parameterization in a Cloud Resolving Model. *Terr. Atmos. Ocean. Sci.* **2021**, *32*, 361–374. [CrossRef]
21. Wang, P.K.; Cheng, K.-Y.; Setvak, M.; Wang, C.-K. The Origin of the Gullwing-Shaped Cirrus above an Argentinian Thunderstorm as Seen in CALIPSO Images. *J. Geophys. Res. Atmos.* **2016**, *121*, 3729–3738. [CrossRef]
22. Klemp, J.B.; Skamarock, W.C.; Dudhia, J. Conservative Split-Explicit Time Integration Methods for the Compressible Nonhydrostatic Equations. *Mon. Weather. Rev.* **2007**, *135*, 2897–2913. [CrossRef]
23. Straka, J.M. *Hail Growth in a Highly-Glaciated Central High Plains Multi-Cellular Hailstorm*; The University of Wisconsin-Madison: Madison, WI, USA, 1989.
24. Johnson, D.E.; Wang, P.K.; Straka, J.M. A Study of Microphysical Processes in the 2 August 1981 CCOPE Supercell Storm. *Atmos. Res.* **1994**, *33*, 93–123. [CrossRef]

25. Lin, H.-M.; Wang, P.K. A Numerical Study of Microphysical Processes in the 21 June 1991 Northern Taiwan Mesoscale Precipitation System. *Terr. Atmos. Ocean. Sci.* **1997**, *8*, 385–404. [CrossRef]
26. Wang, P.K. Moisture Plumes above Thunderstorm Anvils and Their Contributions to Cross-Tropopause Transport of Water Vapor in Midlatitudes. *J. Geophys. Res.* **2003**, *108*, 4194. [CrossRef]
27. Morrison, H.; Curry, J.A.; Khvorostyanov, V.I. A New Double-Moment Microphysics Parameterization for Application in Cloud and Climate Models. Part I: Description. *J. Atmos. Sci.* **2005**, *62*, 1665–1677. [CrossRef]
28. Knight, C.A. The Cooperative Convective Precipitation Experiment (CCOPE), 18 May–7 August 1981. *Bull. Am. Meteorol. Soc.* **1982**, *63*, 386–398. [CrossRef]
29. Wade, C.G. A Preliminary Study of an Intense Thunderstorm Which Moved across the CCOPE Research Network in Southeastern Montana. In Proceedings of the Ninth Conference on Weather Forecasting and Analysis, Seattle, WA, USA, 28 June–1 July 1982; American Meteorological Society: Seattle, WA, USA, 1982; pp. 388–395.
30. Miller, L.J.; Tuttle, J.D.; Knight, C.A. Airflow and Hail Growth in a Severe Northern High Plains Supercell. *J. Atmos. Sci.* **1988**, *45*, 736–762. [CrossRef]
31. Wang, P.K. The Thermodynamic Structure atop a Penetrating Convective Thunderstorm. *Atmos. Res.* **2007**, *83*, 254–262. [CrossRef]
32. Fernández-González, S.; Wang, P.K.; Gascón, E.; Valero, F.; Sánchez, J.L. Latent Cooling and Microphysics Effects in Deep Convection. *Atmos. Res.* **2016**, *180*, 189–199. [CrossRef]
33. Morrison, H.; Thompson, G.; Tatarskii, V. Impact of Cloud Microphysics on the Development of Trailing Stratiform Precipitation in a Simulated Squall Line: Comparison of One- and Two-Moment Schemes. *Mon. Weather. Rev.* **2009**, *137*, 991–1007. [CrossRef]
34. Reisner, J.; Rasmussen, R.M.; Bruintjes, R.T. Explicit Forecasting of Supercooled Liquid Water in Winter Storms Using the MM5 Mesoscale Model. *Q. J. R. Meteorol. Soc.* **1998**, *124*, 1071–1107. [CrossRef]

 *atmosphere*

Article

Role of the Observability Gramian in Parameter Estimation: Application to Nonchaotic and Chaotic Systems via the Forward Sensitivity Method [†]

John M. Lewis [1,2,*] and Sivaramakrishnan Lakshmivarahan [3]

[1]   National Severe Storms Laboratory, Norman, OK 73072, USA
[2]   Desert Research Institute, Reno, NV 89512, USA
[3]   School of Computer Science, University of Oklahoma, Norman, OK 73069, USA
*   Correspondence: john.lewis@dri.edu
[†]   This paper is dedicated to the memory of Sherry Lynn Lewis.

Abstract: Data assimilation in chaotic regimes is challenging, and among the challenging aspects is placement of observations to induce convexity of the cost function in the space of control. This problem is examined by using Saltzman's spectral model of convection that admits both chaotic and nonchaotic regimes and is controlled by two parameters—Rayleigh and Prandtl numbers. The problem is simplified by stripping the seven-variable constraint to a three-variable constraint. Since emphasis is placed on observation positioning to avoid cost-function flatness, forecast sensitivity to controls is needed. Four-dimensional variational assimilation (4D-Var) is silent on this issue of observation placement while Forecast Sensitivity Method (FSM) delivers sensitivities used in placement. With knowledge of the temporal forecast sensitivity matrix V, derivatives of the forecast variables to controls, the cost function can be expressed as a function of the observability Gramian $V^T V$ using first-order Taylor series expansion. The goal is to locate observations at places that force the Gramian positive definite. Further, locations are chosen such that the condition number of $V^T V$ is small and this guarantees convexity in the vicinity of the cost function minimum. Four numerical experiments are executed, and results are compared with the structure of the cost function independently determined though arduous computation over a wide range of the two nondimensional numbers. The results are especially good based on reduction in cost function value and comparison with cost function structure.

Keywords: Rayleigh–Bénard convection; data assimilation; observation placement; forecast sensitivity; chaotic regimes; low-order modeling

Citation: Lewis, J.M.; Lakshmivarahan, S. Role of the Observability Gramian in Parameter Estimation: Application to Nonchaotic and Chaotic Systems via the Forward Sensitivity Method. *Atmosphere* **2022**, *13*, 1647. https://doi.org/10.3390/atmos13101647

Academic Editors: Bo-Wen Shen, Roger A. Pielke Sr. and Xubin Zeng

Received: 16 August 2022
Accepted: 28 September 2022
Published: 10 October 2022

1. Introduction

In his pioneering work, Saltzman [1] derived the well-known spectral model of Rayleigh–Bénard convection [2,3]. It was a 7-variable double Fourier series representation of the convection herein called Saltzman's low-order model and denoted by S-LOM (7). His model included a single parameter, the Rayleigh number R. In the governing equations, the ratio of the Rayleigh number to the critical Rayleigh number R_c, $\lambda = \frac{R}{R_c}$, becomes the key parameter. He assumed the other nondimensional number associated with this convection problem, the Prandtl number σ, was set equal to 10. Historically, Saltzman's [1] model achieved fame by providing Lorenz [4] with a nonlinear deterministic model that exhibits nonperiodic solution to explore the limitations of extended-range predictability. Within a decade, Lorenz's three-variable version of Saltzman's model became known as the chaotic-butterfly model [5], so-named because the phase-space depiction of model evolution resembled butterfly wings.

The goal of the research is to explore data assimilation [6] for the Rayleigh–Bénard convection in both chaotic and nonchaotic regimes. Two three-variable stripped-down versions of S-LOM (7) are used as dynamical constraints: (1) S-LOM (3) based on Saltzman's scaling, and L-LOM (3) based on Lorenz's scaling. These truncated systems contain the ingredients central to defining observation placement that yield a cost function minimum. The strength of this methodology rests on expressing the cost function (to a first-order accuracy around the true but unknown minimum) as a quadratic function of the increment in the control variable with (the symmetric and positive semi-definite) observability Gramian as the matrix of this approximate quadratic function. It is shown that by choosing the observation sites that maximize the square of the forecast sensitivities to control, we can indeed force the observability Gramian to be a positive definite matrix. This in turn avoids the occurrence of flat patches of the cost function in the control space. Refer to Lakshmivarahan et al. [7,8] and Lewis et al. [9] for further details and applications. An alternate strategy would be to place observations at locations that would lead to a small condition number for the observability Gramian.

Following this Introduction, we summarize S-LOM (7) and its reduction to the three-variable form [10] in Section 2. In Section 2, we also give attention to the scaling of S-LOM (3) and L-LOM (3). Discussion of the three-variable spectral dynamics is presented alongside graphics for nonchaotic and chaotic regimes in Section 3. Mechanics for data assimilation follow in Section 4. Data assimilation experiments are executed in Section 5, and discussion along with conclusions end the paper in Section 6.

2. Three-Variable Forms of Saltzman's Model

2.1. Spectral Form of Solution

The laboratory experiments of Bénard [2] were conducted with a rigid surface (a metal plate) below and a free upper surface while later experimental investigations used rigid surfaces at top and below (Chandrasekhar 1961 [11], Ch. 2, sect. 18c; Turner 1973 [12], Ch. 7, sect. 7.2). Saltzman's [1] theoretical development assumed two free surfaces. The onset of convection has different critical Rayleigh numbers R_c associated with the various boundary conditions—roughly 600 for free surfaces, 1800 for rigid surfaces, and 1200 for one-free and one-rigid surface. We use the free-surfaces value of $R_c = 657.5$ to develop the data assimilation model.

In an earlier technical report by the authors (Lewis and Lakshmivarahan [13]), a re-derivation of S-LOM (7) led to functional forms of coefficients in the equations that governed the Fourier amplitudes. As it stood in Saltzman's paper [1], the coefficients were expressed in numerical values only. Since our goal was to use Saltzman's equations in FSM data assimilation with the two nondimensional numbers as parameters, we needed to know the coefficients in terms of the wavenumbers and variable nondimensional numbers.

The three-variable form of Saltzman's equations used in this study is a subset of his seven-variable form. By setting the initial condition on the 3-horizontal wave temperature departure to 1 while the other initial conditions are set to zero delivers the three-variable form involving the (horizontal wavenumber 3/vertical wavenumber 1) streamfunction, the (horizontal wavenumber 3/vertical wavenumber 1) temperature departure, and the (vertical wavenumber 2) temperature departure. Let us label the three amplitudes in S-LOM (3) as follows: x_1 for the streamfunction (corresponding to amplitude A in Saltzman's notation). Temperature-departure components x_4 and x_7 (corresponding to amplitudes D and G in Saltzman' notation). This revised notation is consistent with the ordering of coefficients from 1 to 7 in Lewis and Lakshmivarahan [13] where x_1, x_2, ..., x_7 replaced A, B, ... , G.

The streamfunction ψ in S-LOM (3) is given by

$$\psi(x,z,t) = -4x_1(t)\sin(\pi ax)\sin(\pi z) \tag{1}$$

where a, the nondimensional wavenumber, is given by $a = \frac{1}{\sqrt{2}}$ which defines the 3-horizontal wavenumber wave, and the temperature departure θ is given by

$$\theta(x,z,t) = 4x_4(t)\cos(\pi ax)\sin(\pi z) + 4x_7(t)\sin(2\pi z) \tag{2}$$

2.2. Amplitude Equations

When Equations (1) and (2) are substituted into the partial differential equations that govern Rayleigh–Bénard convection, the two-dimensional (x, z) vorticity equation and the temperature departure equation, followed by imposition of integral orthogonality conditions for the double Fourier series, ordinary differential equations governing the time-dependent amplitudes follow where we have expressed coefficients to two decimal places (same accuracy as used by Saltzman [1]):

Saltzman's S-LOM (3) equations:

$$
\begin{aligned}
\dot{x}_1 &= -\sigma\pi^2(1+a^2)x_1 - \sigma\frac{ax_4}{\pi(1+a^2)} \\
&= -14.80\sigma\,x_1 - 0.15\sigma x_4
\end{aligned}
\tag{3}
$$

$$
\begin{aligned}
\dot{x}_4 &= -4\pi^2 ax_1x_4 - \pi aR_c\lambda x_1 - \pi^2(1+a^2)x_4 \\
&= -27.92x_1x_7 - 1460.62\lambda x_1 - 14.80x_4
\end{aligned}
\tag{4}
$$

$$
\begin{aligned}
\dot{x}_7 &= 2\pi^2 ax_1x_4 - 4\pi^2 x_7 \\
&= 13.96x_1x_4 - 39.48x_7
\end{aligned}
\tag{5}
$$

where

$$
\lambda = \frac{R}{R_c} \quad (R : \textit{Rayleigh number})
\tag{6}
$$

Lorenz [4] used another scaling for the derivation of his L-LOM (3) that took the form:

Lorenz's L-LOM (3) equations:

$$
\dot{x}_1 = -\sigma x_1 + \sigma x_4
\tag{7}
$$

$$
\dot{x}_4 = -x_1 x_7 + \lambda x_1 - x_4
\tag{8}
$$

$$
\dot{x}_7 = x_1 x_4 - b x_7
\tag{9}
$$

where

$$
b = \frac{4}{1+a^2} = \frac{8}{3}
\tag{10}
$$

Table 1 shows non-dimensional and scaling for S-LOM (3) and L-LOM (3). The different signs on the right-hand sides of S-LOM (3) and L-LOM (3) stem from the fact that Lorenz assumed double Fourier representation of temperature departure and streamfunction as follows:

$$
\theta(x,z,t) = x_4(t)\cos(\pi ax)\sin(\pi z) - x_7(t)\sin(2\pi z)
\tag{11}
$$

$$
\psi(x,z,t) = x_1(t)\sin(\pi ax)\sin(\pi z)
\tag{12}
$$

where the sign of the x_7-component term in Equation (11) is opposite the corresponding term in Equation (2) and the sign of the streamfunction in Equation (12) is opposite the sign in Equation (1). These sign differences lead to a shift of temperature departure structure relative to the streamfunction structure in the convective regime compared to Saltzman's results. However, temperature–motion structures in S-LOM (3) and L-LOM (3) are both faithful to Rayleigh–Bénard cellular convection.

$$
\frac{1}{(1+a^2)\pi^2} = 0.068; \quad \frac{1+a^2}{a} = 2.121; \quad \frac{[\pi(1+a^2)]^3}{a^2} = 209.292
\tag{13}
$$

If the physical parameters are taken to be those associated with water at 20 °C (Chandrasekhar [11] (1961, Ch. 2, Table V1), the physical parameters are given by

$$
\begin{aligned}
\nu &= 1.006{\cdot}10^{-2}\ \mathrm{cm^2\ sec^{-1}} \\
\kappa &= 1.433{\cdot}10^{-3}\ \mathrm{cm^2\ sec^{-1}} \\
\epsilon &= 2.0{\cdot}10^{-3}\ {}^\circ\mathrm{C^{-1}}
\end{aligned}
\tag{14}
$$

Table 1. Nondimensional forms of Saltzman and Lorenz variables. The superscript asterisk (*) denotes the nondimensional variables where physical parameters for the model follow: g, H, κ, ν, ϵ: acceleration of gravity, depth of the fluid, coefficient of thermal diffusivity, kinematic viscosity, and coefficient of volume expansion, respectively.

Saltzman	Lorenz
$t = \frac{H^2}{\kappa} t^*$	$t = \frac{H^2}{\kappa} t^* \frac{1}{(1+a^2)\pi^2}$
$x = Hx^*, z = Hz^*$	$x = Hx^*, z = Hz^*$
$\Psi = \kappa \psi^*$	$\Psi = \kappa \psi^* \frac{1+a^2}{a}$
$\Theta = \frac{\kappa \nu}{g\epsilon H^3} \theta^*$	$\Theta = \frac{\kappa \nu}{g\epsilon H^3} \theta^* \frac{[\pi(1+a^2)]^3}{a^2}$

In laboratory experiments, the fluid height falls between 2 mm and 1 cm (Chandrasekhar 1961 [11], Ch. 2, sect. 18). Let us assume $H = 2$ mm. Then, the relative values of Saltzman's and Lorenz's nondimensional variables follow:

- Lorenz's nondimensional t^* is roughly 15 times greater than Saltzman's nondimensional t^*,
- Lorenz's nondimensional x^* and z^* are equal to Saltzman's nondimensional x^* and z^*,
- Lorenz's nondimensional ψ^* is roughly 2 times less than Saltzman's nondimensional ψ^*, and
- Lorenz's nondimensional θ^* is roughly 200 times less than Saltzman's nondimensional θ^*

Thus, for example, if the nonchaotic convective regime comes to equilibrium at Lorenz's $t^* = 15$, it comes to equilibrium at Saltzman's $t^* = 1$. Similarly, a temperature departure of $\theta^* = 200$ in Saltzman's case is associated with Lorenz's value of $\theta^* = 1$.

The equations governing the evolution of forecast sensitivities are found by taking partial derivatives of S-LOM (3) and L-LOM (3) with respect to λ and σ separately. Thus, six equations are added to each set, S-LOM (3) and S-LOM (3).

We demonstrate the process of determining evolution of forecast sensitivities by deriving the equation governing for $\frac{\partial x_4}{\partial \lambda}$:

$$\dot{x}_4 = \frac{dx_4}{dt} = -x_1 x_7 + \lambda x_1 - x_4$$
$$\frac{d}{dt}\left(\frac{\partial x_4}{\partial \lambda}\right) = -x_1 \frac{\partial x_7}{\partial \lambda} - x_7 \frac{\partial x_1}{\partial \lambda} + x_1 + \lambda \frac{\partial x_1}{\partial \lambda} - \frac{\partial x_4}{\partial \lambda} \tag{15}$$

Similar ordinary differential equations are derived for the other forecast sensitivities. The initial conditions for the six sensitivity equations are zero, i.e., change in controls λ and σ affect the sensitivity solutions for $t > 0$, but not at $t = 0$. The 3-variable amplitude solutions feed into the forecast sensitivity equations as shown in Equation (15), but the forecast sensitivity equations do not feedback into the equations governing x_1, x_4, and x_7. The set of nine differential equations are solved simultaneously as a coupled set.

3. Dynamics of 3-Mode Systems: S-LOM (3) and L-LOM (3)

3.1. Overview

The 3-mode truncated systems that describe Rayleigh–Bénard convection have similar strength (they capture the interaction between temperature and vorticity that produce the physically meaningful convective motions) and similar weakness (inability to depict turbulent cellular motion with limited wave structures in the presence of large Rayleigh numbers). However, it is the difference in scaling and nondimensional forms that highlights the dynamical difference in the equation sets. It would be interesting to ask Lorenz why he refused to follow Saltzman's scaling and associated coefficients for the differential equations plainly set forth in Saltzman's Table 2 [1] One can guess that he wanted differential equations where the coefficients exhibited comparable magnitudes that lead to ease in understanding interaction between terms in the equation set. However, beyond this difference, the model forecast sensitivities to controls are drastically different between the two sets—relatively large model forecast sensitivity to controls in S-LOM (3)'s case compared to L-LOM (3)'s case. These differences lead to poor data assimilation results when slight changes in ideal observation times are used in the S-LOM (3) experiments. On the other hand, L-LOM (3) tolerated these slight changes very well. Results of these experiments are discussed in Section 5.1 (L-LOM (3) dynamics) and Section 5.2 (S-LOM (3) dynamics). All calculations in Sections 3.2 and 3.3 use L-LOM (3).

3.2. Nonchaotic Regime: $\lambda = 12$, $\sigma = 7$ ($\sigma = \frac{\nu}{\kappa} = 7.03$ *for Water at* 20 °C)

When $\lambda = 12$, the Rayleigh number R is given by

$$R = \lambda R_c = 12\,(657.5) = 8.106 \cdot 10^3 \tag{16}$$

With this value of R and $H = 2$ mm for water at 20 °C, Rayleigh's [3] formula for the temperature difference ΔT between the upper and lower boundaries of the fluid is given by

$$\Delta T = \frac{\kappa \nu R}{g \epsilon H^3} = 7.11\ ^\circ\text{C} \tag{17}$$

In the first three panels of Figure 1, amplitudes are depicted in pairs between $t = 0$ and $t = 10$ (5 min dimensionally): $x_1 x_4$ (first panel, top), $x_4 x_7$ (second panel), and $x_1 x_7$ (third panel). In each of these panels, amplitudes are oscillatory and exhibit convergence to $x_1 \cong 5.50$, $x_4 = 5.75$, and $x_7 = 12.0$. In the bottom panel of Figure 1, the 3D trajectory of the amplitudes in phase space—$(\lambda,\ \sigma)$ space—is shown over the period $t = 0$ to $t = 10$. This plot indicates convergence of L-LOM (3) to the x_1, x_4, x_7 values stated above.

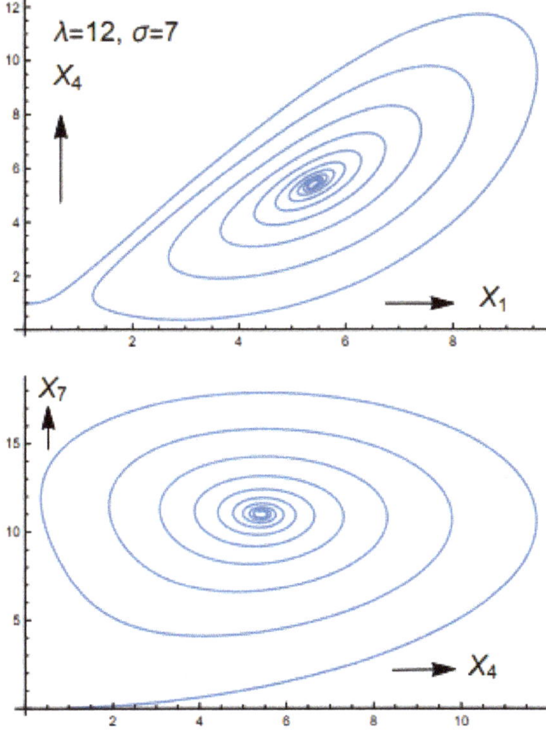

Figure 1. *Cont.*

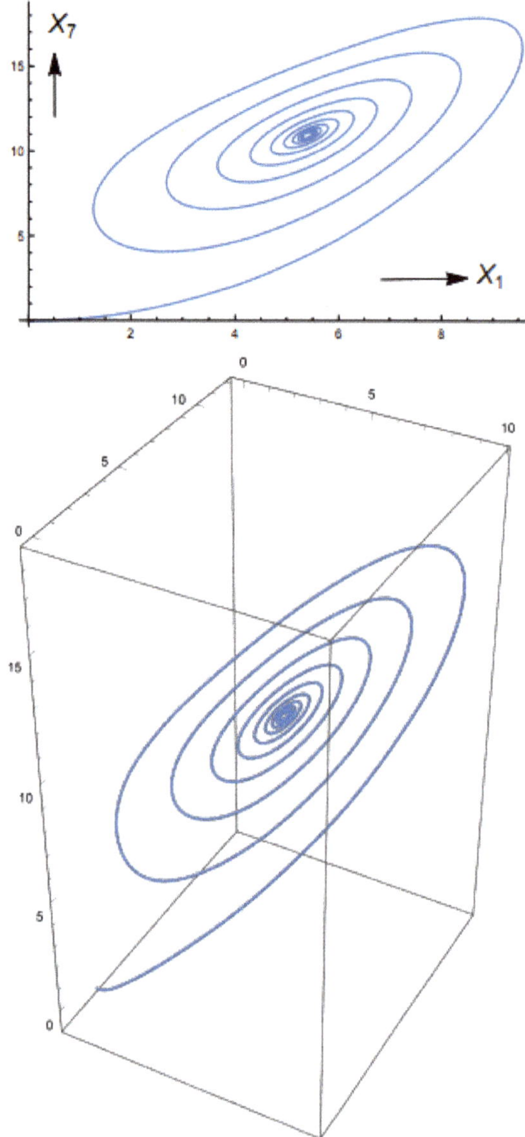

Figure 1. Phase diagrams for the nonchaotic regime with controls $\lambda = 12$, $\sigma = 7$ over time span $[0, 10]$: Panel 1 (x_1, x_4), Panel 2 (x_4, x_7), Panel 3 (x_1, x_7), and Panel 4 (x_1, x_4, x_7) with axes: horizontal $x_1[0, 10]$, $x_4[0, 12]$, $x_7[0, 20]$.

The forecast sensitivities of x_4 and x_7 with respect to λ and σ are shown in Figure 2. These two amplitude components are used to represent the temperature departure, and since only temperature departure observations will be used in the data assimilation experiments, we confine our forecast sensitivity analysis to these two components. These sensitivities exhibit oscillatory decay as steady state is approached.

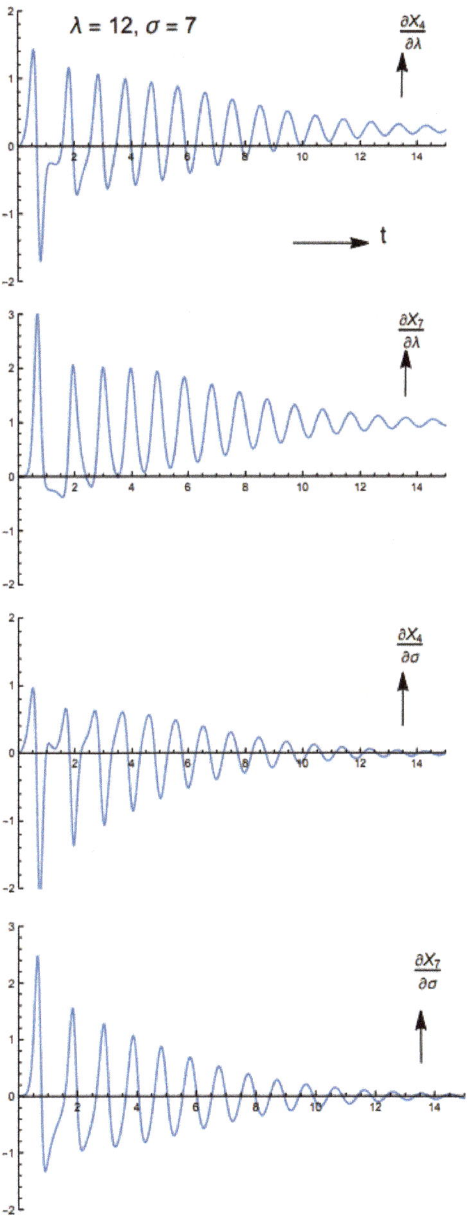

Figure 2. Forecast sensitivities of x_4 and x_7 with respect to (λ, σ) over time span $[0, 14]$ for the non-chaotic regime with controls $\lambda = 12$, $\sigma = 7$.

The structure of the convection at $t = 10$ is shown in Figure 3—two panels, the temperature departure and streamfunction. The maximum/minimum values of temperature departure are $\sim \pm 15$ ($\pm 3\,^{\circ}\mathrm{C}$ dimensionally for water at $20\,^{\circ}\mathrm{C}$) where the maximum values are in the top portion of the water and the minimum values in the bottom portion. This leads to a spurious stable stratification first noted by Kuo [14] when convection is controlled by a truncated low-order spectral system. Within the cells, the maximum ($\frac{\partial \psi}{\partial x} > 0$) and minimum ($\frac{\partial \psi}{\partial x} < 0$) upward/downward vertical velocities, respectively, are the order of $1\,\mathrm{mm \cdot s^{-1}}$ so roughly 50 circuits of a fluid particle around the cell takes place between the onset of convective motion ($t = 0$) and steady state ($t = 10$).

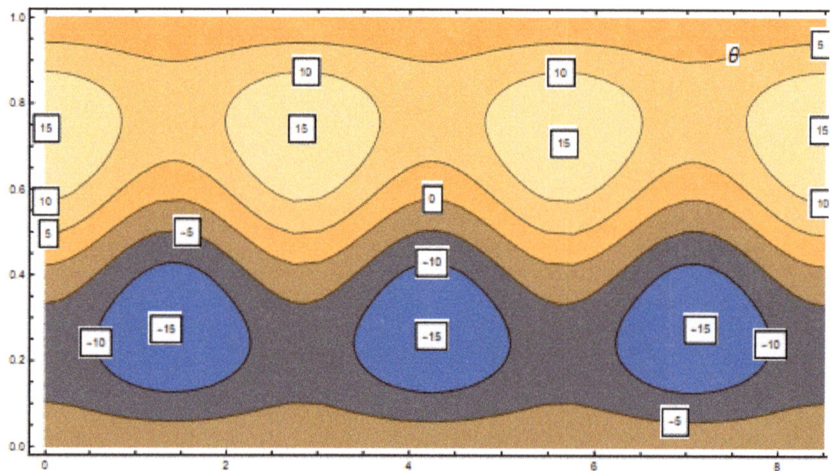

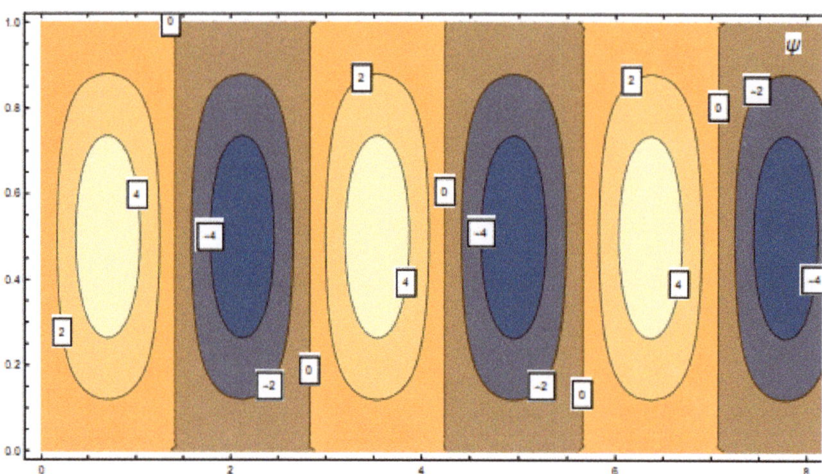

Figure 3. Display of convection pattern in the $x - z$ plane at time $= 10$ for the nonchaotic regime with controls $\lambda = 12$, $\sigma = 7$.

3.3. Chaotic Regime: $\lambda = 28$, $\sigma = 10$

The Prandtl number $\sigma = 10$ was assumed by Saltzman [1] in nonchaotic regimes and by Lorenz [4] in chaotic regimes. With these nondimensional numbers, Lorenz [4] found that instability of steady convection occurred when $\lambda = 24.74$ and beyond. The slightly supercritical value of $\lambda = 28$ was used by Lorenz in his numerical experiments [4]. The states of steady convection are then represented by amplitude points (x_1, x_4, x_7): $(6\sqrt{2}, 6\sqrt{2}, 27)$ and $(-6\sqrt{2}, -6\sqrt{2}, 27)$, while the state of no convection corresponds to the origin in phase space $(0, 0, 0)$.

In the first three panels of Figure 4, amplitudes are depicted in pairs between $t = 0$ and $t = 30$: $x_1 x_4$, $x_4 x_7$, and $x_1 x_7$. The well-known butterfly pattern appears for each pair over the time period $t = 0$ to $t = 30$. The convergence to (x_1, x_4, x_7): $(6\sqrt{2}, 6\sqrt{2}, 27)$ and $(-6\sqrt{2}, -6\sqrt{2}, 27)$ does not occur in finite time, only as $t \longrightarrow \infty$. The 3-D phase plot of amplitudes is shown in the lower right corner of Figure 4 where the two steady-state solutions are located at the centroids of the circular/elliptical trajectories.

The time evolution of sensitivities $\frac{\partial x_7}{\partial \lambda}$, $\frac{\partial x_7}{\partial \sigma}$ is shown in Figure 5. Here, we have broken the plots into two time zones: t [0, 5] and t [5, 10] for $\frac{\partial x_7}{\partial \lambda}$ and t [0, 3] and t [3, 10] for $\frac{\partial x_7}{\partial \sigma}$. Note that the coordinate axes are scaled differently in these plots. Plots for $\frac{\partial x_4}{\partial \lambda}$, $\frac{\partial x_4}{\partial \sigma}$ exhibit similar structures. The common thread between sensitivity plots is that the magnitudes oscillate between $+$ and $-$ values and increase ad infinitum.

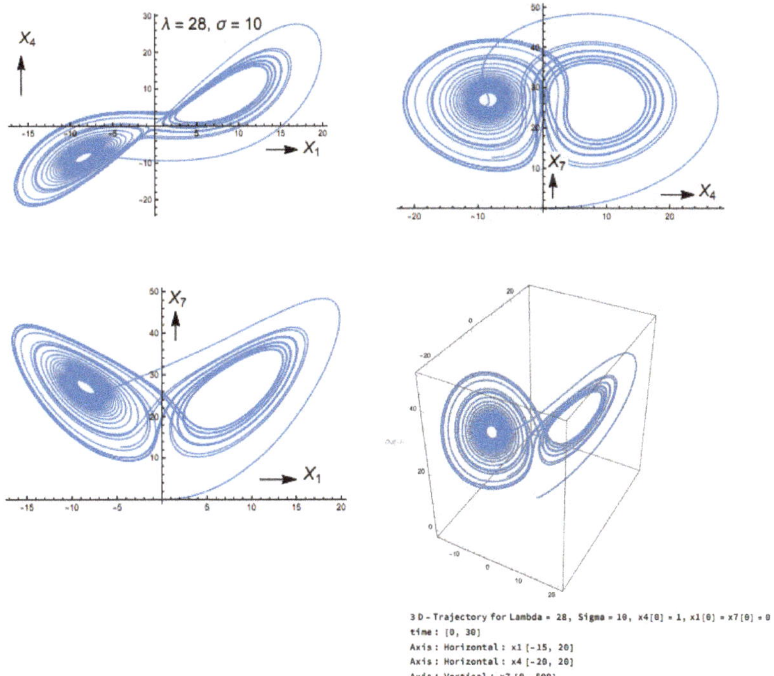

Figure 4. Phase plots of amplitudes for the chaotic regime with controls $\lambda = 28$, $\sigma = 10$.

The sensitivity of the $x_1(t)$ solution to a small change in the initial condition is shown in Figure 6. In this figure, a 5% change in the initial condition leads to significant change in $x_1(t)$ after t∼20.

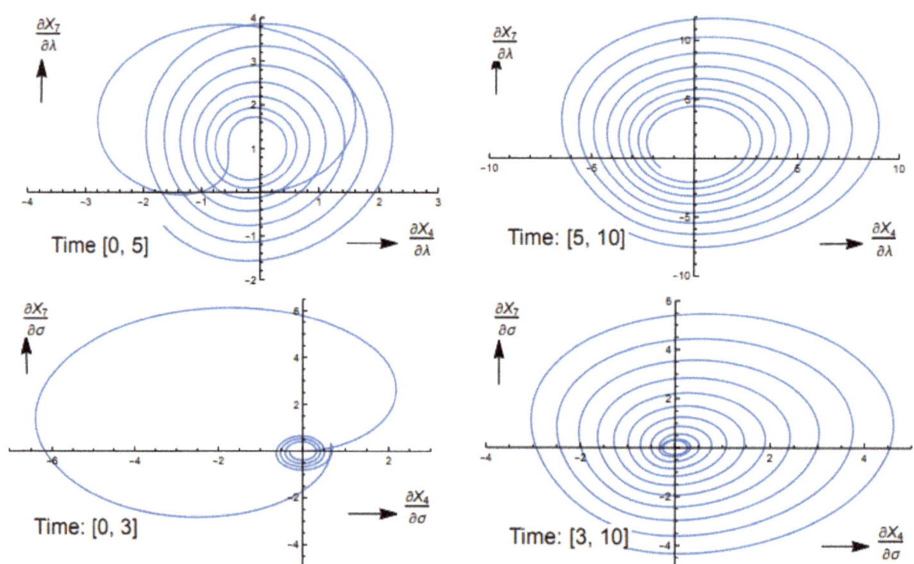

Figure 5. Phase plots of x_4 and x_7 forecast sensitivity over time span [0, 10] in the chaotic regime with controls $\lambda = 28$, $\sigma = 10$.

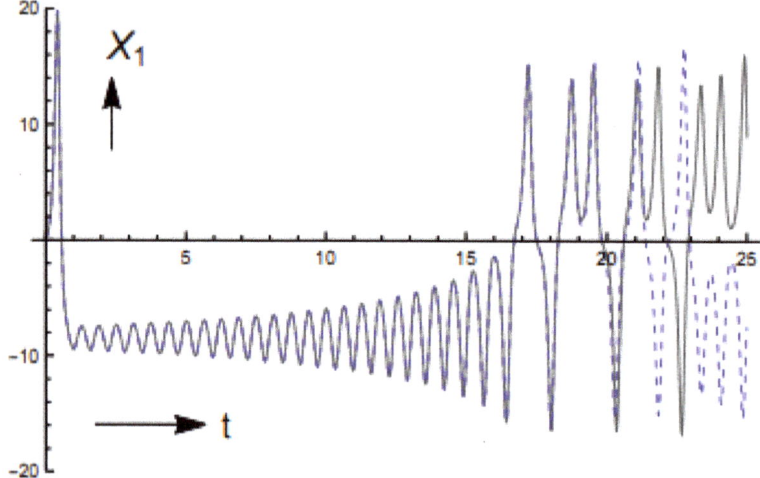

Figure 6. Extreme sensitivity of streamfunction amplitude x_1 to slight change in $x_4(0)$ from 1.00 to 1.05 for the chaotic regime—solid line ($x_4(0) = 1.00$) and dashed line ($x_4(0) = 1.05$).

4. Design of Observation Network

4.1. Model and Forecast Sensitivities

Let $X(k) = (x_1(k),\ x_4(k),\ x_7(k)) \in R^3$ be the state of the S-LOM (3) or L-LOM (3) model at times $k = 0, 1, 2, 3\ldots$ given by

$$X(k+1) = M(X(k), \alpha) \in R^3 \tag{18}$$

with $X(0) \, \epsilon R^3$ as initial condition, $\alpha = (\alpha_1, \alpha_2) = (\lambda, \sigma)^T \epsilon \, R^2$, the parameter vector, and $X(k) = X(k, X(0), \alpha)$ the solution.

Define forecast sensitivity to parameters as

$$
\begin{aligned}
V(k) &= \frac{\partial X(k)}{\partial \alpha} \, \epsilon \, R^{3x2} \\
&= \begin{bmatrix} \frac{\partial x_1}{\partial \lambda} & \frac{\partial x_1}{\partial \sigma} \\ \frac{\partial x_4}{\partial \lambda} & \frac{\partial x_4}{\partial \sigma} \\ \frac{\partial x_7}{\partial \lambda} & \frac{\partial x_7}{\partial \sigma} \end{bmatrix}
\end{aligned}
\tag{19}
$$

It can be shown that $V(k)$ evolves according to the discrete time dynamics given by

$$
\begin{aligned}
V(k+1) &= D_M(k)V(k) + D_M^\alpha(k) \\
&\quad with \; V(0) = 0
\end{aligned}
\tag{20}
$$

where

$$
D_M(k) = \frac{\partial}{\partial x_j(0)}[M_i(X(k), \, \alpha)] \, \epsilon \, R^{3x3}
\tag{21}
$$

$$
D_M^\alpha(k) = \frac{\partial}{\partial \alpha_j}[M_i(X(k), \, \alpha)] \, \epsilon \, R^{3x2}
\tag{22}
$$

with $\alpha = (\alpha_1, \alpha_2) = (\lambda, \sigma)$ [15,16].

An example of forecast sensitivity to parameter λ has been shown earlier in Equation (15).

4.2. Forecast Error

Let $\bar{c} = \bar{\alpha}$ be the true control and $\overline{X}(k)$ be the solution starting from $\bar{c}$. Let $c = \alpha$ be the arbitrary control and let $X(k)$ be the model solution starting from c.

Define

$$
\delta c = \bar{c} - c = \delta \alpha = \bar{\alpha} - \alpha
\tag{23}
$$

where $\bar{\alpha} - \alpha = (\overline{\lambda} - \lambda, \overline{\sigma} - \sigma)^T = \begin{bmatrix} \delta\lambda \\ \delta\sigma \end{bmatrix} = \delta\alpha$.

Let $\delta X(k)$ be the change in the state $X(k)$ resulting from change δc. Then, from Equation (19) and using first-order Taylor expansion, we obtain

$$
\begin{aligned}
\delta X(k) &= \overline{X}(k) - X(k) \\
&= X(k, X(0), \bar{\alpha}) - X(k, X(0), \alpha) \\
&= \frac{\partial X(k)}{\partial \alpha} \delta\alpha \\
&= V(k)\delta\alpha
\end{aligned}
\tag{24}
$$

4.3. Observations and Cost Function

We work with L-LOM (3) in this subsection.

Consider the physical domain defined by the rectangle

$$
D = \left\{ \eta = (x, z)^T \right\}, \quad 0 \leq x \leq 6\sqrt{2}, \; 0 \leq z \leq 1
\tag{25}
$$

where x denotes the lengthwise coordinate and z the breadthwise coordinate of a point η in D. Convection is confined to this domain.

Define a matrix

$$
H(\eta) = [f_4(\eta), \, f_7(\eta)] \epsilon \, R^{1x2}
\tag{26}
$$

where

$$
\begin{aligned}
f_4(\eta) &= \cos(\pi a x) \sin(\pi z) \\
f_7(\eta) &= -\sin(2\pi z)
\end{aligned}
\tag{27}
$$

Let

$$
\bar{\theta}(k, \eta) = H(\eta) \begin{bmatrix} \overline{x}_4 \\ \overline{x}_7 \end{bmatrix} = H(\eta)\hat{\overline{X}} + \xi(k)
\tag{28}
$$

be the actual observation where $\hat{\overline{X}}$ is the projection of $\overline{X}$ onto its last two elements and $\xi(k)$ is the observation noise with variance σ^2.

Let

$$\theta^M(k,\,\eta) = H(\eta)\begin{bmatrix} x_4 \\ x_7 \end{bmatrix} = H(\eta)\hat{X} \tag{29}$$

where $\theta^M(k,\eta)$ is the model counterpart to the observation based on perturbed control. The difference is called the forecast error or innovation (*innov*).

$$e(k) = \overline{\theta}(k,\eta) - \theta^M(k,\,\eta) \tag{30}$$

Using Equations (28) and (29), we rewrite Equation (30) as

$$\begin{aligned} e(k) &= H(\eta)\left(\hat{\overline{X}} - \hat{X}\right) \\ &= H(\eta)\delta\hat{X} \end{aligned} \tag{31}$$

For $\alpha \in R^2$ and $\eta \in R^2$, we define the cost function for a given $(k,\,\eta)$ as

$$J(\alpha, k,\,\eta) = \frac{1}{2\sigma^2}\left(\overline{\theta}(k,\eta) - \theta^M(k,\,\eta)\right)^2 \tag{32}$$

The squared difference between the observation $\overline{\theta}(k,\eta)$ and its model counterpart $\theta^M(k,\,\eta)$. This cost function in Equation (32) is additive in the number of observations. Using Equations (31) and (32), we get

$$J(\alpha,\,\eta) = \frac{1}{2\sigma^2}(H(\eta)(\delta\hat{X})^T H(\eta)(\delta\hat{X}) = \frac{1}{2\sigma^2}(\delta\hat{X})^T H^T(\eta)H(\eta)\,(\delta\hat{X}) \tag{33}$$

Using first-order Taylor expansion, the vector of temperature departure error components, $\begin{bmatrix} \delta x_4 \\ \delta x_7 \end{bmatrix}$, can be represented by

$$\delta\hat{X} = \begin{bmatrix} \delta x_4 \\ \delta x_7 \end{bmatrix} = \begin{bmatrix} \frac{\partial x_4(k)}{\partial\lambda} & \frac{\partial x_4(k)}{\partial\sigma} \\ \frac{\partial x_7(k)}{\partial\lambda} & \frac{\partial x_7(k)}{\partial\sigma} \end{bmatrix}\begin{bmatrix} \delta\lambda \\ \delta\sigma \end{bmatrix} = \hat{V}(k)\delta\alpha, \tag{34}$$

where $\hat{V}(k)$ is a submatrix of the forecast sensitivity matrix $V(k)$; namely, the last two rows of that matrix. Arguments related to the derivation of Equation (34) are found in Lakshmivarahan et al. [7].

Substituting Equation (34) into Equation (33), we get

$$J(\alpha,\,\eta) = \frac{1}{2\sigma^2}(\delta\alpha)^T\left[\hat{V}^T(k)\left(H^T(\eta)H(\eta)\right)\hat{V}(k)\right]\delta\alpha = \frac{1}{2\sigma^2}(\delta\alpha)^T G\,(\delta\alpha) = \frac{1}{2\sigma^2}(\delta\alpha)^T G\,(\delta\alpha) \tag{35}$$

where

$$G(k) \equiv \hat{V}^T(k)\left(H^T(\eta)H(\eta)\right)\hat{V}(k)] = \hat{V}^T(k)\overline{H}(\eta)\hat{V}(k)\Big] \tag{36}$$

and

$$\overline{H}(\eta) \equiv H^T(\eta)H(\eta) \tag{37}$$

Remark 1. *It is important to note that $G(k)$ applies to a single time k. Since the cost function is additive in the number of observations, i.e., if there is more than one observation, a separate matrix exists for each time. Take the case when observations are available at three observation times with G matrices G_1, G_2, and G_3, then the appropriate combined matrix will be $G = G_1 + G_2 + G_3$.*

In the presence of two forms of the cost function, namely Equation (32) and (35), we use Equation (35) to determine observation placement. We use Equation (32) to display the structure of the cost function in the space of controls.

From Equation (35), the gradient of $J(\alpha,\,\eta)$, $\nabla J(\alpha,\eta)$, is given by

$$\nabla J(\alpha, \eta) = \begin{bmatrix} \frac{\partial J}{\partial \lambda} \\ \frac{\partial J}{\partial \sigma} \end{bmatrix} = G\,(\delta\alpha) \tag{38}$$

To determine the optimal adjustment to control through FSM, we write the cost function as

$$J(\alpha, \eta) = \frac{1}{2\sigma^2}\left(H\hat{X} + \frac{\partial H(\eta)\hat{X}}{\partial \alpha}\delta\alpha - \overline{\theta}\right)^T \left(H\hat{X} + \frac{\partial H(\eta)\hat{X}}{\partial \alpha}\delta\alpha - \overline{\theta}\right) \tag{39}$$

where

$$\frac{\partial H(\eta)\hat{X}}{\partial \alpha} = H(\eta)\,\hat{V} \tag{40}$$

Expanding Equation (40) and setting its gradient to zero yields

$$\delta\alpha = \left(G^T G\right)^{-1} G^T innov \tag{41}$$

where the innovation *innov* is given by

$$innov = H\hat{X} - \overline{\theta} \tag{42}$$

4.4. Observation Placement

In this paper we consider two structures for the observations: (1) time only, and (2) time and space. Our strategy for observation placement is to render G positive definite. However, not only render it positive definite, but also reduce its condition number so that a "bowl-like" cost function's structure is present in the vicinity of its true minimum located at $\overline{\alpha} = (\overline{\lambda}, \overline{\sigma})$ for error-free observations. Two types of observations are considered: (a) observations in time alone (Section 4.4.1) and (b) observations in space and time (Section 4.4.2).

4.4.1. Observations in Time Alone

When we consider observations of spectral components $\overline{X}(k)$ as a function of time alone in our data assimilation problem, the $H\,(\eta)$ matrices in Equation (36) are replaced by identity matrices and the cost function then becomes

$$J(\alpha) = \frac{1}{2\sigma^2}(\delta\alpha)^T \left[\hat{V}^T(k)\,\hat{V}(k)\right]\delta\alpha \tag{43}$$

where $G = \hat{V}^T(k)\hat{V}(k)$ is positive definite as shown below. In matrix form

$$\hat{V}^T(k)\hat{V}(k) = \begin{bmatrix} \left(\frac{\partial x_4(k)}{\partial \lambda}\right)^2 + \left(\frac{\partial x_7(k)}{\partial \lambda}\right)^2 & \left(\frac{\partial x_4(k)}{\partial \lambda}\frac{\partial x_4(k)}{\partial \sigma} + \frac{\partial x_7(k)}{\partial \lambda}\frac{\partial x_7(k)}{\partial \sigma}\right) \\ \left(\frac{\partial x_4(k)}{\partial \lambda}\frac{\partial x_4(k)}{\partial \sigma} + \frac{\partial x_7(k)}{\partial \lambda}\frac{\partial x_7(k)}{\partial \sigma}\right) & \left(\frac{\partial x_4(k)}{\partial \sigma}\right)^2 + \left(\frac{\partial x_7(k)}{\partial \sigma}\right)^2 \end{bmatrix} \tag{44}$$

or alternatively represented by

$$\hat{V}(k) = (\varepsilon_1(k),\ \varepsilon_2(k)) \qquad \hat{V}^T = \begin{bmatrix} \varepsilon_1^T(k) \\ \varepsilon_2^T(k) \end{bmatrix} \tag{45}$$

$$\hat{V}^T(k)\hat{V}(k) = \begin{bmatrix} \varepsilon_1^T(k)\varepsilon_1(k) & \varepsilon_1^T(k)\varepsilon_2(k) \\ \varepsilon_2^T(k)\varepsilon_1(k) & \varepsilon_2^T(k)\varepsilon_2(k) \end{bmatrix} \tag{46}$$

From this point onwards, we simplify notation by omitting the parenthetical expression "(k)" following forecast sensitivities.

From the physics of our problem, $V(k)$ is sensitive to λ and σ. The dot products "$\langle a, b \rangle$" are expressed in matrix-vector notation where $\varepsilon_1(k)$ and $\varepsilon_2(k)$ are the first and second columns of $\hat{V}(k)$.

$$\langle \varepsilon_1,\ \varepsilon_2 \rangle = \varepsilon_1^T \varepsilon_2 = \varepsilon_2^T \varepsilon_1 = \langle \varepsilon_2,\ \varepsilon_1 \rangle \tag{47}$$

$$\left\langle \varepsilon_1^T,\ \varepsilon_1 \right\rangle = \|\varepsilon_1\|^2, \quad \left\langle \varepsilon_2^T,\ \varepsilon_2 \right\rangle = \|\varepsilon_2\|^2$$

Thus,

$$\hat{V}^T(k)\hat{V}(k) = \begin{bmatrix} \|\varepsilon_1\|^2 & \langle \varepsilon_1,\ \varepsilon_2 \rangle \\ \langle \varepsilon_2,\ \varepsilon_1 \rangle & \|\varepsilon_2\|^2 \end{bmatrix} \tag{48}$$

Further, $\|\varepsilon_1\|^2 > 0$ and $\|\varepsilon_2\|^2 > 0$, and the determinant of $\hat{V}^T(k)\hat{V}(k)$ is given by

$$\det\left(\hat{V}^T\hat{V}\right) = \|\varepsilon_1\|^2\,\|\varepsilon_2\|^2 - <\varepsilon_1,\ \varepsilon_2>^2 \tag{49}$$

and

$$\langle \varepsilon_1,\ \varepsilon_2 \rangle = \|\varepsilon_1\|\,\|\varepsilon_2\| \cos\theta \tag{50}$$

so

$$<\varepsilon_1,\ \varepsilon_2>^2 = \|\varepsilon_1\|^2\|\varepsilon_2\|^2\ \cos^2\theta \tag{51}$$

The determinant of $\hat{V}^T\hat{V}$ can then be written as

$$\det\left(\hat{V}^T\hat{V}\right) = \|\varepsilon_1\|^2\|\varepsilon_2\|^2 - \|\varepsilon_1\|^2\|\varepsilon_2\|^2\ \cos^2(\theta) = \|\varepsilon_1\|^2\|\varepsilon_2\|^2\left(1 - \cos^2\theta\right) > 0\ (if\ \theta \neq 0) \tag{52}$$

So long as ε_1 and ε_2 are not colinear or $\theta = 0$ or $180°$, then

$$\cos\theta \neq 0,\ \text{and} \det\left(\hat{V}^T\hat{V}\right) > 0 \tag{53}$$

and the following inequalities hold

$$\|\varepsilon_1\|^2 > 0,\ \|\varepsilon_2\|^2 > 0,\ \text{and}\ \det\left(\hat{V}^T\hat{V}\right) > 0 \tag{54}$$

These inequalities imply that $\hat{V}^T\hat{V}$ is a positive definite matrix.

The strategy for choosing observations is therefore:

Pick observation $z(k) = \hat{X}(k) + v(k)$ when $z(k) \in R^2$ and noise $v(k) \in R^2$ such that $z(k_1)$ at time k_1 makes $\|\varepsilon_1\|^2$ a maximum and $z(k_2)$ makes $\|\varepsilon_2\|^2$ a maximum.

Then, at

$$k_1 : V(k_1) = [\varepsilon_1(k_1),\ \varepsilon_2(k_1)]$$

and at

$$k_2 : V(k_2) = [\varepsilon_1(k_2),\ \varepsilon_2(k_2)]$$

and

$$G = V^T(k_1)V(k_1) + V^T(k_2)V(k_2) = G(k_1) + G(k_2) \tag{55}$$

4.4.2. Observations Taken in Space and Time

When observation locations are to be determined in space and time, the G matrix takes the form

$$G(k,\eta) = \hat{V}^T(k)\left(H^T(\eta)H(\eta)\right)\hat{V}(k)] = \hat{V}^T(k)\overline{H}(\eta)\hat{V}(k)]. \tag{56}$$

The space element $\overline{H}(\eta)$ is separable from the time elements $\hat{V}^T(k)$ and $\hat{V}^T(k)$ or in combination $\hat{V}^T(k)\ \hat{V}(k)$. In the presence of this separability, the strategy reduces to first determining desired times as explained above. This is followed by determining desired positions in $(x,\ z)$ space.

Recall that $\hat{x} = (x_4,\ x_7)^T$ and $\alpha = (\lambda,\ \sigma)^T$ and represent $\hat{V}$ as a column partition

$$\hat{V} = \begin{bmatrix} \frac{\partial x_4}{\partial \lambda} & \frac{\partial x_4}{\partial \sigma} \\ \frac{\partial x_7}{\partial \lambda} & \frac{\partial x_7}{\partial \sigma} \end{bmatrix} = \begin{bmatrix} \frac{\partial \hat{x}}{\partial \lambda}, & \frac{\partial \hat{x}}{\partial \sigma} \end{bmatrix} = [V_1,\ V_2] \tag{57}$$

Define vectors

$$g_4 = \begin{bmatrix} \frac{\partial x_4}{\partial \alpha} \end{bmatrix} = \left(\frac{\partial x_4}{\partial \lambda},\ \frac{\partial x_4}{\partial \sigma}\right)^T \in R^2 \tag{58}$$

and

$$g_7 = \left[\frac{\partial x_7}{\partial \alpha}\right] = \left(\frac{\partial x_7}{\partial \lambda}, \frac{\partial x_7}{\partial \sigma}\right)^T \in R^2 \tag{59}$$

It can be verified that the row partition of $\hat{V}$ is given by

$$\hat{V} = \begin{bmatrix} g_4^T \\ g_7^T \end{bmatrix} \tag{60}$$

Recall that H is given by

$$H = [f_4, \ f_7] \in R^{1x2} \tag{61}$$

where $f_4 = \cos(\pi ax)\sin(\pi z)$ and $f_7 = -\sin(2\pi z)$.

Using Equations (60) and (61), we obtain a row vector $w(k, \ \eta) = (w_1(k, \ \eta), \ w_2(k, \ \eta))$ given by

$$w(k, \ \eta) = H\hat{V} = [f_4, \ f_7]\begin{bmatrix} g_4^T \\ g_7^T \end{bmatrix} = (f_4 g_4^T + f_7 g_7^T) \in R^{1x2} \tag{62}$$

Consequently, we can express the Gramian $G(k, \ \eta)$ as a rank-one symmetric product matrix

$$G(k, \ \eta) = w^T(k, \ \eta)w(k, \eta) = \begin{bmatrix} w_1^2(k, \ \eta) & w_1(k, \ \eta)w_2(k, \ \eta) \\ w_2(k, \ \eta)w_1(k, \ \eta) & w_2^2(k, \ \eta) \end{bmatrix} \tag{63}$$

Strategy for determining observation placement:

(1) Pick $\eta(x, \ z)$ such that $\| \ f_4 \ \|$ and $\| \ f_7 \ \|$ are maximum
(2) Let $k_1, \ k_2, \ k_3, \ k_4$ be four time instances where the squares of $\frac{\partial x_4}{\partial \lambda}, \ \frac{\partial x_4}{\partial \sigma}, \ \frac{\partial x_7}{\partial \lambda}, \ \frac{\partial x_7}{\partial \sigma}$ are maximum. Determine the respective $G(k_i, \ \eta)$'s. Then, the final Gramian is given by

$$G = G(k_1, \eta) + G(k_2, \eta) + G(k_3, \eta) + G(k_4, \eta) \tag{64}$$

Remark 2. *In principle we can use just two observations at times k_1 and k_2 where $\left(\frac{\partial x_4}{\partial \lambda}\right)^2$ and $\left(\frac{\partial x_7}{\partial \sigma}\right)^2$ attain their maxima.*

5. Data Assimilation Experiments

Once the observations and their locations have been determined, the FSM method of data assimilation is used to determine optimal values of controls. Development of the FSM was first shown in Lakshmivarahan and Lewis [15] and explored with numerous examples in the textbook Lakshmivarahan et al. [16]. In our case, the optimal values of control are found through the iterative methodology captured by Equation (41). Starting from guess control, solution to Equation (41) delivers the adjustment to guess control. The adjustment is added to the guess and the second operating point in space of control is determined. New forecasts and new forecast sensitivities are found from the updated controls. The process continues until the cost function's change falls within a specified tolerance.

5.1. Experiment I: Nonchaotic Data Assimilation Process (Lorenz Scaling)

True controls are taken to be $\lambda = 12.0$, $\sigma = 7.0$. The forecasts of x_1, x_4, x_7 based on guess control ($\lambda = 12.6$, $\sigma = 4.0$) are shown in Figure 7. The amplitudes reach steady state near t = 10. Graphs of the diagonal elements of $V^T V$ that dictate observations to be made at t = 1 are shown in Figure 8. The observations at this time are found by adding $\approx 10\%$ random error to the forecast based on true controls. The observations of x_4 (Obx_4) and x_7 (Obx_7) at t = 1 are given by

$$(Obx_4, \ Obx_7) = (0.371, \ 11.378). \tag{65}$$

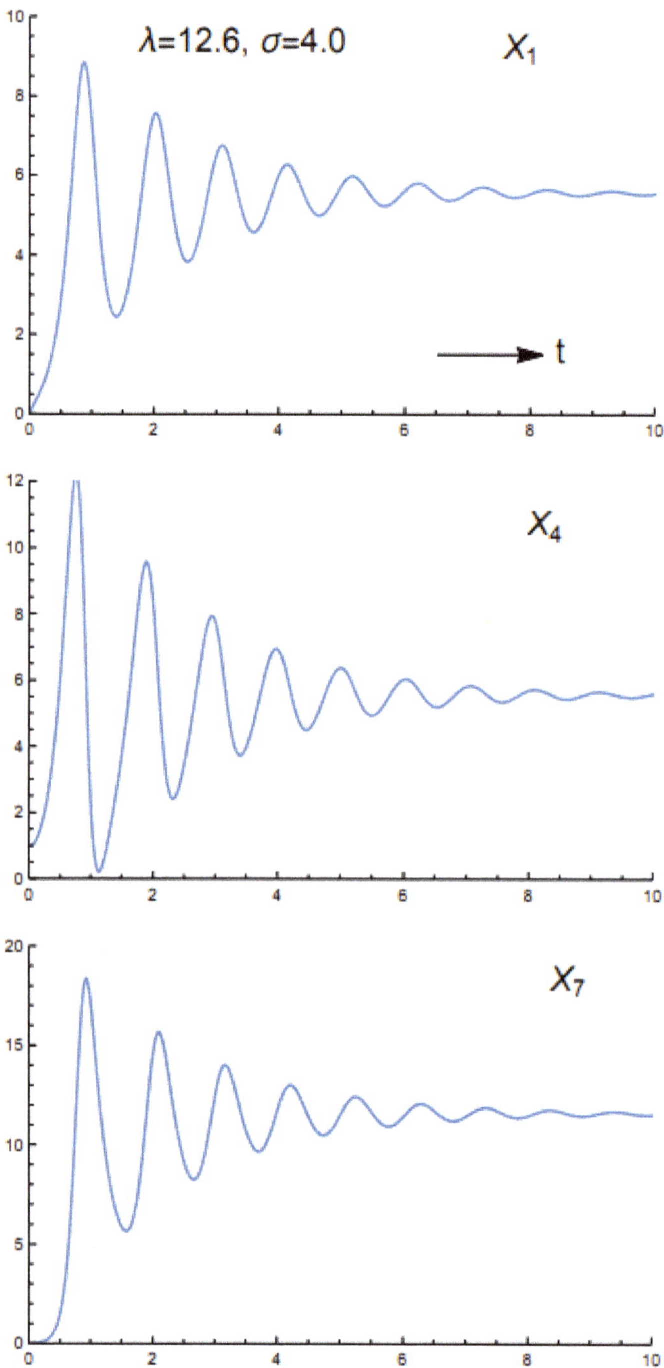

Figure 7. Forecast of the three amplitudes x_1, x_4, x_7 for the nonchaotic regime using L-LOM (3) over the time span [0, 10] with guess controls $\lambda = 12.6$, $\sigma = 4.0$.

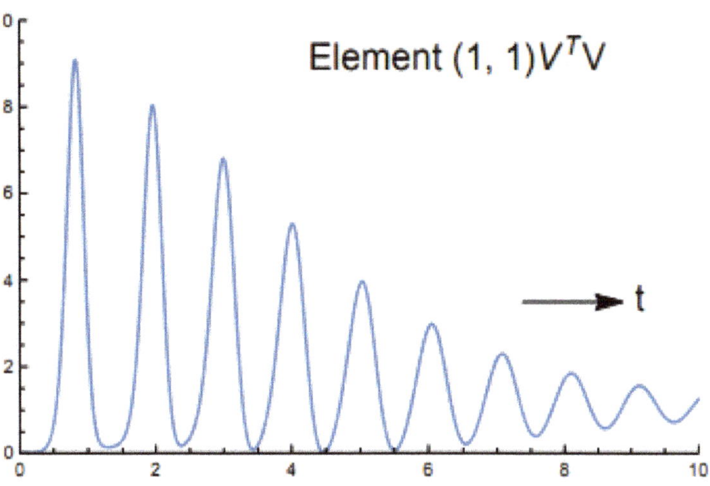

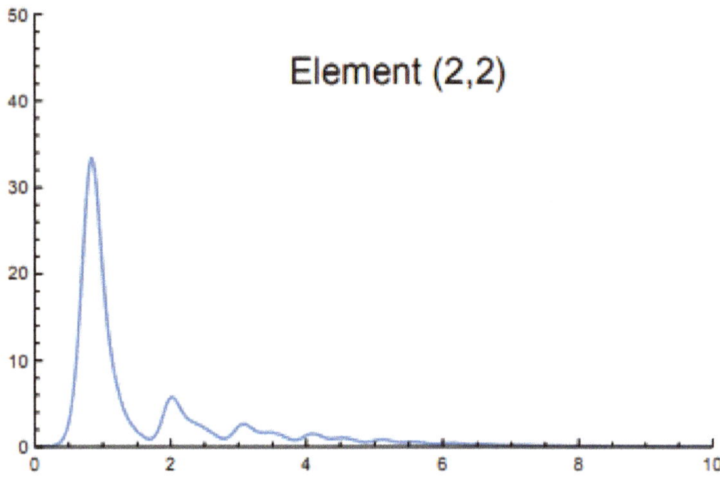

Figure 8. Plots of the diagonal elements of the product sensitivity matrix $V^T V$—elements (1,1) and (2,2)—for the nonchaotic regime using L-LOM (3) when guess controls are given by $\lambda = 12.6$, $\sigma = 4.0$.

Table 2 summarizes results from FSM. The assimilation process converges to a solution in 4 steps. The last column of Table 2 displays the sequential reduction in cost function values from ~19 to $3 \cdot 10^{-5}$, reduction by a factor of 10^{-6}. FSM determines values of $\Delta\lambda$ and $\Delta\sigma$ at each of the 4 iterations. The plot of these values is shown in Figure 9. The contours are always elliptical since the FSM cost function at each iteration is quadradic in $\Delta\lambda$ and $\Delta\sigma$. Display of cost function Equation (32) is shown in Figure 10 and the minimum is found to be located at $\lambda = 12$, $\sigma = 7$.

In a test that changed observation time from $t_{obs} = 1.00$ to $t = 1.05$, the data assimilation algorithm converged to $(\lambda, \sigma) = (11.995, 6.968)$ in 4 iterations.

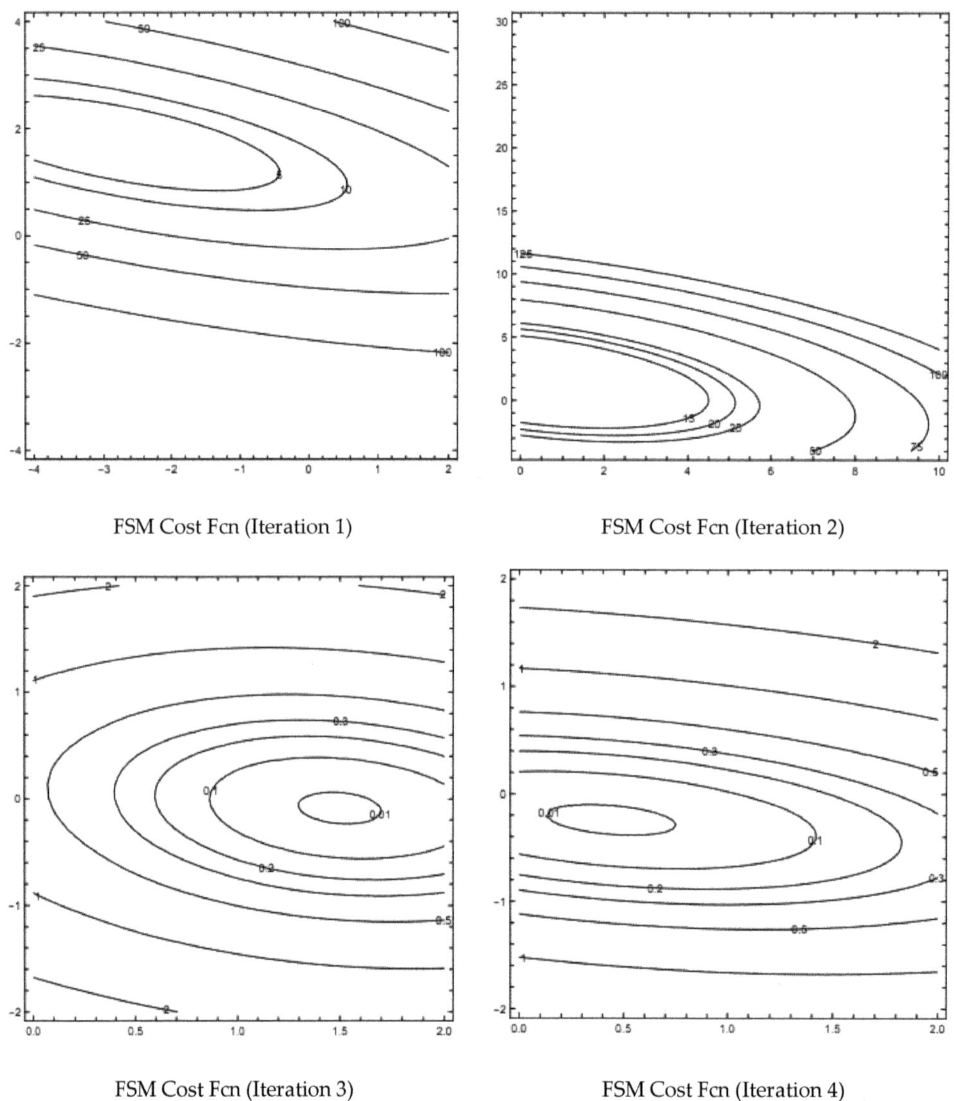

Figure 9. Contour plots of the FSM cost function in the space of control increments ($\delta\lambda$ along the horizontal axis and $\delta\sigma$ along the vertical axis). The values of $\delta\lambda$ and $\delta\sigma$ at the cost function minimum are the incremental adjustments to controls at the operating point.

Table 2. FSM Iterates for λ.

i	$\Delta\lambda^{(i)}$	$\Delta\sigma^{(i)}$	$\lambda^{(i)}$	$\sigma^{(i)}$	$Inov\ (x_4)$	$Inov\ (x_7)$	Cost Fcn
0	—	—	12.600	4.000	-1.987	-5.831	18.957
1	-2.787	1.735	9.813	5.735	-2.245	-1.456	3.580
2	0.252	1.579	10.065	7.309	-1.048	0.113	0.555
3	1.499	-0.087	11.559	7.213	-0.210	0.172	0.037
4	0.439	-0.243	11.997	6.969	-0.002	-0.005	$3\cdot10^{-5}$

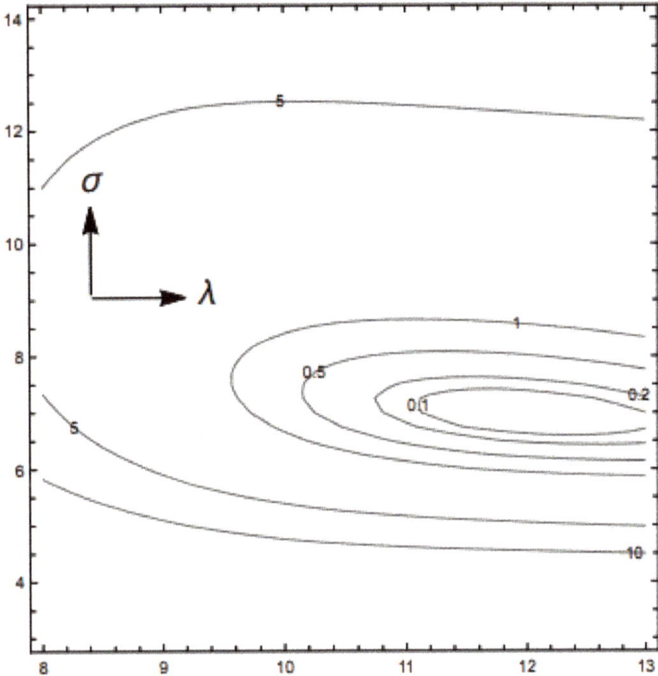

Figure 10. Plot of the cost function given by Equation (32). The minimum is located at $\lambda = 12.0$, $\sigma = 7.0$ (λ the horizontal axis and σ the vertical axis).

5.2. Experiment II: Nonchaotic Data Assimilation Process (Saltzman Scaling)

This data assimilation experiment follows the exact steps for the nonchaotic data assimilation process as shown in Section 5.1 except that the governing 3-mode equations are scaled according to Saltzman's form of the equations as found in Equations (3)–(5).

Figure 11 displays the time-dependent solutions to the 3-mode model. The solid curves are derived from the true controls ($\lambda = 12$, $\sigma = 7$) and the dashed curves from forecast controls (12.6, 4.0). The primary difference between this forecast and the one shown earlier for Lorenz's scaling is the much greater magnitudes of the variables and the shorter time required to reach equilibrium. Otherwise, the time-dependent oscillatory nature of the components is the same. The forecast sensitivity of x_4 and x_7 to λ and σ are shown in Figure 12—again the sensitivities are much greater than those associated with Lorenz's scaling. Based on the elements of the sensitivity product matrix $V^T V$, $t_{obs} = 0.16$. The observations at this time are found by adding $\approx 10\%$ random error to the forecast based on true controls. The observations of x_4 (Obx_4) and x_7 (Obx_7) at t = 0.16 are given by

$$(Obx_4,\ Obx_7) = (427.5 - 260.6) \tag{66}$$

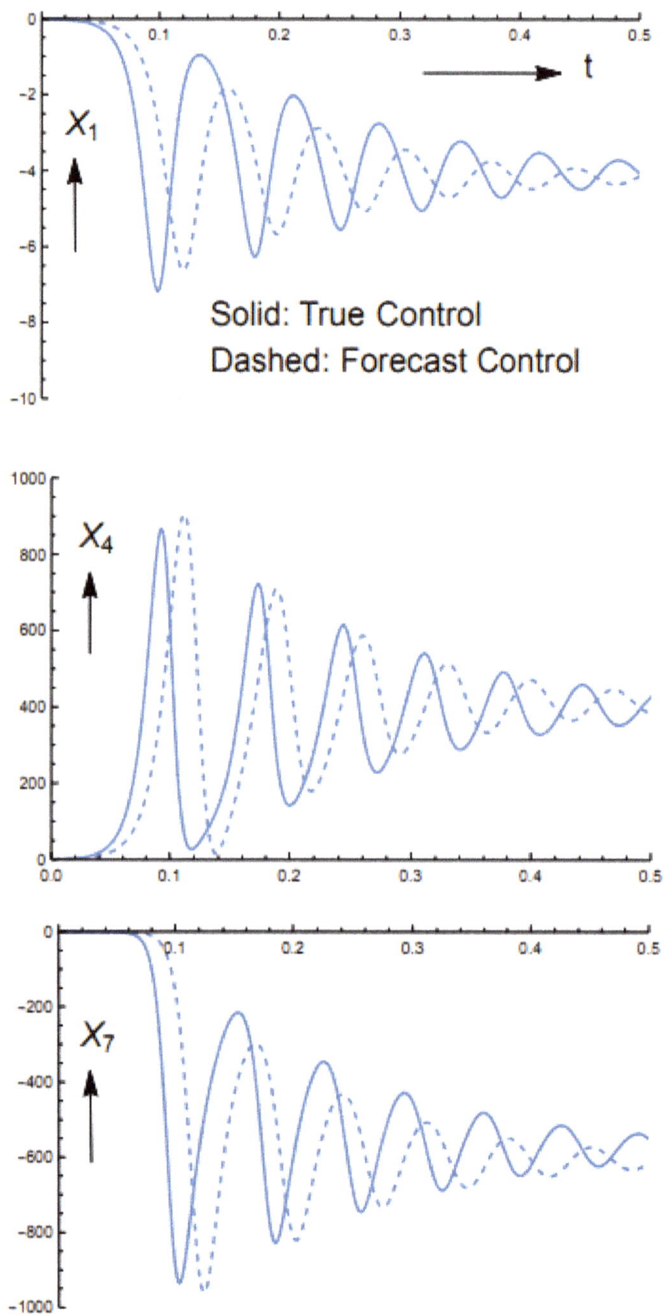

Figure 11. Non-chaotic regime solutions x_1, x_4, x_7 for S-LOM (3) using true control (solid curves: $\lambda = 12.0$, $\sigma = 7.0$) and perturbed/forecast control (dashed curves: $\lambda = 12.6$, $\sigma = 4.00$).

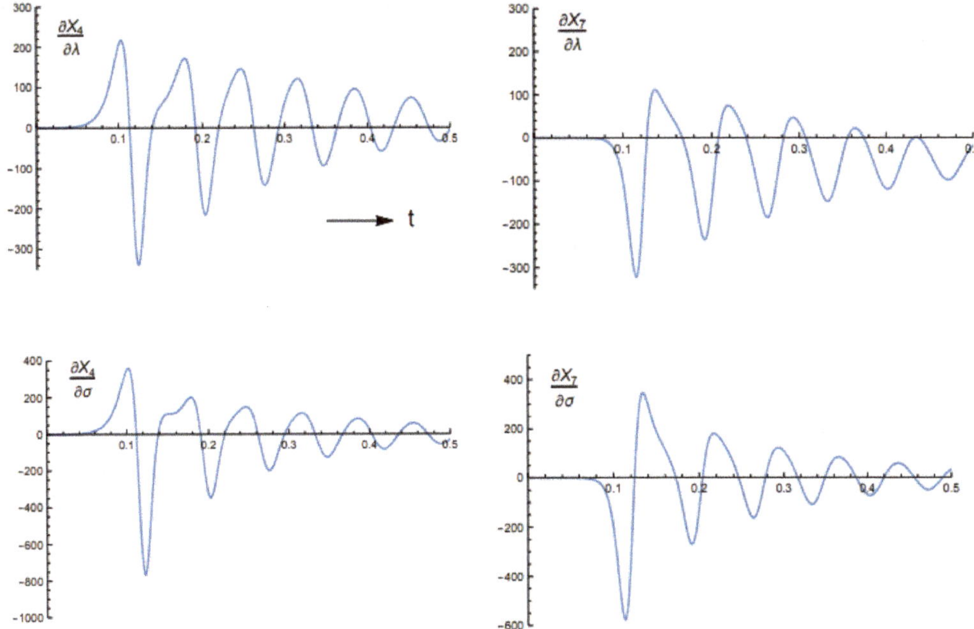

Figure 12. S-LOM (3) forecast sensitivity of x_4 and x_7 to controls for the non-chaotic regime.

Results from the data assimilation experiment are shown in Table 3. The data assimilation process converges to $(\lambda, \sigma) = (13.8, 4.8)$ in 4 iterations. The cost function is reduced 9 orders of magnitude. The cost function for this case is shown in Figure 13. The converged values of control shown in Table 3 are consistent with this cost function's minimum.

In a test that changed observation time from $t_{obs} = 0.16$ to $t = 0.10$, FSM did not converge to an optimal state. Instead, after a single iteration the result was $(\lambda, \sigma) = (-46.3, 26.8)$, unreasonable physically.

Table 3. FSM Iterates for $\lambda = 12$, Saltzman Scaling, Observations taken at $t = 0.16$.

i	$\Delta\lambda^{(i)}$	$\Delta\sigma^{(i)}$	$\lambda^{(i)}$	$\sigma^{(i)}$	*Inov* (x_4)	*Inov* (x_7)	**Cost Fcn**
0	—	—	12.600	4.000	191.857	63.905	$2\cdot10^4$
1	2.397	−0.102	15.000	3.900	−52.480	58.896	$3\cdot10^3$
2	−1.054	0.638	13.943	4.536	4.627	10.944	$7\cdot10^1$
3	−0.135	0.218	13.807	4.754	0.750	0.742	$6\cdot10^{-1}$
4	−0.006	0.018	13.801	4.771	0.006	0.004	$3\cdot10^{-5}$

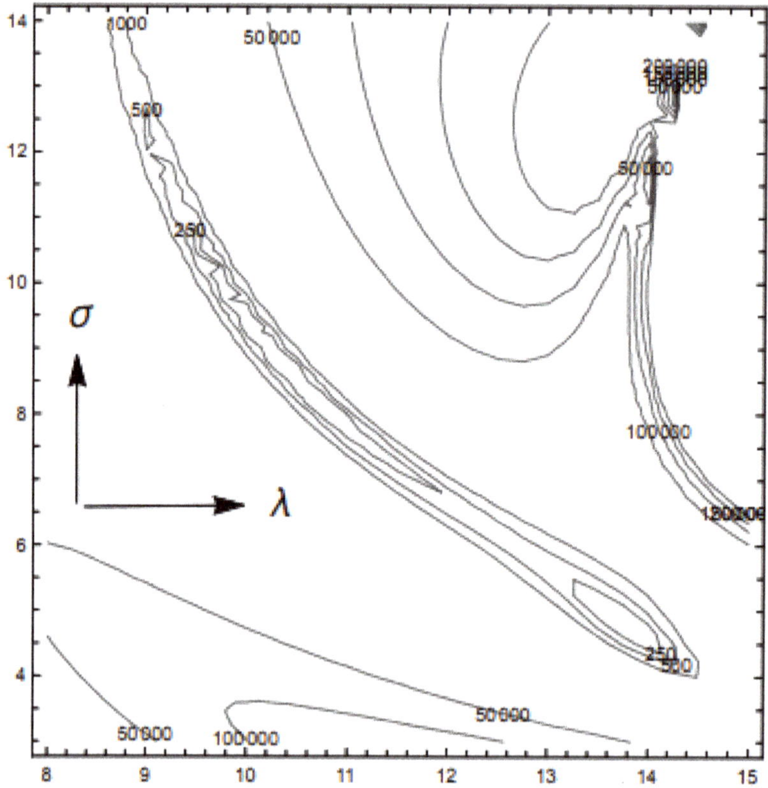

Figure 13. Plot of the cost function based on Equation (32) for the nonchaotic regime using S-LOM (3). The minimum is located at $\lambda = 13.8$, $\sigma = 4.8$ (λ the horizontal axis and σ the vertical axis).

5.3. Experiment III: Chaotic Data Assimilation Process (Lorenz Scaling)

True controls are taken to be $\lambda = 28.0$, $\sigma = 10.0$. The forecasts of x_1, x_4, x_7 based on guess control ($\lambda = 29.0$, $\sigma = 8.0$) are shown in Figure 14. Based on the elements of the product sensitivity matrix $V^T \cdot V$, observations are collected at $t = 0.4$. The observations at this time are found by adding $\approx 10\%$ random error to the forecast based on true controls. The observations of x_4 (Obx_4) and x_7 (Obx_7) at $t = 0.4$ are given by

$$(Obx_4,\ Obx_7) = (22.832,\ 40.194). \tag{67}$$

Table 4 summarizes results from FSM. Assimilation process converges to a solution in 4 steps. The last column of the table displays the sequential reduction in cost function values from $\sim$74 to $\sim 10^{-4}$, reduction by a factor of 10^6. The cost function is shown in Figure 15 and the minimum is found to be located at $\lambda = 28$, $\sigma = 10$, nearly identical to the solution found by FSM.

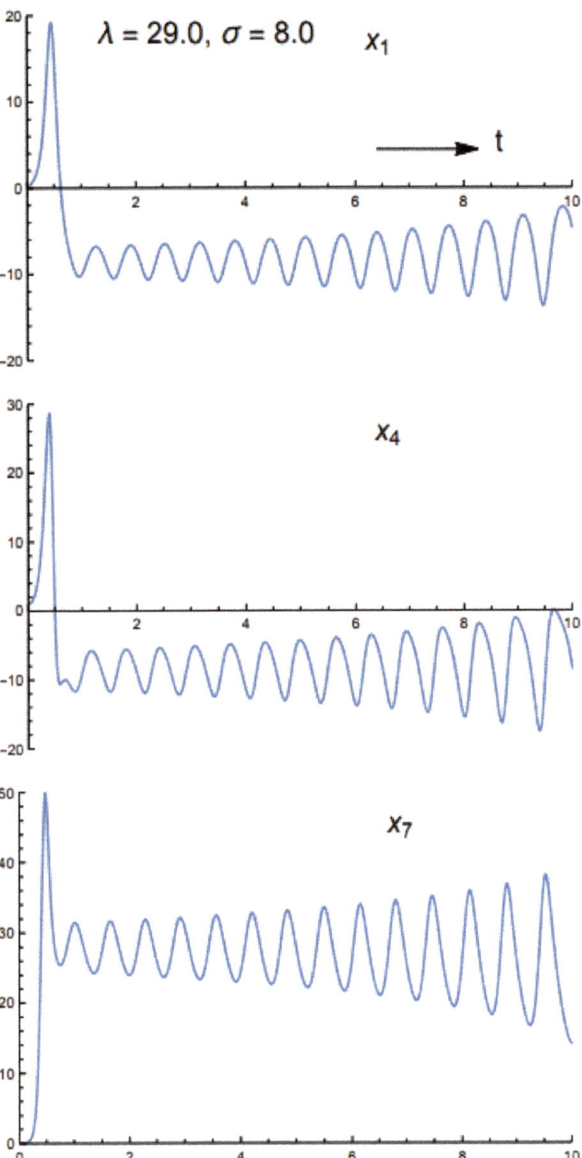

Figure 14. Solution to the amplitude equations for the chaotic regime using L-LOM (3) with guess control given by $\lambda = 29.0$, $\sigma = 8.0$.

Table 4. FSM Iterates for $\lambda = 28$.

i	$\Delta\lambda^{(i)}$	$\Delta\sigma^{(i)}$	$\lambda^{(i)}$	$\sigma^{(i)}$	$In\ (x_4)$	$In\ (x_7)$	**Cost Fcn**
0	—	—	29.000	8.000	−5.732	10.712	73.801
1	−4.419	3.130	24.581	11.130	0.773	7.839	31.024
2	3.003	−0.815	27.584	10.315	0.294	0.258	$8\cdot10^{-2}$
3	0.251	−0.136	27.835	10.178	−0.008	0.011	$9\cdot10^{-5}$
4	0.000	0.002	27.835	10.180	$3\cdot10^{-6}$	$3\cdot10^{-6}$	$9\cdot10^{-12}$

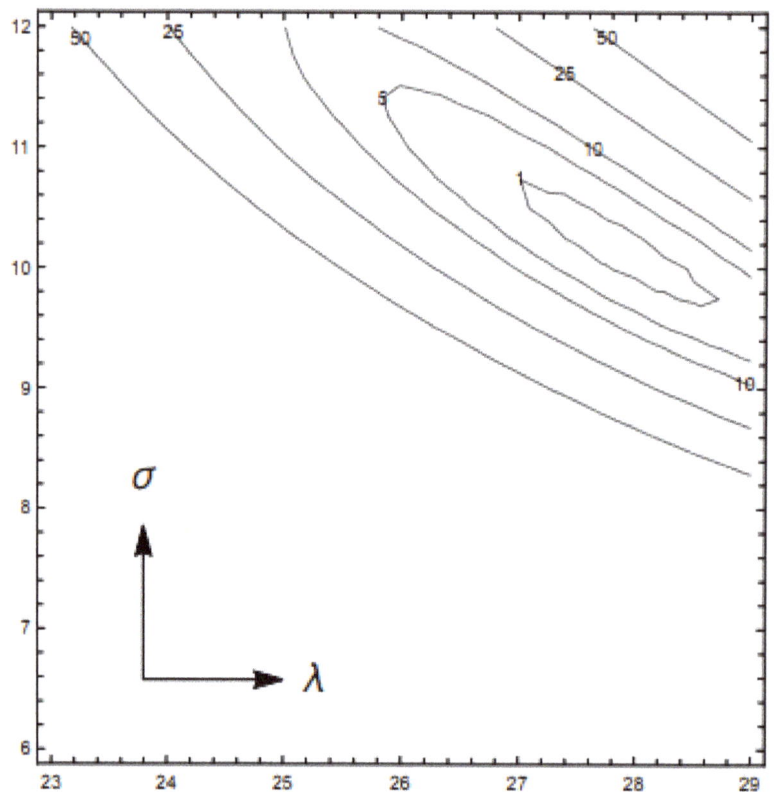

Figure 15. Plot of the cost function based on Equation (32) for the chaotic regime using L-LOM (3). The minimum is located at $\lambda = 27.6$, $\sigma = 10.3$ (λ the horizontal axis and σ the vertical axis).

It is interesting to consider cases where the optimal observation time was greater than $t = 0.4$, times where the model forecast sensitivities are extremely large as shown earlier in Figure 5. There is no reason to believe that observations at these times would not give a reasonable result. The reasonableness of the result, however, would be dependent on the smallness of the condition number.

5.4. Experiment IV: Data Assimilation in Time and Space (x, z)

We confine our attention to a single numerical experiment that combines space and time. Our experiment focuses on true controls $\lambda = 2$ and $\sigma = 7$, a nonchaotic regime. The perturbed controls are $\lambda = 2.5$ and $\sigma = 3.0$. The time of observation is determined from the structure of the sensitivity product matrix $V^T V$. Based on the trace of $V^T V$, observations are taken at $t_{obs} = 2$. The spatial observation sites are determined from the trace of $H^T H$ at $t = 2$. These spatial locations are

$$
\begin{aligned}
x_1 &= 0.10 \quad z_1 = 0.25 \\
x_2 &= 0.10 \quad z_2 = 0.75 \\
x_3 &= 2.55 \quad z_3 = 0.23 \\
x_4 &= 3.11 \quad z_4 = 0.68
\end{aligned}
\tag{68}
$$

In this case, the condition number for the G matrix is 3.8 which builds convexity into the cost function as shown in Figure 16.

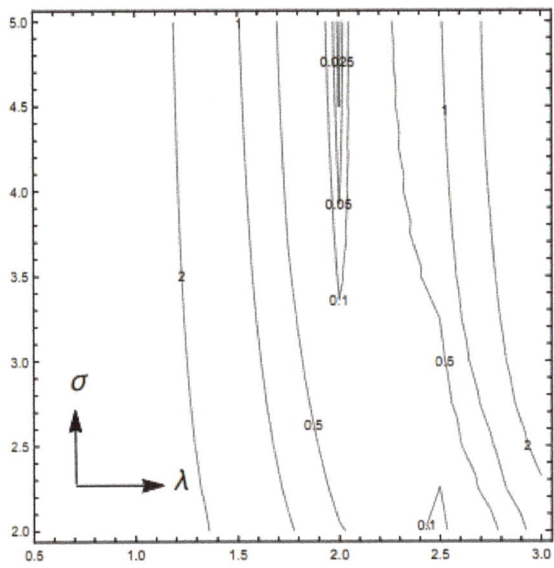

Figure 16. Plot of the cost function based on Equation (32) for the nonchaotic regime using L-LOM (3) with observations in space-time. The minimum is located at $\lambda = 2.0$, $\sigma = 4.4$ (λ the horizontal axis and σ the vertical axis).

Table 5 summarizes to convergence to optimal solution. The innovations at the observation locations exhibit a reduction of 10^7 in the 3 iterations and the optimal minimum is very close to the minimum shown in Figure 16.

Table 5. FSM Iterates for the space-time constraint.

i	$\Delta\lambda^{(i)}$	$\Delta\sigma^{(i)}$	$\lambda^{(i)}$	$\sigma^{(i)}$	*Innov* (x_1,z_1)	*Innov* (x_2,z_2)	*Innov* (x_3,z_3)	*Innov* (x_4,z_4)
0	—	—	2.500	3.000	0.076	-0.638	0.138	-0.599
1	-0.412	0.824	2.087	3.824	-0.032	0.056	-0.034	0.051
2	0.004	0.334	2.091	4.158	-0.002	0.006	-0.002	0.005
3	0.002	0.009	2.093	4.167	$5\cdot10^{-6}$	$7\cdot10^{-7}$	$4\cdot10^{-6}$	$9\cdot10^{-7}$

6. Discussion and Conclusions

S-LOM (3) and L-LOM (3), low-order forms of spectral equations that govern Rayleigh–Bénard convection developed by Saltzman [1] and used by Lorenz [4], respectively, have been used to study advantages of observation placement in dynamic data assimilation. Observation placement is controlled by forecast sensitivity to model control; in this case the control has two elements, Rayleigh and Prandtl numbers. We have used FSM data assimilation since it delivers forecast sensitivities to control that are used to determine observation placement.

By examination of the cost function in terms of forecast sensitivities, a Gramian matrix G becomes the centerpiece to determine observation placement. For time-only observations, the Gramian matrix reduces to the product of $V^T V$ where V is the time-dependent forecast sensitivity to control. This matrix is symmetric positive definite and the maximum values of the trace or norm of this matrix determine the best observation locations. Observation locations that yield the matrix's smallest condition number build convexity into the cost function's minimum. For our case study, the contours of the cost function in the space of the two controls are ellipses. As the condition number approaches the number 1, the major and minor axes of the ellipses approach the same number and the ellipses approach circles. In this case, the path to the minimum is along a straight line. For time-space observations, the Gramian matrix is symmetric semi positive definite in the form $V^T H^T H V$ where H is the spatial-dependent matrix that accounts for the double Fourier series representation of the observations. This product exhibits separability in space and time and therefore permits determination of observation-time placement as a first step to be followed by spatial placement of observations.

We tested the observation placement strategy for both nonchaotic and chaotic regimes. Results from the numerical experiments with FSM were especially good based on (1) iterative reduction of the cost functions' values, and (2) comparison of the optimal parameters with the cost function minimum in the space of controls.

We also wanted to test the role of equation scaling in the data assimilation process. The two models have different scaling where S-LOM (3) follows the standard fluid dynamical scaling based on the fluid's physical parameters while L-LOM (3) incorporates various forms of the non-dimensional wavenumber along with the physical parameters in its scaling. The scaling for L-LOM (3) leads to coefficients of comparable magnitude in the governing amplitude differential equations while scaling for S-LOM (3) leads to order of magnitude differences in the coefficients. In turn, the forecast sensitivity equations for S-LOM (3) exhibit large differences in their coefficients. Nevertheless, the FSM data assimilation method delivered good results for both models. On the other hand, when the ideal observation time is changed slightly, S-LOM (3) results were poor while L-LOM (3) were excellent. It thus appears that scaling plays an important role in data assimilation, at least for our limited number of experiments.

Author Contributions: Formal analysis, J.M.L. and S.L.; Writing—original draft, J.M.L. and S.L.; Writing—review & editing, J.M.L. and S.L. All authors have read and agreed to the published version of the manuscript.

Funding: This research received no external funding.

Institutional Review Board Statement: Not applicable.

Informed Consent Statement: Not applicable.

Acknowledgments: We thank Bo-Wen Shen for encouraging us to explore application of FSM to chaotic systems. The authors also thank: (1) Kevin Fisher-Mack and Junjun Hu for writing the computer program that delivered the cost function display in the space of controls, and (2) the three anonymous reviewers who made cogent suggestions that were followed and this led to clarification of discussions in the paper.

Conflicts of Interest: The authors declare no conflict of interest.

References

1. Saltzman, B. Finite Amplitude Free Convection as an Initial Value Problem-I. *J. Atmos. Sci.* **1962**, *19*, 329–341. [CrossRef]
2. Bénard, M. Les Tourbillions cellulaires dans une nape liquide. *Rev. Générale Des Sci. Pures Et Appl.* **1990**, *11*, 1261–1271, 1309–1328.
3. Rayleigh, L. Convection currents in a horizontal layer of fluid, when higher temperature is on the underside. *Philos. Mag.* **1916**, *32*, 529–546. [CrossRef]
4. Lorenz, E.N. Deterministic Nonperiodic Flow. *J. Atmos. Sci.* **1963**, *20*, 220–241. [CrossRef]

5.   Saravanan, R. *The Climate Demon: Past, Present, and Future of Climate Change*; Cambridge University Press: Cambridge, UK, 2022; 379p.
6.   Lewis, J.M.; Lakshmivarahan, S.; Dhall, S.K. *Dynamic Data Assimilation: A Least Squares Approach*; Cambridge University Press: Cambridge, UK, 2006; 654p.
7.   Lakshmivarahan, S.; Lewis, J.M.; Hu, J. On Controlling the Shape of the Cost Functional in Dynamic Data Assimilation: Guidelines for Placement of Observations and Application to Saltzman's Model of Convection. *J. Atmos. Sci.* **2020**, *77*, 2969–2989.
8.   Lakshmivarahan, S.; Lewis, J.M.; Maryada, S.K.R. Observability Gramian and its Role in the Placement of Observations in Dynamic Data Assimilation. In *Data Assimilation for Atmospheric, Oceanic and Hydrographic Applications*; Park, S.K., Xu, L., Eds.; Springer: Cham, Switzerland, 2022; Volume IV, pp. 215–257.
9.   Lewis, J.M.; Lakshmivarahan, S.; Maryada, S.K.R. Placement of observations for variational data assimilation: Application to Burgers' Equation and seiche phenomenon. In *Data Assimilation for Atmospheric, Oceanic and Hydrographic Applications*; Park, S.K., Xu, L., Eds.; Springer: Cham, Switzerland, 2022; Volume IV, pp. 215–257.
10.   Lakshmivarahan, S.; Lewis, J.M.; Hu, J. Saltzman's Model: Complete Characterization of Solution Properties. *J. Atmos. Sci.* **2019**, *76*, 1587–1608. [CrossRef]
11.   Chandrasekhar, S. *Hydrodynamic and Hydromagnetic Stability*; Oxford at the Clarendon Press: Oxford, UK, 1961; 654p.
12.   Turner, J.S. *Buoyancy Effects in Fluids*; Cambridge University Press: Cambridge, UK, 1973; 368p.
13.   Lewis, J.M.; Lakshmivarahan, S. Derivation of Saltzman's spectral equations for Rayleigh-Bénard convection in terms of two variable parameters, Rayleigh and Prandtl numbers. *Tech. Rep.* 2019; *Unpublished.*
14.   Kuo, H.-L. Solution of the nonlinear equations of cellular convection and heat transport. *J. Fluid Mech.* **1961**, *10*, 611–634. [CrossRef]
15.   Lakshmivarahan, S.; Lewis, J.M. Forward Sensitivity Based Approach to Dynamic Data Assimilation. *Adv. Meteorol.* **2010**, *2010*, 375615. [CrossRef]
16.   Lakshmivarahan, S.; Lewis, J.M.; Jabrzemski, R. *Forecast Error Correction Using Dynamic Data Assimilation*; Springer: Berlin/Heidelberg, Germany, 2017; 270p.

Article

One Saddle Point and Two Types of Sensitivities within the Lorenz 1963 and 1969 Models

Bo-Wen Shen [1,*], Roger A. Pielke, Sr. [2] and Xubin Zeng [3]

[1] Department of Mathematics and Statistics, San Diego State University, San Diego, CA 92182, USA
[2] Cooperative Institute for Research in Environmental Sciences, University of Colorado Boulder, Boulder, CO 80203, USA; pielkesr@colorado.edu
[3] Department of Hydrology and Atmospheric Science, The University of Arizona, Tucson, AZ 85721, USA; xubin@arizona.edu
* Correspondence: bshen@sdsu.edu

Abstract: The fact that both the Lorenz 1963 and 1969 models suggest finite predictability is well known. However, less well known is the fact that the mechanisms (i.e., sensitivities) within both models, which lead to finite predictability, are different. Additionally, the mathematical and physical relationship between these two models has not been fully documented. New analyses, along with a literature review, are performed here to provide insights regarding similarities and differences for these two models. The models represent different physical systems, one for convection and the other for barotropic vorticity. From the perspective of mathematical complexities, the Lorenz 1963 (L63) model is limited-scale and nonlinear; and the Lorenz 1969 (L69) model is closure-based, physically multiscale, mathematically linear, and numerically ill-conditioned. The former possesses a sensitive dependence of solutions on initial conditions, known as the butterfly effect, and the latter contains numerical sensitivities due to an ill-conditioned matrix with a large condition number (i.e., a large variance of growth rates). Here, we illustrate that the existence of a saddle point at the origin is a common feature that produces instability in both systems. Within the chaotic regime of the L63 nonlinear model, unstable growth is constrained by nonlinearity, as well as dissipation, yielding time varying growth rates along an orbit, and, thus, a dependence of (finite) predictability on initial conditions. Within the L69 linear model, multiple unstable modes at various growth rates appear, and the growth of a specific unstable mode (i.e., the most unstable mode during a finite time interval) is constrained by imposing a saturation assumption, thereby yielding a time varying system growth rate. Both models were interchangeably applied for qualitatively revealing the nature of finite predictability in weather and climate. However, only single type solutions were examined (i.e., chaotic and linearly unstable solutions for the L63 and L69 models, respectively), and the L69 system is ill-conditioned and easily captures numerical instability. Thus, an estimate of the predictability limit using either of the above models, with or without additional assumptions (e.g., saturation), should be interpreted with caution and should not be generalized as an upper limit for atmospheric predictability.

Keywords: Lorenz model; chaos; instability; saddle point; SDIC; sensitivities; finite predictability; ill-conditioned

Citation: Shen, B.-W.; Pielke, R.A., Sr.; Zeng, X. One Saddle Point and Two Types of Sensitivities within the Lorenz 1963 and 1969 Models. *Atmosphere* **2022**, *13*, 753. https://doi.org/10.3390/atmos13050753

Academic Editors: Anthony R. Lupo and Jimy Dudhia

Received: 5 March 2022
Accepted: 23 April 2022
Published: 7 May 2022

1. Introduction

Two pioneering studies by Dr. Lorenz [1,2] changed our view on the predictability of weather, and turned our attention from regularity and unlimited predictability associated with Laplace's view of determinism to the irregularity and finite predictability associated with Lorenz's view of deterministic chaos. Chaos is defined as the sensitive dependence of solutions on initial conditions (SDIC), known as the butterfly effect [3]. The feature of SDIC reveals the difficulty in obtaining accurate, long-term predictions, suggesting a finite predictability (e.g., [4]).

Over the past several years, pioneering, yet incomprehensive, theoretical results derived from a limited collection of the above, and other studies, have led to an increased understanding (or misconceptions, to be illustrated) that the conventional view of "weather is chaotic" and that the so-called theoretical limit of predictability of two weeks are well supported by Lorenz's 1963 (L63) and 1969 (L69) studies [1,5]. As a result, when real-world global models produced encouraging simulations at extended-range (15–30 day) time scales [6–9], people who believe in the predictability limit of two weeks have interpreted these new results as inconsistent with chaos theory (e.g., [10]). While some researchers have applied nonlinear L63-type models for understanding the chaotic nature of weather and climate, other researchers have applied the major findings of linear L69-type models for estimating a predictability horizon for the atmosphere. Over the past several decades, researchers in the field of nonlinear dynamics have continuously improved our understanding of nonlinear responses, as well as local and global stability, within Lorenz-type models. Some recent theoretical studies may potentially provide justifications for promising extended-range simulations. However, due to ineffectiveness and difficulties in exchanging ideas and sharing results in different disciplines, related findings are, unfortunately, not fully known within the Earth science community. In this study, dynamical systems methods that are now fairly standard in recent nonlinear studies are applied in order to reveal the unreported features of the classical Lorenz model(s) (in particular, the L69 model).

Based on a comprehensive literature review, we believe current barriers to the advancement of weather and short-term climate predictions originate from gaps between the "improved understanding" of predictability with multistability, derived from advanced theoretical models, and the current approach, based on the conventional, yet incomplete, understanding of predictability with only SDIC and monostability. In contrast to the monostability that allows single type solutions (i.e., chaotic solutions in [1] and unstable solutions in [5]), the concept of multistability that contains coexisting chaotic and non-chaotic solutions has been emphasized (e.g., [11–14]). Recently, based on the concept of time varying multistability, refs. [15,16] provided a revised view that "weather possesses chaos and order; it includes emerging organized systems (such as tornadoes) and recurrent seasons". Such a revised view that suggests distinct predictability for chaotic and non-chaotic systems may lay a foundation for a potential predictability beyond Lorenz's predictability limit of two weeks. The revised view was proposed based on the classical and generalized L63 models that are well studied in nonlinear dynamics (e.g., [11–14,17–31]).

By comparison, it is the L69 model that has mainly been applied for addressing predictability in meteorology (e.g., [32,33]). Since statistical methods were applied to derive the L69 model, which is not a turbulence model but laid a foundation for the development of turbulence models (e.g., [34,35]), the L69 model has been analyzed using methods different from those of dynamical systems. On the other hand, the advantages of applying dynamical systems methods for investigating turbulence models have been documented (e.g., [36,37]). Although eigenvalue problems associated with L69-type models have previously been solved, only features for the largest eigenvalues have been documented. From the perspective of dynamical systems, an analysis of both positive and negative eigenvalues, as well as the corresponding exponential and oscillatory solutions, is helpful for understanding predictability within a multiscale system. Thus, by applying a unified, although simple, approach in a reanalysis of the models, fundamental questions can be revisited, including: (1) what type of mathematical, as well as physical, relationship exists between the two models and (2) what are their similarities and difference in spatial and temporal scale interactions. We hope that such an analysis will help researchers who are familiar with one type of model to quickly capture the major features and findings of the other type of model.

In this study, we perform new analyses, together with a literature review, to provide insights on the L63 and L69 models in terms of two types of sensitivities and the common feature of a saddle point. We then discuss how specific features of the two models were previously applied to determine finite predictability. Section 2 reviews general features of the L63 and L69 models. Similarities and differences of the two models are presented in Section 3. Concluding remarks are provided at the end. Appendix A discusses the full L63 model and its simplified versions, including the non-dissipative L63 model [29]. Part I of Supplementary Materials includes the following two sections: Section (A): a simple illustration of ill-conditioning, and Section (B): an Illustration of a stiff ODE. Part II of Supplementary Materials summarizes additional features of the L63 model regarding SDIC and finite predictability.

2. The Lorenz 1963 and 1969 Models

In this section, through new analyses and a literature review, we discuss general features of the L63 and L69 models in order to propose a simple, 2nd-order ODE that retains the common properties of the two models. Based on the proposed 2nd-order ODE, specific features of the L63 and L69 models, including two types of sensitivities and the impact of a saddle point, are discussed in Section 3.

2.1. The L63 Limited-Scale, Nonlinear Model

The L63 model has been discussed in numerous studies [17,19–21], including in our studies [11,12,25–31] and references therein. A brief summary of the model is provided here, along with mathematical equations for the full and simplified versions found in Appendix A. Based on a system of partial differential equations that describe the time evolution of vorticity and temperature for the Rayleigh–Benard convection [38]), a system of three, 1st-order ODEs (e.g., Equations (A1)–(A3)) were obtained for rediscovering the SDIC [1]. Three major physical processes are heating, dissipation, and nonlinear processes. Such a model is called the Lorenz 1963 (L63) model.

As discussed in Appendix A, the system of three, 1st-order ODEs can be transformed into a system containing one 2nd-order and one 1st-order ODEs (e.g., Equations (A6) and (A7); also see [31]). Such a system with (or without) dissipations can easily be compared to the Pedlosky model [39–43] (or the Duffing Equation) to reveal mathematical universality amongst the systems (e.g., [30,31]). By ignoring dissipative terms, the 1st- and 2nd-order ODEs become uncoupled. The uncoupled, 2nd-order ODE is referred to as the non-dissipative L63 model (e.g., Equation (A11)), which is written as:

$$\frac{\mathrm{d}^2 X}{\mathrm{d}\tau^2} - \lambda^2 X + \frac{X^3}{2} = 0. \tag{1}$$

Here, τ is dimensionless time. The constant λ^2 is proportional to the product of σ and r that represent the Prandtl number and the normalized Rayleigh number, respectively. In general, λ^2 is also a function of initial conditions and, thus, $\lambda^2 = \sigma r + constant$.

Without a loss of generality, λ^2 is assumed to be positive (i.e, $\lambda^2 = \sigma r > 0$) within the non-dissipative L63 model. Thus, the linear version of the system produces both stable and unstable modes (i.e., $e^{-\gamma t}$, $e^{\gamma t}$, and $\gamma = \sqrt{\lambda^2}$). The role of nonlinearity is discussed in Section 3.1.

To facilitate discussions below, the energy function of the above system is obtained by multiplying both sides of Equation (1) by $dX/d\tau$ and by performing an integration with respect to τ (e.g., [23,31]):

$$TE = \left[\frac{1}{2}\left(\frac{dX}{d\tau}\right)^2 \right] + \left[-\frac{1}{2}\lambda^2 X^2 + \frac{1}{8}X^4 \right] = C. \tag{2a}$$

Here, TE indicates the total energy. The first and second pairs of brackets represent the kinetic energy and potential energy, respectively (see [31] for details). C is a constant that is determined by an initial condition. Within a linear system (i.e., no cubic term in Equation (1)), we have:

$$TE = \left[\frac{1}{2}\left(\frac{dX}{d\tau}\right)^2 \right] + \left[-\frac{1}{2}\lambda^2 X^2 \right] = C \quad \text{(for a linear system)}. \tag{2b}$$

Later, to illustrate solution characteristics, we compute the contour lines of total energy for both linear and nonlinear systems. We specifically examine the special case of $TE = 0$ (i.e., $C = 0$).

2.2. The L69 Multiscale, Linear Model

By applying a modified, quasi-normal (QN) closure which assumes the 4th cumulants are zero (e.g., [44]; i.e., relating the fourth-order velocity correlations to second-order velocity correlations), Lorenz proposed the L69 model based on the following two-dimensional, conservative vorticity equation:

$$\frac{d\nabla^2 \Psi}{d\tau} = 0, \tag{3}$$

where Ψ and $\nabla^2 \Psi$ represent the 2D (x, y) stream function and the vorticity, respectively.

Such an approach was extended to yield the Eddy-Damped QN(EDQN) approximation, that replaces the 4th cumulants by a linear damping term, and then the EDQN Markovian (EDQNM) approximation using a minor modification called the Markovianization [34,44,45]. While Leith, 1971 [34] applied a model with the EDQNM approximation in order to study the predictability of 2D turbulence, ref. [35] proposed the test-field mode, overcoming the issue using an artificial cutoff in nonlinear interactions in the Leith 1971 model, for determining the predictability of turbulent flows. In regards to the above models [5,34,35], a common assumption was that: *"Estimates of the predictability of the planetary-scale motions of the atmosphere have been based on turbulence models in which the atmosphere is treated as an isotropic homogeneous two-dimensional turbulent fluid."* [35].

Since many Fourier modes were used for derivations, the L69 model contains multiple, physical modes. While used to illustrate the statistics of predictability within a multiscale framework, the model is not a turbulence model as compared to the models of Leith, 1971 and Leith and Kraichan, 1972 [34,35].

For revealing major features of solutions, linearization was applied to yield the following model in matrix form (e.g., Equation (43) of Lorenz, 1969 [5]):

$$\frac{d^2 \vec{W}}{d\tau^2} = \mathbf{A}\vec{W}. \tag{4}$$

Here, $\mathbf{A}$ is a $N \times N$ time-independent matrix, and N is the total number of wave modes. Each element within the matrix $\mathbf{A}$ represents the scale interactions of two Fourier modes (Equations (2a) and (2b) of Durran and Gingrich, 2014 [33]). $\vec{W}$ represents a column vector consisting of N state variables W_k, $k = 1, 2, \cdots, N$. k is the wavenumber, and W_k represents the ensemble mean of the kinetic energy of the perturbations for the wave mode k. Equation (4) indicates that the L69 model is physically multiscale with $k = 1, 2, \cdots, N$, and also mathematically linear. For a comparison to Equation (2b), the energy function for Equation (4), which contains a symmetric matrix, is written as follows:

$$TE = \left[\frac{1}{2}\left(\frac{d\vec{W}}{d\tau}\right)^2\right] + \left[-\frac{1}{2}\vec{W}^T \mathbf{A}\vec{W}\right] = C.$$

Here, $\vec{W}^T$ is the transpose of the column vector $\vec{W}$. The above equation becomes (2b) when $N = 1$. In general, since A is not a symmetric matrix, the so-called similarity transformation is applied for diagonalization to obtain the above equation. Thus, the vector $\vec{W}$ is replaced by the corresponding transformed column vector.

Since the L69 model is linear, it is important to understand how the model can help reveal the features of the original system (i.e., a partial differential equation for the conservation of vorticity). From a dynamical system perspective, the foundation of linearization is rooted in the linearization theorem, also known as the Hartman–Grobman Theorem (e.g., [22,46]): Suppose the N-dimensional system has an equilibrium point at X_c that is hyperbolic (i.e., with non-zero growth or decay rates). If so, nonlinear flow is then conjugate (i.e., dynamically equivalent) to the flow of the linearized system in the neighborhood of X_c. Within the L69 model, the origin is a saddle point and, thus, a hyperbolic point. As a result, the L69 model may describe the solution of the corresponding nonlinear system near the origin. Due to the limit of a linear system, a saturation assumption was imposed to prevent the unbounded growth of perturbations, as follows: when an unstable mode grows to reach its maximum, as determined by the selected spectrum, it is removed from the system (e.g., [32,47]. Then, a predictability horizon for a specific unstable mode represents the time when the mode becomes saturated. Thus, the impact of the saturation assumption on the estimate of predictability should be examined, and is discussed in Section 3.

From a dynamical perspective, Equation (4) can be transformed into a system of $2N$, 1st order ODEs by introducing additional N state variables that represent 1st-order time derivatives of the original N state variables. The $2N$ state variables can be used as coordinates for constructing the so-called phase space for analysis, as discussed in Section 3. Critical points are defined when the right hand side of the system of $2N$ ODEs becomes zero.

3. Discussions

In the discussion, we first analyze the saddle point and instability within a simple, linear, 2nd-order ODE, then turn the system into the non-dissipative L63 model by including a cubic term, and then present periodicity and limit chaos within the non-dissipative L63 model. We then illustrate specific features of the L69 model, including a saddle point, numerical instability, and numerical sensitivities; and provide comments on a L69-based conceptual model for a chain process. Based on the SDIC of the L63 model and the multiscale instability of the L69 model, we also discuss how to estimate predictability.

3.1. Features of the L63 Model

3.1.1. Physical vs. Numerical Instability within a Linear 2nd-Order ODE

Based on Equations (1) and (4), which represent the L63 and L69 models, respectively, a simple linear, 2nd-order ODE is proposed that is, in reality, Equation (1) without the cubic term, as follows:

$$\frac{\mathrm{d}^2 X}{\mathrm{d}\tau^2} = \lambda^2 X. \tag{5}$$

λ^2 is positive within the L63 model and is either positive or negative within the L69 model. A system that contains a positive value for λ^2 possesses both stable an unstable modes (i.e., $e^{-\gamma\tau}$ and $e^{\gamma\tau}$ for a positive $\gamma = \sqrt{\lambda^2}$). A general solution can be expressed as the linear combination of these two modes (i.e., $c_1 e^{-\gamma\tau} + c_2 e^{\gamma\tau}$, c_1 and c_2 are determined by initial conditions). As a result, given the same model parameter, solutions may contain a stable mode, an unstable mode, or both, depending on the initial condition.

While an unstable mode may be physically interesting and important (e.g., [48]), in the real world, a stable mode likely represents a more realistic solution in response to initial small-scale perturbations. For example, given a tiny perturbation (e.g., any human's flap) as an initial condition, we may select a stable solution if the corresponding response can be described by Equation (5). However, as illustrated by an example that produces an analytical, stable solution [49], an unstable mode may be incorrectly captured by a numerical method. Such a feature, referred to as numerical instability, is reviewed below using Equation (5).

Given $\lambda^2 = 10\pi^2$ (~98.7) and an initial condition of $X(0) = 1$ and $X'(0) = -\sqrt{10}\pi$, an analytical, stable solution of $X = e^{-\sqrt{10}\pi\tau}$ can be obtained. Such an initial condition yields $TE = 0$ in Equation (2b) for the linear system. Values for the solution that is a monotonically decreasing function of time are listed in the 3rd column of Table 8.6.4 of Boyce and DePrima, 2012 [49]. By comparison, when the Runge–Kutta scheme is applied, numerical solutions within the 2nd column of their Table increase with time, indicating an instability. A detailed calculation (not shown) indicates that the growth rate of the numerical, unstable solution is $\sqrt{10}\pi$. Therefore, the occurrence of such an instability appears because: (1) given the same parameters, the model (Equation (5)) allows both stable and unstable modes, and numerical errors produce a tiny, but non-zero, coefficient for the unstable mode; and (2) the error amplifies at the growth rate of the unstable mode and quickly dominates due to a large growth rate (i.e., $\sqrt{10}\pi$). Below, we analyze the solution within a 2D phase space for comparison to a nonlinear solution, and then illustrate that such numerical instability can easily be captured within the L69 model.

3.1.2. A Perspective of Dynamical Systems: Phase Space and a Saddle Point

From the perspective of dynamical systems, the above feature can be qualitatively illustrated using the following system of 1st-order ODEs derived from Equation (5):

$$\frac{dX}{d\tau} = Y, \tag{6a}$$

$$\frac{dY}{d\tau} = \lambda^2 X. \tag{6b}$$

A new variable $Y = X'$ is introduced. Using X and Y (i.e., X and X') as coordinates, a two-dimensional phase space can be constructed. As previously discussed, a solution with a positive value of λ^2 is written as:

$$X = c_1 e^{-\gamma\tau} + c_2 e^{\gamma\tau}, \tag{7a}$$

yielding:

$$\begin{pmatrix} X \\ Y \end{pmatrix} = c_1 e^{-\gamma\tau} \begin{pmatrix} 1 \\ -\gamma \end{pmatrix} + c_2 e^{\gamma\tau} \begin{pmatrix} 1 \\ \gamma \end{pmatrix}. \tag{7b}$$

Here, $Y = X'$ and $\gamma = \sqrt{\lambda^2} = \sqrt{10}\pi$. The first and second vectors on the right hand side of the equation are referred to as the stable and unstable eigenvectors, respectively, within a phase space. Thus, when an orbit begins near the origin $(X, Y) = (0, 0)$, it may move away from the origin in the direction of the unstable eigenvector or approach the origin in the direction of the stable eigenvector. The origin $(X, Y) = (0, 0)$ is called a saddle point. Stable and unstable eigenvectors within a linear system can be generalized to become stable and unstable manifolds of a saddle point within a nonlinear system, respectively: the stable (unstable) eigenvector is tangent to the stable (unstable) manifold at the saddle point (e.g., [50]).

Based on the total energy in Equation (2b), which describes the relationship between X and Y (i.e., X'), a solution trajectory can be analyzed using the contour lines of total energy. As shown in Figure 1, two straight lines with $TE = C = 0$ (i.e., $X' = \sqrt{10}\pi X$ and $X' = -\sqrt{10}\pi X$) indicate the unstable and unstable eigenvectors, respectively. In the previous example, the stable solution, starting at the initial condition of $(X, Y) = (1, -\sqrt{10}\pi)$, moves towards the origin along the zero contour line (Figure 1a,b). Ideally, the orbit should eventually arrive at the origin and stay there forever. However, since the origin is a saddle point, it is numerically challenging to simulate that type of evolution. When the orbit moves very close to the origin where the value of both X and X' are small, any tiny noise (caused by round-off errors) may lead the orbit to move away from the saddle along the direction of the unstable eigenvector (i.e., the line of $X' = \sqrt{10}\pi X$). A similar challenge within a nonlinear system, as documented in Shen, 2020 [30], is reviewed in the next subsection.

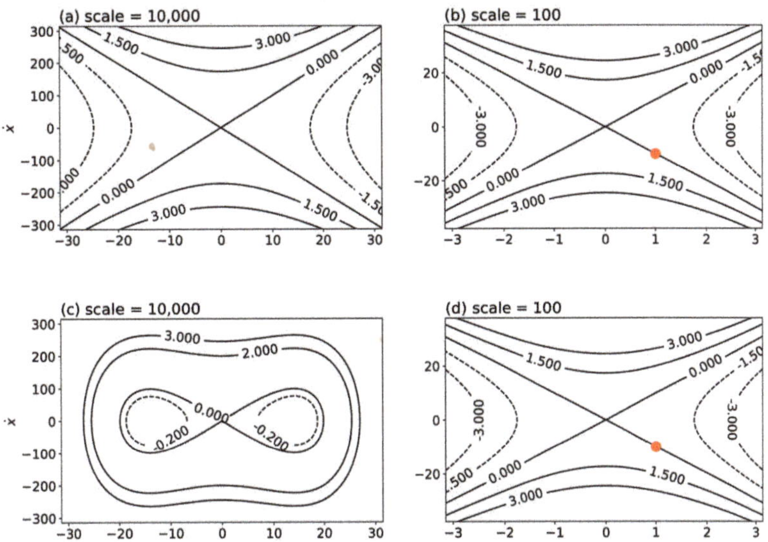

Figure 1. The contour lines of total energy in Equations (2a) and (2b) for the linear (**a,b**) and nonlinear (**c,d**) systems of $X'' - 10\pi^2 X + X^3/2 = 0$. Panels (**b,d**) provide a zoomed-in view of panels (**a,c**), respectively. A large red dot at $(X, X') = (1, -\sqrt{10}\pi)$ displays a starting point of an orbit that moves along the zero contour line towards the origin.

By comparison, a negative eigenvalue ($\lambda^2 < 0$) produces a general periodic solution with trigonometric functions of *sin* and *cos*. The corresponding solution displays a closed curve within the 2D $X - X'$ phase space, and the critical point $(0,0)$ is called a center (Table 1).

Table 1. The classification of various types of solutions for a linear 2D system with real coefficients. We assumed an arbitrary $\alpha, \beta > 0$, and $\gamma > 0$.

Characteristics	Solutions	Critical Points	Remarks
non-oscillatory	$c_1 e^{\gamma\tau} + c_2 e^{-\gamma\tau}$	saddle	monotonic
oscillatory ($\beta \neq 0$)	$c_1 e^{(\alpha+i\beta)\tau} + c_2 e^{(\alpha-i\beta)\tau}$		
	$\alpha = 0$	center	periodic
	$\alpha > 0$	spiral source	
	$\alpha < 0$	spiral sink	

3.1.3. Periodicity and Centers Enabled by Nonlinearity

Mathematically, it is natural to add nonlinear terms, X^2 or X^3, into the linear system in Equation (5) in order to constrain growth of the unstable mode. As previously illustrated, the inclusion of the quadratic and cubic terms yields systems comparable to the Korteweg–de Vries (KdV) and nonlinear Schrodinger equations, respectively, in a traveling-wave coordinate [30,51–55]. The latter can be mathematically written as the non-dissipative L63 model in Equation (1), as follows:

$$\frac{d^2 X}{d\tau^2} - \lambda^2 X + \frac{X^3}{2} = 0,$$

which also represents an unforced Duffing Equation (e.g., [29,56]).

Inclusion of the nonlinear cubic term leads to two, non-trivial critical points at $X = \pm 2\lambda$ and $\lambda > 0$ (e.g., [31]). To reveal the role of the nonlinear term, we can simply compare the contour lines of the energy functions, Equations (2a) and (2b), for the nonlinear and linear systems, Equations (1) and (5), respectively. As shown in Figure 1c, two centers appear, as indicated by a family of nearby closed contour lines that enclose one of the centers. Each of the closed curves represents one periodic solution with a specific initial condition.

As a result of the two centers and the saddle point at the origin, small- and large-cycle oscillations, and homoclinic orbits appear for $C < 0$, $C > 0$, and $C = 0$ in Equations (2a) and (2b), respectively. As shown in Figure 1c, for $C \neq 0$, each closed contour line indicates an oscillatory solution (see [29,30,57,58]). For $C = 0$, an orbit may move in the direction of the unstable manifold of a saddle point and return back along the direction of the stable manifold of the same saddle point. Such an orbit connecting the unstable and stable manifolds of the same saddle point is called a homoclinic orbit. The number "8" shape in Figure 1c indicates two homoclinic orbits. Thus, all three types of solutions reveal how nonlinearity limits the growth of the unstable solution, producing bounded solutions and global stability.

On the other hand, while an unstable eigenvector can be easily captured using a numerical method even though the initial conditions only allow solutions with a stable eigenvector in a linear system, such a feature also appears in association with the homoclinic orbit within a nonlinear system. Below, an analysis of the homoclinic orbit in Shen, 2020 [30] is briefly reviewed for a comparison.

3.1.4. (Computational) Limit Chaos Associated with a Homoclinic Orbit

The association of a saddle point with SDIC has been illustrated using the so-called linear geometric model [59] and the nonlinear non-dissipative L63 model [11,29]. Although the two simplified models may not exactly produce the same type of SDIC as the full L63 model, they may serve as a baseline system for revealing solution sensitivities. Sensitivities associated with the special homoclinic orbit were previously documented in Figure 4 of Shen, 2020 [30]. Most solutions within the non-dissipative L63 model are oscillatory and only two of them are homoclinic orbits. As a result, such sensitivities may be referred to as limit chaos [3]. Since it is challenging to avoid limit chaos in numerical integrations of the non-dissipative L63 model, we may simply call the sensitivity computational limit chaos.

3.1.5. Spiral Sinks Associated with an Additional Dissipative Term $(-\epsilon Y)$

In general, saturation of an unstable mode requires the existence of a stable critical point. However, the above discussions suggest that the appearance of two centers cannot enable unstable modes to become saturated (i.e., the solution reaches a constant). To obtain non-trivial stable critical points that can help yield steady-state solutions, one dissipative term $(-\epsilon Y)$ is added into the non-dissipative L63 model (e.g., Equation (A2)), yielding:

$$\frac{d^2 X}{d\tau^2} + \epsilon \frac{dX}{d\tau} - \lambda^2 X + \frac{X^3}{2} = 0, \tag{8}$$

which can also be obtained from Equation (A10) that represents a simplified Lorenz model with only one dissipative term $(-Y)$ (i.e., without two dissipative terms, $-\sigma X$ and $-bZ$).

Following the derivations for Equations (2a) and (2b), the energy function for Equation (8) indicates that the system energy decreases with time, producing steady state solutions. This is consistent with the analysis of critical points, showing two, stable, non-trivial critical points as spiral sinks [31]. A mathematical solution for a general spiral sink is listed in Table 1. As summarized in Table 3 of Shen, 2021 [31], the locations of the critical points for cases $\epsilon = 0$ and $\epsilon \neq 0$ in Equation (8) are the same. Therefore, it is relatively easy to make a comparison. While the model for $\epsilon = 0$ displays regular oscillatory solutions (e.g., Figure 7 of Shen, 2018 [29]), the model for $\epsilon \neq 0$ simulates steady-state solutions, as shown in Figure 2. For steady-state solutions that indicate a saturation, an orbit moves from the saddle point and towards one spiral sink. However, the time evolution of the solution is not a monotonic function in time.

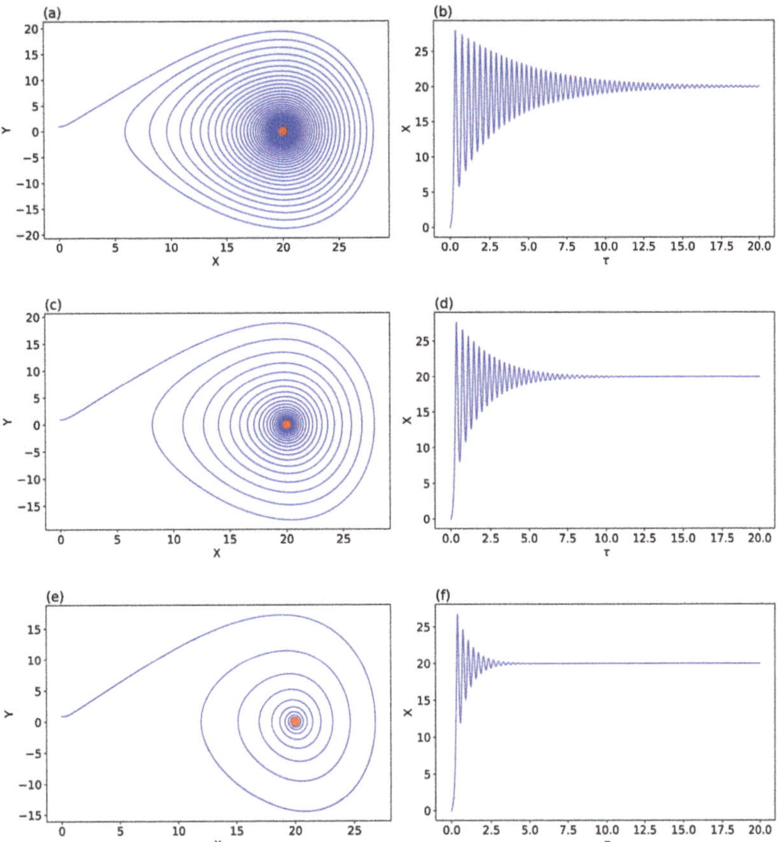

Figure 2. Solutions for the simplified Lorenz model that only contains one dissipative term, $-\epsilon Y$, and $\epsilon = 0.5, 1$, and 2.5, from top to bottom. Left panels (**a,c,e**) display orbits within the $X - Y$ space, while right panels (**b,d,f**) depict the time evolution of X. A stable non-trivial critical point is indicated by a large red dot.

3.1.6. SDIC and Finite Predictability Within the L63 Model

The full L63 model has been widely applied for illustrating the SDIC, as follows. Two runs, the control and parallel runs, are performed using the same model and the same parameters, but slightly different initial conditions (ICs). For example, a tiny perturbation of $\epsilon = 10^{-10}$ is added into the initial condition for the parallel run. Then, the time evolution for the two solutions is analyzed. As a result of the tiny perturbations, the two runs almost produce identical results during an initial period of time. As shown in Figure 3, this feature is called continuous dependence on ICs [50]. However, in spite of the initial tiny perturbations, two runs produce very different solutions after a certain period of time. Such a feature with very different results is referred to as SDIC. Similar results have been documented within the scientific literature (e.g., Figure 2 of Shen, 2019b [8] and Figure 1 of Shen et al., 2021b [16]). Combined CDIC and SDIC determine the predictability horizon. Namely, the predictability limit is determined before the onset of SDIC, and the predictability horizon displays a dependence on initial conditions that are close to or away from one of three critical points within the L63 model [15,60–62].

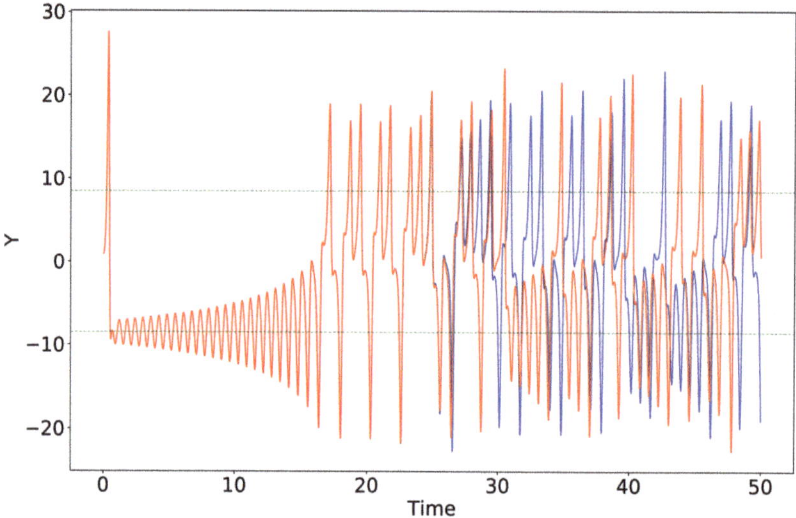

Figure 3. Two nearby trajectories within the L63 model with $r = 28$ and $\sigma = 10$. Solutions, in different colors, were obtained from the control and parallel runs. The only difference between the two solutions is that the parallel run adds a small perturbation (1×10^{-10}) into the initial value of Y. The sensitive dependence of solutions on the initial conditions is indicated by the divergence of two orbits in blue and red. The two horizontal lines indicate the location of the non-trivial critical points.

The above discussions qualitatively reveal the finite predictability of the L63 model. Within the nonlinear L63 model, its chaotic feature is indicated by the appearance of one positive Lyapunov exponent (e.g., [25,26] and references therein). The Lyapunov exponent (LE) is defined as the long-term average exponential rate of divergence of nearby trajectories [63–65] (i.e., the long-term average growth rate of an infinitesimal error [66]). The LE is calculated by evaluating the derivative along the direction of maximum expansion and averaging its logarithm along the trajectory (e.g., [67,68]). Although earlier studies made an attempt to apply the largest Lyapunov exponent for estimating predictability, a general average predictability is less interesting as compared to short-term behavior [66]. Additionally, the L63 limited-scale model with one positive LE is too simplified to estimate predictability in weather and climate. On the other hand, a challenge in extending the predictability horizon by improving the accuracy of initial conditions can be illustrated by assuming the error (i.e., separations of nearby trajectories) to be proportional to $e^{L\tau}$, here L and τ represent a LE and time, respectively (see [23] for details). However, such a rough formula of $e^{L\tau}$ that does not take SDIC into consideration should be applied with caution.

As discussed, when using the linear and nonlinear versions of the non-dissipative L63 model (e.g., Equation (1)), it is challenging to simulate the evolution of an orbit that moves towards the saddle point in a two-dimensional or higher phase space. By comparing the geometric and Lorenz models, such a feature is shown to be related to SDIC.

3.2. Features of the L69 Model

Based on the L63 model, the previous discussions illustrate the different roles of nonlinearity, which are missing in the L69 model. In comparison, within or based on the L69 model, the following features are discussed: (1) eigenvalues and eigenvectors; (2) a conceptual model for a chain process; (3) numerical instability associated with large eigenvalues; (4) ill-conditioning associated with large condition numbers; and (5) monotonicity within the L69 model. Based on the L69 and L63 models, we then provide comments on the estimate of predictability.

3.2.1. Eigenvalues and Eigenvectors

As shown in Equation (4) of this study or Equation 43 of Lorenz, 1969 [5], the L69 model consists of a linear system of N, 2nd-order ODEs, here $N = 21$. Such a system may be transformed to form an eigenvalue problem for an analysis. By assuming a solution vector in the form of $\overrightarrow{W} = e^{\lambda \tau} \overrightarrow{V}$, we can convert Equation (4) into the following equation:

$$\mathbf{A}\overrightarrow{V} = \lambda^2 \overrightarrow{V}. \tag{9}$$

Here, $\overrightarrow{V}$ and λ^2 represent the eigenvector and eigenvalue of the matrix $\mathbf{A}$, respectively. This approach is made possible because Equation (4) does not contain 1st-order derivatives. Here, λ^2, instead of λ, represents an eigenvalue. A system of N, 2nd-order ODEs produces N eigenvalues using this approach. As compared to Equation (5) that contains "one" eigenvalue, each of the N eigenvalues in Equation (9) is associated with two components, $e^{\gamma \tau}$ and $e^{-\gamma \tau}$ for $\gamma = \lambda^2$. Details are provided below.

To simplify discussions, we consider distinct eigenvalues that can be positive or negative. Positive and negative eigenvalues yield real and pure imaginary values for γ, representing exponential modes and oscillatory modes, respectively. Here, we compute the eigenvalues and eigenvectors for the six systems available from Rotunno and Synder, 2008 and Durran and Gingrich, 2014 [32,33]. For these systems, only 9×9 matrices were listed in the studies. However, as discussed below, this limitation does not have a significant impact on our major conclusion since the 9×9 matrices represent sub-matrices of the original 21×21 matrices at larger scales. Therefore, the eigenvalues are relatively small as compared to the eigenvalues of the corresponding full matrices. References to the six matrices are provided in Table 2.

Table 2. Condition numbers in six L69-type systems from Rotunno and Synder (2008, RS08) and Durran and Gingrich (2014, DG14) [32,33]. Both Python and Matlab were used for calculations and displayed similiar results.

	Python	Matlab	Remarks
Table 1 of RS08	8.319352×10^5	8.3194×10^5	2DV dynamics
Table 2 of RS08	8.446532×10^5	8.4465×10^5	vs. Lorenz (1969)
Table 3 of RS08	2.791518×10^4	2.7915×10^4	"unlimited predictability"
Table 4 of RS08	2.146269×10^9	2.1463×10^9	SQG dynamics
Table A1 of DG14	7.967004×10^5	7.9670×10^5	vs. Table 1 of RS08
Table A2 of DG14 *	9.767672×10^6	9.7677×10^6	vs. Table 4 of RS08

* There may be a typo in $C_{1,2}$ in Table A2 of DG14. A revised condition # is $O(10^9)$.

Figure 4 displays the nine eigenvalues of the matrix from Table 4 of Rotunno and Synder, 2008 [32]. Here, we observe positive and negative eigenvalues, both of which are large in magnitude. The first three positive eigenvalues are 457.0, 3.42, and 0.000472, respectively. A positive eigenvalue produces a pair of stable and unstable components, while a negative eigenvalue yields oscillatory components. Additionally, the system also produces eigenvalues with very small magnitudes. Due to the limit of finite precision, here, we do not make an attempt to determine whether (or not) these eigenvalues are zeros or even double zeros. Instead, we focus on eigenvectors with large eigenvalues.

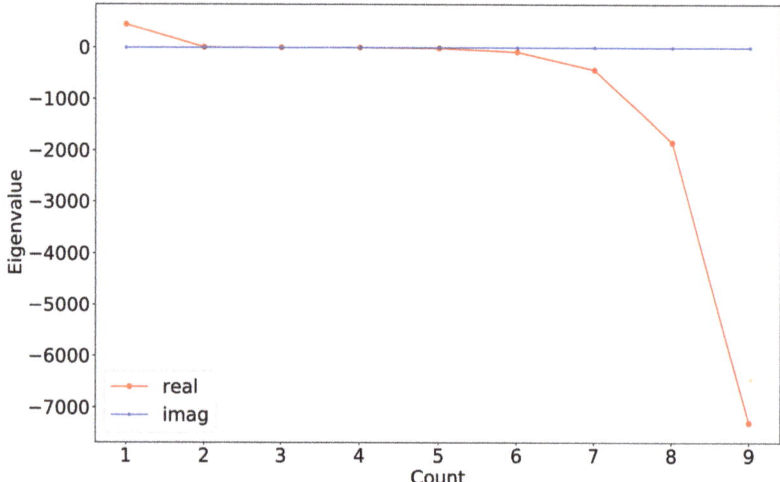

Figure 4. Nine eigenvalues for the 9×9 matrix from Table 4 of Rotunno and Synder, 2008 [32]. As discussed in the main text, each eigenvalue is associated with a pair of solution components. A positive eigenvalue yields a pair of exponentially growing and decaying components, while a negative eigenvalue leads to oscillatory components.

3.2.2. A Conceptual Model for a Chain Process

By analyzing the eigenvalues of the system in Equation (9), growth rates and predictability on various scales can be determined. Since the most unstable mode with the largest exponential growth rate should dominate, a system's predictability is solely determined by the largest growth rate (i.e., the largest $\sqrt{\lambda^2}$). On the other hand, the saturation assumption has been applied to extend a system's predictability as follows: when the unstable mode at the smallest scale reaches its saturated value, as determined by the energy spectrum, it is removed from the system. The time required for the most unstable component to become saturated defines a predictability horizon for the specific unstable mode. Afterwards, a subsystem with a sub-matrix (e.g., a $(N-1) \times (N-1)$ matrix) is obtained. Within the subsystem, different unstable modes appear, and the most unstable mode determines the predictability horizon. After becoming saturated, the mode is removed from the system. Thus, a system's predictability is obtained by adding all of the predictability horizons determined by each of the most unstable modes that sequentially appear, grow, and saturate. Simply speaking, a system's predictability is collectively determined by various eigenvalues at different scales. A mathematical analysis for an estimate of predictability is given near the end of this section. The above procedure relies on the successive eigenvalue calculation of matrices whose elements are continuously removed as a result of the saturation assumption (e.g., [32,47]). The procedure implies a conceptual model for a chain process that consists of repeated removal of the current most unstable mode and the appearance of a new most unstable mode.

The above conceptual model for the chain process is analyzed using Figure 5. Panels (a)–(f) display the first eigenvalues and the corresponding eigenvectors within a 9×9, 8×8, 7×7, 6×6, 5×5, and 4×4 matrix, respectively. Based on Figure 5, we can observe that the first eigenvector contains several non-zero components at smaller scales. Secondly, the first eigenvalue becomes smaller within smaller sub-matrices. Therefore, we may say that smaller scale modes possess larger growth rates. Since the first eigenvalue is always positive and since the corresponding eigenvector contains components at smaller scales, Figure 5 indicates the possibility for a chain process.

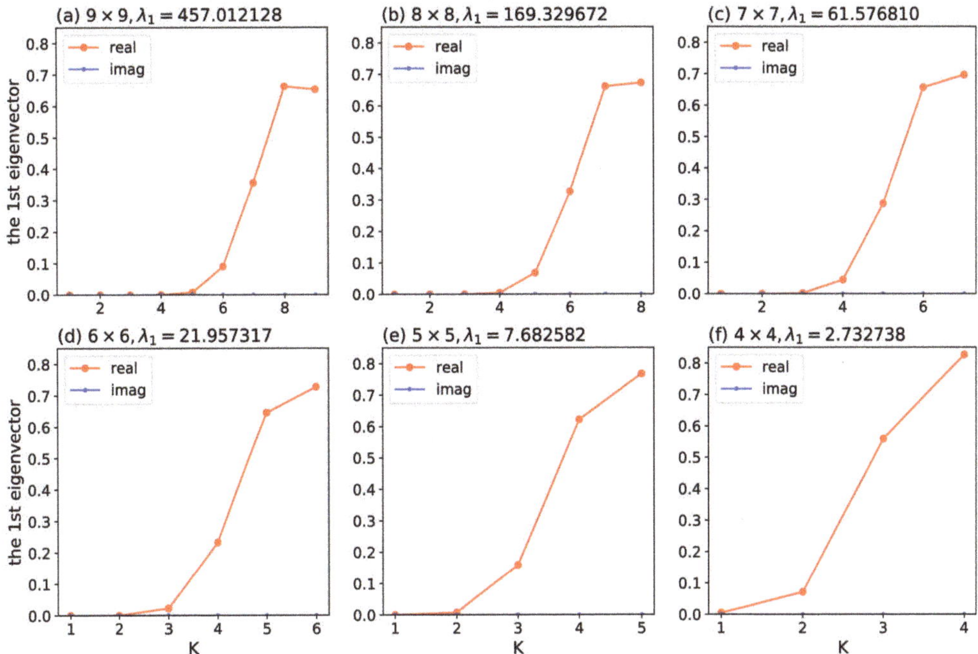

Figure 5. The first eigenvector of the 9×9 matrix from Table 4 of Rotunno and Synder, 2008 [32], and the 8×8, 7×7, 6×6, 5×5, and 4×4 sub-matrices, respectively.

While a specific eigenvector may contain more than one state variable, a specific state variable may appear in more than one eigenvector, depending on the ICs. Mathematically, we say that a general solution can be expressed in terms of a linear combination of all eigenvectors. As a result, a specific state variable (i.e., a specific wave mode) may concurrently grow at more than one growth rate. This feature complicates the process for estimating system predictability.

3.2.3. Numerical Instability Associated with Large Eigenvalues

Here, it should be noted that we cannot exclude the possibility for a specific state variable that appears as a decaying or an oscillatory component. Such a state variable should be more predictable. On the other hand, from a numerical perspective, such a solution with better theoretical predictability may be incorrectly simulated, appearing as an unstable mode. As discussed using Equation (5) in Section 3.1, it is easy for numerical methods to capture unstable modes even though a specific initial condition only allows a stable solution. In the example in Section 3.1, an eigenvalue of $\Lambda = \lambda^2 = 10\pi^2 \approx 98.7$ was used. As a result, numerical instability similar to that in Equation (5) may be easily found within the L69 system that possesses several large eigenvalues, including the largest eigenvalue $\Lambda_1 = 457.0$ within the 9×9 matrix and $\Lambda_1 = 169.3$ within the 8×8 matrix. Thus, it is reasonable to conclude that such numerical instability can be found within the 21×21 matrix.

On the other hand, as discussed in Section 3.1, nonlinearity may constrain the growth of unstable modes, and dissipation may reduce growth rates. Therefore, due to the lack of dissipation and nonlinearity within the L69 model, an estimate of predictability based on the growth rates of unstable modes should be interpreted with caution.

3.2.4. Ill-Conditioning Associated with Large Condition Numbers

In addition to potential numerical instability associated with large eigenvalues, here, we also reveal ill-conditioning with numerical sensitivities within the L69 model. Ill-conditioning is measured by a condition number ($\kappa(\mathbf{A})$), as follows:

$$\kappa(\mathbf{A}) = \|\mathbf{A}\| \|\mathbf{A}^{-1}\|, \tag{10}$$

where $\|\mathbf{A}\|$ represents the matrix norm of $\mathbf{A}$ and $\mathbf{A}^{-1}$ is the inverse matrix of $\mathbf{A}$. The above can be written as:

$$\kappa(\mathbf{A}) = \left|\frac{\Lambda_{max}}{\Lambda_{min}}\right|, \tag{11}$$

here, $\Lambda = \lambda^2$ represents an eigenvalue of matrix $\mathbf{A}$. Ill-conditioning occurs when the condition number ($\kappa(\mathbf{A})$) is large. Equation (11) suggests that a system with large variances in growth rates (i.e., containing a large ratio between the largest and smallest eigenvalues) is ill-conditioned. Such an expression is effective in revealing potential issues in sophisticated model systems that contain multiple scales and, thus, various growth rates. Within an ill-conditioned system, numerical results are sensitive to small changes, including round-off errors within the coefficient matrix. Sensitivity to small changes can be illustrated using a system of equations in Supplementary Materials modified from Kreyszig, 2011 [69], that contains a 2×2 matrix with a condition number of $20,001$. The system has a unique solution of $(0,0)$. However, when the system is perturbed by a tiny noise of δ, its solution is shifted to $(5000.5\delta, 4999.5\delta)$ (e.g., $(0.50005, 0.49995)$ for $\delta = 10^{-4}$).

Table 2 displays large condition numbers for all of the six systems from Rotunno and Synder, 2008 and Durran and Gingrich, 2014 [32,33]. The condition numbers are on the order of 10^4 or higher. As a result, we may expect large uncertainties in determining the location of a saddle point. On the other hand, from the perspective of transient solutions, a large condition number also indicates a large stiffness. As discussed using an example in Supplementary Materials, modified from [49] a system of stiff ODEs requires very small step sizes to obtain stable numerical solutions.

3.2.5. Solutions in Terms of Eigenvalues and Eigenvectors of the L69 Model

As a brief summary, the L69 model can be described as follows:

(1) The model is closure-based, physically multiscale, mathematically linear, and numerically ill-conditioned.
(2) The model possesses multiple positive and negative eigenvalues, and, thus, produces growing and decaying components and oscillatory components. However, the model may easily capture unstable modes due to numerical errors and large growth rates.
(3) Since the system is linear and homogeneous, the only equilibrium point is a trivial equilibrium (or critical) point at $\overrightarrow{W} = 0$. The critical point is a saddle point that contains multiple pairs of stable and unstable eigenvectors associated with multiple positive eigenvalues.

Mathematically, solutions of the L69 linear system can be obtained by performing numerical integration or using a linear combination of system eigenvectors and eigenvalues. Under the saturation assumption, a system's predictability is collectively determined by various eigenvalues at different scales. To illustrate this idea, we provide an analysis of the general solution using eigenvalues (γ_k^2) and eigenvectors ($\overrightarrow{V_k}$), as follows:

$$\overrightarrow{W} = C_1 e^{-\gamma_1 \tau}\overrightarrow{V_1} + D_1 e^{\gamma_1 \tau}\overrightarrow{V_1} + C_2 e^{-\gamma_2 \tau}\overrightarrow{V_2} + D_2 e^{\gamma_2 \tau}\overrightarrow{V_2} + \cdots + C_n e^{-\gamma_n \tau}\overrightarrow{V_n} + D_n e^{\gamma_n \tau}\overrightarrow{V_n}. \tag{12}$$

When $N = 1$, this leads to $\overrightarrow{W} = W_1$ and $\overrightarrow{V_1} = 1$, and Equation (12) yields:

$$\begin{pmatrix} W_1 \\ W_1' \end{pmatrix} = C_1 e^{-\gamma \tau}\begin{pmatrix} 1 \\ -\gamma \end{pmatrix} + D_1 e^{\gamma \tau}\begin{pmatrix} 1 \\ \gamma \end{pmatrix},$$

which is the same as Equation (7b). Here, we, once again, bring the reader's attention to the fact that the L69 model consists of N, 2nd-order ODEs and, thus, $2N$, 1st-order ODEs. When a saturation assumption is applied for determining the predictability horizon, an initial condition for each of the unstable modes is required.

3.2.6. Finite Predictability within the L69 Model

As discussed using Figure 3, the combined CDIC and SDIC suggest finite predictability within the L63 model. While SDIC is indicated by the divergence of two nearby trajectories that start moving towards different non-trivial critical points, two trajectories may closely move around one of the non-trivial critical points during the epoch of CDIC. In Figure 3, such is indicated by oscillatory solutions centered at the value shown in green between $\tau = 3$ and $\tau = 15$. A key message is that when two nonlinear solutions leave the saddle point and move towards one of the non-trivial critical points, both may still closely move and revisit the neighbors of the saddle point several times (e.g., between $\tau = 18$ and $\tau = 22$ in Figure 3) prior to their separation (e.g., after $\tau = 26$ in Figure 3). Revisiting the neighborhood of the saddle is not allowed within a linear system.

By comparison, the L69 multiscale model that has the advantage of including a realistic energy spectrum has been applied to estimate multiscale predictability in weather. On the other hand, due to the limit of linearity, two nearby unstable orbits within the L69 model monotonically diverge, which is different from time varying changes within the L63 model. Additionally, the monotonic increase of errors may not be applicable to oscillatory or decaying solutions.

In Equation (12), a specific state variable (i.e., a specific wave mode, W_k) may concurrently grow at more than one growth rate. This feature complicates the process for estimating system predictability. Below, to simplify discussions without the need for determining "initial conditions", the e-folding time is used in order to estimate the predictability horizon of the most unstable mode (e.g., [70]). The e-folding time is the time interval in which an exponentially growing quantity increases by a factor of e. Thus, given a specific mode k, with a dominant growth rate of γ_k, the predictability limit (or horizon) (i.e., an e-folding time) for the specific mode is inversely proportional to its growth rate, as follows:

$$T_k = \frac{1}{\gamma_k} \tag{13}.$$

Thus, a system's predictability horizon is proportional to the time required for systems at all scales to become saturated (i.e., to increase by a factor of e):

$$T \approx \frac{1}{\gamma_{9\times9}} + \frac{1}{\gamma_{8\times8}} + \frac{1}{\gamma_{7\times7}} \cdots \tag{14}$$

Namely, a system predictability horizon is determined by the sum of the reciprocals for all growth rates. In general, due to 21 modes within the L69 model, Equation (14) does not represent an infinite series.

We also may want to know the condition under which the predictability is finite. If Equation (14) is a geometric series with a factor of $1/2$, the sum is finite, as shown below:

$$T \approx 1 + \frac{1}{2} + \frac{1}{4} + \frac{1}{8} + \cdots = 2. \tag{15}$$

The above example with a common factor of $1/2$ is consistent with discussions of Palmer et al., 2014, Palmer, 2017 and Lorenz, 1969 [5,71,72] in regards to Lorenz's empirical formula:

Except for the smallest scales retained, where the effect of omitting still smaller scales is noticeable, and the very largest scales, where X_k does not conform to a $-2/3$ law, successive differences $t_k - t_{k+1}$ differ by a factor of about $2^{-2/3}$. If one chooses to *reevaluate t_1 by summing the terms of the sequence $t_1 - t_2, t_2 - t_3, \cdots$, one is effectively summing a truncated geometric series.*

The above calculation applied several assumptions, including assumptions that: (1) the series in Equation (14) can be extended to have an infinite number of terms, and (2) the series contains a common factor of $2^{-2/3}$, even for different slopes of the kinetic energy spectrum. On the other hand, the following two series are not convergent:

$$\sum_{n=1} \frac{1}{n} = 1 + \frac{1}{2} + \frac{1}{3} + \frac{1}{4} + \cdots = \infty, \tag{16a}$$

$$\sum_{pprime} \frac{1}{p} = \frac{1}{2} + \frac{1}{3} + \frac{1}{5} + \frac{1}{7} + \cdots = \infty. \tag{16b}$$

The two series do not produce a finite number. Therefore, whether (or not) an extension of Equation (14) with an infinite number of terms produces a finite predictability is still a challenging question. On the other hand, since Equation (14) is based on several assumptions, its validity should be examined. Other than the above, the L69 model is not a turbulence model, and all weather systems cannot be turbulent forever. As a result, it may be legitimate to conclude that "practical predictability" within the L69 is finite. Here, the practical predictability indicates a dependence of predictability on mathematical formulas and ICs, in contrast to intrinsic predictability that only depends on a flow itself [73].

3.2.7. A Comparison of Monostability and Multistability

Given model parameters, the L63 model produces single type solutions, including steady-state, chaotic, and limit cycle solutions. Such a feature is referred to as monostability. By comparison, generalized Lorenz models possess coexisting chaotic and non-chaotic attractors that appear with the same model parameters but different initial conditions [11–14]. The feature of attractor coexistence is called multistability. Although the L69 model is linear, it does allow various types of solutions associated with positive or negative eigenvalues. Thus, the L69 model may be viewed to possess multistability, despite the fact that only unstable solutions have been a focus.

Based on the results with multistability, we recently suggested a revised view that weather possesses both chaos and order, in contrast to the conventional view of "weather is chaotic". Such a view is additionally supported by this study. While chaotic solutions of the L63 model or linearly unstable solutions of the L69 model produce finite predictability, non-chaotic regular solutions may have unlimited predictability (up to the lifetime of the system or the duration of forcing). Such a revised view that turns our attention from monostability to multistability is neither too optimistic nor pessimistic as compared to the Laplacian view of deterministic predictability and the Lorenz view of finite predictability.

4. Concluding Remarks

Both the Lorenz 1963 (L63) and 1969 (L69) models [1,5] have been applied in the past to illustrate finite predictability. An estimate of a predictability limit of two weeks was initially obtained using the L69 model. In this study, new analyses along with a literature review provide insights on the mathematical and physical relationship, two types of sensitivities, and the impact of a saddle point on the two types of sensitivities. The L63 and L69 models are derived from different partial differential equations. One system is for convection, and the other system is for the conservation of barotropic vorticity. The L63 model is limited-scale and nonlinear; and the L69 model is closure-based, physically multiscale, mathematically linear, and numerically ill-conditioned. The former possesses a sensitive dependence of solutions on initial conditions, known as the butterfly effect, and the latter contains numerical sensitivities resulting from an ill-conditioned matrix with a large condition number (i.e., a large variance of growth rates). A common feature that produces unstable components in both systems is the existence of a saddle point at the origin. A saddle point provides an essential ingredient for chaos within the L63 model and for linear instability within the L69 model. Table 3 provides a summary for the L63 model, the geometric model, the non-dissipative L63 model, and the L69 model. All of the listed models reveal the common feature of a saddle point.

Table 3. The Lorenz 1963 (L63) model [1], the geometric model [59], the non-dissipative L63 model [29], and the Lorenz 1969 (L69) model [5]). The non-dissipative L63 model is derived from the L63 model without dissipative terms shown in red. For an extreme case of $N = 1$ and a positive eigenvalue, the L69 model is dynamically comparable to the non-dissipative L63 model without the cubic term shown in blue. A saddle point is a common feature amongst the four models.

(1) The L63 Model	(2) The Geometric Model
$\frac{dX}{d\tau} = \sigma Y - \sigma X,$ $\frac{dY}{d\tau} = -XZ + rX - Y,$ $\frac{dZ}{d\tau} = XY - bZ.$	$\frac{dX}{d\tau} = -3X,$ $\frac{dY}{d\tau} = 2Y,$ $\frac{dZ}{d\tau} = -Z.$
(3) The Non-dissipative L63 Model	(4) The L69 Model
$\frac{d^2 X}{d\tau^2} = (\Omega_0 + \sigma r)X - \frac{X^3}{2},$ Ω_0: constant.	$\frac{d^2 \vec{W}}{d\tau^2} = \mathbf{A}\vec{W},$ $\mathbf{A}$: $N \times N$ matrix, $\vec{W}$: a vector for N state variables.

Within the chaotic regime of the L63 nonlinear model, unstable growth is constrained by nonlinearity, as well as dissipation, yielding bounded solutions and time varying growth rates along an orbit. The appearance of SDIC suggests a finite predictability that displays a dependence on initial conditions. Within unstable solutions of the L69 linear model, multiple growth rates at various scales exist. Unlimited growth of the most unstable mode is suppressed by artificially imposing the saturation assumption: a saturated unstable mode that reaches its upper limit within a finite interval is removed from the system, enabling the appearance of a new most unstable mode. Thus, a system's growth rate appears to be time varying and a system's predictability is collectively determined by various eigenvalues at different scales.

While both the L63 and L69 models suggest finite predictability, only single type solutions (e.g., chaotic solutions within the L63 model and linearly unstable solutions within the L69 model) were considered. The SDIC of the L63 model leads to finite predictability. By comparison, ill-conditioning and the appearance of numerical instability are likely responsible for the "finite predictability" of the L69 study, as suggested by the following quote from the L69 study: *two states of the system differing initially by a small "observational error" will evolve into two states differing as greatly as randomly chosen states of the system within a finite time interval, which cannot be lengthened by reducing the amplitude of the initial error."* Thus, the mechanisms for finite predictability within the L63 and L69 model are not exactly the same, although both models contains a saddle point.

Based on the L69 linear model with a saturation assumption, the conceptual model for a chain process possesses a collection of unstable modes that sequentially appear, grow, and saturate. Although the L69 model effectively describes the phenomena of instability, it cannot precisely reveal the true nature of chaos. In contrast to growing solutions, the L69 linear model also produces decaying components (with positive eigenvalues) and oscillatory components (with negative eigenvalues). As documented in [11,12,15,16], the L63 model and its generalized version (e.g., [11]) can produce non-chaotic solutions and coexisting chaotic and non-chaotic solutions. Therefore, a realistic model should possess both chaotic and non-chaotic processes (as well as both unstable and stable solutions) and, thus, distinct predictability. An estimate of a predictability limit using either the (classical) L63 or L69 model, with or without additional assumptions (e.g., saturation), should be interpreted with caution and should not be generalized as an upper limit for predictability.

Supplementary Materials: Supplementary Materials can be downloaded at: https://www.mdpi.com/article/10.3390/atmos13050753/s1. Table S1 displays the definitions of CDIC and SDIC. Lists

1–12 are provided for the analysis of the L63 model by Palmer et al., 2014 [71], the fundamental local theorem of ODEs, CDIC, Lipschitz constant, CDIC and the Lipschitz constant, SDIC and chaos, a definition of chaos, volume contraction within the L63 model, a definition of the Lyapunov function, existence of the Lyapunov function and global stability within the L63 model for $r \leq 1$. Figures S1–S7 display: a relationship between c_2 in [71] and the Lipschitz constant L, an illustration of CDIC and SDIC, numerical experiments of control and parallel runs within the L63 model, time-varying local Lyapunov exponents, dependence of predictability on initial conditions, and qualitative description of local predictability on the Lorenz attractor [74].

Author Contributions: B.-W.S. designed and performed research; B.-W.S., R.A.P.S. and X.Z. wrote the paper. All authors have read and agreed to the published version of the manuscript.

Funding: This research received no external funding.

Institutional Review Board Statement: Not applicable.

Informed Consent Statement: Not applicable.

Data Availability Statement: Not applicable.

Acknowledgments: We thank anonymous reviewers, D. Durran, R. Rotunno, and H.-C. Kuo for valuable comments and discussions.

Conflicts of Interest: The authors declare no conflict of interest.

Appendix A. The Lorenz 1963 Model and Its Simplified Systems

Here, the Lorenz 1963 model [1] is rewritten by introducing two additional parameters (σ_1 and ϵ) as the coefficients of dissipative terms, as follows:

$$\frac{\mathrm{d}X}{\mathrm{d}\tau} = \sigma Y - \sigma_1 X, \tag{A1}$$

$$\frac{\mathrm{d}Y}{\mathrm{d}\tau} = -XZ + rX - \epsilon Y, \tag{A2}$$

$$\frac{\mathrm{d}Z}{\mathrm{d}\tau} = XY - bZ. \tag{A3}$$

Here, τ is dimensionless time. The above three, first-order ODEs describe the time evolution of three state variables X, Y, and Z. Two parameters, σ and r, are called the Prandtl number and the normalized Rayleigh number, respectively. The parameter b is a function of the ratio of the horizontal and vertical scales of a convection cell. The term $-bZ$ introduces dissipation. Compared to the classical Lorenz model, Equations (A1)–(A3) introduce two additional parameters, σ_1 and ϵ, in order to trace the impact of each of the three dissipative terms. When $\sigma_1 = \sigma$ and $\epsilon = 1$, Equations (A1)–(A3) represents the classical L63 model.

By introducing $\Omega = X^2/2 - \sigma Z$, one can transform Equations (A1)–(A3) into the following 2nd- and 1st-order ODEs that produce the same form as the Pedlosky model [31]:

$$\frac{\mathrm{d}^2 X}{\mathrm{d}\tau^2} + (\sigma_1 + \epsilon)\frac{\mathrm{d}X}{\mathrm{d}\tau} - (\Omega + \sigma r - \sigma_1 \epsilon)X + \frac{X^3}{2} = 0, \tag{A4}$$

$$\frac{\mathrm{d}\Omega}{\mathrm{d}\tau} + b\Omega + (\sigma_1 - \frac{b}{2})X^2 = 0. \tag{A5}$$

From left to right, the four terms of Equation (A4) represent acceleration, linear dissipation, linear forcing, and a nonlinear cubic restoring force, respectively.

The above system in Equations (A4) and (A5) has been compared to the Pedlosky model for revealing their mathematical universality, identifying two crucial dissipative terms, and illustrating the physical relevance of related findings to predictability problems.

Equations (A4) and (A5) can be simplified into the following systems:

(a) $\sigma_1 = \sigma$ and $\epsilon = 1$ (i.e., the L63 model):

$$\frac{d^2X}{d\tau^2} + (\sigma+1)\frac{dX}{d\tau} - (\Omega + \sigma r - \sigma)X + \frac{X^3}{2} = 0, \tag{A6}$$

$$\frac{d\Omega}{d\tau} + b\Omega + (\sigma - \frac{b}{2})X^2 = 0. \tag{A7}$$

The system has a $r_c = \frac{\sigma(\sigma+b+3)}{\sigma-b-1} = 24.74$ for the onset of chaos.

(b) $\sigma_1 = \sigma$ and $\epsilon = 0$ (i.e., the simplest Lorenz-type model for chaos):

$$\frac{d^2X}{d\tau^2} + (\sigma)\frac{dX}{d\tau} - (\Omega + \sigma r)X + \frac{X^3}{2} = 0, \tag{A8}$$

$$\frac{d\Omega}{d\tau} + b\Omega + (\sigma - \frac{b}{2})X^2 = 0. \tag{A9}$$

The system has a $r_c = \frac{\sigma(\sigma+b)}{\sigma-b} = 17.27$ for the onset of chaos.

(c) $\sigma_1 = b = 0$ (an uncoupled 2D system with $d\Omega/d\tau = 0$):

$$\frac{d^2X}{d\tau^2} + \epsilon\frac{dX}{d\tau} - (\Omega + \sigma r)X + \frac{X^3}{2} = 0. \tag{A10}$$

The above system is briefly analyzed in the main text, yielding spiral sink solutions.

(d) $\sigma_1 = b = \epsilon = 0$ (i.e., the non-dissipative L63 model with $d\Omega/d\tau = 0$):

$$\frac{d^2X}{d\tau^2} - (\Omega + \sigma r)X + \frac{X^3}{2} = 0. \tag{A11}$$

As discussed in Shen, 2018, 2020 [29,30] and reviewed in the main text, the above system produces two types of oscillatory solutions and homoclinic orbits.

(e) No nonlinear term in Equation (A11) (i.e., a linear system with $d\Omega/d\tau = 0$):

$$\frac{d^2X}{d\tau^2} - (\Omega + \sigma r)X = 0. \tag{A12}$$

The above system represents the most fundamental 2nd order ODE with stable and unstable solutions.

References

1. Lorenz, E.N. Deterministic nonperiodic flow. *J. Atmos. Sci.* **1963**, *20*, 130–141. [CrossRef]
2. Lorenz, E.N. Predictability: Does the flap of a butterfly's wings in Brazil set off a tornado in Texas? In Proceedings of the 139th Meeting of AAAS Section on Environmental Sciences, New Approaches to Global Weather, GARP, AAAS, Cambridge, MA, USA, 29 December 1972; 5p.
3. Lorenz, E.N. *The Essence of Chaos*; University of Washington Press: Seattle, WA, USA, 1993; 227p.
4. Lighthill, J. The recently recognized failure of predictability in Newtonian dynamics. *Proc. R. Soc. Lond. A* **1986**, *407*, 35–50.
5. Lorenz, E.N. The predictability of a flow which possesses many scales of motion. *Tellus* **1969**, *21*, 289–307. [CrossRef]
6. Shen, B.-W.; Tao, W.-K.; Wu, M.-L. African Easterly Waves in 30-day High-resolution Global Simulations: A Case Study during the 2006 NAMMA Period. *Geophys. Res. Lett.* **2010**, *37*, L18803. [CrossRef]
7. Shen, B.-W.; Tao, W.-K.; Green, B. Coupling Advanced Modeling and Visualization to Improve High-Impact Tropical Weather Prediction(CAMVis). *IEEE Comput. Sci. Eng. (CiSE)* **2011**, *13*, 56–67. [CrossRef]
8. Shen, B.-W. On the predictability of 30-day global mesoscale simulations of multiple African easterly waves during summer 2006: A view with a generalized Lorenz model. *Geosciences* **2019**, *9*, 281. [CrossRef]
9. Judt, F. Atmospheric Predictability of the Tropics, Middle Latitudes, and Polar Regions Explored through Global Storm-Resolving Simulations. *J. Atmos. Sci.* **2020**, *77*, 257–276. [CrossRef]
10. Carroll, M. Predictability Limit: Scientists Find Bounds of Weather Forecasting. 2019. Available online: https://www.sciencedaily.com/releases/2019/04/190415154722.htm (accessed on 26 April 2022).
11. Shen, B.-W. Aggregated negative feedback in a generalized Lorenz model. *Int. J. Bifurc. Chaos* **2019**, *29*, 1950037. [CrossRef]

12. Shen, B.-W.; Reyes, T.; Faghih-Naini, S. Coexistence of chaotic and non-chaotic orbits in a new nine-dimensional Lorenz model. In *11th Chaotic Modeling and Simulation International Conference*; Springer Proceedings in Complexity; Skiadas, C., Lubashevsky, I., Eds.; Springer: Cham, Switzerland, 2019. [CrossRef]

13. Reyes, T.; Shen, B.-W. A recurrence analysis of chaotic and non-chaotic solutions within a generalized nine-dimensional Lorenz model. *Chaos Solitons Fractals* **2019**, *125*, 1–12. [CrossRef]

14. Cui, J.; Shen, B.-W. A Kernel Principal Component Analysis of Coexisting Attractors within a Generalized Lorenz Model. *Chaos Solitons Fractals* **2021**, *146*, 110865. [CrossRef]

15. Shen, B.-W., Pielke, R.A., Sr.; Zeng, X.; Baik, J.-J.; Faghih-Naini, S.; Cui, J.; Atlas, R. Is weather chaotic? coexistence of chaos and order within a generalized Lorenz model. *Bull. Am. Meteorol. Soc.* **2021**, *2*, E148–E158. Available online: https://journals.ametsoc.org/view/journals/bams/102/1/BAMS-D-19-0165.1.xml (accessed on 29 January 2021). [CrossRef]

16. Shen, B.-W., Pielke, R.A., Sr.; Zeng, X.; Baik, J.-J.; Faghih-Naini, S.; Cui, J.; Atlas, R.; Reyes, T.A. Is Weather Chaotic? Coexisting Chaotic and Non-Chaotic Attractors within Lorenz Models. In *The 13th Chaos International Conference CHAOS 2020*; Springer Proceedings in Complexity; Skiadas, C.H., Dimotikalis, Y., Eds.; Springer: Cham, Switzerland, 2021. [CrossRef]

17. Curry, J.H. Generalized Lorenz Systems. *Commun. Math. Phys.* **1978**, *60*, 193–204. [CrossRef]

18. Curry, J.H.; Herring, J.R.; Loncaric,J.; Orszag, S.A. Order and disorder in two- and three-dimensional Benard convection. *J. Fluid Mech.* **1984**, *147*, 1–38. [CrossRef]

19. Sparrow, C. *The Lorenz Equations: Bifurcations, Chaos, and Strange Attractors*; Applied Mathematical Sciences; Springer: New York, NY, USA, 1982; 269p.

20. Guckenheimer, J.; Holmes, P. *Nonlinear Oscillations, Dynamical Systems, and Bifurcations of Vector Fields*; Springer: New York, NY, USA, 1983; 459p.

21. Roy, D.; Musielak, Z.E. Generalized Lorenz models and their routes to chaos. I. energy-conserving vertical mode truncations. *Chaos Soliton. Fract.* **2007**, *32*, 1038–1052. [CrossRef]

22. Hirsch, M.; Smale, S.; Devaney, R.L. *Differential Equations, Dynamical Systems, and an Introduction to Chaos*, 3rd ed.; Academic Press: Waltham, MA, USA, 2013; 432p.

23. Strogatz, S.H. *Nonlinear Dynamics and Chaos: With Applications to Physics, Biology, Chemistry, and Engineering*; Westpress View: Boulder, CO, USA, 2015; 513p.

24. Faghih-Naini, S.; Shen, B.-W. Quasi-periodic in the five-dimensional non-dissipative Lorenz model: The role of the extended nonlinear feedback loop. *Int. J. Bifurc. Chaos* **2018**, *28*, 1850072. [CrossRef]

25. Shen, B.-W. Nonlinear feedback in a five-dimensional Lorenz model. *J. Atmos. Sci.* **2014**, *71*, 1701–1723. [CrossRef]

26. Shen, B.-W. Nonlinear feedback in a six-dimensional Lorenz Model. Impact of an additional heating term. *Nonlin. Processes Geophys.* **2015**, *22*, 749–764. [CrossRef]

27. Shen, B.-W. Hierarchical scale dependence associated with the extension of the nonlinear feedback loop in a seven-dimensional Lorenz model. *Nonlin. Processes Geophys.* **2016**, *23*, 189–203. [CrossRef]

28. Shen, B.-W. On an extension of the nonlinear feedback loop in a nine-dimensional Lorenz model. *Chaotic Modeling Simul. (CMSIM)* **2017**, *2*, 147–157.

29. Shen, B.-W. On periodic solutions in the non-dissipative Lorenz model: The role of the nonlinear feedback loop. *Tellus A* **2018**, *70*, 1471912. [CrossRef]

30. Shen, B.-W. Homoclinic Orbits and Solitary Waves within the non-dissipative Lorenz Model and KdV Equation. *Int. J. Bifurc. Chaos* **2020**, *30*, 15. [CrossRef]

31. Shen, B.-W. Solitary Waves, Homoclinic Orbits, and Nonlinear Oscillations within the non-dissipative Lorenz Model, the inviscid Pedlosky Model, and the KdV Equation. In *The 13th Chaos International Conference CHAOS 2020*; Springer Proceedings in Complexity; Skiadas, C.H., Dimotikalis, Y., Eds.; Springer: Cham, Switzerland, 2021. [CrossRef]

32. Rotunno, R.; Snyder, C. A generalization of Lorenz's model for the predictability of flows with many scales of motion. *J. Atmos. Sci.* **2008**, *65*, 1063–1076. [CrossRef]

33. Durran, D.R.; Gingrich, M. Atmospheric predictability: Why atmospheric butterflies are not of practical importance. *J. Atmos. Sci.* **2014**, *71*, 2476–2478. [CrossRef]

34. Leith, C.E. Atmospheric predictability and two-dimensional turbulence. *J. Atmos. Sci.* **1971**, *28*, 145–161. [CrossRef]

35. Leith, C.E.; Kraichnan, R.H. Predictability of turbulent flows. *J. Atmos. Sci.* **1972**, *29*, 1041–1058. [CrossRef]

36. Holmes, P. Can dynamical systems approach turbulence. In *Whither Turbulence? Turbulence at the Crossroads*; Lecture Notes in Physics; Lumley, J.L., Ed.; Springer: Berlin/Heidelberg, Germany, 1990; Volume 357.

37. Bohr, T.; Jensen, M.; Paladin, G.; Vulpiani, A. *Dynamical Systems Approach to Turbulence*; Cambridge Nonlinear Science Series; Cambridge University Press: Cambridge, CA, USA, 1998. [CrossRef]

38. Saltzman, B. Finite amplitude free convection as an initial value problem. *J. Atmos. Sci.* **1962**, *19*, 329–341. [CrossRef]

39. Pedlosky, J. Finite-amplitude baroclinic waves with small dissipation. *J. Atmos. Sci.* **1971**, *28*, 587–597. [CrossRef]

40. Pedlosky, J. Limit cycles and unstable baroclinic waves. *J. Atmos. Sci.* **1972**, *29*, 53–63. [CrossRef]

41. Pedlosky, J. *Geophysical Fluid Dynamics*, 2nd ed.; Springer: New York, NY, USA, 1987; 710p.

42. Pedlosky, J. The Effect of Beta on the Downstream Development of Unstable, Chaotic Baroclinic Waves. *J. Phys. Oceanogr.* **2019**, *49*, 2337–2343. [CrossRef]

43. Pedlosky, J.; Frenzen, C. Chaotic and periodic behavior of finite-amplitude baroclinic waves. *J. Atmos. Sci.* **1980**, *37*, 1177–1196. [CrossRef]
44. Lesieur, M. *Turbulence in Fluids*, 4th ed.; Springer: Berlin, Germany, 2008; 558p.
45. Orszag, S.A. Analytical theories of turbulence. *J. Fluid Mech.* **1970**, *41*, 363–386. [CrossRef]
46. Shen, B.-W. Lecture Slides for Linearization Theorems. *Course Mater. Math.* **2017**. [CrossRef]
47. Leung, T.Y.; Leutbecher, M.; Reich, S.; Shepherd, T.G. Atmospheric Predictability: Revisiting the Inherent Finite-Time Barrier. *J. Atmos. Sci.* **2019**, *76*, 3883–3892. [CrossRef]
48. Lucarini, V.; Bodai, T. Global stability properties of the climate: Melancholia states, invariant measures, and phase transitions. *Nonlinearity* **2020**, *33*, R59. [CrossRef]
49. Boyce, W.E.; Diprima, R.C. *Elementary Differential Equations*, 10th ed.; John Wiley & Sons, Inc.: Hoboken, NJ, USA, 2012; ISBN1 13: 978-0470458327. ISBN2 10: 0470458321.
50. Alligood, K.; Saucer, T.; Yorke, J. *Chaos An Introduction to Dynamical Systems*; Springer: New York, NY, USA, 1996.
51. Korteweg, D.J.; de Vries, G. On the change of long waves advancing in a rectangular canal, and on a new type of long stationary waves. *Phil. Mag. (Ser. 5)* **1895**, *39*, 422–433. [CrossRef]
52. Lighthill, J. *Waves in Fluids*; Cambridge University Press: Cambridge, UK, 1978; 504p.
53. Whitham, G.B. *Linear and Nonlinear Waves*; John Wiley & Sons, Inc.: New York, NY, USA, 1974; 636p.
54. Haberman, R. *Applied Partial Differential Equations, with Fourier Series and Boundary Valule Problems*, 5th ed.; Pearson Education, Inc.: Hoboken, NJ, USA, 2013; 756p.
55. Jordan, D.W.; Smith, S. *Nonlinear Ordinary Differential Equations. An Introduction for Scientists and Engineers*, 4th ed.; Oxford University Press: New York, NY, USA, 2007; 560p.
56. Shen, B.-W. *Is Weather Chaotic? Multistability, Multiscale Instability, and Predictability within Lorenz Models*; Oxford University: Oxford, UK, 11 October 2021. [CrossRef]
57. Balmforth, N.J. Solitary waves and homoclinic orbits. *Annu. Rev. Fluid Mech.* **1995**, *27*, 335–373. [CrossRef]
58. Boyd, J.P. Dynamical methodology: Solitary waves. In *Encyclopedia of Atmospheric Sciences*, 2nd ed.; Academic Press: San Diego, CA, USA, 2015; pp. 417–422.
59. Guckenheimer, J.; Williams, R.F. Structural stability of Lorenz attractors. *Publ. Math. IHES* **1979**, *50*, 59–72. [CrossRef]
60. Slingo, J.; Palmer, T. Uncertainty in weather and climate prediction. *Philos. Trans. R. Soc.* **2011**, *369A*, 4751–4767. 0161. [CrossRef]
61. Nese, J.M. Quantifying local predictability in phase space. *Physica D* **1989**, *35*, 237–250. [CrossRef]
62. Zeng, X.; Pielke, R.A., Sr.; Eykholt, R. Chaos theory and its applications to the atmosphere. *Bull. Am. Meteor. Soc.* **1993**, *74*, 631–644. [CrossRef]
63. Aurell, E.; Boffetta, G.; Crisanti, A.; Paladin, G.; Vulpiani, A. Predictability in Systems with Many Characteristic Times: The Case of Turbulence. *Phys. Rev. E* **1996**, *53*, 2337–2349. [CrossRef] [PubMed]
64. Drazin, P.G. *Nonlinear Systems*; Cambridge University Press: New York, NY, USA, 1992; p. 333.
65. Eckhardt, B.; Yao, D. Local Lyapunov exponents in chaotic systems. *Physica D* **1993**, *65*, 100–108. [CrossRef]
66. Lorenz, E.N. 1996 "Predictability—A Problem Partly Solved" (PDF). Seminar on Predictability, Vol. I, ECMWF. Available online: https://www.ecmwf.int/en/elibrary/10829-predictability-problem-partly-solved (accessed on 20 April 2022).
67. Benettin, G.; Galgani, L.; Giorgilli, A.; Strelcyn, J.M. Lyapunov Characteristic Exponents for Smooth Dynamical Systems and for Hamiltonian Systems; A Method for Computing All of Them. Part 1: Theory. 1980. Available online: https://link.springer.com/article/10.1007/BF02128236 (accessed on 20 April 2022)
68. Sprott, J.C. *Chaos and Time-Series Analysis*; Oxford University Press: Oxford, UK, 2003; 528p. Available online: http://sprott.physics.wisc.edu/chaos/lyapexp.htm (accessed on 26 April 2022).
69. Kreyszig, E. *Advanced Engineering Mathematics*, 10th ed.; John Wiley & Sons, Inc.: Hoboken, NJ, USA, 2011; p. 1113
70. Lewis, J.; Lakshmivarahan, S.; Dhall, S. *Dynamic Data Assimilation: A Least Squares Approach*; Cambridge University Press: Cambridge, UK, 2006; 654p.
71. Palmer, T.N.; Doring, A.; Seregin, G. The real butterfly effect. *Nonlinearity* **2014**, *27*, R123–R141. [CrossRef]
72. Palmer, T. The Butterfly Effect—What Does It Really Signify? 2017. Available online: https://www.youtube.com/watch?v=vkQEqXAz44I&t=1711s (accessed on 26 April 2022)
73. Lorenz, E. The predictability of hydrodynamic flow. *Trans. N. Y. Acad. Sci.* **1963**, *25*, 409–432. [CrossRef]
74. Schuster, H.G.; Just, W. *Deterministic Chaos: An Introduction*, 4th ed.; John Wiley & Sons, Inc.: Weinheim, Germany, 2005; p. 287. [CrossRef]

Article

Atmospheric Instability and Its Associated Oscillations in the Tropics

Xiping Zeng

DEVCOM Army Research Laboratory, Adelphi, MD 20783, USA; xiping.zeng.civ@army.mil

Abstract: The interaction between tropical clouds and radiation is studied in the context of the weak temperature gradient approximation, using very low order systems (e.g., a two-column two-layer model) as a zeroth-order approximation. Its criteria for the instability are derived in the systems. Owing to the connection between the instability (unstable fixed point) and the oscillation (limit cycle) in physics (phase) space, the systems suggest that the instability of tropical clouds and radiation leads to the atmospheric oscillations with distinct timescales observed. That is, the instability of the boundary layer quasi-equilibrium leads to the quasi-two-day oscillation, the instability of the radiative convective equilibrium leads to the Madden–Julian oscillation (MJO), and the instability of the radiative convective flux equilibrium leads to the El Niño–southern oscillation. In addition, a linear model as a first-order approximation is introduced to reveal the zonal asymmetry of the atmospheric response to a standing convective/radiative heating oscillation. Its asymmetric resonance conditions explain why a standing ~45-day oscillation in the systems brings about a planetary-scale eastward travelling vertical circulation like the MJO. The systems, despite of their simplicity, replicate the oscillations with the distinct timescales observed, providing a novel cloud parameterization for weather and climate models. Their instability criteria further suggests that the models can successfully predict the oscillations if they properly represent cirrus clouds and convective downdrafts in the tropics.

Keywords: tropical clouds; radiation; instability; oscillations

Citation: Zeng, X. Atmospheric Instability and Its Associated Oscillations in the Tropics. *Atmosphere* **2023**, *14*, 433. https://doi.org/10.3390/atmos14030433

Academic Editor: Jimy Dudhia

Received: 10 January 2023
Revised: 13 February 2023
Accepted: 16 February 2023
Published: 21 February 2023

1. Introduction

The views of atmospheric predictability changed from Laplace's determinism to the deterministic chaos of Lorenz (1963), and recently to the coexistence of determinism and chaos (Shen et al. 2021; see Shen et al. 2022 for review) [1–3]. Since the previous studies focused on the Bénard convection and the geostrophic baroclinic model in the middle latitudes (Lorenz 1963, 1969, 1990) [1,4,5], the present study focuses on tropical convection, the engine of atmospheric circulations (Riehl and Malkus 1958) [6].

Lorenz (1963) used a nonlinear system of three differential equations to study the predictability of dry convection with a timescale of minutes [1], referring to the predictability of weather and climate change. Since the Lorenz system included a convective instability, its prognostic variables changed periodically and/or non-periodically. Once the system had no instability, its prognostic variables eventually became steady, which exhibited *the importance of the convective instability in the predictability*.

In this paper, simple systems like the Lorenz system were constructed to address the predictability of weather and climate change in the tropics. The systems incorporated two key components of weather and climate change: moisture (or clouds), which involves the release of potential energy and radiation, which controls the generation of potential energy and is affected by the clouds. In addition, the systems use the instability of tropical clouds and large-scale circulation (Raymond and Zeng 2000) in place of the convective instability in the Lorenz system, where the instability of Raymond and Zeng (2000) has a much longer timescale than the convective instability (see Section 2) [7].

In the proposed systems, the instability was usually connected to oscillations, where one of them is illustrated as

$$\frac{dX_2}{d\tau} = -70X_2 - 35(X_2 - 6.3)(X_2 + 1)(X_2 - Y_2) \text{ and} \tag{1a}$$

$$\frac{dY_2}{d\tau} = 140(X_2 + 1)(X_2 - Y_2), \tag{1b}$$

where X_2 and Y_2 represent the dimensionless air surface entropy (or energy) in one region and the air saturation moist entropy (or temperature) at an altitude of 2 km in another neighboring region, respectively. The system of Equation (1) describes the charge and discharge of entropy in the planetary boundary layer (PBL) (see Appendix C for details). It has an unstable fixed point of $X_2 = Y_2 = 0$ in the phase space (X_2, Y_2), which corresponds to the instability of the boundary layer quasi-equilibrium. It also has a limit cycle that corresponds to a quasi-two-day oscillation (see Figure 1). Once its coefficients are changed so that the fixed point becomes stable, its limit cycle expires. This connection between the unstable fixed point and the limit cycle is generalized to all the continuous dynamical systems on the plane by the Poincaré—Bendixson theorem [8], suggesting that the quasi-two-day oscillation is connected to the instability of the boundary layer quasi-equilibrium.

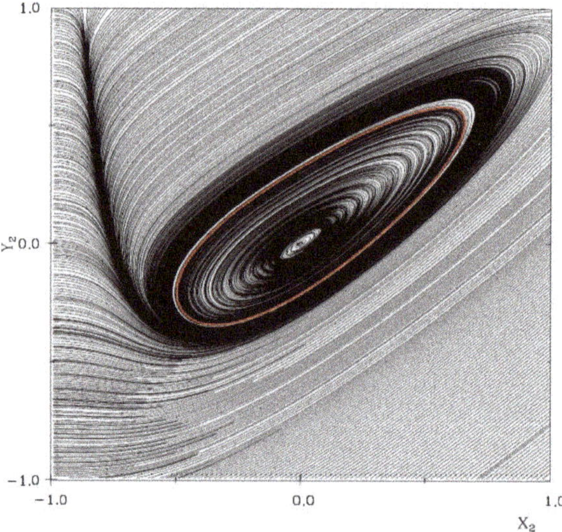

Figure 1. (black line) Streamlines of Equation (1) in the phase space (X_2, Y_2). The fixed point $(0, 0)$, which corresponds to the boundary layer quasi-equilibrium, is unstable. The red line represents a limit cycle that corresponds to a quasi-two-day oscillation in the tropics.

The instability of the boundary layer quasi-equilibrium in the system functions similarly to the convective instability in the Lorenz system, and, therefore, is key to understand the quasi-two-day oscillation. In this paper, the connection between the instability and the oscillation is extended to other two cases: the Madden–Julian oscillation (MJO) and the El Niño–southern oscillation (ENSO).

The paper is organized as follows. Section 2 uses the moist entropy to distinguish two opposite views on convective clouds and large-scale circulations. Section 3 illustrates the instability of the radiative convective equilibrium in the context of the weak temperature gradient approximation. Section 4 introduces a two-column two-layer model of the interaction between tropical clouds and large-scale circulations. Section 5 extends the model to the other two cases with distinct timescales observed. Section 6 concludes the paper.

2. Are Convective Clouds Active or Passive?

There are two opposite views on atmospheric convection in the tropics. Deep convective clouds, as an active factor, penetrate through the level of the minimum moist entropy into the upper troposphere (UT) as hot towers and subsequently drive large-scale vertical circulations (e.g., the Hadley circulation, hurricanes) (Riehl and Malkus 1958) [6]. On the other hand, convective clouds occur as a passive factor to balance atmospheric radiation losses and consequently their spectrum is controlled by radiative cooling (Manabe and Strickler 1964) [9]. In this section, the two opposite views are distinguished via a series of ideal experiments on the radiative convective equilibrium (RCE), beginning with the expression of the moist entropy.

2.1. Expression of Moist Entropy

Consider an air parcel with water vapor, liquid drops and ice particles. Its moist entropy per unit mass of dry air equals the sum of the entropy of dry air and the entropies of vapor, liquid and solid water. The moist entropy thus is expressed as [10]

$$s = (C_p + c_l q_t) \ln \frac{T}{T_{ref}} - R_d \ln \frac{p-e}{p_{ref}} + \left[\frac{L_v(T)}{T} - R_v \ln f \right] q_v - \left[\frac{L_f(T)}{T} - R_v \ln \frac{E_{sw}(T)}{E_{si}(T)} \right] q_i \qquad (2)$$

where T is the air temperature, p is the total pressure of the moist air, e is the partial pressure of the water vapor, q_t is the total mixing ratio of the airborne water (i.e., $q_t = q_v + q_c + q_i$), $q_v/q_c/q_i$ is the mixing ratio of the water vapor/cloud water/cloud ice, L_v/L_f is the latent heat of the vaporization/freezing, C_p/c_l is the specific heat of the dry air/liquid water, R_d/R_v is the gas constant of the dry air/water vapor, E_{sw}/E_{si} is the saturation vapor pressure over water/ice and $f = e/E_{sw}$ is the relative humidity of the air. In addition, $T_{ref} = 273.15$ K and $p_{ref} = 10^5$ pa are the reference temperature and pressure, respectively.

When an air parcel ascends or descends adiabatically, its moist entropy changes with time t as

$$ds/dt = Q_s, \qquad (3)$$

where Q_s represents the increase in entropy from the irreversible processes (e.g., the heat flow from high to low temperature, liquid drop evaporation and ice particle melting; see [10] for its expression). When $T_{ref} = 273.15$ K (i.e., at the triple point of pure water), Q_s is small. Hence $Q_s \approx 0$ (or the conservation of moist entry) is usually assumed in a conceptual model.

The moist entropy is related to the equivalent potential temperature θ_{se} by $s = C_p \ln(\theta_{se}/T_{ref})$ [11]. It functions similar to the moist static energy (or $C_p T + gz + L_v q_v$ where z is the height and g is the acceleration due to gravity), although the moist static energy is not conserved in the convective regions with a strong vertical velocity. In addition, Equation (3) can be used to simulate an air parcel with the same accuracy as the energy equation represented in terms of the (potential) temperature [12].

The moist entropy yields conceptual models of atmospheric convection [6]. In contrast, other variables (e.g., temperature or its variation) often yield ill-posed models by artificially separating the water and energy cycles in the atmosphere. One ill-posed model, for example, is the convective instability of the second kind (CISK) [13], which is reasoned as follows. Consider a horizontally uniform atmosphere with convective available potential energy (CAPE). No matter how large the low-level horizontal convergence is and the strength of the convective clouds it induces, *the mid-tropospheric temperature never surpasses the temperature of an air parcel that moves adiabatically from the surface to the middle troposphere.* In other words, the mid-tropospheric temperature has an upper limit, which is contrary to the prediction of the unlimited mid-tropospheric temperature in the CISK [14,15]. Since the moist entropy does not need an artificial separation of the water and energy cycles, it cannot yield an ill-posed model of moist convection like the CISK.

2.2. Approach of the RCE

Consider a tropical atmosphere subject to radiative cooling over the oceanic equator. Since the radiative cooling brings about the CAPE, convection occurs and transports the energy upward, leading to a balance between the radiative cooling and the convective heating, which is referred to as the RCE [16–19].

The RCE has three forcing factors: the radiative cooling rate, air surface wind speed and sea surface temperature (SST). If the forcing factors are fixed, coupling does not occur between the convection, atmospheric radiation and surface wind in the atmosphere, providing a simple case to determine the timescales in the approach of the RCE.

Suppose that the radiative cooling rate is 1 °C/day below the height z = 9 km and decreases linearly with height to zero at z = 15 km. In addition, the surface wind speed w_s = 4 m s^{-1} and the SST = 302 K are set, which are used for determining the sensible heat flux from the sea to the air and the latent heat flux due to sea surface evaporation.

The RCE is simulated with a cloud resolving model (CRM), using the periodic lateral boundary conditions and the vertical velocity w = 0 at the sea surface [19]. Figure 2 displays the time series of the three modeled variables: the flux of entropy from the sea to the air due to sea surface evaporation and the sensible heat transfer from the sea to the air, the flux of entropy from the air to space due to the radiative cooling, and the internal generation of entropy due to cloud microphysics (e.g., rainwater evaporation) and the heat transfer from high to low temperature (or mixing). In the figure, the surface entropy flux from the sea to the air approaches the top entropy flux from the air to space, arriving at the RCE eventually, where the internal entropy generation is negligible.

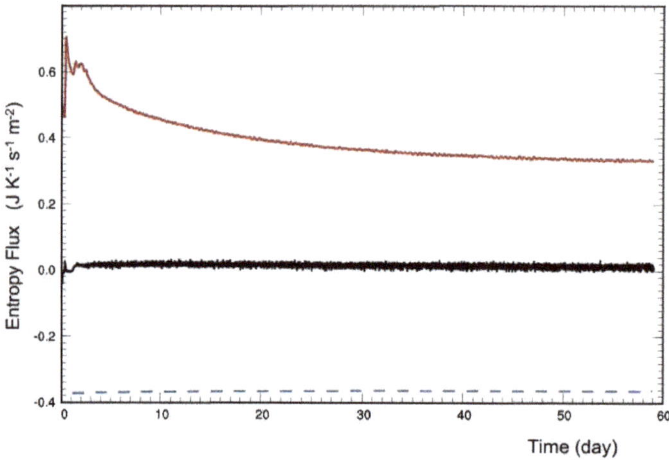

Figure 2. Time series of three entropy fluxes in the approach of the RCE: (red) the flux of entropy from the sea to air, (blue) the flux of entropy from the air to space due to radiative cooling and (black) the total internal generation of entropy due to mixing and rainwater evaporation.

Figure 3 displays the time series of the air temperature and the relative humidity at z = 0 and 4.5 km, showing two timescales of the temperature in the approach of the RCE. The short timescale is ~2 days and is associated with the timescale of the relative humidity. Hence, the short timescale measures the approach of the water quasi-balance between the sea surface evaporation and precipitation (figure omitted). In contrast, the long timescale measures the approach of the RCE while the atmosphere remains at the water quasi-balance.

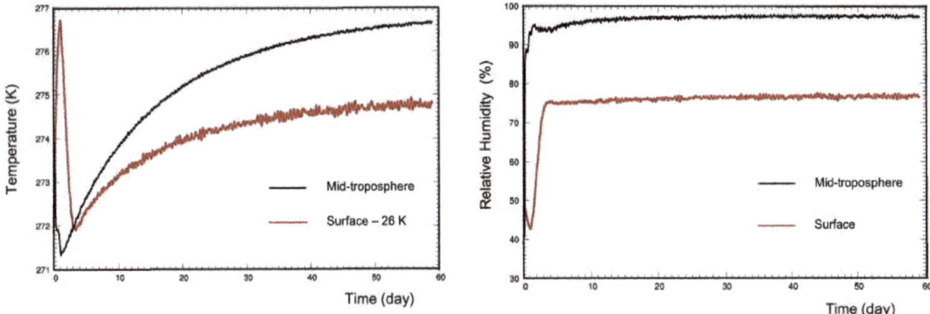

Figure 3. Time series of (**left**) the horizontally averaged air temperature and (**right**) the relative humidity near (red) the sea surface and (black) at a 4.5 km altitude. The air surface temperature is displayed with its value minus 26 K.

The long timescale of the temperature τ_t= 29.2 days, obtained by fitting the mid-tropospheric temperature after day four in Figure 3 with $\exp(-2t/\tau_t)$, is explained with the budget equation of the tropospheric entropy. Consider a small perturbation of the troposphere from the RCE whose deviation of the moist entropy of the air at $z = 0$ is denoted as Δs_s. Thus, $\mu(p_s - p_t)g^{-1}\Delta s_s$ represents the deviation of the total entropy of the troposphere, where p_s and p_t are the pressure at the sea surface and tropopause, respectively; and μ is the vertically averaged deviation of the moist entropy relative to that at $z = 0$. After overlooking the internal generation of entropy, the budget equation of the moist entropy of the troposphere becomes

$$\frac{\mu(p_s - p_t)d\Delta s_s}{gdt} = \rho_s c_E w_s (s_{sea} - s_{s0} - \Delta s_s) - F_R, \tag{4}$$

where the first and second terms on the right-hand side represent the surface and top entropy fluxes, respectively; ρ_s is the air density near the sea surface, $c_E = 1.1 \times 10^{-3}$, w_s is the surface wind speed, s_{sea} is the entropy of the sea surface water as a function of the SST, s_{s0} is the surface moist entropy at the RCE; and F_R is the entropy flux from the air to space due to radiative cooling.

Since the top and surface entropy fluxes are balanced at the RCE or $\rho_s c_E w_s (s_{sea} - s_{s0}) = F_R$, Equation (4) is solved with $\Delta s_s \propto \exp(-2t/\tau_t)$, where the following timescale

$$\tau_t = \frac{2\mu(p_s - p_t)}{\rho_s c_E g w_s} \tag{5}$$

shows that τ_t is inversely proportional to w_s.

The CRM simulations support the inverse relationship between τ_t and the surface wind speed [19]. Figure 4 displays the modeled τ_t against the dimensionless number

$$N_w = \rho_s c_E L_v w_s / \overline{Q_R}, \tag{6}$$

where $\overline{Q_R}$ is the total loss of energy of the troposphere due to radiative cooling. The figure also displays the theoretical timescale estimated with Equation (5) and $\mu = 3.6$ for comparison. Generally speaking, Equation (5) agrees with the CRM results except at some low values of N_w ~100 where the initial status is so close to the RCE that the long timescale in the approach of the RCE is not clear and, consequently, its estimation is "polluted" by the short timescale.

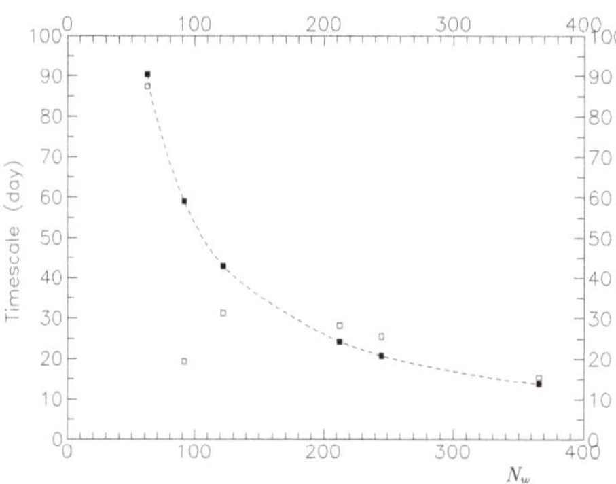

Figure 4. Long timescale versus the dimensionless number N_w when the radiative cooling rate is 1 °C/day. Symbol $\square$ represents the CRM simulation results; symbol $\blacksquare$ and the dashed line represent the analytic result from Equation (5).

In addition, the modeled τ_t decreases with an increase in the radiative cooling rate, which is attributed to the sensitivity of μ to the radiative cooling rate. If the CAPE = 0 were maintained strictly, $\mu = 1$. However, convection usually lags the generation of the CAPE, bringing about $\mu \neq 1$. When the radiative cooling rate varies with a very short forcing period, for example, $\mu \gg 1$ because the surface entropy does not have sufficient time to respond to the cooling of the troposphere. The CRM simulations show that τ_t at the radiative cooling rate of 2 °C/day is about one half of that at the radiative cooling rate 1 °C/day [19]. Hence, $\mu = 1.8$ at the radiative cooling rate of 2 °C/day is obtained that is about one half of $\mu = 3.6$ at the radiative cooling rate 1 °C/day.

2.3. Ideal Experiments of the RCE

To show whether the convective clouds are active or passive, two ideal experiments of the RCE were designed with prescribed forcing. The first experiment used a fixed radiative cooling rate of

$$w_s = w_{s0}[1 + A\sin(\omega t)], \tag{7}$$

where w_{s0} is the mean surface wind speed, A is the dimensionless amplitude and ω is the angular frequency of the forcing. Substituting Equation (7) for Equation (4) and then solving the resulting equation yields

$$\Delta s_s = \frac{A\tau_t}{4 + (\omega\tau_t)^2} \frac{gF_R}{\mu(p_s - p_t)}[2\sin(\omega t) - \omega\tau_t\cos(\omega t)] \tag{8}$$

and the surface entropy flux with the aid of Equation (5).

Figure 5 displays the amplitudes of the air surface temperature and the surface energy flux from Equation (8) against the forcing period of the surface wind. For the forcing period $2\pi/\omega \ll \tau_t$, Equation (8) degenerates into the term of $\omega\tau_t\cos(\omega t)$, which is referred to here as the active mode. In this mode, w_s leads to a sine variation of the surface entropy flux and a cosine variation of the surface moist entropy (or the CAPE). The convective clouds then distribute the cosine variation of the surface moist entropy into the mid-troposphere, bringing about a cosine variation of the mid-tropospheric moist entropy (or temperature). Hence, *the convective clouds are active, leading to the variation of the mid-tropospheric moist entropy (or temperature).*

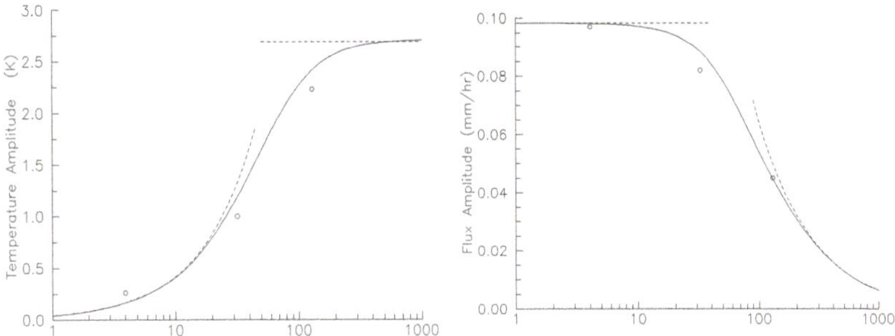

Forcing period of the surface wind (day)

Figure 5. Amplitude of (**left**) the air surface temperature and (**right**) the surface energy flux against the forcing period of the surface wind $2\pi/\omega$, where the energy flux takes the unit of the equivalent water evaporation. The circles denote the CRM results, and the solid lines denote the analytic results of Equation (8). The dashed lines over small period denote the active mode where the difference between the sea surface entropy and the air surface entropy is fixed; the dashed lines over long periods signify the passive mode where the surface entropy flux is fixed.

For $2\pi/\omega \gg \tau_t$, Equation (8) degenerates into the term of $2\sin(\omega t)$, which is referred to as the passive mode. In this mode, the atmosphere is always close to the RCE and, thus, the surface entropy flux is controlled by the fixed radiative cooling rate. Since the surface entropy flux or $w_s(s_{sea} - s_s)$ is constant, $s_{sea} - s_s$ is inversely proportional to w_s. As a result, the surface entropy s_s changes with w_s but the surface entropy flux does not, indicating that *the convective clouds are passive and controlled by the radiative cooling rate.*

As a general case, *Equation (8) possesses both the active and passive modes.* When the forcing period $2\pi/\omega$ is close to τ_t or

$$2\pi/\omega = \pi\tau_t, \tag{9}$$

the two terms of $\omega\tau_t\cos(\omega t)$ and $2\sin(\omega t)$ in Equation (8) have the same amplitude, indicating the active and passive modes have the same importance. Their relative importance is measured by the phase delay of the variation in the mid-tropospheric entropy (or temperature) with respect to the variation in the forcing.

To test the analytic solution of Equation (8), three CRM simulations were carried out [19]. The simulations used the forcing period $2\pi/\omega = 4$, 32 and 128 days in Equation (7), respectively. In addition, the simulations set $w_{s0} = 6$ m s^{-1}, $A = 1/3$, and the radiative cooling rate at 2 K/day. Their results are displayed in Figure 5, supporting the analytic solution of Equation (8).

The second ideal experiment, in contrast to the first one, was carried out to test the response of the convective clouds to the radiative cooling. The experiment used a fixed surface wind speed w_s, but the radiative cooling rate was

$$Q_r = Q_{r0}[1 + A\sin(\omega t)], \tag{10}$$

where $Q_{r0}(z)$ equals 1.5 °C/day below the height $z = 9$ km and decreases linearly with the height to zero at $z = 15$ km. Substituting Equation (10) for Equation (4) yields

$$\Delta s_s = \frac{A\tau_t}{4 + (\omega\tau_t)^2}\frac{gF_{R0}}{\mu(p_s - p_t)}[-2\sin(\omega t) + \omega\tau_t\cos(\omega t)] \tag{11}$$

and the surface entropy flux, where F_{R0} denotes the entropy flux from the air to space with the radiative cooling rate Q_{r0}.

The amplitudes of the air surface temperature and the surface energy flux were computed using Equation (11). They were displayed against the forcing period of the radiative cooling in Figure 6. To support the analytic solution of Equation (11), three CRM simulations were carried out with the forcing period $2\pi/\omega$ = 4, 32 and 128 days in (10), respectively [19]. In addition, the CRM simulations used w_s = 4 m s^{-1} and A = 1/3. Their results are displayed in Figure 6 for comparison, supporting the analytic solution of Equation (11).

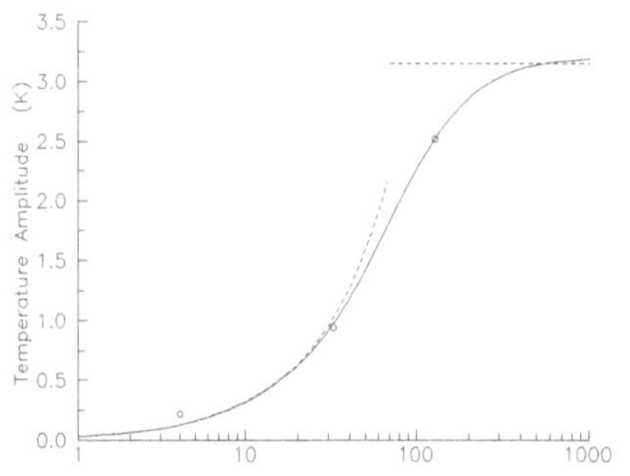
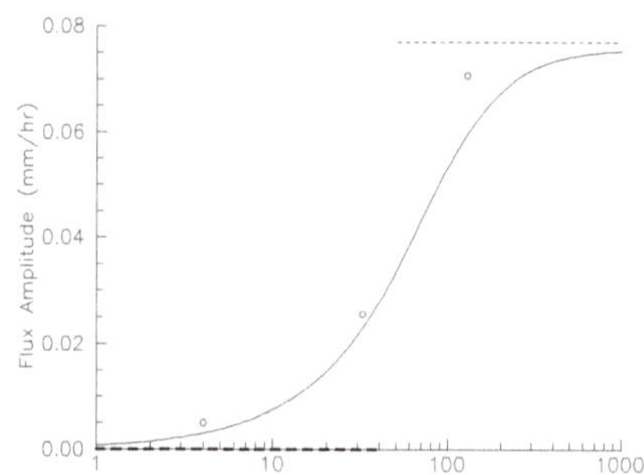

Forcing period of the radiative cooling (day)

Figure 6. Same as Figure 5 except against the forcing period of the radiative cooling. The dashed lines over the small period denote the active mode where the top entropy flux has no effect on the surface entropy flux; the dashed lines over the long period denote the passive mode where the top entropy flux controls the surface entropy flux via convective clouds.

The second experiment supported the first one regarding the separation between the active and passive modes (see Figures 5 and 6). When the forcing period was long (or $2\pi/\omega \gg \tau_t$), *the top entropy flux determined the surface entropy flux whether the surface wind w_s changed or not,* indicating the radiative cooling rate controlled the convective clouds. In contrast, when the forcing period was short (or $2\pi/\omega \ll \tau_t$), the top entropy flux had almost no effect on the surface entropy flux.

In summary, the active and passive modes were separated at the forcing period of the surface wind and radiative cooling at $2\pi/\omega = \pi\tau_t \approx 45$ days (if τ_t = 14 days; see Figure 4 for the variation of τ_t), which corresponds to the period of the MJO [20–22]. Roughly speaking, the convective clouds were active in the weather systems with a timescale shorter than ~ 45 days, they are passive in the climate systems with a timescale longer than ~45 days and they possessed dual roles in the systems with a timescale of ~45 days (e.g., the MJO).

3. Instability of the RCE

The RCE in a wide region is unstable once its clouds are organized via a large-scale vertical circulation into two sub-regions: one with deep convective clouds and the other with shallow (or no) clouds (Raymond and Zeng 2000; Sessions et al. 2010) [7,23], which explains the strong Hadley circulation observed (Raymond 2000) [24]. In this section, the instability of the RCE is discussed with its key processes, beginning with its foundation in the weak temperature gradient (WTG) approximation (Bretherton and Smolarkiewicz 1989; Sobel and Bretherton 2000; Bretherton and Sobel 2002) [25–27].

3.1. The WTG and Its Application

The WTG is the foundation for understanding the interaction between tropical convection and large-scale vertical circulations (Nilsson and Emanuel 1999; Raymond and Zeng 2000; Sobel and Bretherton 2000) [7,26,28]. In a three-dimensional stratified atmosphere with a large Rossby number (e.g., in the tropics), gravity waves damp the temperature perturbations excited by clouds so that the perturbations disappear eventually, which is referred to as buoyancy damping [19]. On the other hand, the imbalance of the convective heating and radiative cooling generates a forced vertical circulation. Since the horizontal scale of convective clouds is small compared to the Rossby radius of deformation, the amplitude of the forced circulation and the horizontal temperature gradient in the mid-troposphere are small, which is referred to as the WTG [26,27,29]. In brief, buoyancy damping and the WTG represent the process and result of gravity waves, respectively. They are analogous to the damping of the surface waves and the little slope of the pond surface, respectively, when pebbles are tossed into a pond (see Zeng et al. 2020 for the connection between the WTG and buoyancy damping [30]).

Consider two adjacent regions: one with convective clouds and the other with a clear sky. Let $T_e(z)$ denote the vertical profile of the temperature in the clear sky (or environment) region and let $\Delta T_e(z)$ denote the difference in the temperature between the convective region and the clear sky region. Obviously, $\Delta T_e(z)$ is caused by the imbalance between the convective heating and radiative cooling in the convective region. The analytic model of gravity waves showed that $\Delta T_e(z)$ was inversely (directly) proportional to the gravity wave phase speed (the heating intensity and the distance between the two regions) [19,31]. Since many gravity waves with different vertical wavelengths (or phase speeds) work at the same time, their ensemble effect on $\Delta T_e(z)$ can be obtained. Generally speaking, *$\Delta T_e(z)$ is small in the mid-troposphere, working as the WTG. In addition, $\Delta T_e(z)$ is relatively large and thus the WTG is not suitable in the PBL, indicating the PBL can maintain its local characteristics. Additionally, $\Delta T_e(z)$ is large in the UT, too,* because the tropopause with a strong static stability and the sea surface trap internal gravity waves in the troposphere as two "rigid" boundaries.

3.2. Origin of the Instability

Consider the individual clouds in the convective region. Let $\Delta T_c^{(i)}(z)$ denote the temperature of a cloud marked with superscript i. Since $\Delta T_e(z)$ represents the difference in the temperature between the two regions, the temperature of cloud i at the mid-tropospheric height z_m is expressed as

$$T_c^{(i)}(z_m) = \Delta T_c^{(i)}(z_m) + \Delta T_e(z_m) + T_e(z_m). \tag{12}$$

Replacing $\Delta T_c^{(i)}(z)$ approximately with its ensemble average $\Delta \overline{T}_c(z)$ yields

$$T_c^{(i)}(z_m) = \Delta \overline{T}_c(z_m) + \Delta T_e(z_m) + T_e(z_m), \tag{13}$$

where the left-hand side has superscript i but the right-hand side does not.

Equation (13) has two different applications. (a) Given the surface entropy s_s, an air parcel near the sea surface ascends adiabatically to the mid-troposphere. Thus, $\Delta T_c^{(i)}(z_m)$ is determined using the entropy conservation and Equation (12). (b) Since the surface entropy s_s below a cloud is changed by the local factors of the cloud itself as well as its neighboring clouds (e.g., convective downdrafts, surface wind gusts), the surface entropy s_s cannot be known at first. Instead, given the right-hand side of Equation (13), s_s is known using the entropy conservation or

$$s_s = s^*\left(\Delta \overline{T}_c(z_m) + \Delta T_e(z_m) + T_e(z_m), p_m\right), \tag{14}$$

where p_m denotes the pressure at the mid-troposphere and $s^*(T, p)$ is the saturation moist entropy of the air with temperature T and pressure p (or the entropy of the air that is saturated with respect to water at temperature T and pressure p).

Equation (14) resembles the model of Emanuel et al. (1994) in that the vertical temperature profile itself is controlled by convection and tied directly to the surface entropy [14]. The model of Emanuel et al. (1994) is based on the idea that convection is nearly in statistical equilibrium with its environment or $\Delta T_c^{(i)}(z_m) \approx 0$. After taking the WTG (i.e., $\Delta T_e(z_m) \approx 0$) and $\Delta T_c^{(i)}(z_m) \approx 0$, Equation (14) is approximated as a concept or

$$s_s \approx s^*(T_e(z_m), p_m). \tag{15}$$

That is, *the surface entropy in the convective region is determined by the mid-tropospheric temperature in the clear sky region.*

Other models of the convective atmosphere in a tropical β channel complement Equation (15) with higher-order accuracy. The first model with Equation (15) in Section 4, as the zeroth-order approximation, produces a small variation of the imbalance between convective heating and radiative cooling and, thus, a small variation of $T_e(z_m)$ in a region. The second model in Appendix D, as the first-order approximation, represents a global response (i.e., vertical circulations) to the small thermal forcing in the first model. The third model, as the second-order approximation, represents the feedback between the forced vertical circulations and convective clouds especially those beyond the region of the first model.

If a large-scale vertical circulation was not present between the two regions, the atmosphere in the convective region would stay at the RCE with the surface entropy s_{s0}. Since s_{s0} is determined by the convective downdrafts, SST, surface wind and other local factors (see Section 2),

$$s_{s0} \neq s^*\left(\Delta \overline{T}_c(z_m) + \Delta T_e(z_m) + T_e(z_m), p_m\right) \tag{16}$$

because the right-hand side is determined by a non-local factor, the mid-tropospheric temperature in the clear sky region. The inequality (16) leads to the instability of the RCE. Specifically, *the large-scale vertical circulation rises to modulate deep convective clouds that, in turn, change s_s until s_s satisfies Equation (16).* In other words, the instability exhibits a movement of s_s from s_{s0} to $s^*(T_e(z_m), p_m)$ with a growing large-scale vertical circulation.

3.3. Convective Downdrafts

Convective downdrafts embody the movement of s_s from s_{s0} to $s^*(T_e(z_m), p_m)$ by redistributing entropy vertically. Figure 7 displays the vertical profile of the buoyancy of an air parcel that ascends adiabatically from the surface to the UT, where the buoyancy is directly proportional to the difference in entropy between the surface entropy (green/black line) and the saturation moist entropy of the environmental air (red line). During their ascent from the surface to the UT, the air parcels detrain into the UT and then mix with their environment. Since the entropy of the mixed parcels falls between the blue and black lines, the cloud detrainment increases the UT entropy. If some mixed parcels have an entropy between the blue and red lines, their buoyancy becomes negative, forming convective downdrafts. Once the lower tropospheric downdrafts penetrate into the PBL, they bring down the mid-tropospheric air with low entropy into the PBL and, consequently, decrease the surface entropy.

Observations show that convective downdrafts are common in the tropics [32–35]. Figure 8, for example, displays the vertical distribution of downdrafts in a mature mesoscale convective system (MCS) observed by airborne nadir pointing Doppler radar during the Tropical Composition, Cloud and Climate Coupling (TC4) Experiment [36]. The figure displays the vertical cross sections of the observed radar reflectivity and Doppler velocity using the ER-2 Doppler radar (EDOP) on the NASA high-altitude (~20 km) ER-2 aircraft [37,38]. After taking the fall speed of the hydrometeors (ranging from 2.5 to 10 m s^{-1} with a reflectivity from 0 to 50 dBZ for the drops; 1.2 to 1.8 m s^{-1} with a reflectivity increasing from −10 to 30 dBZ for snow and 1.8 to 5.4 m s^{-1} with a reflectivity from 30 to 45 dBZ for graupel) off the Doppler velocity [37], the figure clearly shows that *convective downdrafts, especially those in the lower troposphere, are quite common.*

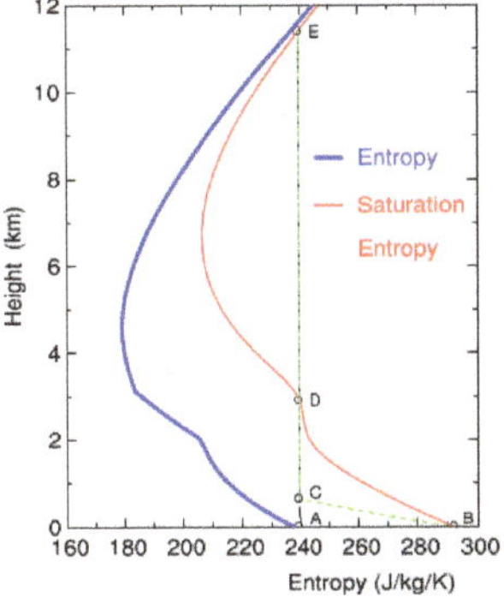

Figure 7. Diagrammatical analysis of the buoyancy of an air parcel that ascends adiabatically from the surface. The blue and red solid lines represent the vertical profiles of the mean moist entropy and the saturation moist entropy in the tropics, respectively. If a surface air parcel ascends adiabatically, its moist entropy and saturation moist entropy follow the black dashed line and the green dashed line, respectively, from the surface level A/B to the lifted condensation level C, the level of free convection D and finally to the cloud top E. Its buoyancy is directly proportional to the difference in entropy between the green line B–C–E and the red line.

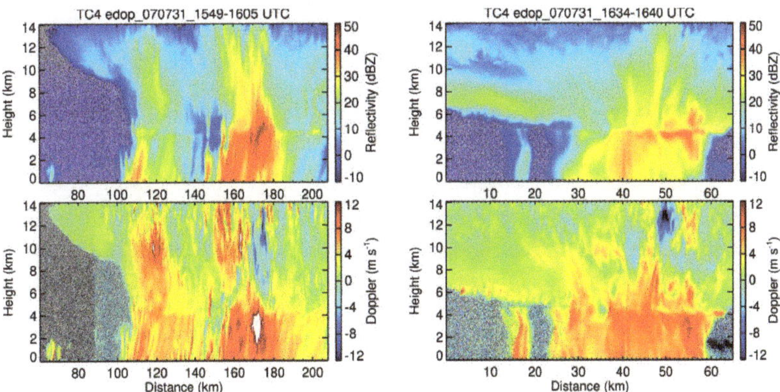

Figure 8. Vertical cross sections of (**top**) the radar reflectivity and (**bottom**) the Doppler velocity of an MCS observed on 31 July 2007 over the Pacific Ocean south of Costa Rica during the TC4 experiment along two EDOP flight tracks. In the bottom panes, red (or positive) and blue (or negative) represent the downward and upward motions, respectively; the white areas surrounded by the red ones indicate the Doppler folding (or the Doppler velocity larger than 34 m s^{-1}). (**Left**) The flight track over 1549–1605 UTC is parallel to the leading edge of the MCS, and (**right**) the track over 1634–1640 UTC is perpendicular to the leading edge of the MCS and right across the center of the MCS.

The movement of s_s from s_{s0} to $s^*(T_e(z_m), p_m)$ is affected indirectly by a large-scale vertical circulation via deep convective clouds and their downdrafts. To be specific, the large-scale ascent (descent) raises (lowers) the top of deep convective clouds, which is discussed with the aid of Figure 7. A UT large-scale descent brings about an increase in entropy by vertical advection and an increase in the temperature due to the air compression in the UT. As a result, the descent leads to a shift of the blue and red lines in the UT to the right in Figure 7. Since the descent has no impact on the surface entropy (i.e., no impact on the green line in Figure 7), the descent lowers the top of the deep convective clouds (or the height with zero buoyancy). If the descent continues, the deep convective clouds decrease in population and, eventually, shallow ones take their place, which are simulated by the CRM simulations of Sessions et al. (2010) [23]. Similarly, the UT large-scale ascent decreases the UT entropy by vertical advection and, thus, leads to a shift of the blue and red lines in the UT to the left in Figure 7, raising the top of the convective clouds. In short, the large-scale ascent/descent affects the deep convective clouds that in turn vertically redistribute entropy, impacting s_s indirectly.

3.4. Cirrus Clouds and Their Radiation

Cloud radiation, as a complement of the aforementioned dynamical factors, maintains the instability by fueling deep convective clouds. Cirrus clouds, including the cloud anvil from the MCSs, emit infrared radiation into space, bringing about a decrease in the UT temperature and entropy [39]. Meanwhile, the cirrus cloud amount is affected by large-scale vertical velocity. In general, the amount is increased (decreased) by the large-scale ascent (descent) [7]. In addition, the amount is changed indirectly by the large-scale ascent via UT convective downdrafts (see Figures 8 and 9) [40–43].

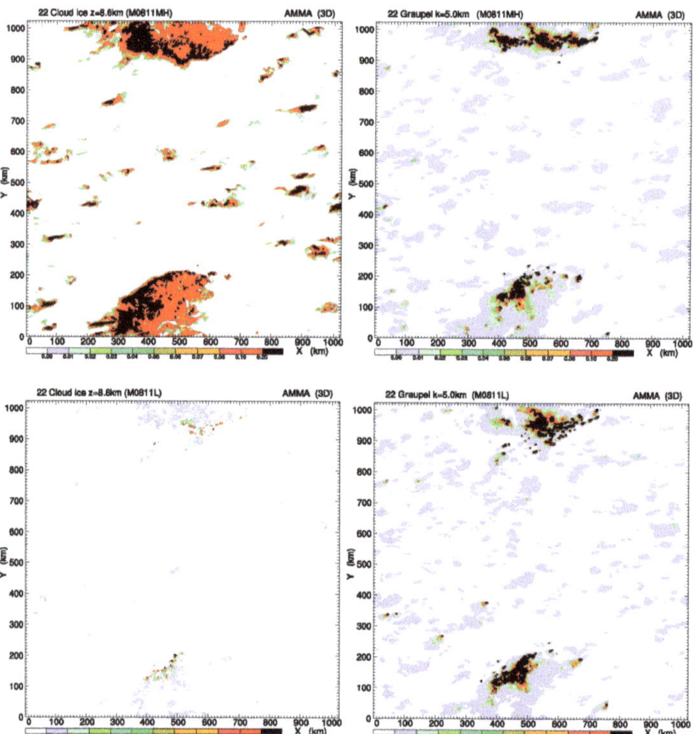

Figure 9. Horizontal distribution of the cloud–ice mixing ratio at 8.6 km (**left**; representing cloud anvil) and the graupel mixing ratio at 5 km (**right**; representing convective cores) at 0400 UTC 11 August 2006. The results come from the two numerical experiments with a (**top**) high and (**bottom**) low number concentration of activated IN. The broad anvil in the experiment with the plentiful activated IN (top/left) agrees with the in-situ satellite and radar observations during the field campaign of the AMMA (the African Monsoon Multidisciplinary Analysis) over West Africa. Adapted with permission from Zeng et al. (2013) [43].

The indirect effect of the large-scale ascent on cirrus clouds via UT downdrafts is discussed as follows. Consider an air parcel with super-cooled drops that undergoes a vertical circulation. When the parcel ascends, its temperature decreases first to activate many ice nuclei (IN), and then ice crystals form on the activated IN [44,45]. With the parcel descending, its super-cooled droplets evaporate first and then a part of the ice crystals disappear via sublimation. The sublimation of the ice crystals leaves new solid particles as residues that can be activated at warmer temperatures than the original IN [46,47]. As a result, the parcel has significantly more activated IN than the same parcel with no vertical circulation. In short, *the recycling of IN brings about plentiful ice crystals (or activated IN)*. Roughly speaking, the number of the activated IN depends on the descent distance of convective downdrafts; it increases by 10 times when the descent distance increases by ~0.5 km [41–43].

The downdraft-induced activated IN lead to an increase in the ice crystal number, then a decrease in the precipitation efficiency and eventually an increase in cloud anvils (or cirrus clouds) in the UT [41,43]. This process is shown by two CRM simulations in Figure 9: one with a high number concentration of activated IN and the other with a low number concentration. The broad anvil of the MCSs in the CRM simulation with the plentiful activated IN agrees with the field observations, suggesting that the UT convective downdrafts play an important role in cloud anvils.

In summary, UT convective downdrafts are common in a convective region with a large-scale ascent but rare in a clear sky region with a large-scale descent, indicating that large-scale vertical circulation can significantly change the UT cirrus cloud amount via UT downdrafts. Since the cirrus cloud amount changes the radiative cooling of the UT, it modulates the population of deep convective clouds as feedback.

4. A Two-Column Two-Layer Model

A highly simplified model was constructed to assemble the aforementioned processes. In this section, the model is introduced with its structure first, then the criteria for stable RCE and, finally, a connection between the RCE's instability and a 40-day oscillation of deep convective clouds.

4.1. Model Structure

Consider a two-column two-layer model on the interaction between convective clouds and large-scale vertical circulations in the tropics (see its schematic in Figure 10). The model, similar to Raymond and Zeng (2000) [7], consisted of two adjacent columns: one for testing and the other for reference. For simplicity, the two columns had the same area with width L. They used the same SST and surface wind speed so that the model focused on the interaction between cloud radiation and large-scale vertical circulation. Each column was characterized by two layers: one for the PBL and the other for the UT with cirrus clouds or cloud anvils that detrained from deep convective clouds (i.e., the layer from 7 to 12 km in Figures 7–9). The two layers were bridged by deep convective clouds.

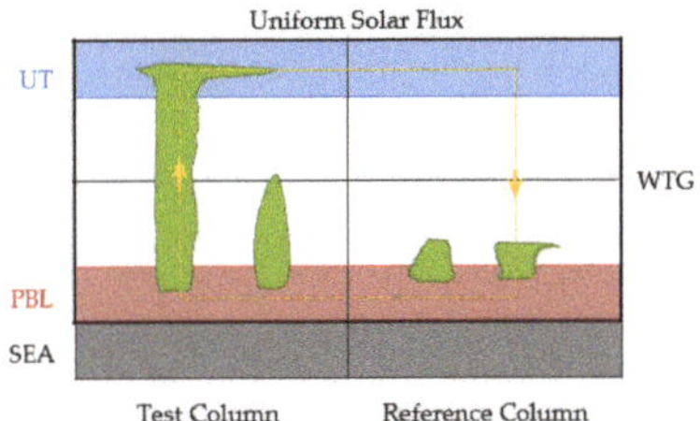

Figure 10. Schematic showing (green area) the deep and shallow convective clouds in two adjacent columns are connected by (golden line) a large-scale vertical circulation. The two columns maintain the WTG in the mid-troposphere, and each column is characterized with the three layers of (blue) the UT, (red) PBL and (gray area) sea surface.

The two columns were connected by a large-scale vertical circulation with a mid-tropospheric vertical velocity w_m (or $-w_m$) in the test (or reference) column. Suppose the two columns stay at the RCE initially with entropy s_{s0} in the PBL, s_{u0} in the UT layer and $w_m = 0$. Once a vertical circulation occurs between them, it induces a perturbation of the surface entropy, denoted by Δs_s (or $-\Delta s_s$) in the test (or reference) column, and a perturbation of the UT entropy, denoted by Δs_u (or $-\Delta s_u$) in the test (or reference) column. Since the perturbation of the total entropy of the two columns was zero, only the entropy budget of the test column is discussed except for specification.

In the test column, w_m is associated with the deep convective clouds. To be specific, w_m is affected by two factors: $\Delta s_s + s_{s0} - s^*(T_m, p_m)$ and $\Delta s_s - \Delta s_u$, where $s^*(T_m, p_m)$ denotes the saturation moist entropy with a mid-tropospheric temperature T_m and a pressure p_m. The first factor, $\Delta s_s + s_{s0} - s^*$, represents the average mid-tropospheric buoyancy of the convective clouds (Figure 7). The second factor, $\Delta s_s - \Delta s_u$, represents the deviation of the average UT buoyancy of the convective clouds. Owing to a low water vapor content in the UT, the UT moist entropy is close to the saturation moist entropy (Figure 7) and, thus, can approximately represent its saturation moist entropy. Hence, $\Delta s_s - \Delta s_u$ approximately represents the deviation of the average UT buoyancy of the convective clouds.).

When $\Delta s_s > \Delta s_u$, for example, more convective clouds reach the UT than at the RCE and, subsequently, $w_m > 0$. Hence, the mid-tropospheric vertical velocity is expressed as

$$w_m = \frac{H}{C_p^2 \tau_c}(\Delta s_s - \Delta s_u)[\Delta s_s + s_{s0} - s^*(T_m, p_m)], \qquad (17)$$

where $H = 12$ km is the depth of the troposphere and τ_c is a timescale to measure the upward transportation of the air mass into the UT induced by the deep convective clouds. The factor of $\Delta s_s + s_{s0} - s^*(T_m, p_m)$ is key for the model to embody the WTG, because $\Delta s_s + s_{s0} - s^*(T_m, p_m) \geq 0$ represents an essential condition that a cumulus cloud becomes a deep convective cloud.

Deep convective clouds decrease the entropy in the PBL via their downdrafts. Some of the downdrafts penetrate into the PBL (Figure 8) and, thus, bring the mid-tropospheric air with the minimum entropy s_{min} into the PBL (Figure 7). As a result, the vertical entropy flux from the lower troposphere to the PBL is proportional to $\Delta s_s + s_{s0} - s_{min}$. Since the deviation of the deep convective clouds from the RCE is proportional to w_m, the deviation of the vertical entropy flux from the lower troposphere to the PBL is also proportional to w_m. Hence, the entropy budget of the PBL is expressed as

$$\frac{d\Delta s_s}{dt} = -\frac{\Delta s_s}{2\tau_t} - U_s\frac{2\Delta s_s}{L} - \frac{k_d w_m}{H}(\Delta s_s + s_{s0} - s_{min}), \qquad (18)$$

where the three terms on the right-hand side, in turn, correspond to the entropy flux from sea, the horizontal entropy flux from the reference column and the vertical entropy flux from the lower troposphere to PBL. U_s represents the horizontal wind speed of environmental air in PBL with respect to the test column. The dimensionless coefficient k_d represents the effect of the downdrafts on the surface entropy.

Deep convective clouds transport the surface air with high entropy into UT and consequently increase the entropy. Hence, the entropy budget of the UT layer is expressed as

$$\frac{d\Delta s_u}{dt} = -U_u\frac{2\Delta s_u}{L} + \frac{k_u w_m}{H}(\Delta s_s + s_{s0} - \Delta s_u - s_{u0}) - \frac{g\sigma T_u^3 \Delta \sigma_c}{p_m - p_t}, \qquad (19)$$

where the three terms on the right-hand side, in turn, correspond to the horizontal entropy flux from the reference column, the upward entropy flux from the PBL and the entropy flux to space via infrared radiation. U_u represents the UT horizontal wind speed of the environmental air with respect to the test column, and the dimensionless coefficient k_u represents the effect of deep convective cores on the UT entropy. σ is the Stefan–Boltzmann constant,

T_u is the UT temperature, p_m/p_t are the pressure of the mid-troposphere/tropopause and $\Delta\sigma_c$ is the deviation of the UT cloud fraction from σ_{c0}, the UT cloud fraction at the RCE.

The UT cloud fraction, $\Delta\sigma_c + \sigma_{c0}$, depends on the detrainment of the deep convective clouds in the UT and the precipitation formation [48]. Suppose $\sigma_{c0} = 0.5$. Since $0 < \Delta\sigma_c + \sigma_{c0} < 1$, $-0.5 < \Delta\sigma_c < 0.5$, hence, the UT cloud fraction is described by

$$\frac{d\Delta\sigma_c}{dt} = k_c\frac{w_m}{H}[1 - (2\Delta\sigma_c)^2] - \frac{\Delta\sigma_c}{\tau_p}, \tag{20}$$

where the two terms on the right-hand side are attributed to the detrainment of deep convective clouds and the sink of cirrus clouds due to precipitation formation, respectively. The dimensionless coefficient k_c represents the effect of the detrainment of deep convective clouds on cirrus cloud fraction, and τ_p is the timescale of the sink of UT cirrus clouds.

4.2. Stability Criteria of the RCE

The two-column two-layer model is simplified into a very low order system of Equations (18)–(20). The system has three prognostic variables: Δs_s, Δs_u and $\Delta\sigma_c$. It is studied in the phase space (Δs_s, Δs_u, $\Delta\sigma_c$) similar to Lorenz (1963) [1]. Obviously, the system has a fixed point of the origin in the phase space that corresponds to the RCE, as shown as

$$\Delta s_s = \Delta s_u = \Delta\sigma_c = 0. \tag{21}$$

The criteria for the stability of the fixed point (or the RCE) are derived in Appendix A. They contained two factors of cloud radiation: the detrainment of deep convective clouds (or k_c) and the timescale of cirrus cloud sink (or τ_p). When the detrainment of deep convective clouds is strong and the cirrus cloud sink is slow, the RCE is unstable. In addition, the criteria also contain the width L of convective region. When L is small (e.g., ~100 km), the RCE is stable; when L is large (e.g., ~3,000 km), then the RCE is unstable. In other words, the criteria predict that the RCE is unstable over a wide convective region where the horizontal exchange in entropy between the columns is negligible.

4.3. Oscillation and an Unstable RCE

A specific case with an unstable RCE is discussed in Appendix A, where $L = 3,000$ km and $\tau_p = 10$ days. Once its initial status was slightly away from the RCE, its status deviated from the RCE permanently and eventually approached a limit cycle (see Figure A1). In addition, its prognostic variables (e.g., surface entropy) oscillated with a period of 40 days (Figure 11). Since the peak of the UT cloud fraction corresponded to the onset of deep convective clouds, Figure 11 shows that the deep convective clouds also oscillated in number with a period of 40 days, providing a prototype of the MJO.

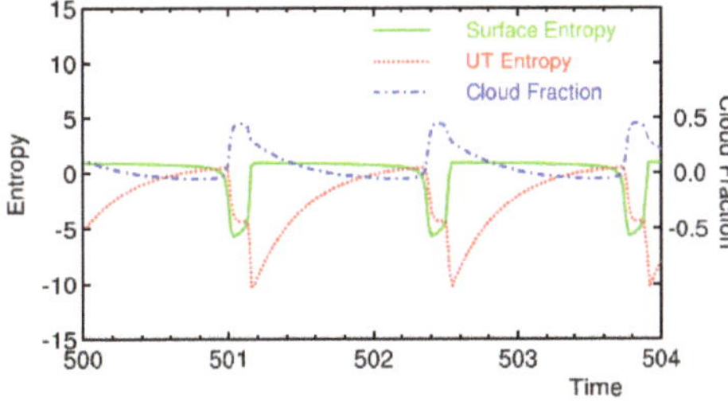

Figure 11. Time series of (green) the dimensionless surface entropy X, (red) the UT entropy Y and (blue line) the UT cloud fraction Z in the system of Equations (18)–(20) (see Appendix A for the definitions and parameters used). The horizontal axis of time is scaled with $2\tau_t = 28$ days (or ~ 1 month) as the unit.

If L is decreased to 100 km, the RCE becomes stable and the oscillation disappears, indicating that the instability of the RCE is an essential condition of the oscillation. If τ_p is decreased from 10 to 5 days while maintaining $L = 3{,}000$ km, the RCE is still unstable, but the oscillation disappears. The system, instead, approaches a steady state where one column shows a large-scale ascent and deep convection and the other shows a large-scale descent and shallow clouds, which resemble the model of Raymond and Zeng (2000) [7]. This sensitivity of the system target to τ_p (or bifurcation in the phase space) indicates that the slow sink of UT cirrus clouds is another essential condition of the oscillation.

Recently, Zeng et al. (2022) showed the possibility of a slow sink of cirrus clouds in the tropics with a new precipitation process—the radiative effect on microphysics [48]. Consider relatively large ice crystals near cloud top. Owing to their radiative cooling, they grow to precipitating particles at the expense of small ones and then fall off their parent clouds. As a result, thick cirrus clouds (detrained from deep convective clouds) quickly become thin with a timescale of hours. Once the thin cirrus clouds form, their ice crystals undergo radiative warming (see Figure 3 of Zeng et al. 2022 [48]) and, consequently, survive with a slow gravitational deposition of ice crystals and/or a slow crystal sublimation in a large-scale descent.

The surface wind, of course, can be incorporated into the two-column model to represent its effects on the instability and the oscillation (Fuchs and Raymond 2005, 2017; Emanuel 2020; Wang and Sobel 2022) [49–52]. It is related to a large-scale vertical circulation via the surface wind gusts driven by cloud systems (see Section 5.1) and large-scale surface wind [50,53–55].

5. Other Forms of Instability

Instability behaves differently when its timescale is much shorter or much longer than ~45 days. In this section, the two-column two-layer model was extended to two cases with a timescale ~2 days and ~2 years, respectively. To minimize its complexity, the model was tuned for each case by adding or deleting processes.

5.1. Boundary Layer Quasi-Equilibrium

The boundary layer quasi-equilibrium was proposed to describe the quasi-balance in the entropy flux between the surface and the PBL top (Emanuel 1995; Raymond 1995) [56,57]. It arrives with a timescale of ~1 day (or the short timescale of the temperature and relative humidity in Figures 2 and 3). This subsection deals with the instability of the quasi-equilibrium in the context of the WTG, beginning with CRM simulations.

Consider a two-column model that represents clouds in each column with a three-dimensional CRM [19]. The two columns take the same forcing factors (e.g, the SST, radiative cooling rate) similar to the experiment in Figures 2 and 3, except for maintaining the WTG via a large-scale vertical circulation between them. Surprisingly, the model replicates a quasi-two-day oscillation, such as those observed in the western Pacific [58–61].

To show how the oscillation works, Figure 12 displays the modeled time series of the surface entropy and the entropy flux from the sea to the air. In the test column with the large-scale ascent, the deep convective clouds spend and, thus, rapidly decrease the surface entropy while the sea slowly supplies entropy to the PBL.

The time series in Figure 12 are explained by the evolution of the horizontally averaged surface wind speed and the entropy flux at the PBL top in Figure 13. In the test column, the deep convective clouds generated strong surface gusts and subsequently increased the entropy flux from the sea to the air, providing energy to maintain the deep convective clouds. On the other side, in the reference column with the large-scale descent, few to no deep convective clouds brought about the small upward entropy flux at the PBL top and subsequently a high surface entropy. Once the surface entropy in the reference column becomes very high, deep convective clouds occur and, subsequently, invert the large-scale circulation between the two columns. Hence, the modeled two-day oscillation can be treated as an alternation of the charge and discharge of entropy in the PBL.

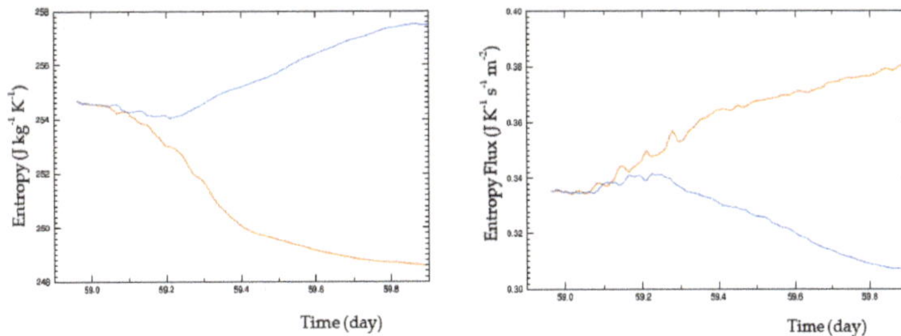

Figure 12. Modeled time series of (**left**) the surface moist entropy and (**right**) the entropy flux from the sea to the air in the (red) test and (blue line) reference columns.

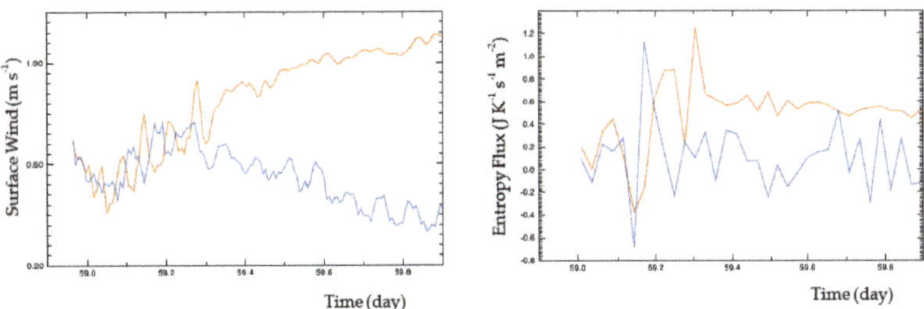

Figure 13. Modeled time series of (**left**) the horizontally averaged surface wind speed and (**right**) the entropy flux at $z = 0.75$ km in the (red) test and (blue line) reference columns.

In the modeled two-day oscillation, the surface wind gusts induced by convective downdrafts changed the entropy flux from the sea to the air, bringing about the deviation of the boundary layer quasi-equilibrium, which resembled the wind-induced surface heat exchange (WISHE) model (Emanuel 1986; Rotunno and Emanuel 1987; Yano and Emanuel 1991) [53–55], to some extent. On the other hand, the population of convective downdrafts depended on the population of deep convective cores that, in turn, depended on the buoyancy at the cloud base. Generally speaking, the higher the buoyancy at the cloud base, the more deep convective cores were initiated, given the same external perturbations. Figure 14 displays the vertical profiles of the moist entropy and the saturation moist entropy near the MCS shown in Figure 8. An ascending surface air parcel, as shown by Figure 14, had a strong negative buoyancy at ~2 km, which explained the lack of scattered convective cores around the MCS.

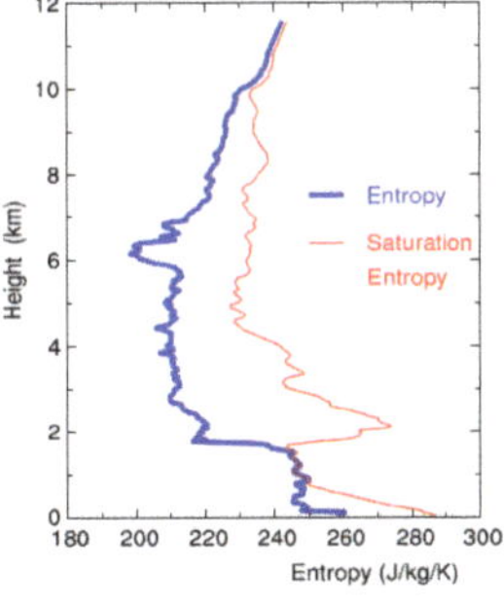

Figure 14. Vertical profiles of the moist entropy (blue line) and the saturation moist entropy (red line) obtained from the dropsonde at ~1551 UTC on 31 July 2007 during the TC4 experiment. The dropsonde was released from the DC-8 aircraft flying at an altitude of 12 km to measure the atmospheric temperature, pressure, wind and relative humidity near the MCS shown in Figure 8.

To explain the modeled oscillation, the two-column two-layer model was modified based on the CRM simulation of Figures 12 and 13. The modified model represents the surface wind gusts by introducing Δw_s (or $-\Delta w_s$), the deviation of the surface wind speed from w_{s0} the surface wind speed at the RCE in the test (or reference) column. Since Δw_s changes the entropy flux from the sea to the PBL, the PBL entropy budget in the test column (18) becomes

$$\frac{d\Delta s_s}{dt} = \frac{c_E}{H_{PBL}}(s_{sea} - s_{s0})\Delta w_s - \frac{\Delta s_s}{2\tau_t} - U_s\frac{2\Delta s_s}{L} - \frac{k_d w_m}{H}(\Delta s_s + s_{s0} - s_{min}), \qquad (22)$$

where H_{PBL} is the depth of the PBL.

To complement the thermodynamic Equation (22), the model introduced a dynamic equation on ΔT_{2km}, the deviation of the temperature or Δs^*_{2km}, the deviation of the saturation moist entropy from the RCE at $z = 2$ km in the reference column, where Δs^*_{2km} corresponds to the peak of red line at $z = 2$ km in Figure 14. Since Δs^*_{2km} is a function of ΔT_{2km}, differentiating Equation (2) with respect to time yields

$$\frac{d\Delta s^*_{2km}}{dt} = (C_p + \frac{L_v^2 q_{vs}}{T_{2km}^2})\frac{d\Delta T_{2km}}{T_{2km}dt} \qquad (23)$$

with the aid of the Clausius—Clapeyron equation, where q_{vs} is the saturation mixing ratio of water vapor at temperature T_{2km} and pressure p_{2km}. In addition, ΔT_{2km} is proportional to $(\Gamma_d - \Gamma_s)/2$ times the large-scale vertical velocity at $z = 2$ km, where Γ_d and Γ_s denote the dry and saturated adiabatic lapse rates in the reference and test columns, respectively. Thus, Equation (23) is simplified to

$$\frac{d\Delta s^*_{2km}}{dt} = b_1(C_p + \frac{L_v^2 q_{vs}}{T_{2km}^2})\frac{\Gamma_d - \Gamma_s}{2T_{2km}}w_m, \qquad (24)$$

where b_1 is the ratio between the large-scale vertical velocities at $z = 2$ km and the mid-troposphere.

Clearly, $\Delta s_s - \Delta s^*_{2km}$ represents the average buoyancy of the convective clouds at $z = 2$ km in the test column and, therefore, approximately describes whether the surface parcels can pass the lifted condensation level to form clouds. Hence, the large-scale vertical velocity at the mid-troposphere is related to $\Delta s_s - \Delta s^*_{2km}$ by

$$w_m = \frac{H}{C_p^2 \tau_c}[\Delta s_s + s_{s0} - s^*(T_m, p_m)](\Delta s_s - \Delta s^*_{2km}). \qquad (25)$$

Since the deep convective clouds generate downdrafts that, in turn, generate surface wind gusts, Δw_s is directly proportional to w_m (see Figure 13), which is expressed as

$$\Delta w_s = b_2 w_m, \qquad (26)$$

where b_2 is a dimensionless coefficient. In addition, $\Delta s_u = 0$ is set for simplicity because the radiative cooling slightly impacts the boundary layer quasi-equilibrium (see Section 2.3).

The system of Equation (22) and Equation (24) has a fixed point of $\Delta s_s = \Delta s^*_{2km} = 0$ in the phase space that corresponds to the boundary layer quasi-equilibrium. The criteria for the stability of the fixed point of the origin are derived in Appendix C. The criteria showed that the boundary layer quasi-equilibrium is unstable when b_2 is large (or the downdraft-induced surface gusts are strong), which is supported by the CRM result in Figure 13. The criteria also show that the boundary layer quasi-equilibrium is unstable when L is large (or over a large convective domain), which is supported by the modeling of the MCS onset in horizontally uniform environments with no wind shear [43,62,63].

A specific case with an unstable boundary layer quasi-equilibrium is discussed in Appendix C, where $L = 350$ km. Once its initial status was away from the origin, its status deviated from the origin and eventually approached a limit cycle (see Figure 1). In addition, its prognostic variables (e.g., surface entropy) oscillated with a period of 2 days (Figure 15). This standing two-day oscillation can lead to a westward travelling two-day oscillation via the zonal asymmetry of the atmospheric response to the zonally symmetric convective heating (see Appendix D). Hence, the system of Equation (22) and Equation (24) provided a prototype of the observed quasi-two-day oscillation.

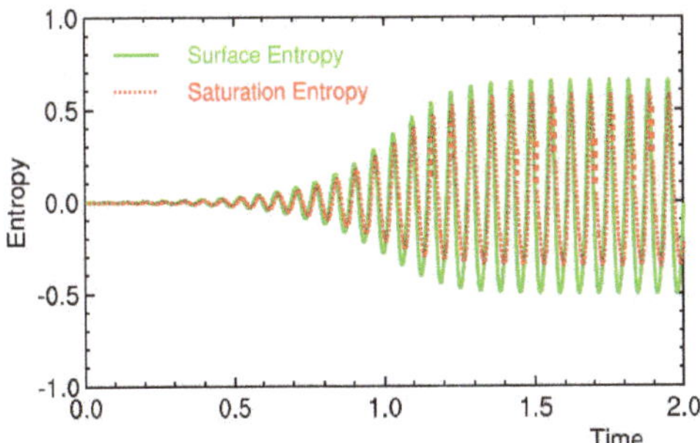

Figure 15. Time series of (green) X_2 the dimensionless surface entropy in the test column and (red) Y_2 the saturation moist entropy at $z = 2$ km in the reference column in the system of Equation (22) and Equation (24) (see Appendix C for the definitions and parameters used). The horizontal axis of time is scaled with $2\tau_t = 28$ days as the unit.

5.2. Radiative Convective Flux Equilibrium

Consider a two-column model with the SST as a prognostic variable (Nilsson and Emanuel 1999) [28]. If a large-scale vertical circulation is not present between the two columns, the atmosphere approaches a balance between the radiative cooling, convective heating and surface energy flux in each column, which is referred to as the radiative convective flux equilibrium (RCFE). In contrast to the RCE, the RCFE focuses on the balance between the atmospheric radiative cooling and the surface energy flux, where the surface energy flux is related to the SST that is, in turn, related to the UT cirrus clouds via solar and infrared radiation.

The surface energy flux possesses three timescales: two timescales in the approach of the RCE (see Figure 3) and an additional timescale in the approach of the RCFE. Consider a mixed ocean layer with depth $H_{mix} = 100$ m that is situated below the atmosphere. Its thermal capacity is about 25 times as much as that of the atmosphere. Since the atmosphere has a long timescale τ_t of ~14 days, the surface energy flux has a very long timescale of 25 × 14 = 350 days (or ~1 year) in the approach of the RCFE. This third timescale of ~1 year measures the approach of the RCFE while the atmosphere remains at the RCE.

The two-column two-layer model in Section 4 was extended to incorporate the mixed ocean layer as the third layer (see Figure 10), addressing the instability of the RCFE and its connection to the ENSO [64]. Suppose that the test column sat over the central and eastern Pacific and the reference column over the maritime continent. The two columns would be connected by a large-scale vertical circulation (or Walker circulation).

Let Δs_{sea} (or $-\Delta s_{sea}$) denote the deviation of the sea entropy from s_{sea}, the sea entropy at the RCFE in the test (or reference) column. Since s_{sea} is related to the sea surface temperature T_{sst} by $s_{sea} = c_l \ln\left(T_{sst}/T_{ref}\right)$ [10], Δs_{sea} is related to ΔT_{sst}, the derivation of the SST, in the test column by

$$\Delta s_{sea} = c_l \Delta T_{sst}/T_{sst}. \tag{27}$$

In the test column, Δs_{sea} is affected by three factors: infrared and solar radiation, the horizontal exchange in entropy between the columns induced by ocean circulations and the vertical exchange between the sea and air. Hence, its tendency is governed by

$$\frac{d\Delta s_{sea}}{dt} = \frac{p_s - p_t}{\rho_{sea} g H_{mix}} \frac{\Delta s_s - \Delta s_{sea}}{2\tau_t} - U_{sea} \frac{2\Delta s_{sea}}{L} - \frac{\sigma}{\rho_{sea} H_{mix} T_{sst}} \left[(T_{sst} + \Delta T_{sst})^4 - T_{sst}^4 - \Delta\sigma_c(\sigma^{-1}F_{sun} - T_u^4) \right], \quad (28)$$

where ρ_{sea} is the density of the sea water, U_{sea} is the flow speed of the sea surface with respect to the test column and F_{sun} is the average flux of the solar radiation absorbed by the sea with a clear sky.

The model introduces Δs_{sea} into Equation (18), yielding a new PBL entropy budget equation

$$\frac{d\Delta s_s}{dt} = -\frac{\Delta s_s - \Delta s_{sea}}{2\tau_t} - U_s \frac{2\Delta s_s}{L} - \frac{k_d w_m}{H}(\Delta s_s + s_{s0} - s_{min}). \quad (29)$$

On the other hand, the model still uses Equation (19) and Equation (20) to describe the UT entropy and cloud fraction, respectively. The model overlooks the feedback of water vapor on radiation for simplicity [28].

In summary, the system of the RCFE represents a two-column three-layer model. It has four prognostic Equations (28), (29), (19) and (20). It has a fixed point of the origin in the phase space (Δs_s, Δs_u, $\Delta\sigma_c$, Δs_{sea}) that corresponds to the RCFE. Its criteria for the stability of the fixed point of the origin are presented in Appendix B. The criterion (A8) shows that the fixed point for the RCFE is unstable when $(\sigma^{-1}F_{sun} - T_u^4)/T_{sst}^4$ is large.

A specific case with an unstable RCFE is discussed in Appendix B, where $L = 10,000$ km. Once its initial status was slightly away from the RCFE, the system deviated from the RCFE permanently and eventually approached a limit cycle (Figure A2). In addition, its prognostic variables (e.g., SST, UT cloud fraction) oscillated with a period of ~20 months (Figure 16). Since τ_t varies with the surface wind speed and radiative cooling rate (Figure 4), the oscillation in Figure 16 possesses a period close to that of the ENSO, providing a prototype of the ENSO.

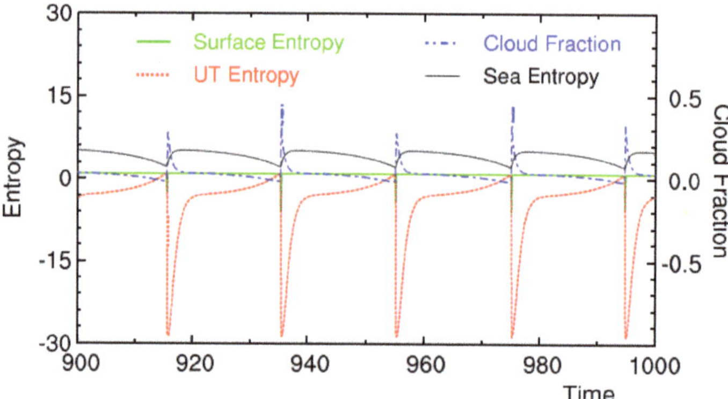

Figure 16. Time series of (green) the dimensionless air surface entropy X, (red) the UT entropy Y, (blue) the UT cloud fraction Z and (black) the sea surface entropy Ψ in the system of Equations (28), (29), (19) and (20) (see Appendix B for the definitions and parameters used). The horizontal axis of time is scaled with $2\tau_t = 28$ days (or ~1 month) as the unit.

6. Conclusions and Discussion

Convective clouds rapidly consume the CAPE and, thus, are self-destructive. Meanwhile, they modulate the external energy fluxes from the sea and to space, increasing the CAPE slowly for future convective clouds. If the convective clouds are treated as one group, they change themselves via the sea surface flux and radiative cooling. This nonlinear feedback between convective clouds and their external energy fluxes is discussed in this paper with three very low order systems.

The three systems are so simple that their criteria for a stable fixed point are derived. Furthermore, their connection between an unstable fixed point and a limit cycle provides a new clue for understanding the observed atmospheric oscillations. To be specific, the fixed point of the origin in the three systems corresponds to the boundary layer quasi-equilibrium, the RCE and the RCFE, respectively, and the limit cycle for three standing oscillations with a period of ~2 days, ~40 days and ~2 years, respectively. The three standing oscillations, due to the zonal asymmetry of the atmospheric response to the zonally symmetric convective/radiative heating (see Appendix D), bring about a westward travelling quasi-two-day oscillation, eastward travelling MJO and standing ENSO, respectively.

The systems show that the atmosphere is predictable via limit cycles if their parameters are known. On the other side, their parameters, especially the local variable of the surface wind speed, vary greatly. Since the parameters vary, the systems may frequently swing across the criteria for the stable equilibrium, bringing about bifurcations in the phase space. If so, their predictability is limited by their accuracy of process representation and the introduction of other processes, such as the effect of vertical wind shear on cloud anvil [65,66].

The current weather and climate models do not represent clouds well because of their common biases of "excessive water vapor" and "too dense clouds" [67–70]. Thus, they cannot represent the cloud-radiation interaction accurately and, consequently, cannot simulate the oscillations well [64,71]. The present theoretical study of the instability criteria suggests it is imperative to properly represent the feedback of clouds on radiation as well as convective downdrafts in the tropics. Recently, a new precipitation mechanism in cirrus clouds—the radiative effect on microphysics—has been proposed to remove the biases [48,72], raising a prospect that the models can predict the oscillations well.

Institutional Review Board Statement: Not applicable.

Informed Consent Statement: Not applicable.

Data Availability Statement: Not applicable.

Acknowledgments: The author wishes to thank Bo-Wen Shen, Roger A. Pielke and Xubin Zeng for organizing this interesting issue on atmospheric predictability. He also wishes to thank Lin Tian and Gerald Heymsfield at NASA for providing the airborne Doppler radar data used in Figures 8 and 14. He greatly appreciated the three anonymous reviewers for their kind comments and Sandra Montoya for reading the manuscript. This research was supported by the NSF (via PI David Raymond), NASA and the DoD over the past decades. The article is a continuation of the work presented at the 98th American Meteorological Society Annual Meeting and the David Raymond Symposium in Austin on 11 January 2018.

Conflicts of Interest: The author declares no conflict of interest.

Appendix A. Stability Analysis of the RCE

The stability criteria of the RCE are derived from the ordinary differential Equations (18)−(20). After introducing the four dimensionless variables

$$X = \frac{\Delta s_s}{s^*(T_m, p_m) - s_{s0}}, \ Y = \frac{\Delta s_u}{s^*(T_m, p_m) - s_{s0}}, \ Z = \Delta \sigma_c, \text{ and } \tau = \frac{t}{2\tau_t}, \tag{A1}$$

Equations (18)–(20) are rewritten in a dimensionless form as

$$\frac{dX}{d\tau} = -c_1 X - c_2 (X-1)(X-Y)(X+a_1), \tag{A2a}$$

$$\frac{dY}{d\tau} = -c_3 Y + c_4 (X-1)(X-Y)(X-Y+a_2) - c_5 Z \text{ and} \tag{A2b}$$

$$\frac{dZ}{d\tau} = c_6(X-1)(X-Y)(1-4Z^2) - c_7 Z \tag{A2c}$$

with the aid of Equation (17), where the dimensionless parameters and coefficients are

$$a_1 = \frac{s_{s0} - s_{min}}{s^* - s_{s0}}, \quad a_2 = \frac{s_{s0} - s_{u0}}{s^* - s_{s0}}, \tag{A3a}$$

$$c_1 = 1 + \frac{4U_s \tau_t}{L}, \quad c_2 = \frac{2k_d \tau_t (s^* - s_{s0})^2}{C_p^2 \tau_c}, \quad c_3 = \frac{4U_u \tau_t}{L}, \quad c_4 = \frac{2k_u \tau_t (s^* - s_{s0})^2}{C_p^2 \tau_c}, \tag{A3b}$$

$$c_5 = \frac{2g \tau_t \sigma T_u^3}{(p_m - p_t)(s^* - s_{s0})}, \quad c_6 = \frac{k_c \tau_t (s^* - s_{s0})^2}{C_p^2 \tau_c} \text{ and } c_7 = \frac{2\tau_t}{\tau_p}. \tag{A3c}$$

When $X = Y = Z = 0$, $dX/d\tau = dY/d\tau = dZ/d\tau = 0$, indicating that $(0, 0, 0)$ is a fixed point in the phase space (X, Y, Z) that corresponds to the RCE in the physics space.

To analyze the stability of the fixed point, Equation (A2) is linearized with the first order around the fixed point, yielding

$$\frac{dX}{d\tau} = (c_2 a_1 - c_1)X - c_2 a_1 Y, \tag{A4a}$$

$$\frac{dY}{d\tau} = -c_4 a_2 X + (c_4 a_2 - c_3)Y - c_5 Z \text{ and} \tag{A4b}$$

$$\frac{dZ}{d\tau} = -c_6 X + c_6 Y - c_7 Z. \tag{A4c}$$

The stability of Equation (A4) is determined by the eigenvalues of its corresponding matrix. Let λ denote an eigenvalue of the matrix. Thus,

$$\begin{vmatrix} \alpha_1 - \lambda & -c_2 a_1 & 0 \\ -c_4 a_2 & \alpha_2 - \lambda & -c_5 \\ -c_6 & c_6 & -c_7 - \lambda \end{vmatrix} = 0 \tag{A5}$$

which is read as

$$\lambda^3 + (c_7 - \alpha_1 - \alpha_2)\lambda^2 - [\alpha_1 c_3 + c_1 c_4 a_2 + c_5 c_6 + (\alpha_1 + \alpha_2)c_7]\lambda, \\ -(\alpha_1 c_3 c_7 + c_1 c_4 c_7 a_2 + c_1 c_5 c_6) = 0 \tag{A6}$$

where

$$\alpha_1 = c_2 a_1 - c_1 \text{ and} \tag{A7a}$$

$$\alpha_2 = c_4 a_2 - c_3. \tag{A7b}$$

The Routh—Hurwitz stability criterion of Equation (A4) is expressed with the polynomial coefficients of Equation (A6) [73,74], showing that λ has a negative real part (or the fixed point of the origin is stable) only when

$$c_1 + c_3 + c_7 - c_2 a_1 - c_4 a_2 > 0, \tag{A8a}$$

$$c_3 - c_4 a_2 - c_2 a_1 \frac{c_3}{c_1} - c_5 \frac{c_6}{c_7} > 0 \text{ and} \tag{A8b}$$

$$(\alpha_1 + \alpha_2)\left[c_1 c_4 a_2 + \alpha_1 c_3 + (\alpha_1 + \alpha_2)c_7 - c_7^2\right] + (c_2 a_1 + \alpha_2 - c_7)c_5 c_6 > 0. \tag{A8c}$$

If one of the (A8) equations is violated, the fixed point (or the RCE) is unstable.

The RCE, as shown by Equations (A8a) and (A8b), is unstable when c_2a_1, c_4a_2 and/or c_6 are large and/or c_7 is small. The RCE is stable when c_1 and c_3 are large. Owing to the inverse relationship of c_1 and c_3 to L, Equation (A8) shows that the RCE is stable over a small convective region. In contrast, when L is very large, c_3 is so small that Equation (A8b) is violated, indicating that the RCE is unstable over a wide convective region.

Consider, for example, a specific case with $L = 3{,}000$ km, $T_{sst} = 300$ K, $T_u = 240$ K, $p_m = 500$ hPa, $p_t = 100$ hPa, $U_s = U_t = 5$ m s^{-1}, $\tau_t = 14$ days, $\tau_c = 1$ h, $\tau_p = 10$ days, $s_{s0} = 236$ J kg^{-1} K^{-1}, $s_{min} = 210$ J kg^{-1} K^{-1}, $s^* = 240$ J kg^{-1} K^{-1}, $s_{u0} = 230$ J kg^{-1} K^{-1}, $k_d = k_u = 500$ and $k_c = 189$. Thus, $a_1 = 6.5$, $a_2 = 1.5$, $c_1 = 9$, $c_2 = 5$, $c_3 = 8$, $c_4 = 5$, $c_5 = 116.2$, $c_6 = 1$ and $c_7 = 2.8$. The case, thus, violates Equations (A8a) and (A8b), showing that the RCE is unstable.

Suppose that Equation (A2) is initially at point $(0.001, 0, 0)$ in the phase space (X, Y, Z), slightly away from the origin (or the RCE). Since the RCE is unstable, its trajectory deviates from the origin first and then approaches a limit cycle. The limit cycle is approximately displayed in Figure A1 with the trajectory extending from $\tau = 500$ to 512.6. The limit cycle is also displayed against time in Figure 11.

The system changes its topology with its coefficients. Suppose that the system takes the same coefficients as in Figure A1, except for $c_7 = 5.6$ (or $\tau_p = 5$ days). Although the system still violates Equations (A8a) and (A8b) (or the RCE is still unstable), the system has no oscillation. Instead, the system approaches a fixed point that corresponds to a steady status with a large-scale ascent (descent) in one column (the other).

Suppose that the system takes the same coefficients as in Figure A1, except for $L = 100$ km (or the convective region is small). Thus, $c_1 = 243$ and $c_3 = 242$. Hence, the system satisfies all the criteria of Equation (A8), showing that the RCE is stable. As a result, the system approaches the origin from its initial point of $(0.001, 0, 0)$.

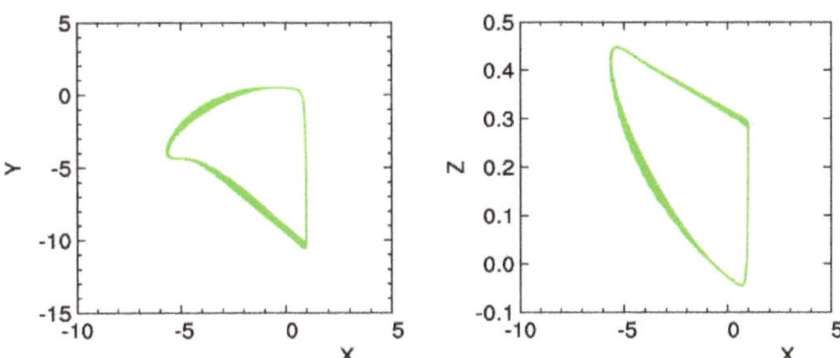

Figure A1. Numerical solution of Equation (A2) with $L = 3{,}000$ km. Projections on the (**left**) X–Y and (**right**) X–Z plane in the phase space of the segment of the trajectory extending from $\tau = 500$ to 512.6.

Appendix B. Stability Analysis of the Radiative Convective Flux Equilibrium

The stability criteria of the RCFE, following the procedure and notation in Appendix A, are derived from Equations (28), (29), (19) and (20). Using the four dimensionless variables in (A1) and

$$\Psi = \frac{\Delta s_{sea}}{s^*(T_m, p_m) - s_{s0}}, \tag{A9}$$

Equations (29), (19), (20) and (28) are changed to

$$\frac{dX}{d\tau} = -c_1 X - c_2(X-1)(X-Y)(X+a_1) + \Psi, \tag{A10a}$$

$$\frac{dY}{d\tau} = -c_3 Y + c_4(X-1)(X-Y)(X-Y+a_2) - c_5 Z, \tag{A10b}$$

$$\frac{dZ}{d\tau} = c_6(X-1)(X-Y)(1-4Z^2) - c_7 Z \quad \text{and} \tag{A10c}$$

$$\frac{d\Psi}{d\tau} = c_8 X + c_9 Z - c_{10}\Psi, \tag{A10d}$$

with the aid of Equation (27) and $(T_{sst} + \Delta T_{sst})^4 - T_{sst}^4 \approx 4T_{sst}^3 \Delta T_{sst}$, where the dimensionless parameters and coefficients are defined in Equation (A3) and

$$\chi = \frac{8\sigma T_{sst}^3 \tau_t}{c_l \rho_{sea} H_{mix}}, \tag{A11a}$$

$$c_8 = \frac{p_s - p_t}{\rho_{sea} g H_{mix}}, \quad c_9 = \frac{c_l \chi (\sigma^{-1} F_{sun} - T_u^4)}{4 T_{sst}^4 [s^*(T_m, p_m) - s_{s0}]}, \quad c_{10} = c_8 + \chi + \frac{4 U_{sea} \tau_t}{L}. \tag{A11b}$$

When $X = Y = Z = \Psi = 0$, $dX/d\tau = dY/d\tau = dZ/d\tau = d\Psi/d\tau = 0$, indicating that $(0, 0, 0, 0)$ is a fixed point in the phase space (X, Y, Z, Ψ) that corresponds to the RCFE in the physics space.

To show the stability of the fixed point, Equation (A10) is linearized with the first order around the fixed point, yielding

$$\frac{dX}{d\tau} = (c_2 a_1 - c_1)X - c_2 a_1 Y + \Psi, \tag{A12a}$$

$$\frac{dY}{d\tau} = -c_4 a_2 X + (c_4 a_2 - c_3)Y - c_5 Z, \tag{A12b}$$

$$\frac{dZ}{d\tau} = -c_6 X + c_6 Y - c_7 Z \quad \text{and} \tag{A12c}$$

$$\frac{d\Psi}{d\tau} = c_8 X + c_9 Z - c_{10}\Psi. \tag{A12d}$$

The stability of the fixed point of the origin is analyzed using the matrix of Equation (A12). The matrix has an eigenvalue λ, satisfying

$$\begin{vmatrix} \alpha_1 - \lambda & -c_2 a_1 & 0 & 1 \\ -c_4 a_2 & \alpha_2 - \lambda & -c_5 & 0 \\ -c_6 & c_6 & -c_7 - \lambda & 0 \\ c_8 & 0 & c_9 & -c_{10} - \lambda \end{vmatrix} = 0 \tag{A13}$$

which is read as

$$\lambda^4 + (c_7 + c_{10} - \alpha_1 - \alpha_2)\lambda^3 - [\alpha_1 c_3 + c_1 c_4 a_2 + c_5 c_6 + (\alpha_1 + \alpha_2)(c_7 + c_{10}) - c_7 c_{10} - c_8]\lambda^2$$
$$- [(\alpha_1 c_3 + c_1 c_4 a_2)(c_7 + c_{10}) + (c_1 + c_{10})c_5 c_6 + (\alpha_1 + \alpha_2)c_7 c_{10} + c_6 c_9 + c_8(\alpha_2 - c_7)]\lambda$$
$$- [(\alpha_1 c_3 + c_1 c_4 a_2)c_7 c_{10} + (c_1 c_{10} - c_8)c_5 c_6 + \alpha_2 c_7 c_8 + c_3 c_6 c_9] = 0 \tag{A14}$$

The Routh—Hurwitz stability criterion of Equation (A12) is expressed with the polynomial coefficients of Equation (A14), showing that λ has a negative real part (or the fixed point of the origin is stable) only when

$$c_1 + c_3 + c_7 + c_{10} - c_2 a_1 - c_4 a_2 > 0, \tag{A15a}$$

$$(c_3 - c_4 a_2)\left(1 + \frac{c_8}{c_1 c_{10}}\right) - c_2 a_1 \frac{c_3}{c_1} - \left(1 - \frac{c_8}{c_1 c_{10}}\right)\frac{c_5 c_6}{c_7} - \frac{c_3}{c_1}\frac{c_6 c_9}{c_7 c_{10}} > 0, \tag{A15b}$$

$$(\alpha_1 + \alpha_2)[\alpha_1 c_3 + c_1 c_4 a_2 + (\alpha_1 + \alpha_2)(c_7 + c_{10}) - (c_7 + c_{10})^2] \text{and}$$
$$+ (c_{10} - \alpha_1)c_8 + (c_7 + c_{10})c_7 c_{10} + (c_2 a_1 + \alpha_2 - c_7)c_5 c_6 + c_6 c_9 > 0 \tag{A15c}$$

$$(c_7 + c_{10} - \alpha_1 - \alpha_2)^2[(\alpha_1 c_3 + c_1 c_4 a_2)c_7 c_{10} + (c_1 c_{10} - c_8)c_5 c_6 + \alpha_2 c_7 c_8 + c_3 c_6 c_9]$$
$$- [(\alpha_1 c_3 + c_1 c_4 a_2)(c_7 + c_{10}) + (c_1 + c_{10})c_5 c_6 + (\alpha_1 + \alpha_2)c_7 c_{10} + c_6 c_9 + c_8(\alpha_2 - c_7)]$$
$$\left\{ (\alpha_1 + \alpha_2)[\alpha_1 c_3 + c_1 c_4 a_2 + (\alpha_1 + \alpha_2)(c_7 + c_{10}) - (c_7 + c_{10})^2] \right.$$
$$\left. + (c_{10} - \alpha_1)c_8 + (c_7 + c_{10})c_7 c_{10} + (c_2 a_1 + \alpha_2 - c_7)c_5 c_6 + c_6 c_9 \right\} > 0 \tag{A15d}$$

If one of the (A15) equations is violated, the fixed point (or the RCFE) is unstable.

The criteria of Equation (A15) are similar to those for the RCE but are more complicated. The RCFE, as shown by Equation (A15b), is unstable when c_6/c_7 and/or c_9/c_{10} are large. Note $c_8/(c_1 c_{10}) < 1$.

Consider, for example, a specific case with $L = 10{,}000$ km, $T_{sst} = 300$ K, $T_u = 240$ K, $p_s = 1013.25$ hPa, $p_t = 100$ hPa, $U_s = 5$ m s^{-1}, $U_t = 2$ m s^{-1}, $U_{sea} = 0$, $\tau_t = 14$ days, $\tau_c = 1$ h, $\tau_p = 21.5$ days, $s_{s0} = 236$ J kg^{-1} K^{-1}, $s_{min} = 210$ J kg^{-1} K^{-1}, $s^* = 240$ J kg^{-1} K^{-1}, $s_{u0} = 230$ J kg^{-1} K^{-1}, $k_d = k_u = 500$, $k_c = 189$ and $H_{mix} = 100$ m. Thus, $a_1 = 6.5$, $a_2 = 1.5$, $c_1 = 3.42$, $c_2 = 5$, $c_3 = 0.97$, $c_4 = 5$, $c_5 = 116.2$, $c_6 = 1$, $c_7 = 1.3$, $c_8 = 0.093$, $c_9 = 25$, $c_{10} = 0.13$. Hence, the case violates Equations (A15a) and (A15b), showing that the RCFE is unstable.

Suppose that Equation (A10) is initially at (0.001, 0, 0, 0) in the phase space (X, Y, Z, Ψ), slightly away from the origin (or the RCFE). Since the RCFE is unstable, its trajectory deviates from the origin first and then approaches a limit cycle. The limit cycle is approximately displayed in Figure A2 with the trajectory extending from $\tau = 8{,}000$ to $10{,}000$. The limit cycle is also displayed against time in Figure 16.

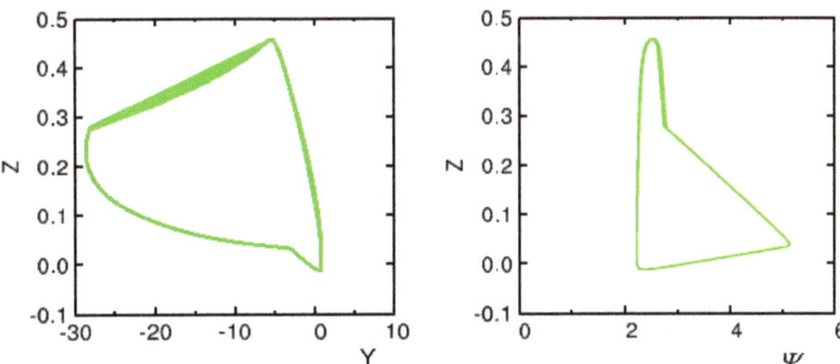

Figure A2. Numerical solution of Equation (A10) with $L = 10{,}000$ km. The projections on the (**left**) Y–Z and (**right**) Ψ–Z plane in the phase space of the segment of the trajectory extending from $\tau = 8{,}000$ to $10{,}000$.

Appendix C. Stability Analysis of the Boundary Layer Quasi-Equilibrium

The stability criteria of the boundary layer quasi-equilibrium are derived from Equations (22) and (24). Using the three dimensionless variables

$$X_2 = \frac{\Delta s_s}{s_{s0} - s^*(T_m, p_m)}, \quad Y_2 = \frac{\Delta s^*_{2km}}{s_{s0} - s^*(T_m, p_m)} \text{ and } \tau = \frac{t}{2\tau_t}, \tag{A16}$$

Equations (22) and (24) are changed to

$$\frac{dX_2}{d\tau} = -c_{21}X_2 - c_{22}(X_2 - a_{21})(X_2 + 1)(X_2 - Y_2) \text{ and} \tag{A17a}$$

$$\frac{dY_2}{d\tau} = c_{23}(X_2 + 1)(X_2 - Y_2) \tag{A17b}$$

with the aid of Equations (25) and (26), where the dimensionless parameters and coefficients are

$$a_{21} = \frac{b_2 c_E k_d^{-1} H H_{PBL}^{-1}(s_{sea} - s_{s0}) - (s_{s0} - s_{min})}{s_{s0} - s^*(T_m, p_m)}, \tag{A18a}$$

$$c_{21} = 1 + \frac{4U_s \tau_t}{L}, \quad c_{22} = \frac{2k_d \tau_t}{C_p^2 \tau_c}[s_{s0} - s^*(T_m, p_m)]^2 \text{ and} \tag{A18b}$$

$$c_{23} = (1 + \frac{L_v^2 q_{vs}}{C_p T_{2km}^2}) \frac{b_1 H \tau_t}{C_p \tau_c T_{2km}}(\Gamma_d - \Gamma_s)[s_{s0} - s^*(T_m, p_m)]. \tag{A18c}$$

When $X_2 = Y_2 = 0$, $dX_2/d\tau = dY_2/d\tau = 0$. Thus, $(0, 0)$ is a fixed point in the phase space (X_2, Y_2) that corresponds to the boundary layer quasi-equilibrium.

To show the stability of the fixed point, Equation (A17) is linearized with the first order around the fixed point, yielding

$$\frac{dX_2}{d\tau} = (c_{22}a_{21} - c_{21})X_2 - c_{22}a_{21}Y_2, \tag{A19a}$$

$$\frac{dY_2}{d\tau} = c_{23}X_2 - c_{23}Y_2. \tag{A19b}$$

Let λ denote an eigenvalue of the matrix of Equation (A19). Thus,

$$\begin{vmatrix} c_{22}a_{21} - c_{21} - \lambda & -c_{22}a_{21} \\ c_{23} & -c_{23} - \lambda \end{vmatrix} = 0, \tag{A20}$$

which is read as

$$\lambda^2 + (c_{21} - c_{22}a_{21} + c_{23})\lambda + c_{21}c_{23} = 0. \tag{A21}$$

The Routh—Hurwitz stability criterion of Equation (A19) shows that λ has a negative real part (or the fixed point of the origin is stable) when

$$c_{21} - c_{22}a_{21} + c_{23} > 0, \text{ and} \tag{A22a}$$

$$c_{21}c_{23} > 0. \tag{A22b}$$

In other words, the boundary layer quasi-equilibrium becomes unstable when

$$a_{21} > (c_{21} + c_{23})/c_{22} \text{ or} \tag{A23a}$$

$$c_{23} < 0. \tag{A23b}$$

Substituting Equation (A18a) into (A23a) yields a criterion on b_2 for an unstable boundary layer quasi-equilibrium. That is, when b_2 is large or the effect of downdrafts on surface wind gusts is strong, the boundary layer quasi-equilibrium is unstable.

Consider, for example, a specific case with $L = 350$ km, $U_s = 5$ m s^{-1}, $\tau_t = 14$ days, $\tau_c = 1$ h, $s_{s0} = 240$ J kg^{-1} K^{-1}, $s_{min} = 210$ J kg^{-1} K^{-1}, $s^* = 236$ J kg^{-1} K^{-1}, $s_{sea} = 393$ J kg^{-1} K^{-1}, $T_{2km} = 290$ K, $q_{vs} = 16$ g kg^{-1}, $\Gamma_d = 0.0098$ K m^{-1}, $\Gamma_s = 0.0065$ K m^{-1}, $k_d = 3280$, $b_1 = 0.7$, $b_2 = 9 \times 10^4$ and $H_{PBL} = 1$ km. Thus, $a_{21} = 6.3$, $c_{21} = 70$, $c_{22} = 35$ and $c_{23} = 140$. Hence, the case satisfies Equation (A23a), showing that the boundary layer quasi-equilibrium is unstable.

Suppose that Equation (A17) is initially at $(0.001, 0)$ in the phase space (X_2, Y_2), slightly away from the origin. Since the fixed point of the origin is unstable, its trajectory deviates from the origin first and then approaches a limit cycle (see Figure 1). In addition, its variables oscillate with a period of two days (see Figure 15).

Appendix D. Zonal Asymmetry of Atmospheric Response to Diabatic Heating

The two-column model in Section 4, as the zeroth-order approximation, produces a standing 40-day oscillation of convective clouds, mimicking the observed oscillations of the MJO's major precipitation events confined between 60° E and 180° E [71]. In this appendix, a first-order approximation is introduced to reveal the zonal asymmetry of the atmospheric response to a standing convective/radiative heating oscillation, explaining the eastward propagation of the MJO.

Consider an atmosphere in the coordinate system (x, y, z) with a diabatic heating rate $Q(x, y, z, t)$. Its velocity vector (u, v, w), potential temperature perturbation θ' and pressure perturbation p' change in response to Q, which is governed by a linear model for a tropical β channel. That is,

$$\partial_t(\rho u) - \beta y \rho v + \partial_x p' = 0, \tag{A24a}$$

$$\partial_t(\rho v) + \beta y \rho u + \partial_y p' = 0, \tag{A24b}$$

$$\partial_x(\rho u) + \partial_y(\rho v) + \partial_z(\rho w) = 0, \tag{A24c}$$

$$\partial_t(\rho \theta') + N^2 g^{-1} \theta_0 \rho w = Q \text{ and} \tag{A24d}$$

$$\partial_z p' - g \rho \theta'/\theta_0 = 0, \tag{A24e}$$

where β is the Rossby parameter, N is the Brunt–Väisälä frequency, ρ is the air density and θ_0 is the potential temperature of the background. In the following four subsections, Equation (A24) is solved to show that a standing 40-day oscillation can lead to an eastward travelling planetary-scale disturbance.

Appendix D.1. Forced Modes versus Free Modes

The forced and free modes of Equation (A24) are distinguished first. For the waves trapped in the troposphere, they are expanded vertically in the Fourier series by

$$\{\rho u, \rho v, p'\} = \sum_{m=1}^{\infty} \left\{ (\rho u)_m, (\rho v)_m, p_l^m \right\} \cos \frac{m\pi}{H} z \text{ and} \tag{A25a}$$

$$\{\rho w, \rho \theta', Q\} = \sum_{m=1}^{\infty} \left\{ (\rho w)_m, (\rho \theta')_m, \frac{N\theta_0}{c_m g} Q_m \right\} \sin \frac{m\pi}{H} z, \tag{A25b}$$

where a symbol with the subscript m denotes the expansion coefficient of its corresponding variable at the vertical mode m, and H is the height of the troposphere. Substituting Equation (A25) into Equation (A24) first and then eliminating $(\rho w)_m$ and $(\rho \theta')_m$ in the resulting equation yields

$$\partial_t(\rho u)_m - \beta y (\rho v)_m + \partial_x p'_m = 0, \tag{A26a}$$

$$\partial_t(\rho v)_m + \beta y (\rho u)_m + \partial_y p'_m = 0 \text{ and} \tag{A26b}$$

$$\partial_t p'_m + c_m^2 \left[\partial_x(\rho u)_m + \partial_y(\rho v)_m \right] = -Q_m, \tag{A26c}$$

where $c_m = HN/m\pi$ is the phase speed of the gravity wave at mode m.

Using a variable array $\mathbf{A} = \{(\rho u)_m, (\rho v)_m, p'_m\}$ and a forcing array $\mathbf{Q} = \{0, 0, -Q_m\}$, the linear Equation (A26) is rewritten as

$$\mathbf{L}(\mathbf{A}) = \mathbf{Q} \text{ and } \mathbf{A}\,|_{t=0} = \mathbf{A}_0, \tag{A27}$$

where $\mathbf{L}$ is a linear operator on $\mathbf{A}$ and $\mathbf{A}_0$ is the initial value of $\mathbf{A}$. The solution of (A26) consists of two kinds of modes: a solution of forced motion $\mathbf{A}_{\text{forced}}$ and a solution of free waves $\mathbf{A}_{\text{free}}$. That is,

$$\mathbf{A} = \mathbf{A}_{\text{forced}} + \mathbf{A}_{\text{free}}. \tag{A28}$$

As a special solution of Equation (A26), $\mathbf{A}_{\text{forced}}$ satisfies

$$\mathbf{L}(\mathbf{A}_{\text{forced}}) = \mathbf{Q}. \tag{A29}$$

Since $\mathbf{L}$ is a linear operator, $\mathbf{L}(\mathbf{A}) = \mathbf{L}(\mathbf{A}_{\text{forced}}) + \mathbf{L}(\mathbf{A}_{\text{free}})$. Thus, substituting this equation and Equation (A29) into Equation (A27) yields

$$\mathbf{L}(\mathbf{A}_{\text{free}}) = 0 \text{ and } \mathbf{A}_{\text{free}}\,|_{t=0} = \mathbf{A}_0 - \mathbf{A}_{\text{forced}}\,|_{t=0}. \tag{A30}$$

As the solution of Equation (A30) with an initial condition of $\mathbf{A}_0 - \mathbf{A}_{\text{forced}}\,|_{t=0}$, $\mathbf{A}_{\text{free}}$ accounts for all the free waves induced by the initial condition of $\mathbf{A}_0$ and the forcing of $\mathbf{Q}$.

The free modes of $\mathbf{A}_{\text{free}}$ were introduced in Matsuno (1966) [75]. The forced modes of $\mathbf{A}_{\text{forced}}$ are presented below, focusing on their resonance conditions. The subscript of $\mathbf{A}_{\text{forced}}$ is omitted hereafter for simplicity.

Appendix D.2. Analytic Solution

Consider a thermal forcing with factor $\exp(ikx-i\omega t)$, where k and ω represent the zonal angular wavenumber and angular frequency, respectively. Given k and ω, the forced motion in Equation (A26) is expressed as

$$(\rho u)_m = \rho \sum_{n=-1}^{\infty} C_n ik\varepsilon^{-1} \hat{u}_n \exp(-\hat{y}^2/2) \exp(ikx - i\omega t), \tag{A31a}$$

$$(\rho v)_m = \rho \sum_{n=-1}^{\infty} C_n \hat{v}_n \exp(-\hat{y}^2/2) \exp(ikx - i\omega t), \tag{A31b}$$

$$p'_m = \sum_{n=-1}^{\infty} C_n i\omega\varepsilon^{-1} \hat{p}_n \exp(-\hat{y}^2/2) \exp(ikx - i\omega t) \text{and} \tag{A31c}$$

$$Q_m = \sum_{n=-1}^{\infty} C_n \varepsilon c_m^2 \hat{Q}_n \exp(-\hat{y}^2/2) \exp(ikx - i\omega t), \tag{A31d}$$

where C_n is the coefficient for the meridional mode n, $\varepsilon \equiv (\beta|k|/\omega)^{1/2}$, $\hat{y} \equiv \varepsilon y$ and $\hat{u}_n, \hat{v}_n, \hat{p}_n$ and $\hat{Q}_n$ are the $\hat{y}$-dependent polynomials. Substituting Equation (A31) into Equation (A26) yields the expressions of $\hat{u}_n, \hat{v}_n, \hat{p}_n$ and $\hat{Q}_n$. Alternatively, the expressions of $\hat{u}_n, \hat{v}_n, \hat{p}_n$ and $\hat{Q}_n$ can be substituted into Equations (A31) and (A26) for a test.

For the westward traveling zonal modes (or $k < 0$),

$$\hat{u}_n = \varepsilon^2 k^{-2}\left[-\hat{y}H_n(\hat{y}) + \hat{y}^2 H_{n-1}(\hat{y}) + 2\hat{y}H_{n-2}(\hat{y}) + 2H_{n-3}(\hat{y})\right] - H_{n-1}(\hat{y}),$$

$$\hat{v}_n = H_n(\hat{y}),$$

$$\hat{p}_n = \varepsilon^2 k^{-2}\left[\hat{y}^2 H_{n-1}(\hat{y}) + 2\hat{y}H_{n-2}(\hat{y}) + 2H_{n-3}(\hat{y})\right] - H_{n-1}(\hat{y}) \text{ and}$$

$$\hat{Q}_n = (1 - \mu_c^2)\left[\hat{y}^2 H_{n-1}(\hat{y}) + 2\hat{y}H_{n-2}(\hat{y}) + 2H_{n-3}(\hat{y})\right] - [2n + 1 - (2n^* + 1)\mu_c]H_{n-1}(\hat{y})$$

when $n \geq 3$, where H_n is the Hermite polynomial of the n'th order, the ratio between the phase speeds of the forced and free modes is defined as

$$\mu_c \equiv \omega / (|k|c_m),\tag{A32}$$

and a meridional number n^* of the forced modes, similar to n of the free modes, is defined as

$$2n^* + 1 \equiv \left[\omega^2 - (kc_m)^2 - \beta k\omega^{-1}c_m^2\right] / (\beta c_m).\tag{A33}$$

The mode with $n = 1$ is expressed as

$$\hat{u}_1 = \varepsilon^2 k^{-2}(1 - \hat{y}^2) - 1,$$

$$\hat{v}_1 = H_1(\hat{y}),$$

$$\hat{p}_1 = \varepsilon^2 k^{-2}(\hat{y}^2 + 1) - 1 \text{ and}$$

$$\hat{Q}_1 = (1 - \mu_c^2)(\hat{y}^2 + 1) - 3 + (2n^* + 1)\mu_c.$$

Clearly, the preceding modes have two resonance conditions of

$$\mu_c = 1 \text{ and } n^* = n.\tag{A34}$$

In addition, the mode with $n = 2$ is expressed as

$$\hat{u}_2 = (3\varepsilon^2 k^{-2} - 2\varepsilon^4 k^{-4} - 2)\hat{y} + \varepsilon^2 k^{-2}(2 - \varepsilon^2 k^{-2})\hat{y}(1 - \hat{y}^2),$$

$$\hat{v}_2 = 2\varepsilon^2 k^{-2} + (2 - \varepsilon^2 k^{-2})(2\hat{y}^2 - 1),$$

$$\hat{p}_2 = (3\varepsilon^2 k^{-2} - 2)\hat{y} + \varepsilon^2 k^{-2}(2 - \varepsilon^2 k^{-2})\hat{y}^3 \text{ and}$$

$$\hat{Q}_2 = [7\varepsilon^2 k^{-2} - 10 + (2n^* + 1)\mu_c(2 - 3\varepsilon^2 k^{-2})]\hat{y} + (1 - \mu_c^2)(2 - \varepsilon^2 k^{-2})\hat{y}^3,$$

which has two resonance conditions of

$$\mu_c = 1 \text{ and } n^* = (4 - 2\varepsilon^2 k^{-2}) / (2 - 3\varepsilon^2 k^{-2}).\tag{A35}$$

For the eastward traveling zonal modes (or $k > 0$),

$$\hat{u}_n = \frac{H_{n+1}(\hat{y})}{2(n+1)} + \frac{\varepsilon^2}{2(n+3)k^2}\left[2(n+2)\hat{y}H_n(\hat{y}) - (\hat{y}^2 - \frac{1}{n+1})H_{n+1}(\hat{y})\right],$$

$$\hat{v}_n = H_n(\hat{y}),$$

$$\hat{p}_n = \frac{H_{n+1}(\hat{y})}{2(n+1)} - \frac{\varepsilon^2}{2(n+3)k^2}\left[2\hat{y}H_n(\hat{y}) + (\hat{y}^2 - \frac{1}{n+1})H_{n+1}(\hat{y})\right] \text{ and}$$

$$\hat{Q}_n = \frac{\mu_c^2 - 1}{2(n+3)}\left[2\hat{y}H_n(\hat{y}) + (\hat{y}^2 - \frac{1}{n+1})H_{n+1}(\hat{y})\right] + \frac{2n + 1 - (2n^* + 1)\mu_c}{2(n+1)}H_{n+1}(\hat{y})$$

when $n \geq 0$, which have the same resonance conditions as Equation (A34). In addition, the mode with $n = -1$ is expressed as

$$\hat{u}_{-1} = -\varepsilon^2 k^{-2}, \ \hat{v}_{-1} = 0, \ \hat{p}_{-1} = -\varepsilon^2 k^{-2} \text{ and } \hat{Q}_{-1} = \mu_c^2 - 1.$$

Obviously, the mode with $n = -1$ has a similar spatial structure to the Kelvin wave and has a sole resonance condition of $\mu_c = 1$.

Since $\hat{v}_n = H_n(\hat{y})$ works in almost all the forced modes and the sequence of $H_n(\hat{y})$ forms an orthogonal basis, any meridional velocity component of the forced motion, thus, can be expanded into the series of Hermite polynomials or $\hat{v}_n$. In other words, any forced motion can be expanded as Equation (A31). Note, there is no westward traveling mode with $n = 0$ or -1 for the restraint of the Coliolis force (i.e., $C_{-1} = C_0 = 0$ when $k < 0$).

Appendix D.3. Resonance in the Westward Travelling Modes

A standing oscillation of convective cloud systems induces a standing thermal forcing oscillation, such as

$$2\sin(\omega t)\cos(kx) = \sin(\omega t - kx) + \sin(\omega t + kx), \tag{A36}$$

that decomposes into eastward and westward travelling forcings with the same amplitude. *If the atmosphere responds to the eastward traveling forcing more effectively than the westward travelling one, its forced motion, thus, has a net eastward travelling component, leading to an eastward propagating disturbance. Otherwise, a standing thermal forcing leads to a westward propagating disturbance.*

To measure the atmospheric response to the thermal forcing, a response factor for the vertical mode m and the meridional mode n is defined as

$$R_{mn} = R_E \beta c_m \max\left| k\varepsilon^{-1}\hat{u}_n \exp(-\hat{y}^2/2)\right| / \max\left|\varepsilon c_m^2 \hat{Q}_n \exp(-\hat{y}^2/2)\right|,$$

where R_E is the radius of the Earth. The response factor of the modes $n = 1$, 2 and 3 is displayed in Figure A3, revealing its asymmetry between the westward and eastward travelling modes. Specifically, the westward travelling mode with $n = 2$ has a resonance at $\omega/(\beta c_m)^{1/2} = -k(c_m/\beta)^{1/2} = 2^{-1/2}$ or (A35). Suppose $c_m = 20$ m s^{-1}. The resonance occurs at a forcing period of ~5 days and, thus, brings about a high response factor near the forcing period, agreeing with the observations of the westward travelling quasi-five-day oscillations [76,77].

To address the origin of the quasi-five-day oscillations, the two-column model in Section 5.1 was modified to cover two layers: the lower troposphere (below 3 km) with a down-gradient entropy and shallow convection and the upper troposphere (above 7 km) with an up-gradient entropy and deep convection (see Figure 7). Since the cloud spectrum is statistically steady at the RCE, the ratio between the numbers of deep and shallow cumulus clouds is a constant at the RCE. However, this constant ratio becomes unstable over a large domain (i.e., in a two-column model) under the constraint of the WTG, bringing about a standing oscillation. Since the thermal capacity of the lower troposphere is about 2.5 times as much as that of the PBL, the oscillation, thus, has a period of 2.5 × 2 = 5 days. Owing to its high response factor, the westward travelling mode with $n = 2$ at the forcing period of ~5 days becomes prominent, explaining the westward travelling quasi-five-day oscillations observed.

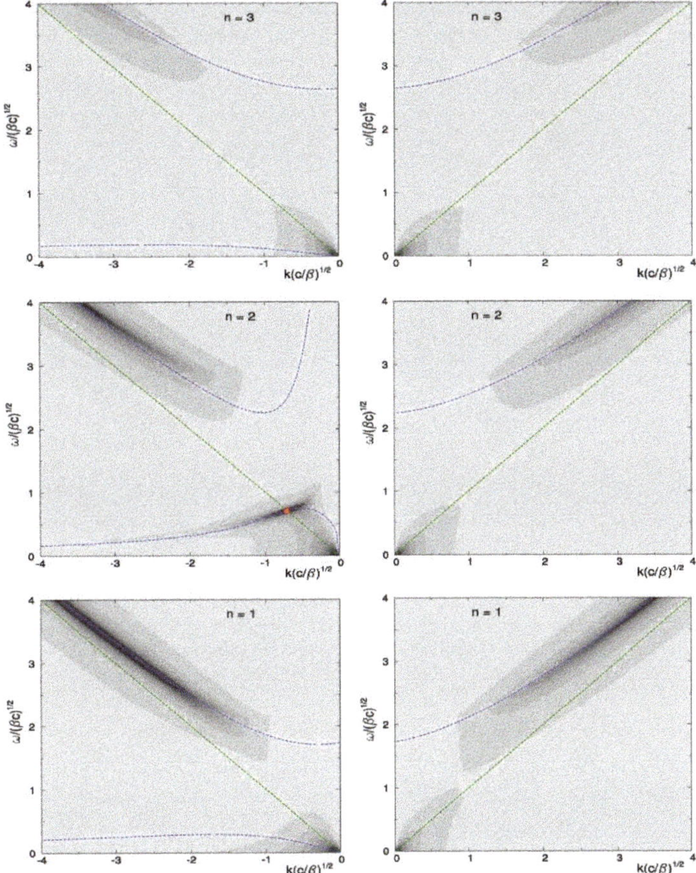

Figure A3. Response factor of (**left**) the westward and (**right**) the eastward travelling modes with n = (**bottom**) 1, (**middle**) 2 and (**top**) 3, where c is the phase speed of the gravity wave. The shading density is directly proportional to the response factor with a contour interval of 4. The green and blue lines represent the zonal and meridional resonance conditions, respectively. The blue lines also represent the dispersion relationship of the free modes except for the westward travelling mode with n = 2. The red spot indicates that both resonance conditions are satisfied.

In addition, the westward travelling mode with n = 1 has no resonance (Figure A3). However, it still has a higher response factor than its eastward travelling counterpart. Hence, the standing two-day oscillation in Section 5.1 can lead to a westward travelling disturbance, which is consistent with the satellite observations of the westward travelling quasi-two-day oscillations with n = 1 [59–61]. Since the eastward traveling mode still exists, it may be amplified through an interaction with the other factors, such as its parent MJO and vertical wind shear [65,66], occasionally leading to a standing or eastward travelling quasi-two-day disturbance [59].

Appendix D.4. Resonance in the Eastward Travelling Modes

Figure A4 displays the response factor of the eastward travelling modes with n = −1 and 0. Clearly, the model with n = −1 has a resonance at the dispersion line of the Kelvin wave, which agrees well with the satellite observations of the plentiful Kelvin waves [77]. Since the response factor for the eastward travelling mode with n = −1 is much higher than that for the mixed westward travelling modes, a standing 45-day oscillation of the forcing leads to a standing oscillation of the vertical circulation plus an eastward travelling oscillation of the vertical circulation, which agrees with the observations of Wheeler and Kiladis (1999) [71,77], with the aid of the reverse direction of Equation (A36).

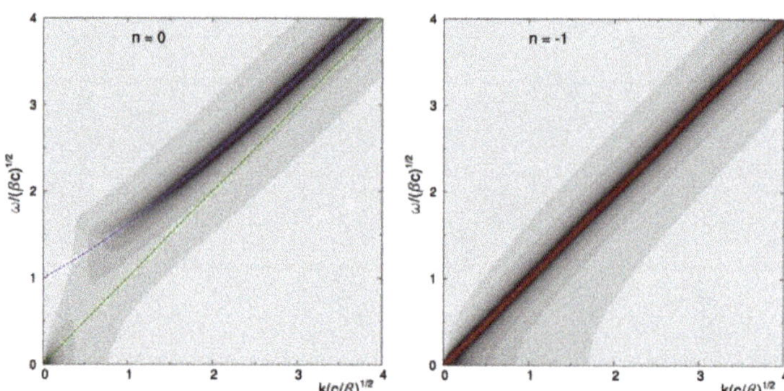

Figure A4. Similar to Figure A3, except for the eastward travelling modes with $n = 0$ (**left**) and -1 (**right**). The red line indicates the sole resonance condition.

Compare multiple cases of the eastward travelling mode with $n = -1$ that have a forcing period of 45 days but a different zonal forcing wavelength. The larger their forcing wavelength, the higher their response factor (Figure A5), which is clear in physics because the case with the largest wavelength is closest to the resonance condition (Figure A4). Hence, *a standing 45-day oscillation of convection can lead to a planetary-scale eastward travelling vertical circulation like the MJO.* Nevertheless, the eastward travelling vertical circulation further interacts with convection, exhibiting a complicated phenomenon of convection propagation [49–52].

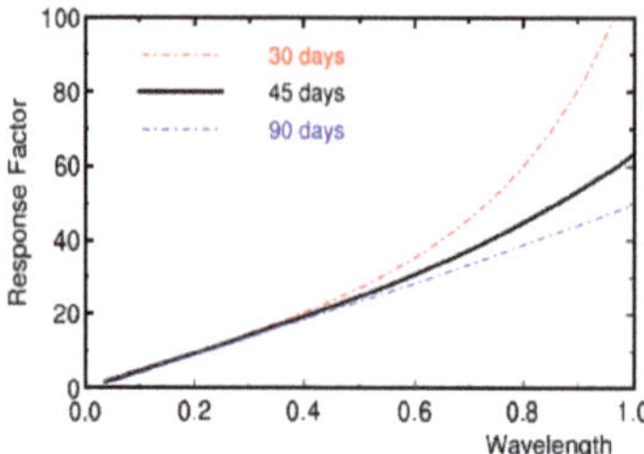

Figure A5. Response factor of the eastward travelling mode with $n = -1$ versus the zonal forcing wavelength (normalized by the circumference of the Earth). The red, black, and blue lines represent the forcing period 30, 45 and 90 days, respectively.

Funding: This research received no external funding.

References

1. Lorenz, E.N. Deterministic nonperiodic flow. *J. Atmos. Sci.* **1963**, *20*, 130–141. [CrossRef]
2. Shen, B.-W.; Pielke, R.A., Sr.; Zeng, X.; Baik, J.-J.; Faghih-Naini, S.; Cui, J.; Atlas, R. Is Weather Chaotic? Coexistence of Chaos and Order within a Generalized Lorenz Model. *Bull. Am. Meteorol. Soc.* **2021**, *2*, E148–E158.
3. Shen, B.-W.; Pielke, R., Sr.; Zeng, X.; Cui, J.; Faghih-Naini, S.; Paxson, W.; Kesarkar, A.; Zeng, X.; Atlas, R. The Dual Nature of Chaos and Order in the Atmosphere. *Atmosphere* **2022**, *13*, 1892. [CrossRef]
4. Lorenz, E.N. The predictability of a flow which possesses many scales of motion. *Tellus* **1969**, *21*, 289–307. [CrossRef]
5. Lorenz, E.N. Can chaos and intransitivity lead to interannual variability? *Tellus* **1990**, *42*, 378–389. [CrossRef]
6. Riehl, H.; Malkus, J.S. On the heat balance in the equatorial trough zone. *Geophysica* **1958**, *6*, 503–537.

7. Raymond, D.J.; Zeng, X. Instability and large-scale circulations in a two-column model of the troposphere. *Q. J. R. Meteorol. Soc.* **2000**, *126*, 3117–3135. [CrossRef]
8. Teschl, G. *Ordinary Differential Equations and Dynamical Systems*; American Mathematical Society: Providence, RL, USA, 2012; ISBN 978-0-8218-8328-0.
9. Manabe, S.; Strickler, R.F. Thermal equilibrium of the atmosphere with a convective adjustment. *J. Atmos. Sci.* **1964**, *21*, 361–385. [CrossRef]
10. Zeng, X.; Tao, W.-K.; Simpson, J. An equation for moist entropy in a precipitating and icy atmosphere. *J. Atmos. Sci.* **2005**, *62*, 4293–4309. [CrossRef]
11. Hauf, T.; Höller, H. Entropy and potential temperature. *J. Atmos. Sci.* **1987**, *44*, 2887–2901. [CrossRef]
12. Zeng, X.; Tao, W.-K.; Simpson, J. A choice of prognostic variables for long-term cloud-resolving model simulations. *J. Meteorol. Soc. Jpn.* **2008**, *86*, 839–856. [CrossRef]
13. Charney, J.G.; Eliassen, A. On the growth of the hurricane depression. *J. Atmos. Sci.* **1964**, *21*, 68–75. [CrossRef]
14. Emanuel, K.A.; Neelin, J.D.; Bretherton, C.S. On large-scale circulations in convecting atmospheres. *Q. J. R. Meteorol. Soc.* **1994**, *120*, 1111–1143. [CrossRef]
15. Raymond, D.J.; Torres, D.J. Fundamental moist modes of the equatorial troposphere. *J. Atmos. Sci.* **1998**, *55*, 1771–1790. [CrossRef]
16. Emanuel, K.A. *Atmospheric Convection*; Oxford University Press: New York, NY, USA, 1994; p. 580.
17. Robe, F.R.; Emanuel, K.A. Moist convective scaling: Some inferences from three-dimensional cloud ensemble simulations. *J. Atmos.* **1996**, *53*, 3265–3275. [CrossRef]
18. Tompkins, A.M.; Craig, G.C. Radiative-convective equilibrium in a three-dimensional cloud ensemble model. *Q. J. R. Meteorol. Soc.* **1998**, *124*, 2073–2098.
19. Zeng, X. Ensemble Simulation of Tropical Convection. Ph.D. Dissertation, New Mexico Institute of Mining and Technology, Socorro, NM, USA, 2001; p. 124.
20. Madden, R.; Julian, P.R. Detection of a 40–50 day oscillation in the zonal wind in the tropical Pacific. *J. Atmos. Sci.* **1971**, *28*, 702–708. [CrossRef]
21. Madden, R.; Julian, P.R. Description of global-scale circulation cells in the Tropics with a 40–50 day period. *J. Atmos. Sci.* **1972**, *29*, 1109–1123. [CrossRef]
22. Madden, R.; Julian, P.R. Observations of the 40–50 day tropical oscillation—A review. *Mon. Wea. Rev.* **1994**, *122*, 814–837. [CrossRef]
23. Sessions, S.L.; Sugaya, S.; Raymond, D.J.; Sobel, A.H. Multiple equilibria in a cloud resolving model using the weak temperature gradient approximation. *J. Geophys. Res.* **2010**, *115*, D12110. [CrossRef]
24. Raymond, D.J. Thermodynamic control of tropical rainfall. *Q. J. R. Meteorol. Soc.* **2000**, *126*, 889–898. [CrossRef]
25. Bretherton, C.S.; Smolarkiewicz, P.K. Gravity waves, compensating subsidence and detrainment around cumulus clouds. *J. Atmos. Sci.* **1989**, *46*, 740–759. [CrossRef]
26. Sobel, A.H.; Bretherton, C.S. Modeling tropical precipitation in a single column. *J. Climate* **2000**, *13*, 4378–4392. [CrossRef]
27. Bretherton, C.S.; Sobel, A.H. A simple model of a convectively coupled Walker circulation using the weak temperature gradient approximation. *J. Climate* **2002**, *15*, 2907–2920. [CrossRef]
28. Nilsson, J.; Emanuel, K.A. Equilibrium atmosphere of a two-column radiative convective model. *Q. J. R. Meteorol. Soc.* **1999**, *125*, 2239–2264. [CrossRef]
29. Raymond, D.J.; Zeng, X. Modeling tropical atmospheric convection in the context of the weak temperature gradient approximation. *Q. J. R. Meteorol. Soc.* **2005**, *131*, 1301–1320. [CrossRef]
30. Zeng, X.; Wang, Y.; MacCall, B.T. A $k-\varepsilon$ turbulence model for the stable atmosphere. *J. Atmos. Sci.* **2020**, *77*, 167–184. [CrossRef]
31. Zeng, X.; Tao, W.-K.; Lang, S.; Hou, A.Y.; Zhang, M.; Simpson, J. On the sensitivity of the atmospheric ensembles to cloud microphysics in long-term cloud-resolving model simulations. *J. Meteorol. Soc. Jpn.* **2008**, *86*, 45–65. [CrossRef]
32. LeMone, M.A.; Zipser, E.J. Cumulonimbus vertical velocity events in GATE. Part I: Diameter, intensity and mass flux. *J. Atmos. Sci.* **1980**, *37*, 2444–2457. [CrossRef]
33. Wei, D.; Blyth, A.M.; Raymond, D.J. Buoyancy of convective clouds in TOGA COARE. *J. Atmos. Sci.* **1998**, *55*, 3381–3391. [CrossRef]
34. Igau, R.C.; LeMone, M.A.; Wei, D. Updraft and downdraft cores in TOGA COARE: Why so many buoyant downdraft cores? *J. Atmos. Sci.* **1999**, *56*, 2232–2245. [CrossRef]
35. Zipser, E.J. Some view on "hot towers" after 50 years of tropical field programs and two years of TRMM data. In *Cloud Systems, Hurricanes, and the Tropical Rainfall Measuring Mission (TRMM)*; Tao, W.-K., Adler, R., Eds.; Meteorological Monographs; American Meteorological Society: Boston, MA, USA, 2003; Volume 29, pp. 49–58.
36. Toon, O.B.; Starr, D.O.; Jensen, E.J.; Newman, P.A.; Platnick, S.; Schoeberl, M.R.; Wennberg, P.O.; Wofsy, S.C.; Kurylo, M.J.; Maring, H.; et al. Planning, Implementation, and First Results of the Tropical Composition, Cloud and Climate Coupling Experiment (TC4). *J. Geophys. Res.* **2010**, *115*, D00J04. [CrossRef]
37. Heymsfield, G.M.; Tian, L.; Heymsfield, A.J.; Li, L.; Guimond, S. Characteristics of Deep Tropical and Subtropical Convection from Nadir-Viewing High-Altitude Airborne Doppler Radar. *J. Atmos. Sci.* **2010**, *67*, 285–308. [CrossRef]
38. Heymsfield, G.M.; Tian, L.; Li, L.; McLinden, M.; Cervantes, J.I. Airborne Radar Observations of Severe Hailstorms: Implications for Future Spaceborne Radar. *J. Appl. Meteorol. Climatol.* **2013**, *52*, 1851–1867. [CrossRef]

39. Houze, R.A., Jr. Cloud clusters and large-scale vertical motion in the tropics. *J. Meteorol. Soc. Japan* **1982**, *60*, 396–410. [CrossRef]
40. Zeng, X.; Tao, W.-K.; Zhang, M.; Hou, A.Y.; Xie, S.; Lang, S.; Li, X.; Starr, D.; Li, X. A contribution by ice nuclei to global warming. *Q. J. R. Meteorol. Soc.* **2009**, *135*, 1614–1629. [CrossRef]
41. Zeng, X.; Tao, W.-K.; Zhang, M.; Hou, A.Y.; Xie, S.; Lang, S.; Li, X.; Starr, D.; Li, X.; Simpson, J. An indirect effect of ice nuclei on atmospheric radiation. *J. Atmos. Sci.* **2009**, *66*, 41–61. [CrossRef]
42. Zeng, X.; Tao, W.-K.; Matsui, T.; Xie, S.; Lang, S.; Zhang, M.; Starr, D.; Li, X. Estimating the Ice Crystal Enhancement Factor in the Tropics. *J. Atmos. Sci.* **2011**, *68*, 1424–1434. [CrossRef]
43. Zeng, X.; Tao, W.-K.; Powell, S.W.; Houze, R.A., Jr.; Ciesielski, P.; Guy, N.; Pierce, H.; Matsui, T. A comparison of the water budgets between clouds from AMMA and TWP-ICE. *J. Atmos. Sci.* **2013**, *70*, 487–503. [CrossRef]
44. Fletcher, N.H. *The Physics of Rain Clouds*; Cambridge University Press: Cambridge, UK, 1962; p. 386.
45. DeMott, P.J.; Hill, T.C.J.; McCluskey, C.S.; Prather, K.A.; Collins, D.B.; Sullivan, R.C.; Ruppel, M.J.; Mason, R.H.; Irish, V.E.; Lee, T.; et al. Sea spray aerosol as a unique source of ice nucleating particles. *Proc. Natl. Acad. Sci. USA* **2016**, *113*, 5797–5803. [CrossRef]
46. Roberts, P.; Hallett, J. A laboratory study of the ice nucleating properties of some mineral particulates. *Q. J. R. Meteorol. Soc.* **1968**, *94*, 25–34. [CrossRef]
47. Edwards, G.R.; Evans, L.F. The mechanism of activation of ice nuclei. *J. Atmos. Sci.* **1971**, *28*, 1443–1447. [CrossRef]
48. Zeng, X.; Heymsfield, A.J.; Ulanowski, Z.; Neely III, R.R.; Li, X.; Gong, J.; Wu, D.L. The radiative effect on cloud microphysics from the Arctic to the Tropics. *Bull. Am. Meteorol. Soc.* **2022**, *103*, E2108–E2129. [CrossRef]
49. Fuchs, Z.; Raymond, D.J. Large-scale modes in a rotating atmosphere with radiative–convective instability and WISHE. *J. Atmos. Sci.* **2005**, *62*, 4084–4094. [CrossRef]
50. Fuchs, Ž.; Raymond, D.J. A simple model of intraseasonal oscillations. *J. Adv. Model. Earth Syst.* **2017**, *9*, 1195–1211. [CrossRef]
51. Emanuel, K.A. Slow modes of the equatorial waveguide. *J. Atmos. Sci.* **2020**, *77*, 1575–1582. [CrossRef]
52. Wang, S.; Sobel, A.H. A unified moisture mode theory for the Madden–Julian oscillation and the boreal summer intraseasonal oscillation. *J. Clim.* **2022**, *35*, 1267–1291. [CrossRef]
53. Emanuel, K.A. An air-sea interaction theory for tropical cyclones. Part I: Steady state maintenance. *J. Atmos. Sci.* **1986**, *43*, 585–604. [CrossRef]
54. Rotunno, R.; Emanuel, K.A. An air-sea interaction theory for tropical cyclones. Part II: Evolutionary study using a non-hydrostatic axisymmetric numerical model. *J. Atmos. Sci.* **1987**, *44*, 542–561. [CrossRef]
55. Yano, J.-I.; Emanuel, K.A. An improved model of the equatorial troposphere and its coupling with the stratosphere. *J. Atmos. Sci.* **1991**, *48*, 377–389. [CrossRef]
56. Emanuel, K.A. The behavior of a simple hurricane model using a convective scheme based on subcloud-layer entropy equilibrium. *J. Atmos. Sci.* **1995**, *52*, 3960–3968. [CrossRef]
57. Raymond, D.J. Regulation of moist convection over the west Pacific warm pool. *J. Atmos. Sci.* **1995**, *52*, 3945–3959. [CrossRef]
58. Takayabu, Y. Large-scale cloud disturbances associated with equatorial waves. Part II: Western-propagating inertio-gravity waves. *J. Meteorol. Soc. Jpn.* **1994**, *72*, 451–465. [CrossRef]
59. Chen, S.S.; Houze, R.A., Jr.; Mapes, B.E. Mutiscale variability of deep convection in relation to large-scale circulation in TOGA COARE. *J. Atmos. Sci.* **1996**, *53*, 1380–1409. [CrossRef]
60. Takayabu, Y.; Lau, K.M.; Sui, C.H. Observation of a quasi-two-day wave during TOGA COARE. *Mon. Wea. Rev.* **1996**, *124*, 1892–1913. [CrossRef]
61. Haertel, P.T.; Johnson, R.H. Two-day disturbances in the equatorial western Pacific. *Q. J. R. Meteorol. Soc.* **1998**, *124*, 615–636. [CrossRef]
62. Bretherton, C.S.; Blossey, P.N.; Khairoutdinov, M. An energy-balance analysis of deep convective self-aggregation above uniform SST. *J. Atmos. Sci.* **2005**, *62*, 4273–4292. [CrossRef]
63. Zeng, X.; Tao, W.-K.; Houze, R.A., Jr. Modeling Mesoscale Convective Systems in a Highly Simplified Environment. The First Science Team Meeting of the Atmospheric System Research (ASR) Program. Bethesda, Maryland. 15–19 March 2010 . Available online: http://asr.science.energy.gov/meetings/stm/posters/poster_pdf/2010/P000163.pdf (accessed on 10 January 2023).
64. Barnston, A.G.; Tippett, M.K.; L'Heureux, M.L.; Li, S.; DeWitt, D.G. Skill of real-time seasonal ENSO model predictions during 2002-2011: Is Our Capability Increasing? *Bull. Am. Meteorol. Soc.* **2012**, *93*, 631–651. [CrossRef]
65. Rotunno, R.; Klemp, J.B. The influence of shear-induced pressure gradient on thunderstorm motion. *Mon. Wea. Rev.* **1982**, *110*, 136–151. [CrossRef]
66. Houze, R.A., Jr. 2004: Mesoscale convective systems. *Rev. Geophys.* **2004**, *42*, RG4003. [CrossRef]
67. Zhang, M.H.; Lin, W.Y.; Klein, S.; Bacmeister, J.T.; Bony, S.; Cederwall, R.T.; Del Genio, A.D.; Hack, J.J.; Loeb, N.G.; Lohmann, U.; et al. Comparing clouds and their seasonal variations in 10 atmospheric general circulation models with satellite measurements. *J. Geophy. Res.* **2001**, *110*, D15S02. [CrossRef]
68. Nam, C.; Bony, S.; Dufresne, J.-L.; Chepfer, H. The 'too few, too bright' tropical low-cloud problem in CMIP5 models. *Geophys. Res. Lett.* **2012**, *39*, L21801. [CrossRef]
69. Jiang, J.H.; Su, H.; Zhai, C.; Perun, V.S.; Del Genio, A.; Nazarenko, L.S.; Donner, L.J.; Horowitz, L.; Seman, C.; Cole, J.; et al. Evaluation of cloud and water vapor simulations in CMIP5 climate models using NASA "A-Train" satellite observations. *J. Geophys. Res.* **2012**, *117*, D14105. [CrossRef]

70. Jiang, J.H.; Su, H.; Zhai, C.; Wu, L.; Minschwaner, K.; Molod, A.M.; Tompkins, A.M. An assessment of upper troposphere and lower stratosphere water vapor in MERRA, MERRA2, and ECMWF re-analyses using Aura MLS observations. *J. Geophys. Res.* **2015**, *120*, 11468–11485. [CrossRef]
71. Zhang, C. Madden-Julian Oscillation. *Rev. Geophys.* **2005**, *43*, RG2003. [CrossRef]
72. Zeng, X.; Gong, J.; Li, X.; Wu, D.L. Modeling the radiative effect on microphysics in cirrus clouds against satellite observations. *J. Geophys. Res.* **2021**, *126*, e2020JD033923. [CrossRef]
73. Henrici, P. *Applied and Computational Complex Analysis, Vol II*; Wiley: New York, NY, USA, 1977.
74. Shen, B.-W. Limit Cycles, Closed Orbits and Poincare-Bendixson Theorem in 2D Systems. *Course Mater.* M537 **2017**, 102. [CrossRef]
75. Matsuno, T. 1966: Quasi-geostrophic motions in the equatorial area. *J. Meteorol. Soc. Jpn.* **1966**, *44*, 25–43. [CrossRef]
76. Yanai, M.; Maruyama, T. Stratospheric wave disturbances propagating over the equatorial Pacific. *J. Meteorol. Soc. Japan* **1966**, *44*, 291–294. [CrossRef]
77. Wheeler, M.; Kiladis, G.N. Convectively coupled equatorial waves: Analysis of clouds and temperature in the wavenumber–frequency domain. *J. Atmos. Sci.* **1999**, *56*, 374–399. [CrossRef]

Article

The Dual Nature of Chaos and Order in the Atmosphere

Bo-Wen Shen [1,*], Roger Pielke, Sr. [2], Xubin Zeng [3], Jialin Cui [4], Sara Faghih-Naini [5,6], Wei Paxson [1], Amit Kesarkar [7], Xiping Zeng [8] and Robert Atlas [9,†]

[1] Department of Mathematics and Statistics, San Diego State University, San Diego, CA 92182, USA
[2] Cooperative Institute for Research in Environmental Sciences, University of Colorado Boulder, Boulder, CO 80203, USA
[3] Department of Hydrology and Atmospheric Science, The University of Arizona, Tucson, AZ 85721, USA
[4] Department of Computer Science, North Carolina State University, Raleigh, NC 27695, USA
[5] Department of Mathematics, University of Bayreuth, 95447 Bayreuth, Germany
[6] Department of Computer Science, Friedrich-Alexander University Erlangen-Nuremberg, 91058 Erlangen, Germany
[7] National Atmospheric Research Laboratory, Gadanki 517112, India
[8] Army Research Laboratory, Adelphi, MD 20783, USA
[9] Atlantic Oceanographic and Meteorological Laboratory, National Oceanic and Atmospheric Administration, Miami, FL 33149, USA
* Correspondence: bshen@sdsu.edu
† The author has retired.

Citation: Shen, B.-W.; Pielke, R., Sr.; Zeng, X.; Cui, J.; Faghih-Naini, S.; Paxson, W.; Kesarkar, A.; Zeng, X.; Atlas, R. The Dual Nature of Chaos and Order in the Atmosphere. *Atmosphere* **2022**, *13*, 1892. https://doi.org/10.3390/atmos13111892

Academic Editors: Ilias Kavouras and Anthony R. Lupo

Received: 10 September 2022
Accepted: 11 November 2022
Published: 12 November 2022

Abstract: In the past, the Lorenz 1963 and 1969 models have been applied for revealing the chaotic nature of weather and climate and for estimating the atmospheric predictability limit. Recently, an in-depth analysis of classical Lorenz 1963 models and newly developed, generalized Lorenz models suggested a revised view that *"the entirety of weather possesses a dual nature of chaos and order with distinct predictability"*, in contrast to the conventional view of *"weather is chaotic"*. The distinct predictability associated with attractor coexistence suggests limited predictability for chaotic solutions and unlimited predictability (or up to their lifetime) for non-chaotic solutions. Such a view is also supported by a recent analysis of the Lorenz 1969 model that is capable of producing both unstable and stable solutions. While the alternative appearance of two kinds of attractor coexistence was previously illustrated, in this study, multistability (for attractor coexistence) and monostability (for single type solutions) are further discussed using kayaking and skiing as an analogy. Using a slowly varying, periodic heating parameter, we additionally emphasize the predictable nature of recurrence for slowly varying solutions and a less predictable (or unpredictable) nature for the onset for emerging solutions (defined as the exact timing for the transition from a chaotic solution to a non-chaotic limit cycle type solution). As a result, we refined the revised view outlined above to: *"The atmosphere possesses chaos and order; it includes, as examples, emerging organized systems (such as tornadoes) and time varying forcing from recurrent seasons"*. In addition to diurnal and annual cycles, examples of non-chaotic weather systems, as previously documented, are provided to support the revised view.

Keywords: dual nature; chaos; generalized Lorenz model; predictability; multistability

1. Introduction

Two studies of Prof. Lorenz (Lorenz 1963, 1972 [1–3]) laid the foundation for chaos theory that emphasized the feature of Sensitive Dependence of Solutions on Initial Conditions (SDIC) [4,5]. Such a feature can be found in earlier studies [6–8] and was rediscovered by Lorenz [1]. Since Lorenz's studies in the 1960s, the concept of SDIC has changed the view on the predictability of weather and climate, yielding a paradigm shift from Laplace's view of determinism with unlimited predictability to Lorenz's view of deterministic chaos with finite predictability. Specifically, the statement "weather is chaotic" has been well accepted [9,10].

Amongst Lorenz models [1,2,11–16], the Lorenz 1963 and 1969 models have been applied to reveal the chaotic nature of the atmosphere and to estimate a predictability horizon. While the Lorenz 1963 (L63) equations represent a limited-scale, nonlinear, chaotic model [1,3,4], the Lorenz 1969 (L69) model is a closure-based, physically multiscale, mathematically linear, and numerically ill-conditioned system [17]. Although the L63 and L69 models have been applied for revealing the chaotic nature and for estimating the predictability limit for the atmosphere, we, in the scientific community, should be fully aware of the strength and weakness of the idealized models. As reviewed in Section 2, both models also produce various types of solutions. However, in past years, only chaotic solutions of the L63 model and linearly unstable solutions of the L69 model have been applied for studying atmospheric predictability. Therefore, asking whether or not other features of the classical Lorenz models and new features of generalized Lorenz models can improve our understanding of the nature of weather and climate is legitimate.

In contrast to single type chaotic solutions applied to reveal the nature of the atmosphere, recent studies using classical, simplified, and generalized Lorenz models [18–20] have emphasized the importance of considering various types of solutions, including solitary type solutions and coexisting chaotic and non-chaotic solutions, for revealing the complexity of the atmosphere and, thus, distinct predictability. For example, the physical relevance of findings within Lorenz models for real world problems has been reiterated by providing mathematical universality between the Lorenz simple weather and the Pedlosky simple ocean models [21–24]; and amongst the non-dissipative Lorenz model, the Duffing, the Nonlinear Schrodinger, the Korteweg–de Vries equations, and the simplified, epidemic SIR model [25–29]. Such an effort illustrated that mathematical solutions of the classical Lorenz model can help us understand the dynamics of different physical phenomena (e.g., solitary waves, homoclinic orbits, epidemic waves, and nonlinear baroclinic waves).

The L63 model mainly produces single types of nonlinear solutions, although it yields coexisting attractors over a very small interval for the heating parameter (e.g., [30]). By comparison, a newly developed generalized Lorenz model (GLM) with nine state variables yields attractor coexistence over a wide range of parameters. Such a feature is defined when chaotic and non-chaotic solutions appear within the same GLM that applies the same model parameters but different initial conditions. Additionally, when a time varying heating function is applied, various types of solutions may alternatively and concurrently appear. Based on findings obtained from classical and generalized Lorenz models [9,10], the conventional view is being revised to emphasize the dual nature of chaos and order, as follows:

> "Weather possesses chaos and order; it includes, as examples, emerging organized systems (such as tornadoes) and time varying forcing from recurrent seasons".

Since 2019 [9,10,18–20], the first portion of the above quote has been explicitly documented in our recent studies using the L63 model and GLM, while the second portion of the quote was implicitly suggested in Shen et al., 2021 [9]. In addition to various types of solutions within Lorenz models, earlier studies that applied regular solution to study non-chaotic weather systems were summarized in Shen et al., 2021 [9,10]. To support the revised view with a focus on the illustration for the second sentence, this article not only provides a brief review but also elaborates on additional details for the following features: (1) an analogy for monostability and multistability using skiing vs. kayaking; (2) single-types of attractors, SDIC, and monostability within the L63 model; (3) coexisting attractors and multistability within the GLM; (4) time varying multistability and slow time varying solutions; (5) the onset of emerging solutions; (6) various types of solutions within the L69 model; (7) distinct predictability within Lorenz models; (8) a list of non-chaotic weather systems; and (9) a short list of suggested future tasks. Table 1 lists the definitions of related concepts.

Table 1. The definitions of concepts related to predictability and multistability.

Name	Definitions	Recommendations
1st kind of attractor coexistence	The coexistence of chaotic and steady-state solutions	[9,10,28]
2nd kind of attractor coexistence	The coexistence of nonlinear oscillatory and steady-state solutions	[9,10]
attractor	The smallest attracting point set that, itself, cannot be decomposed into two or more subsets with distinct basins of attraction.	[31]
autonomous	A system of ODEs is autonomous if time does not explicitly appear within the equations.	[32]
bifurcation	It occurs when the structure of a system's solution significantly changes as a control parameter varies.	[32,33]
butterfly effect	*The phenomenon that a small alteration in the state of a dynamical system will cause subsequent states to differ greatly from the states that would have followed without the alteration.*	[3]
basin of attraction	As time advances, orbits initialized within a basin tend asymptotically to the attractor lying within the basin. The set of initial conditions leading to a given attractor.	[33]
chaos	Bounded aperiodic orbits exhibit a sensitive dependence on ICs.	[3]
final state sensitivity	Nearby orbits settle to one of multiple attractors for a finite but arbitrarily long time.	[34]
hidden attractor	An attractor is called a hidden attractor if its basin of attraction does not intersect with small neighborhoods of equilibria.	[35]
intransitivity	A specific type of solution lasts forever.	[36]
intrinsic predictability	Predictability that is only dependent on flow itself.	[9,37]
limit cycle	A nonlinear oscillatory solution; an isolated closed orbit	[32]
monostability	The appearance of single-type solutions	[9,10,17]
multistability	A system with multistability contains more than one bounded attractor that depends only on initial conditions. For example, the coexistence of two types of solutions.	[9,10,17,38,39]
non-autonomous	Variable time (τ) appears on the right-hand side of the equations.	[32]
phase space	Within a system of the first-order ODEs, a phase space or state space can be constructed using time-dependent variables as coordinates.	[40]
practical predictability	Predictability that is limited by imperfect initial conditions and/or (mathematical) formulas.	[9,37]
recurrence	Defined when a trajectory returns back to the neighborhood of a previously visited state. Recurrence may be viewed as a generalization of "periodicity" that braces quasi-periodicity with multiple frequencies and chaos.	[33]
sensitive dependence	*The property characterizing an orbit if most other orbits that pass close to it at some point do not remain close to it as time advances.*	[3]

2. Analysis and Discussion

As outlined at the end of Section 1, this section first applies skiing and kayaking to provide an analogy for monostability and multistability. The section then discusses major features of the L63 model, the GLM, and the L69 model. To present monostability and multistability, various types of solutions are reviewed. A list for non-chaotic weather systems is provided at the end of the section.

2.1. An Analogy for Monostability and Multistability Using Skiing and Kayaking

Since the SDIC, monostability, and multistability are the most important concepts in this study, to help readers, they are first illustrated using real-world analogies of skiing and kayaking. To explain SDIC, the book entitled "The Essence of Chaos" by Lorenz, 1993 [3] applied the activity of skiing (left in Figure 1) and developed an idealized skiing model for revealing the sensitivity of time-varying paths to initial positions (middle in Figure 1). Based on the left panel, when slopes are steep everywhere, SDIC always appears. This feature with a single type of solution is referred to as monostability.

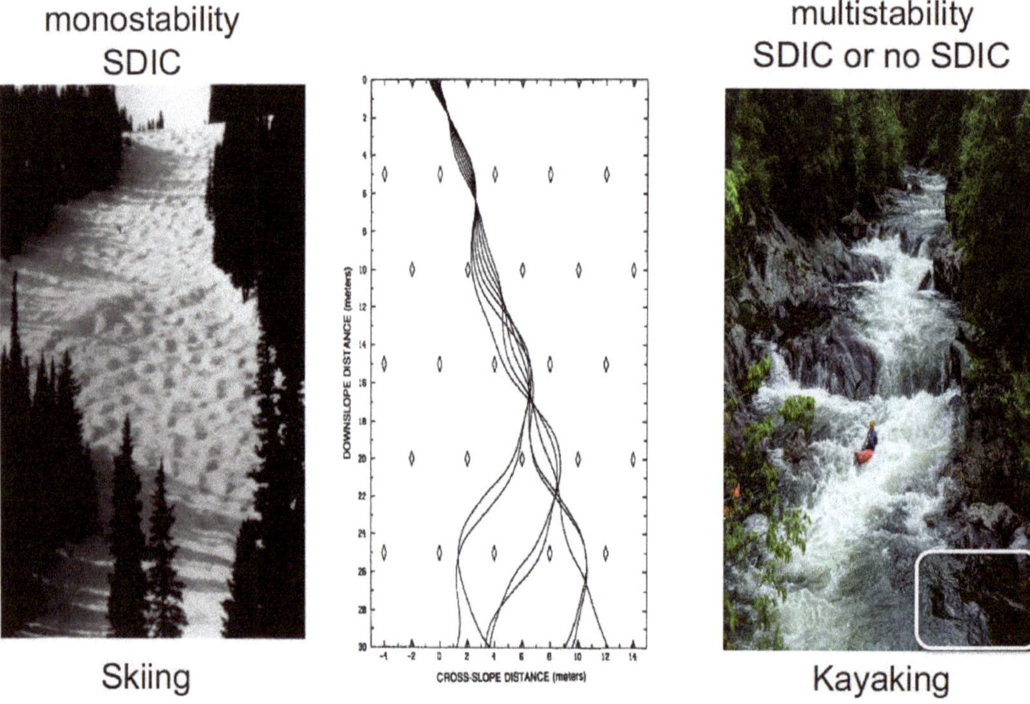

Figure 1. Skiing as used to reveal monostability (left and middle, Lorenz 1993 [3]) and kayaking as used to indicate multistability (right, courtesy of Shutterstock-Carol Mellema https://www.shutterstock.com/image-photo/kayaker-enjoys-whitewater-sinks-smoky-mountains-649533271 (accessed 1 November 2022)). A stagnant area is outlined with a white box.

In comparison, the right panel of Figure 1 for kayaking is used to illustrate multistability. In the photo, the appearance of strong currents and a stagnant area (outlined with a white box) suggests instability and local stability, respectively. As a result, when two kayaks move along strong currents, their paths display SDIC. On the other hand, when two kayaks move into a stagnant area, they become trapped, showing no typical SDIC (although a chaotic transient may occur [31]). Such features of SDIC or no SDIC suggest two types of solutions and illustrate the nature of multistability.

2.2. Single-Types of Attractors, SDIC, and Monostability within the L63 Model

The well-known L63 model that consists of three, first order ordinary differential equations (ODEs) with three state variables and three parameters is written as follows [1,41]:

$$\frac{dX}{d\tau} = \sigma Y - \sigma X,\tag{1}$$

$$\frac{dX}{d\tau} = \sigma Y - \sigma X,\tag{2}$$

$$\frac{dZ}{d\tau} = XY - bZ.\tag{3}$$

Here, τ is dimensionless time. Time-independent parameters include σ, r, and b. The first two parameters are known as the Prandtl number and the normalized Rayleigh number (or the heating parameter), respectively. The heating parameter represents a measure of temperature differences between the bottom and top layers, while the third parameter, b, roughly indicates the aspect ratio of the convection cell. This study keeps $\sigma = 10$ and $b = 8/3$, but varies r in different runs. For the three state variables, X, Y, and Z, the variable X is associated with a stream function that defines velocities, and variables Y and Z are related to temperature. The 2-page Supplementary Materials of Shen et al., 2021 [9] list the equations of the GLM that can be reduced to the L63 model. At small, medium, and large heating parameters, three types of solutions are known to appear [9,10,19]. Below, three numerical experiments, which apply three different values of Rayleigh parameters and keep the other two parameters as constants within the same L63 model, are discussed. The selected values of the Rayleigh parameters are 20, 28, and 350, representing weak, medium, and strong heating, respectively. For each of the three cases, both control and parallel runs were performed. The only difference in the two runs was that a tiny perturbation with $\epsilon = 10^{-10}$ was added into the initial condition (IC) of the parallel run. The SDIC along with continuous dependence of solutions on IC (CDIC) is then discussed. Solutions of the Y component are provided in the top panels of Figure 2. Three panels, from left to right, display long-term time independent responses, irregular temporal variations, and regular temporal oscillations referred to as steady-state, chaotic, and limit cycle solutions, respectively. Corresponding solutions within the X-Y phase space are shown in the bottom panels of Figure 2. Non-chaotic, steady-state and limit cycle solutions become point attractors and periodic attractors in panels 1d and 1f, respectively. As a comparison, all of the three types of attractors within the three-dimensional X-Y-Z phase space are provided in Figure 3. Below, the features of SDIC for chaotic solutions are further discussed.

For chaotic solutions in the middle panels of Figure 2, both the control and parallel runs produced very close responses at an initial stage, but very different results at a later time. Initial comparable results indicate that CDIC is an important feature of dynamical systems. Despite initial tiny differences, large differences at a later time, as indicated by the red and blue curves in Figure 2b, revealed the feature of SDIC. Such a feature suggests that a tiny change in an IC will eventually lead to a very different time evolution for a solution. However, the concept of SDIC does not suggest a causality relationship. Specifically, the initial tiny perturbation should not be viewed as the cause for a specific event (e.g., a maximum or minimum) that subsequently appears or for any transition between different events.

As discussed above, depending on the relative strength of the heating parameter, one and only one type of steady-state, chaotic, and limit cycle solution appears within the L63 model. Such a feature is referred to as monostability, as compared to multistability for coexisting attractors. Over several decades, chaotic solutions and monostability have been a focal point, yielding the statement "weather is chaotic". As discussed below, such an exclusive statement is being revised by taking coexisting attractors and time varying multistability into consideration.

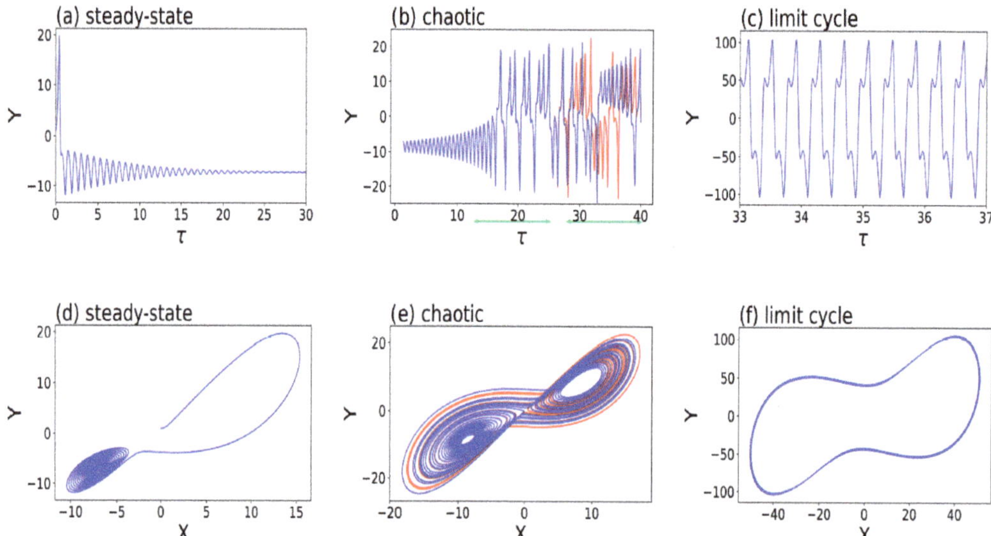

Figure 2. Three types of solutions within the Lorenz 1963 model. Steady-state (**a,d**), chaotic (**b,e**), and limit cycle (**c,f**) solutions appear at small, moderate, and large normalized Rayleigh parameters (i.e., r = 20, 28, and 350), respectively. Control and parallel runs are shown in red and blue, respectively. SDIC is indicated by visible blue and red curves in panel (**b**), where the first and second green horizonal lines indicate CDIC and SDIC, respectively. (**a–c**) depict the time evolution of Y. (**d–f**) show orbits within the X–Y space, appearing as a point attractor (**a,d**), a chaotic attractor (**b,e**), and a periodic attractor (**c,f**), respectively (after Shen et al., 2021 [10]). The other two parameters are kept as constants: σ = 10 and b = 8/3. The initial conditions of (X, Y, Z) for the control and parallel runs are $(0, 1, 0)$ and $(0, 1 + \epsilon, 0)$, respectively.

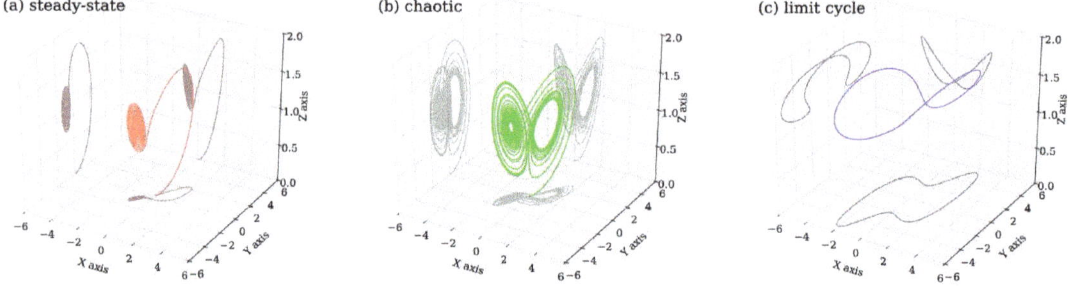

Figure 3. Three types of solutions within the X–Y–Z phase space obtained from the Lorenz 1963 model. Panels (**a–c**) display a steady-state solution, a chaotic solution, and a limit cycle with small, medium, and large heating parameters, respectively. While panels (**a,b**) show the solution for $\tau \in [0, 30]$ panel, to reveal its isolated feature, panel (**c**) displays the limit cycle solution for $\tau \in [10, 30]$. Values of parameters are the same as those in the control run in Figure 2.

2.3. Coexisting Attractors and Multistability within the GLM

Based on various high-dimensional Lorenz models [41–44], a generalized multi-dimensional Lorenz model (GLM) has been developed [18–20]. Mathematical descriptions of the GLM are summarized in the Supplementary Materials of Shen et al., 2021 [9]. Major features of the GLM include: (1) any odd number of state variables greater than three; (2) the aforementioned three types of solutions; (3) hierarchical spatial scale dependence (e.g., [43,44]); and (4) two kinds of attractor coexistence [9,18,20,45,46]. Additionally, aggregated negative feedback appears within high-dimensional LMs when the negative feedback of various smaller scale modes is accumulated to provide stronger dissipations, requiring stronger heating for the onset of chaos in higher-dimensional LMs. Such a finding is indicated in Table 2 of Shen, 2019 [19], which compared the critical values of heating parameters for the onset of chaos within the L63 and GLM that contains 5–9 state variables. Sufficiently large, aggregated negative feedback may cause (some) unstable equilibrium points to become stable and, thus, enable the coexistence of stable and unstable equilibrium points, yielding attractor coexistence, as illustrated below using the GLM with nine state variables.

Amongst the three types of solutions (i.e., steady-state, chaotic, and limit-cycle solutions), two types of solutions may appear within the same model that applies the same model parameters but different initial conditions. Such a feature is known as attractor coexistence. The GLM produces two kinds of attractor coexistence, including coexisting chaotic and steady-state solutions and coexisting limit-cycle and steady-state solutions, referred to as the 1st and 2nd kinds of attractor coexistence, respectively. To illustrate coexisting attractors that display a dependence on initial conditions, Figure 4 displays two ensemble runs, each run using 128 different initial conditions that were distributed over a hypersphere (e.g., Shen et al., 2019 [20]). The 1st and 2nd kinds of attractor coexistence are illustrated using values of 680 and 1600 for the heating parameter, respectively. As shown in Figure 4a, 128 orbits with different starting locations eventually reveal the 1st kind of attractor coexistence for one chaotic attractor, and two point attractors. As clearly seen in Figure 4a, each of the chaotic and non-chaotic attractors occupy a different portion of the phase space. By comparison, in Figure 4b, 128 ensemble members produce the 2nd kind of attractor coexistence, consisting of limit cycle and steady-state solutions.

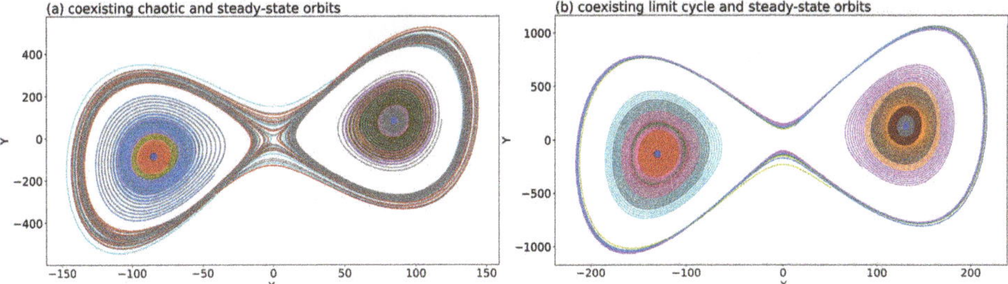

Figure 4. Two kinds of attractor coexistence using the GLM with 9 modes. Each panel displays orbits from 128 runs with different ICs for $\tau \in [0.625, 5]$. Curves in different colors indicate orbits with different initial conditions. (**a**) displays the coexistence of chaotic and steady-state solutions with $r = 680$. Stable critical points are shown with large blue dots. (**b**) displays the coexistence of the limit cycle and steady-state solutions with $r = 1600$.

The above two kinds of attractor coexistence occur in association with the coexistence of unstable (i.e., a saddle point) and stable equilibrium points. As discussed in Shen, 2019 [18] and reviewed above, the appearance of local stable equilibrium points is enabled by the so-called aggregated negative feedback of small-scale convective processes. The feature with coexisting attractors is referred to as multistability, as compared to monostability for single type attractors. As a result of multistability, SDIC does not always appear. Namely, SDIC appears when two orbits become the chaotic attractor that occupies one portion of the phase space; it does not appear when two orbits move towards the same point attractor that occupies the other portion of the phase space.

In a recent study by Shen et al. [47], three major kinds of butterfly effects can be identified: (1) SDIC, (2) the ability of a tiny perturbation in creating an organized circulation at a large distance, and (3) the hypothetical role of small scale processes in contributing to finite predictability. While the first kind of butterfly effect with SDIC is well accepted, the concept of multistability suggests that the first kind of butterfly effect does not always appear.

2.4. Time Varying Multistability and Recurrent Slowly Varying Solutions

Within Lorenz models, a time varying heating function may be applied to represent a large-scale forcing system ([9,36]; and references therein). Since the heating function changes with time, the first and second kinds of attractor coexistence alternatively appear, leading to time varying multistability and transitions from chaotic to non-chaotic solutions. An analysis of the above two features can help reveal the predictable nature of recurrence for slowly varying solutions, and a less predictable (or unpredictable) nature for the onset for emerging solutions (defined as the exact timing for the transition from a chaotic solution to a non-chaotic limit cycle type solution).

Here, by extending the study of Shen et al., 2021 [9], the GLM with a time-dependent heating function is applied in order to revisit time varying multistability. Figure 5a displays three trajectories during a dimensionless time τ between 0 and 35π (i.e., $\tau \in [0, 35\pi]$). As shown in the second panels of Figure 5, these solutions were obtained using tiny differences in ICs and a time varying Rayleigh parameter. The 3rd to 5th panels display the feature of SDIC [3,17,47], the first kind of attractor coexistence (i.e., coexisting steady-state and chaotic solutions), and the second kind of attractor coexistence (i.e., coexisting steady-state and periodic solutions) at different time intervals, respectively. The alternative appearance of two kinds of attractor coexistence suggests time varying multistability. By comparison, the 6th panel indicates that once a "steady state" solution appears (when local time changes for all state variables become zero), it remains "steady" and slowly varies with the time-dependent heating function. Such a recurrent, slowly varying solution may, in reality, be used as an analogy for recurrent seasons.

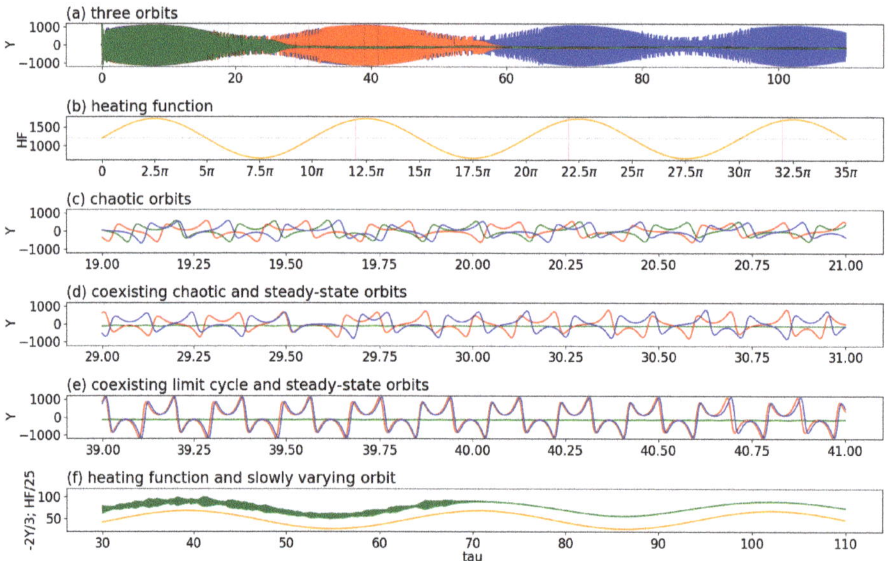

Figure 5. Two kinds of attractor coexistence revealed by three trajectories using a time varying heating parameter (i.e., Rayleigh parameter), $r = 1200 + 520 \sin (\tau/5)$, within a GLM (Shen, 2019 [18]).

The green, blue, and red lines represent the solutions of the control and two parallel runs. The parallel runs include an initial tiny perturbation, $\epsilon = 10^{-8}$ or $\epsilon = -10^{-8}$. The heating function is indicated by an orange line. From top to bottom, panels (**a**,**b**) display the three orbits and the heating parameters for $\tau \in [0, 35\pi]$, respectively. Panel (**c**) for $\tau \in [19, 21]$ displays diverged trajectories, showing SDIC. The first kind of attractor coexistence (i.e., coexisting chaotic and steady-state solutions) is shown in panel (**d**) for $\tau \in [29, 31]$. The green line, indeed, represents a steady state solution. The second kind of attractor coexistence (i.e., coexisting regular oscillations and steady-state solutions) is shown in panel (**e**) for $\tau \in [39, 41]$. Panel (**f**) displays a nearly steady-state solution ($2Y/3$) and the heating function for $\tau \in [30, 110]$. The three vertical lines in panel (**b**) indicate the starting time for the analysis in Figure 6. (After Shen et al., 2021 [9]).

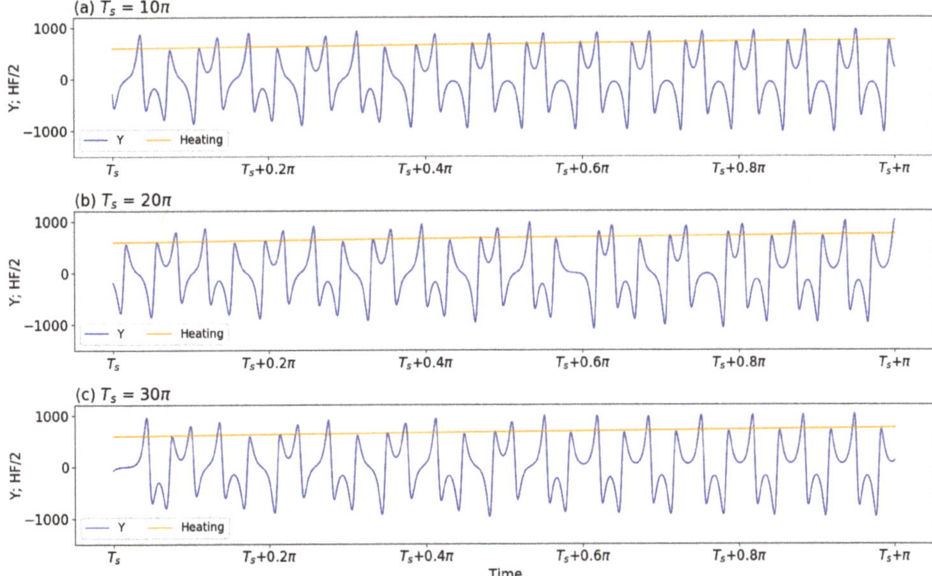

Figure 6. Panels (**a**–**c**) display the same trajectory during three different time intervals of $\tau \in [T_s, T_s + \pi]$, with the starting time T_s equal to 10π, 20π, and 30π, respectively. The orange line in each panel represents the half value of the heating function.

2.5. Onset of Emerging Solutions

Although the GLM with a time dependent heating function produces time varying multistability, as a result of simplicity within the model and the complexity of the problem, here, no attempt is made to address the attractor basin, intra-transitivity, final state sensitivity, or a hidden attractor (e.g., [9,46]; see details in Table 1). Instead, the transition from one type of solution to the other type of solution is a focus. Here, for emulating a real-world scenario for the first appearance of African Easterly waves during a seasonal transition [9,19], the transition from a chaotic (irregular) solution to a limit cycle type (regular) solution is analyzed. Within Figure 5, while two trajectories display a sensitivity to initial conditions after $\tau > 19$ (Figure 5c), they become regularly oscillatory solutions with comparable frequencies and amplitudes for $\tau \in [39, 41]$ (Figure 5e). Reappearance of the regular solution (i.e., limit cycle type solutions) is defined as the "onset of an emerging solution". Below, a challenge in predicting the onset of the transtion from a chaotic solution to a regularly oscillatory solution is illustrated.

Figure 6 displays the same trajectory during three different time intervals of $\tau \in [T_s, T_s + \pi]$. Here, the starting time T_s is equal to 10π, 20π, and 30π in panels (**a**)–(**c**), respectively. An orange line in each panel represents the half value of the heating function. Given any vertical line, its intersection with each of the three orange lines in all panels yields the same value for the heating parameter. In other words, although the starting time is different, the time evolution for the heating functions is exactly the same in all three panels. As a result, if the appearance of regularly oscillatory solutions is solely determined by the values of the heating function, the same time evolution of oscillatory solutions should appear in the three panels of Figure 6. However, differences are observed. For example, panel (a) displays the onset of a regularly oscillatory solution (e.g., the timing of the regular solution) at $\tau = T_s + 0.4\pi = 10.4\pi$. Panels (b) and (c) show an onset around $T_s + 0.8\pi = 20.8\pi$ and $T_s + 0.6\pi = 30.6\pi$, respectively. As a result of the time lag, the correlation coefficient for the solution curves in any two panels is low, although the solutions are regularly oscillatory. The results not only indicate a challenge in predicting the onset of oscillatory solutions but also suggest the importance of removing phase differences for verification (to obtain a better correlation between two solution curves).

For the three solution curves (which represent the same trajectory at different time intervals) in Figure 6, different times for the onset of solutions suggest a time lag (or phase differences). However, the period and amplitude for solutions within the regime of limit-cycle solutions are comparable, suggesting an optimistic view in predictability. Similar to Figure 6, Figure 7a,b provides the same trajectory during three different time intervals of $\tau \in [T_s, T_s + \pi]$, including T_s for 12π, 22π, and 32π. Such a choice is required in order to assure the full development of nonlinear oscillatory solutions. The selected time intervals are referred to as Epoch-1, Epoch-2, and Epoch-3, respectively, and the solution curve from the first epoch (i.e., Epoch-1) is used as a reference curve. The top panels clearly show phase differences amongst the solution curves, although the heating function remains the same for selected epochs. For solution curves during two different epochs, their cross correlation is computed to determine a time lag. Such a time lag is then considered to have a different starting time for the first epoch. For example, in panel (c), a time lag of $\Delta\tau = 0.0522$ is added to become a new starting time for the revised first epoch, referred to as a revised Epoch-1. After the time lag is considered, the solution curves for revised Epoch-1 and Epoch-2 are the same. In a similar manner, panel (d) displays the same evolution for solution curves for another revised Epoch-1 and Epoch-3.

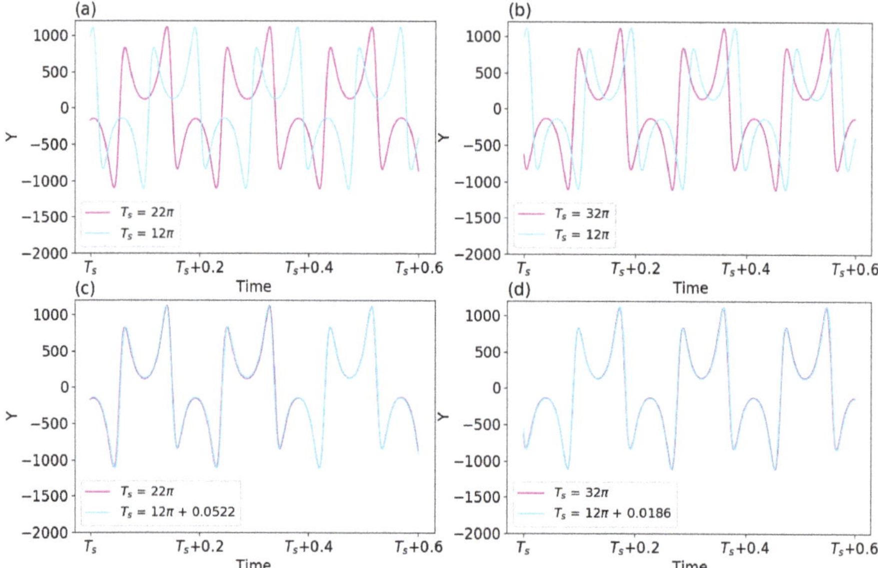

Figure 7. Panels (**a,b**) display the same trajectory during three different time intervals of $\tau \in [T_s, T_s + \pi]$, with the starting time T_s equal to 12π, 22π, and 32π, respectively. These three time intervals are referred to as Epoch-1, Epoch-2, and Epoch-3, respectively. In panels (**c,d**), to adjust the phase differences between two solutions curves, a time lag is added into Epoch-1.

Time varying multistability associated with the alternative occurrence of two kinds of attractor coexistence yields the alternative and concurrent appearance of various types of solutions. Such a feature indicates complexities of weather and climate that possess both chaotic and non-chaotic solutions, and their transitions. While chaotic solutions and their transition to regular solutions possess limited predictability, slowly varying solutions and nonlinear oscillatory solutions (i.e., limit-cycle type solutions) may have better predictability. The relationship between time varying multistability and distinct predictability is further elaborated after a discussion of the L69 model and its solutions.

2.6. Various Types of Solutions within the L69 Model

The L63 model is effective in revealing the chaotic nature of weather and climate and in suggesting a finite predictability [17]. The model has also been applied for estimating the predictability horizon using the global or finite-time Lyapunov exponent [48,49]. However, only a qualitative description of finite predictability has been established. The time varying, finite-time Lyapunov exponents suggest the dependence of finite predictability on initial conditions [9,50,51]. Some initial conditions may lead to better predictability than other initial conditions.

For a real-world application to predictability within the atmosphere, the L69 model was applied in order to produce a predictability limit of two weeks [47,52]. To examine the validity of the findings, a recent study by Shen et al., 2022 [17] summarized the major features of the L69 model, as follows:

- The L69 model is a closure-based, physically multiscale, mathematically linear, and numerically ill-conditioned system.
- As compared to other turbulence models, the L69 model applied the common assumptions of homogeneity and isotropy [53–61]. However, since the L69 model was derived from a conservative partial differential equation without dissipations, it is not a turbulence model.
- The L69 multiscale model has been used for revealing energy transfer and scale interaction.
- The L69 linear model cannot produce chaos.
- Since it possesses both positive and negative eigenvalues with large variances, yielding a large condition number (e.g., Figure 4 and Figure 5 of [17]), the L69 model produces a different kind of sensitivity, as compared to SDIC within the L63 model.
- The model permits the occurrence of linearly stable and unstable solutions as well as oscillatory solutions. However, only unstable solutions have been a focus in predictability studies.

Since the Lorenz 1969 multiscale model consists of 21, 2nd order ODEs (i.e., 42, 1st order ODEs), it is not necessarily viewed as a low-order dynamical system, as compared to many low-order systems in nonlinear dynamics textbooks. As compared to real world weather or climate models [19], it is a low-dimensional system. Given the above features of the L63 and L69 models, estimates for the predictability limit using either of the above models should be cautiously interpreted and should not be generalized as an upper limit for atmospheric predictability. Although these idealized models may not be suitable for quantitative analyses, as discussed below, the abundant features of their solutions may effectively, qualitatively reveal the complicated nature of weather and climate.

2.7. Distinct Predictability within Lorenz Models

As discussed, the L63 and L69 models have been applied for revealing the chaotic and unstable nature of weather and climate, and their intrinsic predictability, as well as the practical predictability of models [17,37]. Major findings regarding predictability and multistability within the two classical models and the GLM are summarized as follows:

- The L63 nonlinear model with monostability is effective for revealing the chaotic nature of weather, suggesting finite intrinsic predictability within the chaotic regime of the system (i.e., the atmosphere).

- The L69 linear model with ill-conditioning easily captures unstable modes and, thus, is effective for revealing the practical finite predictability of the model.
- The GLM with multistability suggests both limited and unlimited (i.e., up to a system's lifetime) intrinsic predictability for chaotic and non-chaotic solutions, respectively.

All of the above three systems in bullets are autonomous. Namely, system parameters are not explicit functions of time. By comparison, when a time varying heating function was applied within the GLM, the system became non-autonomous. Such a system produces a time varying multistability that is modulated by large-scale time varying forcing (heating). As a result, various types of solutions (e.g., steady-state, chaotic, and limit cycle solutions) with distinct predictability appear alternatively and/or concurrently. For example, slowly varying solutions may coexist with chaotic solutions for one period of time, and with limit cycle solutions for another period of time. Additionally, as illustrated in Section 2.5, the onset of an emerging solution (i.e., during a transition from a chaotic solution to a limit-cycle type solution) suggests a different predictability problem. In summary, the above discussions support the revised view that emphasizes the dual nature of chaos and order.

2.8. Non-Chaotic Weather Systems

The L69 model with forty-two, first order ODE was applied in order to study the multiscale predictability of weather. Although the L69 model is neither a low-order system nor a turbulence model (due to the lack of dissipative terms), major findings using the L69 model were indeed supported by studies using turbulence models [17,55,56]. By comparison, the L63 model has been used to illustrate the chaotic nature in weather and climate. Finite-dimensional chaotic responses revealed using the simple L63 model can be captured using rotating annulus experiments in the laboratory [62,63], illustrated by an analysis of weather maps (Figures 10.6 and 10.7 in [63]), and simulated using more sophisticated models (e.g., [64]). Based on ensemble runs using a weather model, the feature of local finite dimensionality [65–67] indicates a simple structure for instability (e.g., within a few dominant state space directions) (personal communication with Prof. Szunyogh) and, thus, suggests the occurrence of finite-dimensional chaotic responses.

As indicated by the title of Chapter 3 in [3], "Our Chaotic Weather", and the title of [63], "Application of Chaos to Meteorology and Climate", applying chaos theory for understanding weather and climate has been a focus for several decades [68,69]. By comparison, non-chaotic solutions have been previously applied for understanding the dynamics of different weather systems, including steady-state solutions for investigating atmospheric blocking (e.g., [70,71]), limit cycles for studying 40-day intra-seasonal oscillations [72], quasi-biennial oscillations [73] and vortex shedding [74], and nonlinear solitary-pattern solutions for understanding morning glory (i.e., a low-level roll cloud, [75]). While additional detailed discussions regarding non-chaotic weather systems are being documented in a separate study, Table 2 provides a summary.

Table 2. Non-chaotic Solutions vs. Weather Systems.

Type	Weather Systems	References
Steady-state Solutions	Atmospheric blocking	[70,71]
Limit Cycles	40-day intra-seasonal oscillations	[72]
	Quasi-biennial oscillations	[73]
	Vortex shedding	[74]
Nonlinear solitary-pattern solutions	Morning glory	[75]

While the conventional view focuses on chaos, as well as monostability, our revised view emphasizes the possibility for coexisting chaotic and non-chaotic weather systems, yielding the concept of multistability. As illustrated using the analogy with skiing and kayaking in Figure 1 (within a physical space), the major difference between monostability and multistability in the acknowledgement of "inhomogeneity" (within a phase space). Namely, SDIC (i.e., chaos) only appears within some physical/phase spaces instead of the entire spaces. Furthermore, the GLM with a time varying heating function, the so-called non-autonomous system [32,36], illustrated not only the temporal evolution but also the onset of various kinds of solutions, better representing the complexities of weather and climate, as compared to classical models with time independent parameters (i.e., autonomous systems).

2.9. Suggested Future Tasks

Numerous interesting studies in nonlinear dynamics exist and many of them have the potential to improve our understanding of weather and climate. Here, since suggested future tasks and additional studies and concepts will be covered in the future, only a few studies are discussed. Based on the above discussions, an effective classification of chaotic and non-chaotic solutions may identify systems with better predictability. Such a goal may be achieved by applying or extending existing tools, including the recurrence analysis method, the kernel principal component analysis method, the parallel ensemble empirical mode decomposition method, etc. (e.g., [45,46,76,77]). On the other hand, to separate chaotic and non-chaotic attractors, the detailed attractor basin for each attractor should be determined. Then, it becomes feasible to address final state sensitivity and intra-transitivity, as defined in Table 1. Whether or not the number of attractors in our weather is finite is another interesting but challenging question. Given a specific system that possesses infinite attractors, the detection of "megastability" and "extreme multistability" that correspond to countable and uncountable attractors [39], respectively, further increases the level of the challenge. In addition to the above three types of solutions and two kinds of attractor coexistence, homoclinic phenomena (e.g., homoclinic bifurcation, homoclinic tangencies, and homoclinic chaos), which have been intensively studied within low-order systems deserve to be explored in high-dimensional systems (e.g., using the GLM with nine modes or higher) [78–80].

There is no doubt that the "butterfly effect", originally derived from the Lorenz's 1963 study [1,3], is a fascinating idea that has inspired many researchers to devote their time and effort to related research. For example, Lorenz's attractor and Lorenz-type attractors (that are associated with Lorenz and Shilnikov types of saddle points, respectively) [1,3,81–85] have been rigorously examined within dynamical systems (e.g., the L63 model and the Shimizu-Morioka model [81]). Among the three kinds of butterfly effects classified by Shen et al., 2022 [47] and reviewed above, the first kind of butterfly effect with SDIC is well accepted, while the second kind of butterfly effect remains a metaphor. Thus, there is a definite need to finally answer the following question: "Can a butterfly flap cause a tornado in Texas?". Stated more robustly, can a small perturbation create a coherent larger scale feature at large distances? If so, what are the sizes and intensity that can produce such a feature and at what distances? All of the above questions are the subject of future studies.

3. Concluding Remarks

In this study, we not only provided a review of the features of various types of solutions within classical and generalized Lorenz models, but also presented a new analysis for the onset of emerging solutions, as well as a list of non-chaotic weather systems, to propose a refined version of the revised view:

> "The atmosphere possesses chaos and order; it includes, as examples, emerging organized systems (such as tornadoes) and time varying forcing from recurrent seasons",

in contrast to the conventional view of "weather is chaotic". The revised view focuses on: (1) the dual nature of chaos and order with distinct predictability in weather and climate; (2) the role of slowly varying environmental forcing (e.g., heating) in modulating the appearance of chaos and order, and in determining slowly varying systems; and (3) the challenge in determining the exact timing for the onset of a new, recurrent epoch (or regime).

From the perspective of dynamical systems, time varying multistability represents the key concept of the revised view. As compared to monostability that indicates the appearance of single type solutions (such as steady-state, chaotic, or limit cycle solutions), multistability permits the coexistence of chaotic and non-chaotic solutions that occupy different portions of the phase space. To reveal fundamental differences between monostability and multistability, an analogy using skiing and kayaking was provided. In addition to diurnal and annual cycles, a short list of non-chaotic weather systems was provided in Section 2.8. Suggested future tasks were presented in Section 2.9.

Author Contributions: Conceptualization, B.-W.S., R.P.S., X.Z. (Xubin Zeng), and R.A.; methodology, B.-W.S.; software, B.-W.S., J.C., W.P., and S.F.-N.; writing—original draft preparation, B.-W.S., R.P.S., and X.Z. (Xubin Zeng); writing—review and editing, B.-W.S., R.P.S., X.Z., J.C., S.F.-N., W.P., A.K., X.Z. (Xiping Zeng) and R.A. All authors have read and agreed to the published version of the manuscript.

Funding: This research received no external funding.

Institutional Review Board Statement: Not applicable.

Informed Consent Statement: Not applicable.

Data Availability Statement: Not applicable.

Acknowledgments: We thank academic editors, editors, four anonymous reviewers, and Istvan Szunyogh for valuable comments and discussions.

Conflicts of Interest: The authors declare no conflict of interest.

References

1. Lorenz, E.N. Deterministic nonperiodic flow. *J. Atmos. Sci.* **1963**, *20*, 130–141. [CrossRef]
2. Lorenz, E.N. Predictability: Does the Flap of a Butterfly's Wings in Brazil Set off a Tornado in Texas? In Proceedings of the 139th Meeting of AAAS Section on Environmental Sciences, New Approaches to Global Weather, GARP, AAAS, Cambridge, MA, USA, 29 December 1972.
3. Lorenz, E.N. *The Essence of Chaos*; University of Washington Press: Seattle, WA, USA, 1993; 227p.
4. Gleick, J. *Chaos: Making a New Science*; Penguin: New York, NY, USA, 1987; 360p.
5. The Nobel Committee for Physics. *Scientific Background on the Nobel Prize in Physics 2021 For Groundbreaking Contributions to Our Understanding of Complex Physical Systems*; The Nobel Committee for Physics: Stockholm, Sweden, 2021; Available online: https://www.nobelprize.org/prizes/physics/2021/popular-information/ (accessed on 28 June 2022).
6. Maxwell, J.C. *Matter and Motion*; Dover: Mineola, NY, USA, 1952.
7. Poincaré, H. Sur le problème des trois corps et les équations de la dynamique. *Acta Math.* **1890**, *13*, 1–270.
8. Poincaré, H. *Science et Méthode. Flammarion*; Maitland, F., Translator; Science and Method 1908; Thomas Nelson and Sons: London, UK, 1914.
9. Shen, B.-W.; Pielke, R.A., Sr.; Zeng, X.; Baik, J.-J.; Faghih-Naini, S.; Cui, J.; Atlas, R. Is Weather Chaotic? Coexistence of Chaos and Order within a Generalized Lorenz Model. *Bull. Am. Meteorol. Soc.* **2021**, *2*, E148–E158. [CrossRef]
10. Shen, B.-W.; Pielke, R.A., Sr.; Zeng, X.; Baik, J.-J.; Faghih-Naini, S.; Cui, J.; Atlas, R.; Reyes, T.A. Is Weather Chaotic? Coexisting Chaotic and Non-Chaotic Attractors within Lorenz Models. In Proceedings of the 13th Chaos International Conference CHAOS 2020, Florence, Italy, 9–12 June 2020; Skiadas, C.H., Dimotikalis, Y., Eds.; Springer Proceedings in Complexity. Springer: Cham, Switzerland, 2021. [CrossRef]
11. Lorenz, E.N. The predictability of a flow which possesses many scales of motion. *Tellus* **1969**, *21*, 289–307. [CrossRef]
12. Lorenz, E.N. Investigating the Predictability of Turbulent Motion. In *Statistical Models and Turbulence, Proceedings of the symposium held at the University of California, San Diego, CA, USA, 15–21 July 1971*; Springer: Berlin/Heidelberg, Germany, 1972; pp. 195–204.
13. Lorenz, E.N. Low-order models representing realizations of turbulence. *J. Fluid Mech.* **1972**, *55*, 545–563. [CrossRef]
14. Lorenz, E.N. Irregularity: A fundamental property of the atmosphere. *Tellus* **1984**, *36A*, 98–110. [CrossRef]
15. Lorenz, E.N. Predictability—A Problem Partly Solved. In Proceedings of the Seminar on Predictability, Reading, UK, 4–8 September 1995; ECMWF: Reading, UK, 1996; Volume 1.
16. Lorenz, E.N. Designing Chaotic Models. *J. Atmos. Sci.* **2005**, *62*, 1574–1587. [CrossRef]
17. Shen, B.-W.; Pielke, R.A., Sr.; Zeng, X. One Saddle Point and Two Types of Sensitivities Within the Lorenz 1963 and 1969 Models. *Atmosphere* **2022**, *13*, 753. [CrossRef]
18. Shen, B.-W. Aggregated Negative Feedback in a Generalized Lorenz Model. *Int. J. Bifurc. Chaos* **2019**, *29*, 1950037. [CrossRef]
19. Shen, B.-W. On the Predictability of 30-Day Global Mesoscale Simulations of African Easterly Waves during Summer 2006: A View with the Generalized Lorenz Model. *Geosciences* **2019**, *9*, 281. [CrossRef]
20. Shen, B.-W.; Reyes, T.; Faghih-Naini, S. Coexistence of Chaotic and Non-Chaotic Orbits in a New Nine-Dimensional Lorenz Model. In Proceedings of the 11th Chaotic Modeling and Simulation International Conference, CHAOS 2018, Rome, Italy, 5–8 June 2018; Skiadas, C., Lubashevsky, I., Eds.; Springer Proceedings in Complexity. Springer: Cham, Switzerland, 2019. [CrossRef]
21. Pedlosky, J. Finite-amplitude baroclinic waves with small dissipation. *J. Atmos. Sci.* **1971**, *28*, 587–597. [CrossRef]

22. Pedlosky, J. Limit cycles and unstable baroclinic waves. *J. Atmos. Sci.* **1972**, *29*, 53–63. [CrossRef]
23. Pedlosky, J. *Geophysical Fluid Dynamics*, 2nd ed.; Springer: New York, NY, USA, 1987; 710p.
24. Pedlosky, J. The Effect of Beta on the Downstream Development of Unstable, Chaotic BaroclinicWaves. *J. Phys. Oceanogr.* **2019**, *49*, 2337–2343. [CrossRef]
25. Shen, B.-W. On periodic solutions in the non-dissipative Lorenz model: The role of the nonlinear feedback loop. *Tellus A* **2018**, *70*, 1471912. [CrossRef]
26. Faghih-Naini, S.; Shen, B.-W. Quasi-periodic orbits in the five-dimensional non-dissipative Lorenz model: The role of the extended nonlinear feedback loop. *Int. J. Bifurc. Chaos* **2018**, *28*, 1850072. [CrossRef]
27. Shen, B.-W. Homoclinic Orbits and Solitary Waves within the non-dissipative Lorenz Model and KdV Equation. *Int. J. Bifurc. Chaos* **2020**, *30*, 2050257. [CrossRef]
28. Shen, B.-W. Solitary Waves, Homoclinic Orbits, and Nonlinear Oscillations within the non-dissipative Lorenz Model, the inviscid Pedlosky Model, and the KdV Equation. In Proceedings of the 13th Chaos International Conference CHAOS 2020, Florence, Italy, 9–12 June 2020; Skiadas, C.H., Dimotikalis, Y., Eds.; Springer Proceedings in Complexity. Springer: Cham, Switzerland, 2021.
29. Paxson, W.; Shen, B.-W. 2022: A KdV-SIR Equation and Its Analytical Solutions for Solitary Epidemic Waves. *Int. J. Bifurc. Chaos* **2022**, *32*, 2250199. [CrossRef]
30. Yorke, J.; Yorke, E. Metastable chaos: The transition to sustained chaotic behavior in the Lorenz model. *J. Stat. Phys.* **1979**, *21*, 263–277. [CrossRef]
31. Sprott, J.C.; Wang, X.; Chen, G. Coexistence of Point, periodic and Strange attractors. *Int. J. Bifurc. Chaos* **2013**, *23*, 1350093. [CrossRef]
32. Jordan, D.W.; Smith, P. Nonlinear Ordinary Differential Equations. In *An Introduction for Scientists and Engineers*, 4th ed.; Oxford University Press: Oxford, UK, 2007; p. 560.
33. Thompson, J.M.T.; Stewart, H.B. *Nonlinear Dynamics and Chaos*, 2nd ed.; John Wiley & Sons, Ltd.: Hoboken, NJ, USA, 2002; p. 437.
34. Grebogi, C.; McDonald, S.W.; Ott, E.; Yorke, J.A. Final state sensitivity: An obstruction to predictability. *Phys. Lett. A* **1983**, *99*, 415–418. [CrossRef]
35. Leonov, G.A.; Kuznetsov, N.V. Hidden attractors in dynamical systems. from hidden oscillations in Hilbert–Kolmogorov, Aizerman, and Kalman problems to hidden chaotic attractor in Chua circuits. *Int. J. Bifurc. Chaos* **2014**, *23*, 1330002. [CrossRef]
36. Lorenz, E.N. Can chaos and intransitivity lead to interannual variability? *Tellus* **1990**, *42A*, 378–389. [CrossRef]
37. Lorenz, E.N. The predictability of hydrodynamic flow. *Trans. N. Y. Acad. Sci.* **1963**, *25*, 409–432. [CrossRef]
38. Lai, Q.; Chen, S. Coexisting attractors generated from a new 4D smooth chaotic system. *Int. J. Contr. Autom. Syst.* **2016**, *14*, 1124–1131. [CrossRef]
39. Jafari, S.; Rajagopal, K.; Hayat, T.; Alsaedi, A.; Pham, V.-T. Simplest Megastable Chaotic Oscillator. *Int. J. Bifurc. Chaos* **2019**, *29*, 1950187. [CrossRef]
40. Hilborn, R.C. Chaos and Nonlinear Dynamics. In *An Introduction for Scientists and Engineers*, 2nd ed.; Oxford University Press: Oxford, UK, 2000; p. 650.
41. Shen, B.-W. Nonlinear feedback in a five-dimensional Lorenz model. *J. Atmos. Sci.* **2014**, *71*, 1701–1723. [CrossRef]
42. Shen, B.-W. Nonlinear feedback in a six-dimensional Lorenz Model: Impact of an additional heating term. *Nonlin. Processes Geophys.* **2015**, *22*, 749–764. [CrossRef]
43. Shen, B.-W. Hierarchical scale dependence associated with the extension of the nonlinear feedback loop in a seven-dimensional Lorenz model. *Nonlin. Processes Geophys.* **2016**, *23*, 189–203. [CrossRef]
44. Shen, B.-W. On an extension of the nonlinear feedback loop in a nine-dimensional Lorenz model. *Chaotic Modeling Simul.* **2017**, *2*, 147–157.
45. Reyes, T.; Shen, B.-W. A Recurrence Analysis of Chaotic and Non-Chaotic Solutions within a Generalized Nine-Dimensional Lorenz Model. *Chaos Solitons Fractals* **2019**, *125*, 1–12. [CrossRef]
46. Cui, J.; Shen, B.-W. A Kernel Principal Component Analysis of Coexisting Attractors within a Generalized Lorenz Model. *Chaos Solitons Fractals* **2021**, *146*, 110865. [CrossRef]
47. Shen, B.-W.; Pielke, R.A., Sr.; Zeng, X.; Cui, J.; Faghih-Naini, S.; Paxson, W.; Atlas, R. Three Kinds of Butterfly Effects within Lorenz Models. *Encyclopedia* **2022**, *2*, 1250–1259. [CrossRef]
48. Wolf, A.; Swift, J.B.; Swinney, H.L.; Vastano, J.A. Determining Lyapunov exponents from a time series. *Phys. D Nonlinear Phenom.* **1985**, *16*, 285–317. [CrossRef]
49. Eckhardt, B.; Yao, D. Local Lyapunov exponents in chaotic systems. *Phys. D Nonlinear Phenom.* **1993**, *65*, 100–108. [CrossRef]
50. Nese, J.M. Quantifying local predictability in phase space. *Phys. D Nonlinear Phenom.* **1989**, *35*, 237–250. [CrossRef]
51. Slingo, J.; Palmer, T. Uncertainty in weather and climate prediction. *Philos. Trans. R. Soc. A* **2011**, *369A*, 4751–4767. [CrossRef]
52. Lewis, J. Roots of ensemble forecasting. *Mon. Weather. Rev.* **2005**, *133*, 1865–1885. [CrossRef]
53. Orszag, S.A. Analytical theories of turbulence. *J. Fluid Mech.* **1970**, *41*, 363–386. [CrossRef]
54. Orszag, S.A. *Fluid Dynamics*; Balian, R., Peuble, J.L., Eds.; Gordon and Breach: London, UK, 1977.
55. Leith, C.E. Atmospheric predictability and two-dimensional turbulence. *J. Atmos. Sci.* **1971**, *28*, 145–161. [CrossRef]
56. Leith, C.E.; Kraichnan, R.H. Predictability of turbulent flows. *J. Atmos. Sci.* **1972**, *29*, 1041–1058. [CrossRef]
57. Lilly, D.K. Numerical simulation of two-dimensional turbulence. *Phys. Fluids* **1969**, *12* (Suppl. 2), 240–249. [CrossRef]
58. Lilly, D.K. Numerical simulation studies of two-dimensional turbulence: II. Stability and predictability studies. *Geophys. Fluid Dyn.* **1972**, *4*, 1–28. [CrossRef]

59. Lilly, K.D. Lectures in Sub-Synoptic Scales of Motions and Two-Dimensional Turbulence. In *Dynamic Meteorology*; Morel, P., Ed.; Reidel: Boston, MA, USA, 1973; pp. 353–418.

60. Vallis, G. *Atmospheric and Oceanic Fluid Dynamics*; Cambridge University Press: Cambridge, UK, 2006; p. 745.

61. Shen, B.-W.; Pielke, R.A., Sr.; Zeng, X. A Note on Lorenz's and Lilly's Empirical Formulas for Predictability Estimates. *ResearchGate* **2022**, preprint. [CrossRef]

62. Ghil, M.; Read, P.; Smith, L. Geophysical flows as dynamical systems: The influence of Hide's experiments. *Astron. Geophys.* **2010**, *51*, 4.28–4.35. [CrossRef]

63. Read, P. Application of Chaos to Meteorology and Climate. In *The Nature of Chaos*; Mullin, T., Ed.; Clarendo Press: Oxford, UK, 1993; pp. 222–260.

64. Legras, B.; Ghil, M. Persistent anomalies, blocking, and variations in atmospheric predictability. *J. Atmos. Sci.* **1985**, *42*, 433–471. [CrossRef]

65. Patil, D.J.; Hunt, B.R.; Kalnay, E.; Yorke, J.A.; Ott, E. Local low-dimensionality of atmospheric dynamics. *Phys. Rev. Lett.* **2001**, *86*, 5878–5881. [CrossRef]

66. Oczkowski, M.; Szunyogh, I.; Patil, D.J. Mechanisms for the Development of Locally Low-Dimensional Atmospheric Dynamics. *J. Atmos. Sci.* **2005**, *62*, 1135–1156. [CrossRef]

67. Ott, E.; Hunt, B.R.; Szunyogh, I.; Corazza, M.; Kalnay, E.; Patil, D.J.; Yorke, J. Exploiting Local Low Dimensionality of the Atmospheric Dynamics for Efficient Ensemble Kalman Filtering. 2002. Available online: https://doi.org/10.48550/arXiv.physics/0203058 (accessed on 1 November 2022).

68. Zeng, X.; Pielke, R.A., Sr.; Eykholt, R. Chaos theory and its applications to the atmosphere. *Bull. Am. Meteorol. Soc.* **1993**, *74*, 631–644. [CrossRef]

69. Ghil, M. A Century of Nonlinearity in the Geosciences. *Earth Space Sci.* **2019**, *6*, 1007–1042. [CrossRef]

70. Charney, J.G.; DeVore, J.G. Multiple flow equilibria in the atmosphere and blocking. *J. Atmos. Sci.* **1979**, *36*, 1205–1216. [CrossRef]

71. Crommelin, D.T.; Opsteegh, J.D.; Verhulst, F. A mechanism for atmospheric regime behavior. *J. Atmos. Sci.* **2004**, *61*, 1406–1419. [CrossRef]

72. Ghil, M.; Robertson, A.W. "Waves" vs. "particles" in the atmosphere's phase space: A pathway to long-range forecasting? *Proc. Natl. Acad. Sci. USA* **2002**, *99* (Suppl. 1), 2493–2500. [CrossRef]

73. Renaud, A.; Nadeau, L.-P.; Venaille, A. Periodicity Disruption of a Model Quasibiennial Oscillation of Equatorial Winds. *Phys. Rev. Lett.* **2019**, *122*, 214504. [CrossRef]

74. Ramesh, K.; Murua, J.; Gopalarathnam, A. Limit-cycle oscillations in unsteady flows dominated by intermittent leading-edge vortex shedding. *J. Fluids Struct.* **2015**, *55*, 84–105. [CrossRef]

75. Goler, R.A.; Reeder, M.J. The generation of the morning glory. *J. Atmos. Sci.* **2004**, *61*, 1360–1376. [CrossRef]

76. Wu, Y.-L.; Shen, B.-W. An evaluation of the parallel ensemble empirical mode decomposition method in revealing the role of downscaling processes associated with African easterly waves in tropical cyclone genesis. *J. Atmos. Ocean. Technol.* **2016**, *33*, 1611–1628. [CrossRef]

77. Shen, B.-W.; Cheung, S.; Wu, Y.; Li, F.; Kao, D. Parallel Implementation of the Ensemble Empirical Mode Decomposition (PEEMD) and Its Application for Earth Science Data Analysis. *Comput. Sci. Eng.* **2017**, *19*, 49–57. [CrossRef]

78. Shilnikov, L.P. On a new type of bifurcation of multi-dimensional dynamical systems. *Dokl. Akad. Nauk SSSR* **1969**, *10*, 1368–1371.

79. Gonchenko, S.V.; Turaev, D.V.; Shilnikov, L.P. Dynamical phenomena in multidimensional systems with a structurally unstable homoclinic Poincar6 curve. *Russ. Acad. Sci. Dokl. Mat.* **1993**, *47*, 410–415.

80. Belhaq, M.; Houssni, M.; Freire, E.; Rodriguez-Luis, A.J. Asymptotics of Homoclinic Bifurcation in a Three-Dimensional System. *Nonlinear Dyn.* **2000**, *21*, 135–155. [CrossRef]

81. Shimizu, T.; Morioka, N. On the bifurcation of a symmetric limit cycle to an asymmetric one in a simple model. *Phys. Lett. A* **1980**, *76*, 201–204. [CrossRef]

82. Shil'nilov, A.L. On bifurcations of a Lorenz-like attractor in the Shimizu-Morioka system. *Phys. D Nonlinear Phenom.* **1992**, *62*, 332–346.

83. Shil'nikov, A.L.; Shil'nikov, L.P.; Turaev, D.V. Normal Forms and Lorenz Attractors. *Int. J. Bifurc. Chaos* **1993**, *3*, 1123–1139. [CrossRef]

84. Gonchenko, S.; Kazakov, A.; Turaev, D.; Shilnikov, A.L. Leonid Shilnikov and mathematical theory of dynamical chaos. *Chaos* **2022**, *32*, 010402. [CrossRef] [PubMed]

85. Simonnet, E.; Ghil, M.; Dijkstra, H. Homoclinic bifurcation in the quasi-geostrophic double-gyre circulation. *J. Mar. Res.* **2005**, *63*, 931–956. [CrossRef]

 *atmosphere*

 MDPI

Essay

Predictability and Predictions

Richard A. Anthes

COSMIC Program, University Corporation for Atmospheric Research, Boulder, CO 80301, USA; anthes@ucar.edu

Abstract: This essay describes the author's lifetime experiences with predictability theory and weather predictions.

Keywords: predictability; predictions; Laplace's demon; butterfly effect; weather forecasts; MM5

Citation: Anthes, R.A. Predictability and Predictions. *Atmosphere* **2022**, *13*, 1292. https://doi.org/10.3390/atmos13081292

Academic Editors: Bo-Wen Shen, Roger A. Pielke Sr., Xubin Zeng and Ilias Kavouras

Received: 18 July 2022
Accepted: 12 August 2022
Published: 14 August 2022

"Prediction is very difficult, especially if it's about the future." [1]

Niels Bohr (Physics Nobel Prize 1922)

Humans have always been fascinated by foretelling the future, and oracles, shamans, priests, and prophets have held prominent positions in all cultures. Various strategies, often occult, form the basis for astrology, palmistry, numerology, soothsaying, prognostication, fortune telling, and Tarot. Horoscopes remain popular today, while weather, economic, and political forecasters hold lucrative positions in society, in spite of often dubious records.

One famous seer, Michel de Nostredame (1503–1566), known as Nostradamas, was a French astrologer and prophet who published a collection of 942 four-lined prophesies in his book Les Prophéties [2]. While mostly foretelling and commenting on political and religious catastrophes, there is one example of his prophesy of a weather and climate disaster:

"For forty years the rainbow will not be seen.

For forty years it will be seen every day.

The dry earth will grow more parched,

and there will be great floods when it is seen."

The prediction of complex systems based on mathematics and physics rather than the occult also has a long history. In the 17th century, Gottfried Leibnitz (1646–1716) speculated that everything proceeds mathematically, and so someone who had sufficient understanding and could take into account everything, "would be a prophet and see the future in the present as in a mirror" [3]. About a hundred years later, the Marquis de Laplace (1749–1827) dreamed of an intelligent being (an intellect, later dubbed Laplace's Demon) who knew the positions and velocities of every single atom and used Newton's equations of motion to predict the future of the entire universe [4].These ideas of Leibniz and Laplace likely inspired the early ideas associated with numerical prediction of the weather, transforming weather prediction from an arcane art to a science.

"One can't predict the weather more than a few days in advance." [5]

Stephen Hawking (1942–2018)

While still in high school, I began my career in meteorology as a Student Trainee in the U.S. Weather Bureau. I spent summers at the Weather Bureau Airport Stations in Richmond and Norfolk, Virginia and was introduced formally to the challenges of operational weather forecasting. We did not have any numerical model guidance at this time, and our forecasts were entirely qualitative, limited to a day or two. However, I noticed that lines of thunderstorms would often develop in the early afternoon over the Blue Ridge Mountains and propagate eastward, reaching the coast in the evening, and these events appeared to have some predictability. (Here, I use the word predictability in its common lay meaning, "able to be predicted." This definition differs somewhat from the generally accepted meaning of predictability in the sense of nonlinear fluid dynamics.) I dreamed of someday being able to quantitatively predict such events well before they occur. I had not yet heard of the concept of numerical weather prediction (NWP), which was just being developed at Princeton under the direction of Jule Charney, based on the early ideas of Vilhelm Bjerknes and Lewis Fry Richardson.

As an undergraduate student at the University of Wisconsin in Madison (Madison, WI, USA) in the early 1960s, I became interested in numerical simulations, and coded a simple nonlinear one-dimensional gravity wave model on punch cards. Given appropriate initial conditions, the model correctly predicted many cycles of a propagating gravity wave. By this time, the Weather Bureau was distributing 36 h forecasts of 500 mb flows using a barotropic model, which at least to us students was of dubious value, but still captured our imagination as we anxiously awaited the latest model run.

While still at Wisconsin, I moved from the Virginia airport stations to the ESSA (Environmental Science Services Administration, now NOAA) National Hurricane Research Laboratory (Miami, FL, USA) and began developing a nonlinear, baroclinic, three-dimensional model of the tropical cyclone. During this development and debugging, I suffered through various forms of numerical and physical instabilities. The numerical instabilities could be controlled by suitable choices of finite difference schemes and various damping or smoothing mechanisms, but physical instabilities persisted and resulted in the evolution of somewhat realistic mesoscale features such as rainbands and eddies on the outflow layer that were not present in the initial conditions [6].

Around this time, in the early 1960s, Edward Lorenz at MIT noted that small differences in a simple numerical model he was working with doubled in size every four days, and in 1963 he published his classic paper in the Journal of Atmospheric Sciences, "Deterministic nonperiodic flow," which has been the theoretical foundation of modern predictability theory for nonlinear fluid dynamics, including numerical prediction of atmospheres and oceans [7]. In 1972, Lorenz gave a talk on atmospheric predictability with the title "Does the flap of a butterfly's wing in Brazil set off a tornado in Texas?" This rhetorical and provocative question has intrigued scientists and the public ever sense, and "the butterfly effect" has come to mean chaos and the lack of predictability of chaotic nonlinear systems, including human societies [8].

During my time at the National Hurricane Research Laboratory, I became friends with Fred Sanders, a leading synoptic meteorologist from MIT. I sailed with Fred on his sailboat, the *Stillwater*, from Miami to Marblehead, Massachusetts in June 1968. One quiet warm night under the stars during our midnight watch, we had a long philosophical discussion of whether the universe was theoretically predictable. I was of the opinion that it might be deterministic, that if someone knew all the physics, chemistry, and biology and had perfect knowledge of the complete state of the universe at a given time, the future of the universe was knowable. I had been captured by Laplace's demon. Fred must have been aware of Lorenz' work on chaos theory by this time, and he gently argued against this possibility.

When I moved to Penn State as an Assistant Professor in 1971, my PhD student Tom Warner and I began modifying the hurricane simulation model to accept real data and make forecasts, and the model evolved into multiple generations of what we called the Penn State Mesoscale Model, or MM for short. The fifth-generation of MM, MM5, became widely used internationally as a research and forecast model. There were some prominent scientists in the community that were skeptical of this effort, warning us that trying to forecast small-scale (mesoscale) atmospheric phenomena was a fool's errand, making various arguments based on turbulence and predictability theories and lack of observations on the mesoscale. Yet, as our model development and testing continued, we saw many examples of realistic and observed mesoscale features developing in the model forecasts, in spite of only large-scale features present in the initial conditions. These included fronts, jet streaks, precipitation systems, and various features associated with terrain variations, surface conditions such as soil moisture variations, and land-water contrasts. In our first comprehensive paper describing the MM model [9], we noted that nonlinear processes are capable of producing smaller scale information in the forecast than is present in the initial conditions, as long waves interact to produce energy in shorter waves. Frontogenesis through horizontal deformation in a large-scale baroclinic field was a prime example [10]. These nonlinear effects starting with realistic large-scale initial conditions, combined with local forcing associated with variations in terrain and surface conditions, and diabatic effects led us to hypothesize "in many synoptic situations, if the local forcing is modeled correctly, the details of the initial perturbations are not particularly important." Perhaps some predictability of mesoscale features did exist, at least for limited times into the numerical forecast. I was still thinking of those lines of thunderstorms that developed over the Blue Ridge Mountains in summer.

I summarized the above ideas in a 1984 paper, "Predictability of mesoscale meteorological phenomena," classifying the development of mesoscale weather systems into two types [11]: (1) those resulting from forcing by surface inhomogeneities and (2) those resulting from internal modifications of large-scale flow patterns. Land-sea breezes, mountain-valley breezes, mountain waves, heat island circulations, coastal fronts, and dryline and moist convection are often generated by the first mechanisms. Fronts and jet stream phenomena, generated by shearing and stretching deformation associated with large-scale flows, belong to the second class. After showing examples of these two types of mesoscale predictions at a conference on predictability, I recall Michael Ghil, a theoretical climate dynamicist from UCLA, jumped up and exclaimed, tongue in cheek, "maybe we should stop studying predictability and just start making predictions!"

On longer time scales, forcing associated with anomalies such as ocean temperatures and changing greenhouse gas concentrations offered hope for some predictive skill of climate models, as argued by Jagadish Shukla in his 1998 Science article [12].

After I moved to NCAR in 1981 and began the administrative phase of my career, my active research with MM ended. However, the development of the MM was continued very competently by Tom Warner, Bill Kuo, and others, and as the model resolution and physical parameterizations improved, we continued to see examples of successful predictions of mesoscale features. Mesoscale modeling efforts were now going on in many places, and I was excited more than 20 years later when I became aware of some stimulating results from a mesoscale global model at NASA carried out by Bo-Wen Shen and his colleagues. By 2008, global mesoscale models with horizontal resolutions of ~10 km were possible (in 1978 our MM experiments with real data were over a limited area with horizontal resolutions of ~111 km). The NASA simulations of tropical storm genesis were supporting Tom's and my hypothesis from 1978! Large-scale forcing associated with the Madden–Julian Oscillation was responsible for the correct prediction of tropical cyclones such as Nargis [13] in the Indian Ocean, many days before any evidence of the tropical cyclone circulation existed [13]. Based in large part on this successful forecast, I wrote a short essay in UCAR Magazine in 2011 with an optimistic view on predictability [14].

In 2014, Bo-Wen moved to San Diego State University's Department of Mathematics and Statistics (San Diego, CA, USA) where he continued his work on theoretical aspects of atmospheric predictability, extending Lorenz' revolutionary work in the 1960s and 1970s (Ref. [15], which concludes "the entirety of weather possesses both chaos and order.") This work, and the success of high-resolution global models, now supported by mesoscale satellite observations assimilated with sophisticated four-dimensional variational schemes, in predicting useful mesoscale phenomena such as tropical cyclones, provide optimism for continuing to increase the forecast lead time of significant weather. Nevertheless, it seems clear to me now that the butterflies will ultimately prevail, and nothing is predictable forever. The demon has been exorcised.

Funding: This essay received no external funding.

Institutional Review Board Statement: Not applicable.

Informed Consent Statement: Not applicable.

Acknowledgments: I thank Bo-Wen Shen for encouraging me to write this essay.

Conflicts of Interest: The authors declare no conflict of interest.

References

1. Bohr Niels Quote. Available online: https://www.brainyquote.com/authors/niels-bohr-quotes (accessed on 13 August 2022).
2. Nostradamus. Les Prophéties, translated by Edgar Leoni and published in Wikisource free online library. 1555. Available online: https://en.wikisource.org/wiki/Les_Propheties (accessed on 13 August 2022).
3. Alexander, H.G. The Leibniz-Clarke Correspondence. *Philosophy* **1956**, *32*, 365–366. [CrossRef]
4. Laplace, P.S. *A Philosophical Essay on Probabilities 1814*, 6th ed.; translated into English from the original, French; Truscott, F.W., Emory, F.L., Eds.; Dover Publications: New York, NY, USA, 1951; p. 4. Available online: https://en.wikipedia.org/wiki/A_Philosophical_Essay_on_Probabilities (accessed on 13 August 2022).
5. Hawking, Stephen in Stephen Hawking Quotes. Available online: https://www.quotetab.com/quotes/by-stephen-hawking (accessed on 13 August 2022).
6. Anthes, R.A. Development of Asymmetries in a Three-Dimensional Numerical Model of the Tropical Cyclone. *Mon. Weather Rev.* **1972**, *100*, 461–476. [CrossRef]
7. Lorenz, E.N. Deterministic non-periodic flow. *J. Atmos. Sci.* **1963**, *20*, 130–141. [CrossRef]
8. Lorenz, E.N. Predictability: Does the flap of a butterfly's wing in Brazil set off a tornado in Texas? In Proceedings of the Talk at the 139th Meeting of the American Association for the Advancement of Sciences, Washington, DC, USA, 29 December 1972; Available online: http://climate.envsci.rutgers.edu/climdyn2017/LorenzButterfly.pdf (accessed on 13 August 2022).
9. Anthes, R.A.; Warner, T.T. Development of hydrodynamic models suitable for air pollution and other mesometeorological studies. *Mon. Wea. Rev.* **1978**, *106*, 1045–1078. [CrossRef]
10. Hoskins, B.J.; Bretherton, F.P. Atmospheric frontogenesis models: Mathematical formulation and solution. *J. Atmos. Sci.* **1972**, *29*, 11–37. [CrossRef]
11. Anthes, R.A. Predictability of mesoscale meteorological phenomena. In *Predictability of Fluid Motions (La Jolla Institute 1983)*; Holloway, G., West, B.J., Eds.; American Institute of Physics: New York, NY, USA, 1984; pp. 247–270.
12. Shukla, J. Predictability in the Midst of Chaos: A Scientific Basis for Climate Forecasting. *Science* **1998**, *282*, 728–731. [CrossRef] [PubMed]
13. Shen, B.-W.; Tao, W.-K.; Lau, W.K.; Atlas, R. Predicting tropical cyclogenesis with a global mesoscale model: Hierarchical multiscale interactions during the formation of tropical cyclone Nargis (2008). *J. Geophys. Res. Atmospheres.* **2010**, *115*, D14102. [CrossRef]
14. Anthes, R. "Turning the Tables on Chaos: Is the Atmosphere More than We Assume?" UCAR Magazine, spring/summer 2011. 2011. Available online: https://news.ucar.edu/4505/turning-tables-chaos-atmosphere-more-predictable-we-assume (accessed on 13 August 2022).
15. Shen, B.-W.; Pielke, S.R.A.; Zeng, X.; Baik, J.-J.; Faghih-Naini, S.; Cui, J.; Atlas, R. Is Weather Chaotic? Coexistence of Chaos and Order within a Generalized Lorenz Model. *Bull. Am. Meteorol. Soc.* **2021**, *102*, E148–E158. [CrossRef]

Article

Time-Lagged Ensemble Quantitative Precipitation Forecasts for Three Landfalling Typhoons in the Philippines Using the CReSS Model, Part I: Description and Verification against Rain-Gauge Observations

Chung-Chieh Wang [1,*], Chien-Hung Tsai [1], Ben Jong-Dao Jou [2] and Shirley J. David [3]

[1] Department of Earth Sciences, National Taiwan Normal University, Taipei 11677, Taiwan; iamboytimmy@yahoo.com.tw

[2] Department of Atmospheric Sciences, National Taiwan University, Taipei 10617, Taiwan; jouben@ntu.edu.tw

[3] Philippine Atmospheric, Geophysical and Astronomical Services Administration, Quezon City 1100, Philippines; sjdavid@pagasa.dost.gov.ph

* Correspondence: cwang@ntnu.edu.tw

Abstract: In this study, the 2.5 km Cloud-Resolving Storm Simulator was applied to forecast the rainfall by three landfalling typhoons in the Philippines at high resolution: Mangkhut (2018), Koppu (2015), and Melor (2015), using a time-lagged strategy for ensemble. The three typhoons penetrated northern Luzon, central Luzon, and the middle of the Philippine Archipelago, respectively, and the present study verified the track and quantitative precipitation forecasts (QPFs) using categorical statistics against observations at 56 rain-gauge sites at seven thresholds up to 500 mm. The predictability of rainfall is the highest for Koppu, followed by Melor, and the lowest for Mangkhut, which had the highest peak rainfall amount. Targeted at the most-rainy 24 h of each case, the threat score (TS) within the short range ($\leq$72 h) could reach 1.0 for Koppu at 350 mm in many runs (peak observation = 502 mm), and 1.0 for Mangkhut and 0.25 for Melor (peak observation = 407 mm) both at 200 mm in the best member, when the track errors were small enough. For rainfall from entire events (48 or 72 h), TS hitting 1.0 could also be achieved regularly at 500 mm for Koppu (peak observation = 695 mm), and 0.33 at 350 mm for Melor (407 mm) and 0.46 at 200 mm for Mangkhut (786 mm) in the best case. At lead times beyond the short range, one third of these earlier runs also produced good QPFs for both Koppu and Melor, but such runs were fewer for Mangkhut and the quality of QPFs was also not as high due to larger northward track biases. Overall, the QPF results are very encouraging, and comparable to the skill level for typhoon rainfall in Taiwan (with similar peak rainfall amounts). Thus, at high resolution, there is a fair chance to make decent QPFs even at lead times of 3–7 days before typhoon landfall in the Philippines, with useful information on rainfall scenarios for early preparation.

Keywords: quantitative precipitation forecast (QPF); typhoon; time-lagged ensemble; cloud-resolving model; categorical statistics; Philippines

Citation: Wang, C.-C.; Tsai, C.-H.; Jou, B.J.-D.; David, S.J. Time-Lagged Ensemble Quantitative Precipitation Forecasts for Three Landfalling Typhoons in the Philippines Using the CReSS Model, Part I: Description and Verification against Rain-Gauge Observations. *Atmosphere* **2022**, *13*, 1193. https://doi.org/10.3390/atmos13081193

Academic Editors: Bo-Wen Shen, Roger A. Pielke, Sr. and Xubin Zeng

Received: 29 June 2022
Accepted: 25 July 2022
Published: 28 July 2022

1. Introduction

Located at the boundary of tectonic plates in the tropical western North Pacific, the Philippines is among the nations with highest risk of natural disasters and is under constant threat of volcanic eruptions, earthquakes, typhoons, tsunami, and other hazards such as sea level rise, e.g., [1–4]. As the most exposed country in the world to tropical cyclones (TCs) [5], the Philippines often experience floods, inundation, landslides, and storm surges in the category of meteorological hazards, e.g., [6–10], all mainly caused by the frequent landfall of TCs, or typhoons in the western North Pacific, e.g., [9–15]. Therefore, accurate quantitative precipitation forecasts (QPFs) for typhoons are urgently needed in order to help prevent and reduce the impacts of hazards from excess typhoon rainfall in the Philippines, especially under the current trend toward a warmer climate [12,16–20]. At the present time, however, QPFs over heavy-rainfall thresholds are still very challenging around the world, e.g., [21–23], and few studies have been carried out to verify typhoon QPFs objectively in the Philippines [24,25].

Just north of the Philippines across the Luzon Strait, Taiwan has similar geological characteristics and also its fair share of heavy rainfall from TCs [26]. Past studies have shown that due to its steep topography, TC rainfall is enhanced in Taiwan and its distributions over short periods are highly dictated by the location of the storm relative to the island [26–29], and therefore the total rainfall is linked to the track of the TC (including moving speed). Such an orographic phase-locking effect gives the potential for more accurate QPFs if realistic tracks can be predicted, and hence raises the predictability of TC rainfall in Taiwan. Based on this concept, the TC rainfall in Taiwan can be predicted from rainfall climatology constructed from past typhoons using forecast or projected tracks [30,31]. Similarly, the ensemble typhoon QPF model developed in [32] made use of available lagged ensemble predictions for the same TC, and thus can take into account the stochastic nature of model predictions, e.g., [33,34], as well as the specific characteristics of individual storms. In both cases, the predicted TC tracks must be accurate enough for the QPFs to be successful. The performances of such ensemble prediction systems are also evaluated for their probabilistic forecasts and potential applications, e.g., [35–37].

One other requirement for models to improve their typhoon QPFs over complex terrain is high resolution, preferably with a large enough fine domain size, as demonstrated in recent years in Taiwan. At the grid size (Δx) of 2.5 km, close to the recommendation for operational use [38], real-time forecasts by the Cloud-Resolving Storm Simulator (CReSS) [39,40] show improved 24 h QPFs in Taiwan at the short range (within three days) for typhoons from 2010–2015 [41–45], linked to a better capability to resolve both the TC convection and local topography. Using objective categorical measures, for example, the overall threat scores (TSs) at the threshold value of 350 mm (per 24 h), defined as the fraction of correct forecast of events (i.e., hits) in the union area of either the observation or prediction to reach that threshold (thus, $0 \leq TS \leq 1$, see Section 2.2 for details), are 0.28, 0.25, and 0.15 on day 1 (0–24 h), day 2 (24–48 h), and day 3 (48–72 h), respectively [44]. In addition, such scores also exhibit a clear dependency on the observed rainfall magnitude in Taiwan, i.e., the more rain, the higher the TSs (at the same thresholds) [41,43,44], mainly for two reasons: the hits are easier to achieve when the rainfall area is larger in size over the island, and the larger (rainier) events tend to be under stronger forcing that the model can also capture at a sufficient resolution. Therefore, for the most-rainy top 5% of samples, the TSs at 350 mm on days 1–3 are roughly 0.1 higher than their counterparts for all TCs [44]. In some cases where a good track can be predicted, high TSs may be achieved even at a longer lead time, e.g., TS = 0.45 at 350 mm on day 3 for Typhoon (TY) Fanapi (2010), TS = 0.6 at 350 mm on day 3 and 0.87 at 500 mm on day 1 for TY Megi (2010) [41], TS > 0.38 at 750 mm at all three ranges of days 1–3 for TY Soulik (2013) [44], and TS $\geq$ 0.28 at 500 mm on days 1–3 and TS = 0.36 at 750 mm on day 1 for TY Soudelor (2015) [45]. Note that in these examples of individual forecasts, TS must drop to zero once the threshold exceeds the observed or predicted peak rainfall (as no hits can be produced), and all the TCs here were very rainy and produced a peak 24 h amount of 842–1110 mm.

As large computational resources are necessary to achieve high resolution for a sufficiently large domain [46], it is not ideal to divide the resource and run a multi-member ensemble system, e.g., [47–49], to obtain probability information. To cope with this situation, the time-lagged ensemble, e.g., [50–54], using a longer forecast range has been recommended by Wang et al. [55,56], and this allows not only for the ensemble information but also the possibility of realistic, high-quality QPFs at lead times beyond 3 days. This is particularly suitable for TCs, which in general have relatively long lifespans of at least one week, e.g., [57]. Using daily runs initialized at 0000 UTC, the system was able to produce good 24 h QPFs at lead times beyond 72 h in more than half of the six TC examples in 2012–2013 [55]. These included TS = 0.21 at 200 mm on day 8 (168–192 h) for TY Talim (observed peak = 415 mm), TS = 0.5 at 350 mm on day 6 (120–144 h) for TY Saola (peak = 884 mm), TS = 0.5 at 200 mm also on day 6 for TY Jelawat (peak = 216 mm), and TS = 0.25 at 200 mm on day 7 (144–168 h, peak = 426 mm) and also on day 4 (72–96 h, for a different day, peak = 367 mm) for TY Kong-Rey (2013), all better than many QPFs made within the short range [55]. Recently, lagged runs every 6 h were also tested on three rainy TCs [58,59] and TY Morakot (2009), the most hazardous typhoon to hit Taiwan in five decades [60–62]. For two of them, the lead time of decent QPFs, with a similarity skill score (SSS, $0 \leq$ SSS ≤ 1, also, see Section 2.2 for definition) of at least 0.75 against the observed rainfall pattern, can be significantly extended to more than 5.5 days prior to the starting time of rainfall accumulation (typically before landfall) [58,59]. For TYs Morakot (2009) and Soulik (2013), however, a better QPF could not be obtained beyond the short range due to the limitation by track errors [59], consistent with previous studies [61–65]. In any case, the evolution of heavy-rainfall probabilities with time for all typhoons can provide very useful information for hazard preparation at the earliest time possible, at a fraction of computational cost compared to the multi-member ensemble [58,59,62].

Given the above development of high-resolution time-lagged ensemble in recent years and the similarity of TC rainfall in Taiwan and the Philippines, e.g., [66], the purpose of the present study is therefore to apply the same strategy to typhoons making landfall at the Philippines and to evaluate the performance in the QPFs. In order to make a comparison with typhoon QPFs in Taiwan, similar methods of verification will be used, including the categorical matrix. From four seasons of 2015–2018, three typhoons are eventually chosen for study: Mangkhut (2018), Koppu (2015), and Melor (2015), as they were all very rainy ($\geq$300 mm) but made landfall in different parts of the Philippine Archipelago (Figure 1a). Each event also yielded $\geq$200 mm of rainfall in the Manila region, and thus posed high risk of flooding. In fact, the time-lagged ensemble was also applied to forecast the landfall intensity of Supertyphoon Haiyan (2013), and consistently produced peak surface winds reaching 73 m s^{-1} and minimum central mean sea level pressure below 900 hPa at (or near) landfall from about 48 h prior to its occurrence [15]. However, this study focuses on TC intensity instead of QPFs, since the main hazard from Haiyan was the storm surge.

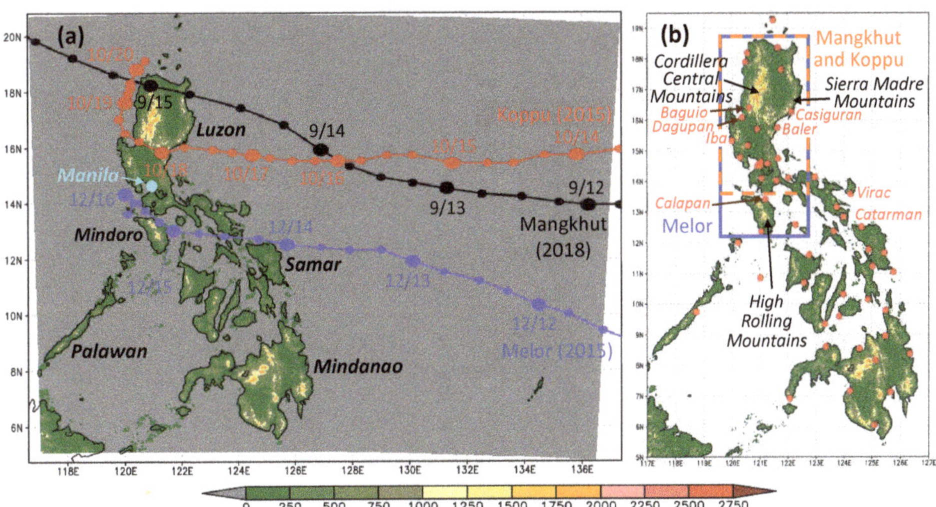

Figure 1. (a) The geography and topography (m, color) of the Philippines, model domain (shaded region), and the Joint Typhoon Warning Center (JTWC) best tracks of the three typhoons in this study

(center positions given every 6 h, with dates marked at 0000 UTC). The major islands and the capital city of Manila are labeled. (**b**) The topography (m, color), major mountains, locations of rain gauges (red dots) around the Philippines, and the small domains used to compute the SSS. Rain-gauge sites mentioned in text are labeled. The color scales for the topography are the same in the two panels.

The remaining part of this paper is arranged as follows. In Section 2, the CReSS model, numerical experiments, and the methodology for the verification of QPFs are described. The results of the time-lagged CReSS ensemble for the three typhoons of Mangkhut (2018), Koppu (2015), and Melor (2015) are presented and discussed in Sections 3–5, respectively. Further comparison with QPF results in Taiwan and discussion are given in Section 6, and the conclusions are given in Section 7.

2. The CReSS Model, Experiments, and Verification of QPFs

2.1. The CReSS Model and Experiments

The CReSS model (version 3.4.2) was used to perform all experiments in this study. Developed at Nagoya University, Japan [39,40], it is a cloud-resolving model with a compressible and non-hydrostatic equation set, a single domain of uniform grid size, and a terrain-following vertical coordinate based on height. The model treats all clouds explicitly using a bulk cold-rain scheme (Table 1) with six water species of vapor, cloud water, cloud ice, rain, snow, and graupel [40,67–71], without the use of cumulus parameterization. The sub-grid scale parameterized processes (Table 1) include the turbulence in the planetary boundary layer [72,73] and the radiation, momentum, and energy fluxes at the surface [74,75], with the aid of a substrate model [40]. Readers are referred to [39,40] and some earlier studies [15,41,44,55,62] for further details.

Table 1. The basic domain configuration and physical packages of the 2.5 km CReSS in this study.

Map projection	Lambert Conformal (center at 123° E, secant at 5° N and 20° N)
Grid spacing (km)	2.5 × 2.5 × 0.1–0.5695 (0.4) *
Grid dimension (x, y, z)	864 × 696 × 50
Domain size (km)	2160 × 1740 × 20
Forecast frequency	Every 6 h (at 0000, 0600, 1200, and 1800 UTC)
Forecast length	8 days (192 h)
IC/BCs	NCEP GFS 0.5° × 0.5° analyses and forecasts (26 levels)
Cloud microphysics	Double-moment bulk cold-rain scheme
PBL parameterization	1.5-order closure with turbulent kinetic energy prediction
Surface processes	Shortwave/longwave radiation and momentum/energy fluxes
Soil model	43 levels, every 5 cm to 2.1 m in depth

* The vertical grid spacing of CReSS is uneven and stretched (smallest at bottom) and the parentheses give the averaged value.

As in [55,62], the horizontal grid size was set to 2.5 km, with a domain size of 2160 × 1740 km^2 and 50 vertical levels (Table 1, Figure 1a). This allows most TCs to enter the high-resolution domain at about 2.5–4 days before landfall, and thus a potential for the model to produce high-quality QPFs at lead times beyond the short range. The Global Forecast System (GFS) gross analyses and forecasts at the National Center for Environmental Prediction (NCEP), USA [76,77], were used as the initial and boundary conditions (IC/BCs) of our CReSS runs (Table 1). With a resolution of 0.5° × 0.5° at 26 levels, these data are freely available in real time and were routinely accessed. During the case periods of the three selected typhoons, CReSS experiments were run every 6 h out to a range of eight days (192 h), but some earlier runs with larger track errors were skipped if deemed unnecessary. Since only data available before each run were used, our results here represent what would have been achieved in real time.

2.2. Observational Data and Verification of Model QPFs

The performance of the model hindcasts were examined in the following aspects. The JTWC best-track data were used to verify the TC tracks. At selected times, the retrievals of rainfall intensity from the Tropical Rainfall Measuring Mission (TRMM) satellites at 3 h intervals [78], available from the Naval Research Laboratory (NRL), USA were also employed to compare with the rainfall structure of model TCs when such observations were available. To verify model QPFs, three-hourly rainfall observations from a total of 56 rain gauges over the Philippines (Figure 1b), provided by the Philippine Atmospheric, Geophysical and Astronomical Services Administration (PAGASA), were also used. The data were summed over the selected accumulation periods of 24, 48, or 72 h for the verification.

In this study, the categorical statistics based on the 2 × 2 contingency table, e.g., [79–81], are adopted, so as to allow easy comparisons with previous studies in Taiwan. As shown in Table 2, the outcome of a rainfall forecast versus observation to reach a specified threshold (called an event) over a given period at any verification point can be one (and only one) of four possibilities: hit (H), miss (M), false alarm (FA), and correct negative (CN). Among a set of N total points ($N = H + M + FA + CN$), the correct forecasts are H and CN, while the incorrect ones are M and FA.

Table 2. The 2 × 2 contingency table for dichotomous (yes/no) forecasts of rainfall reaching a given threshold at a set of verification points (N).

		Observation		
		Yes (event)	**No** (no event)	**Total**
Forecast	**Yes** (event)	Hits (H)	False alarm (FA)	Forecast yes (F)
	No (no event)	Misses (M)	Correct negatives (CN)	Forecast no
	Total	Observation yes (O)	Observation no	Total points (N)

From the table, several widely used skill scores such as the probability of detection (POD, of observed events), success ratio (SR, of forecast events), TS, and bias score (BS) can be computed [79–81], respectively, as

$$POD = H/(H + M) = H/O, \tag{1}$$

$$SR = H/(H + FA) = H/F, \tag{2}$$

$$TS = H/(H + M + FA), \tag{3}$$

$$BS = (H + FA)/(H + M) = F/O. \tag{4}$$

Thus, from the equations, one can see that POD, SR, and TS are all bounded by 0 and 1, and the higher the better. Additionally, as the ratio of correct prediction of events to the union of all those either observed or predicted (or both), TS can be no higher than POD or SR (where (1)—SR is also known as the false alarm ratio, or FAR). On the other hand, the BS is the ratio of predicted to observed event points and thus reflects over-forecasting (if BS > 1) or under-forecasting (if BS < 1) by the model. Hence, its most ideal value is unity. By using the bilinear method, model QPFs were interpolated onto the rain-gauge locations (Figure 1b) in this study, where the above scores in Equations (1)–(4) were computed. Later, the scores will be presented using the performance diagram [82]. In this study, QPFs accumulated over both 24 h and longer periods (48 or 72 h) for entire events are verified, at threshold amounts of 0.05, 10, 50, 100, 200, 350, and 500 mm when applicable.

Besides the categorical matrix, the SSS that measures the overall similarity between two patterns, in this case the observed and predicted rainfall patterns in the Philippines, is defined and used as in [83]. It uses the mean squared error (MSE) as in the Brier score [80] or fractions skill score [84], and is formulated as

$$ \text{SSS} = 1 - \frac{\frac{1}{N} \sum_{i=1}^{N} (f_i - o_i)^2}{\frac{1}{N} \left(\sum_{i=1}^{N} f_i^2 + \sum_{i=1}^{N} o_i^2 \right)}, \tag{5} $$

where f_i and o_i are the forecast and observed rainfall at the ith point among N, respectively. In essence, SSS is the skill measured against the worst MSE possible, i.e., the one when no rainfall in the forecast overlaps with the observation and the second term on the right-hand side (RHS) of Equation (5) equals 1. Thus, $0 \leq \text{SSS} \leq 1$ (the higher the better) and a score of 1 means $f_i = o_i$ at all the points and the two patterns are exactly the same. As the QPFs of time-lagged forecasts are verified, the performance at the short range ($\leq$72 h) and how many runs could produce good QPFs at longer lead times (beyond the short range) in each of the three typhoons are our main interests.

3. Time-Lagged Ensemble for Typhoon Mangkhut (2018)

3.1. Track Forecast and Examples of Rainfall Structure

Using the method described in Section 2, a total of 19 hindcast experiments were carried out for the first case of TY Mangkhut (2018) at initial times of 0000 and 1200 UTC on 9–10 as well as every 6 h from 0000 UTC 11 to 1800 UTC 14 September (Figure 2a). While Mangkhut made landfall near 1800 UTC 14 September and moved across the northernmost part of Luzon at a pace of about 30 km h^{-1} (also Figure 1a), most of the hindcasts prior to 12 September had a northward bias in its track and the TC centers did not make landfall in these runs. Thus, the track errors in these earlier runs were largely cross-track in nature, and the along-track errors (errors in moving speed) were relatively small (Figure 2a). The track errors gradually reduced with time (Figure 2b), and the one from the run made at 1800 UTC 11 September was only about 110 km near landfall, which is in fact very small with a lead time of 72 h. Starting from 12 September, all the track errors further reduced to within 80 km at landfall (Figure 2b), and inside 50 km in runs made at 0000 and 1800 UTC 13 and all those on 14 September. In fact, most runs from 13–14 September exhibited small southward bias in tracks during landfall. Thus, the track predictions of Mangkhut from the hindcasts were reasonable, and track errors were quite small within the short range ($\leq$72 h).

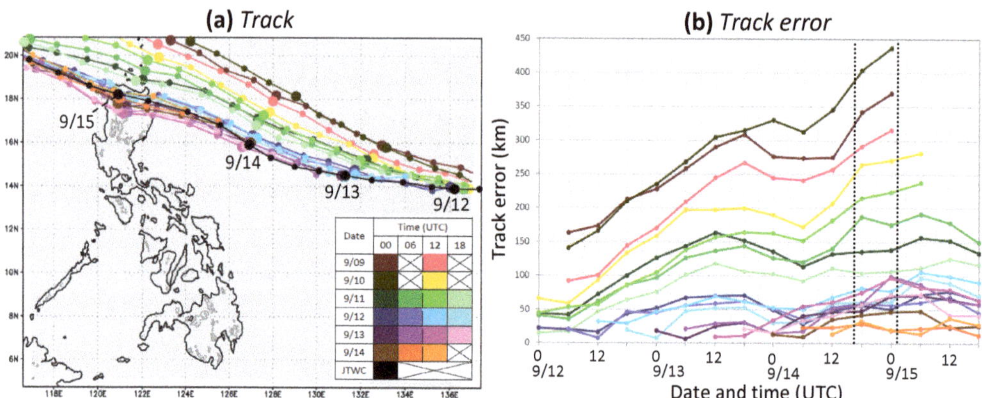

Figure 2. (**a**) The JTWC best track (black) and predicted tracks (colors) of TY Mangkhut (2018) by the time-lagged forecasts at different initial times (see insert), and (**b**) the track errors against time from 0000 UTC 12 to 1800 UTC 15 September, with the two dotted vertical lines indicating the landfall period based on the JTWC best track.

In Figure 3, the rainfall structure of Mangkhut from one of the hindcasts, with t_0 at 0600 UTC 13 September, is compared with TRMM rainrate retrievals at two selected times as examples. With a track error of about 60 km at landfall (Figure 2b), this particular experiment was among the ones that performed better in track prediction. With more details than the TRMM data, the CReSS model at $\Delta x = 2.5$ km can be seen to capture the double eyewall structure and the size of the inner eyewall of Mangkhut near 0944 UTC 14 September before its landfall (Figure 3a,c). Later at 2134 UTC when the storm center was over northern Luzon near 18° N during landfall, the model also reproduced its rainfall distribution quite well, with most intense rainfall to the southeast of the TC center and just off the northeastern tip of Luzon (Figure 3b,d). Although Figure 3 only includes two instances from one hindcast that was selected somewhat arbitrarily, similar quality of realistic TC rainfall structure is confirmed for other runs for Mangkhut (not shown), as also for other cases near Taiwan using comparable model configurations [41,44,45,55].

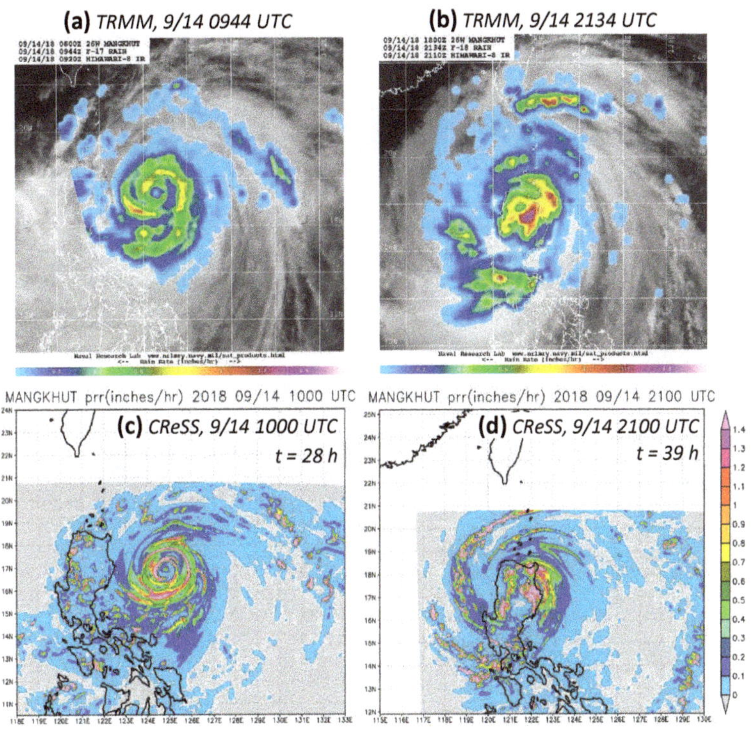

Figure 3. TRMM rainrate retrieval (inch h^{-1}, color), overlaid on infrared cloud imagery by the Himawari-8 satellite at closest time, of TY Mangkhut (2018) at (**a**) 0944 and (**b**) 2134 UTC 14 September (source: NRL), and model rainrate (inch h^{-1}, color) valid at (**c**) 1000 and (**d**) 2100 UTC 14 September from the forecast initialized at 0600 UTC 13 September 2018 (forecast hour labeled). The region plotted and color scale used are similar for each pair.

3.2. Model QPFs for Typhoon Mangkhut (2018)

The daily rainfall distributions on 14 September and those targeted for this day but produced by the 0000 UTC runs at different ranges (from long to short as labeled, header colors as in Figure 2a) are shown in Figure 4a–f (top row), while similar plots for 15 September are shown in Figure 4g–l (second row). With a track through northern Luzon, the northern half of the island received much more rain from Mangkhut than the southern half and the rest of the Philippines on 14 September, with an observed peak daily amount of 250.2 mm at Baguio (Figures 2b and 4f). With reasonable tracks, most predictions within day 4 also captured such characteristics in rainfall, with more rain along the Cordillera Central Mountain (CCM) and the Sierra Madre Mountain (SMM, Figure 4b–e). However, at the longest range on day 6, the experiment starting at t_0 = 0000 UTC 9 September had most of the rainfall occurring offshore and little over Luzon (Figure 4a), obviously due to the relatively large northward track bias of around 350 km. On 15 September when Mangkhut was departing from the northwestern corner of Luzon (Figure 2a), the observed rainfall was mainly concentrated along the CCM in northwestern Luzon, with a peak of 535.6 mm again at Baguio, and diminished in the northeastern part of the island (Figure 4l). This change in rainfall pattern was also well reproduced by all the runs shown within day 4 (Figure 4i–k), but not those at longer lead times (Figure 4g,h), again owing to the larger cross-track errors (Figure 2).

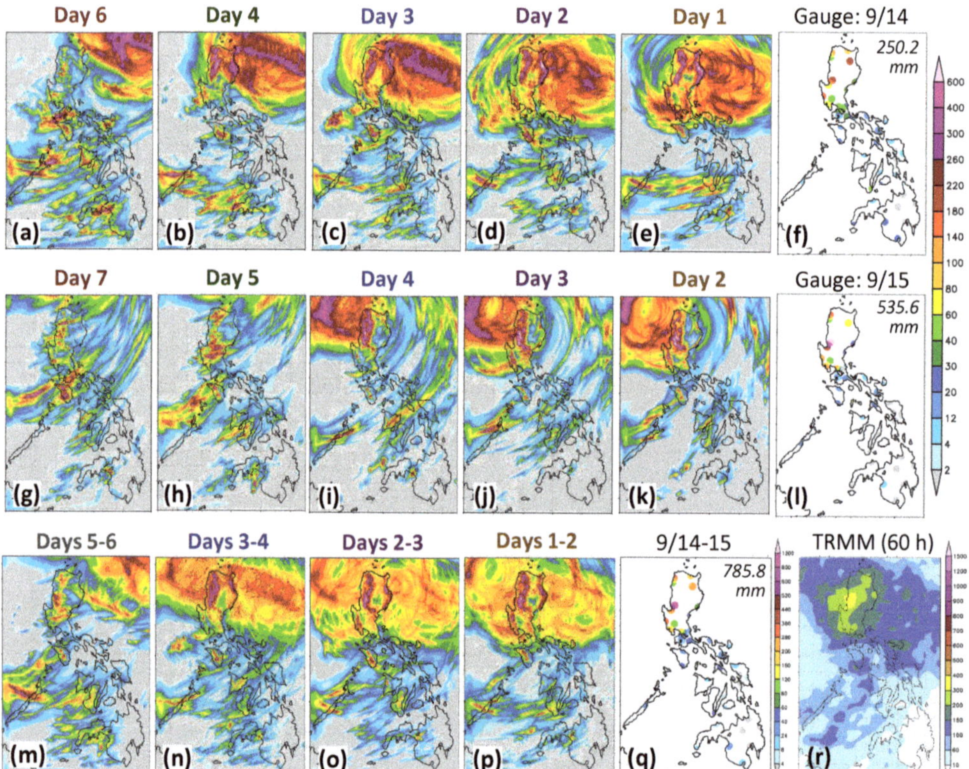

Figure 4. Predicted distributions of the 24 h daily total rainfall (mm, scale to the right) at (**a**–**e**) five different ranges (as labeled) for 14 September and (**f**) the corresponding gauge observations during TY Mangkhut (2018). (**g**–**l**) As in (**a**–**f**), except for 15 September 2018. (**m**–**q**) As in (**a**–**f**), except for 48 h QPFs over 14–15 September (in UTC) at (**m**–**p**) four different ranges and (**q**) from the gauge data. In the headings of model predictions, the same colors as in Figure 2a (for t_0) are used. (**r**) TRMM total rainfall (mm) for the entire typhoon event (60 h starting from 1200 UTC 13 September). The same color scale is used for (**a**–**l**) and doubled for (**m**–**q**). Only (**r**) has its own scale.

For the two-day combined rainfall (Figure 4m–r, bottom row) over 14–15 September, the maximum value reached 785.8 mm (at Baguio) in the gauge observations (Figure 4q). Here, results from four experiments are shown for comparison. Again, the main rainfall area along and near the CCM in northwestern Luzon was well captured by the model within the range of days 3–4 (Figure 4n–p), but the rainfall at Iba (264.6 mm) along the coast of western Luzon near 15.3° N (Figure 2b) could only be better reproduced within day 3 (Figure 4o,p). At longer lead times, for example, on days 5–6 (t_0 = 0000 UTC 10 September), not enough rain fell in Luzon due to the larger track errors too far north (Figure 4m). Plotted in Figure 4r to compare with the rain-gauge observations in Figure 4q, the 60 h total rainfall from the TRMM retrievals for the entire event of Mangkhut also indicates a maximum rainfall along the CCM near Baguio in northwestern Luzon, but the peak value is less than 400 mm and seemingly considerably underestimated. Overall, the visual comparison in Figure 4 suggests a good performance by the 2.5 km CReSS in predicting the rainfall of Mangkhut at ranges within 3–4 days, when the track errors became smaller.

3.3. Categorical Skill Scores and Rainfall Similarity

Through the use of categorical measures as described in Section 2.2, the QPFs for TY Mangkhut (2018) by the model can also be verified in an objective and quantitative fashion. Firstly, the results of all the hindcast runs for the 24 h target periods of 14 and 15 September are presented in Figure 5a,b, respectively, verified at the 56 rain gauges across the Philippines for six rainfall thresholds from 0.05 to 350 mm (per 24 h) using the performance diagram [82]. For the rainfall on 14 September (Figure 5a), one can see that the TS values tend to decrease with increasing thresholds, as they are mostly about 0.75–0.85 at 0.05 mm (per 24 h), 0.45–0.7 at 10 mm, 0.25–0.4 at 50 mm, and 0.1–0.3 at 100 mm, respectively, as typically found in the literature, e.g., [41–45,55]. As the qualified points become fewer toward the higher thresholds, the scores also tend to become more dispersed with a larger spread in the diagram. At 200 mm, a few runs were able to yield fairly high TS values (TS = 0.5 for those with t_0 = 1200 UTC 12, and 0000 and 1800 UTC 13 September; TS = 0.33 for the two with t_0 = 1800 UTC 12 and 0600 UTC 13 September). However, the TSs are zero in all other runs at 200 mm with no hits. As $H > 0$ are required for POD, SR, and TS to be above zero, all three scores must drop to zero in any forecast when the threshold goes beyond either the observed or predicted peak amount. Therefore, 350 mm is not a meaningful threshold to examine in Figure 5a. For the BSs, all data points are within 0.2 and 3.6, while most of them are inside 0.5–2.5. However, the earlier runs tended to have BSs < 1 due to the northward track bias, whereas the later runs tended to have BSs > 1 with some over-prediction associated with slight southward track bias (more inland).

For 15 September (Figure 5b), most points have BS values between 1.25 and 2.5 and none is below 0.8, so the over-prediction became somewhat more apparent for the second day when the observed peak amount (535.6 mm) more than doubled from one day before. The TSs of QPFs targeted on 15 September from many runs are about 0.55–0.65 at the lowest threshold of 0.05 mm (per 24 h), 0.3–0.55 at 10 mm, 0.1–0.35 at 50 mm, 0.1–0.33 at 100 mm, and mostly zero at 200 mm (Figure 5b). Thus, despite a much higher observed peak rainfall amount on 15 September, the overall QPFs for this day appear no better in TS values. However, several experiments did perform quite well in TS toward the high thresholds in Figure 5b, such as the ones at 0600 UTC 13 (TS = 0.75 at 100 mm), 1800 UTC 13 (TS = 0.5 at 200 mm), and 0600 UTC 14 September (TS = 1.0 at 200 mm). Hence, some individual runs produced decent QPFs at high thresholds for 15 September, although the overall TSs did not seem to improve.

Next, the categorical skill scores of the hindcasts are examined in Figure 5c for the two-day total rainfall from Mangkhut over 14–15 September (in UTC), where the observed peak amount was 785.5 mm at Baguio. Many experiments produced TSs of 0.8–0.85 at 0.05 mm (per 48 h), 0.6–0.8 at 10 mm, 0.25–0.55 at 50 mm, and 0.1–0.4 at both 100 and 200 mm, respectively (Figure 5c). Due to a longer accumulation period and higher total rainfall amounts, these TS values are in general higher than those for individual days seen in Figure 5a,b (at the same threshold amounts of accumulation). However, still no hit was produced at 350 mm (per 48 h) by any of the runs, even though some experiments at shorter lead times are certainly able to produce rainfall totals reaching 600 mm near Baguio in Figure 4. Additionally, similar to the situation for 15 September, the data points for the 48 h QPFs again suggest some over-prediction, especially over the threshold range of 50–200 mm (per 48 h) with BS values from 1 to about 2–2.5 (Figure 5c). Nevertheless, at low thresholds of ≤10 mm, most TSs are quite high (above 0.6) and the BSs (roughly from 0.95 to 1.3) are also fairly close to unity.

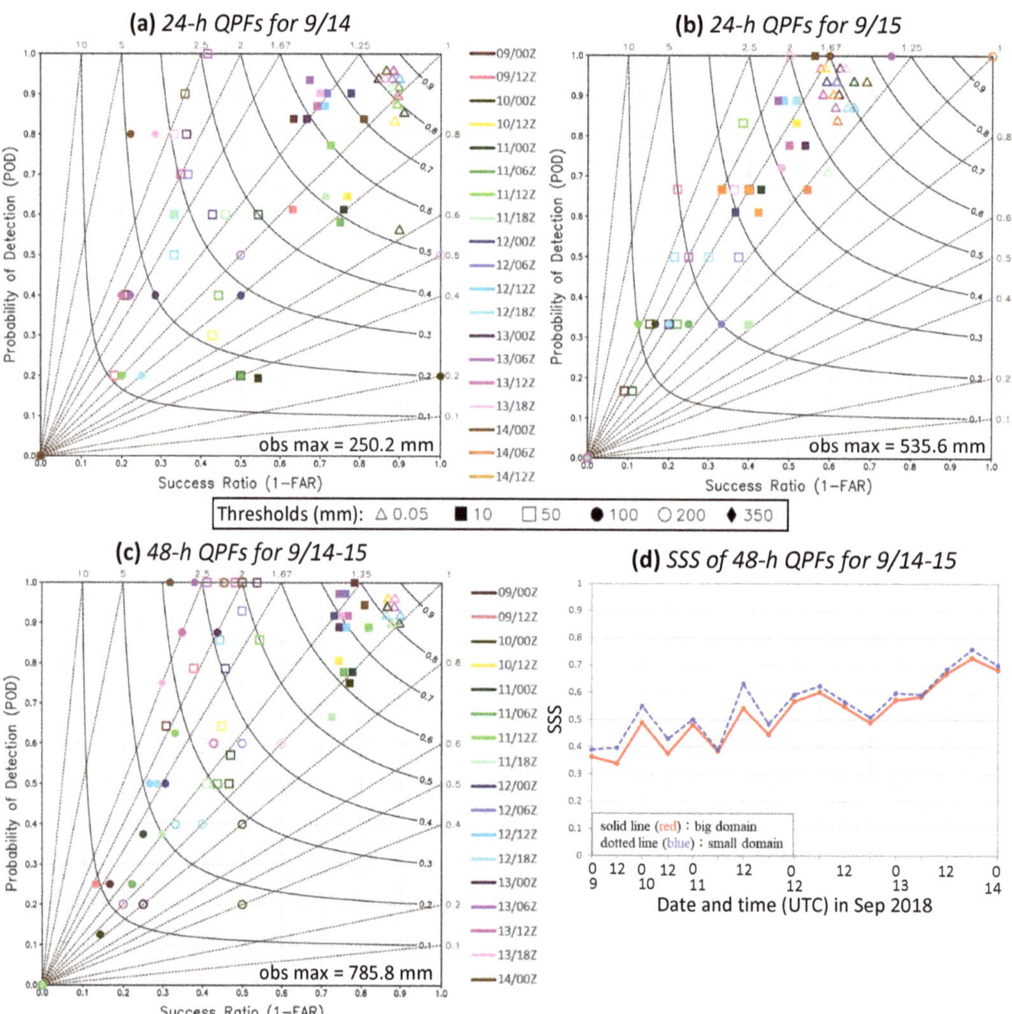

Figure 5. Performance diagrams of 24 h QPFs by forecasts at different initial times (color) targeted for (**a**) 14 and (**b**) 15 September, and of (**c**) 48 h QPFs targeted for 14–15 September 2018 during Mangkhut at six selected thresholds from 0.05 to 350 mm (symbol, see insert). (**d**) The SSS values of 48 h QPFs for 14–15 September as functions of initial time for the big (entire Philippines) and small (Luzon island only) domains. Note that only two runs were executed per day (at 0000 and 1200 UTC) from 9–10 September.

As described in Section 2.2, the SSS measures the overall similarity (from 0 to 1) between the observed and predicted rainfall patterns at the gauge sites over the Philippines ($N = 56$). In Figure 5d, the SSS computed for the patterns of 48 h rainfall accumulation (for the whole event) was about 0.35 in the first two runs and gradually increased to about 0.7 in the last 2–3 runs (red curve), despite some fluctuation among the members. For the smaller domain of Luzon (see Figure 1b), where most of the significant rainfall (say, ≥50 mm) was received (Figure 4q), the SSS values in Figure 5d are slightly higher (blue dashed), roughly by 0.02–0.05. Among all runs, the highest score was achieved by the one with t_0 at 1800 UTC 13 September, with SSS = 0.73 over the entire Philippines (big domain) and SSS = 0.76 for Luzon only (small domain).

4. Time-Lagged Ensemble for Typhoon Koppu (2015)

4.1. Track Forecast and Examples of Rainfall Structure

The second case is Koppu that struck the Philippines in the middle of October in 2015. This typhoon made landfall near 2300 UTC 17 October and penetrated the middle section of Luzon, before turning north over the ocean along its northwestern coast (Figure 1a). A total of 19 hindcast runs were made for Koppu, every 6 h from 0000 UTC 13 to 1200 UTC 17 October (Figure 6a). Compared to Mangkhut, Koppu moved much slower at less than 15 km h^{-1} before landfall, and even only about 6 km h^{-1} off the northwestern coast of Luzon on 19 October after the northward turn (Figures 1a and 6a). At longer lead times, the tracks of Koppu in the predictions from 13–14 October also gradually developed a northward bias during its approach, so that the model TCs made landfall at the northeastern part of Luzon instead (Figure 6a), with track errors of 150–270 km (Figure 6b), and then turned north afterwards. With time, the track errors gradually reduced and improved considerably after 0600 UTC 15 October to all become within 75 km near landfall. However, the northward turn still occurred too early in the model, over land instead of the nearby ocean (Figure 6a). Only until the runs at 0600 and 1200 UTC 17 October did the TC move offshore (with track errors within 60 km during landfall), but only briefly. Thus, although the track errors were not too large and also reduced with time, the slow motion of Koppu off northwestern Luzon on 19–20 October was not well captured by any of the hindcasts, perhaps not too surprising as the last run was at 1200 UTC 17 October.

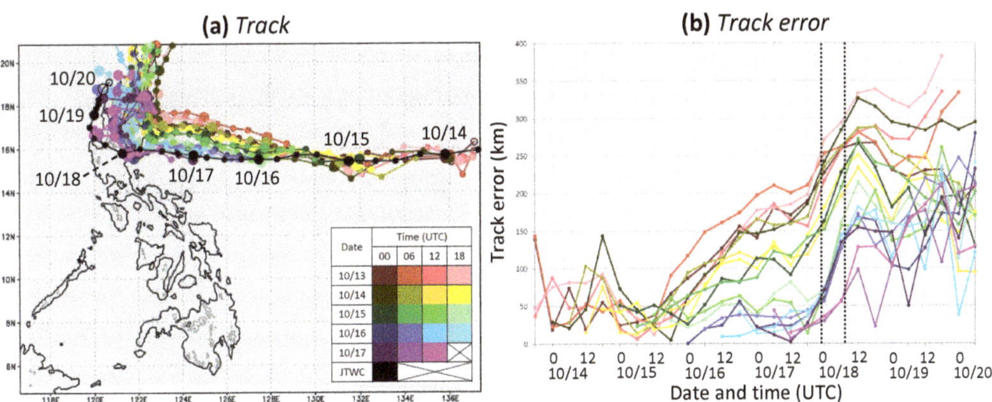

Figure 6. As in Figure 2, but for (**a**) the tracks and (**b**) track errors of TY Koppu (2015) by the time-lagged forecasts, against the JTWC best track.

Similar to Figure 3, the model-predicted rainfall structure of Koppu is compared with the TRMM retrievals in Figure 7, at three different times by the run initialized at 0000 UTC 16 October 2015. The track error in this particular hindcast remained quite small (less than ~65 km) up to landfall, but increased rapidly afterward as the model TC moved north over land (Figure 6b) as mentioned. Therefore, the track was not among the best. Nevertheless, Figure 7 shows that the model was able to reproduce the change in rainfall structure of Koppu, from a more symmetric structure at around 2300 UTC 16 October before landfall (Figure 7a,d) to an more asymmetric pattern on late 17 and early 18 October after landfall (Figure 7b,e), with most of the clouds on the western side of the storm and intense rainfall over Luzon near 15°–16° N. Even near 1100 UTC 19 October when Koppu's center moved offshore and a primary rainband developed and extended northward (Figure 7c), this highly asymmetric rainfall structure was captured quite well, with the rainband at almost the correct location (Figure 7f) despite the track error.

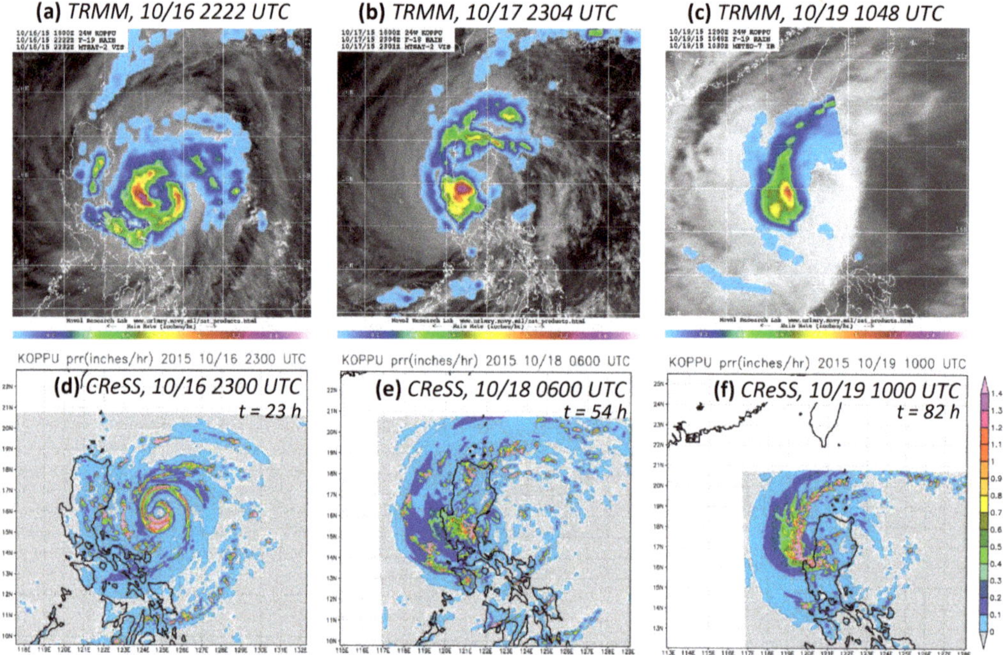

Figure 7. TRMM rainrate retrieval (inch h^{-1}, color), overlaid on visible (or infrared) cloud imagery of the MTSAT-2 (or Meteo-7) satellite at closest time, of TY Koppu (2015) at (**a**) 2222 UTC 16, (**b**) 2304 UTC 17, and (**c**) 1048 UTC 19 October (source: NRL), and model rainrate (inch h^{-1}, color) valid at (**d**) 2300 UTC 16, (**e**) 0600 UTC 18, and (**f**) 1000 UTC 19 October from the forecast initialized at 0000 UTC 16 October 2015 (forecast hour labeled). The region plotted and color scale used are similar for each pair.

4.2. Model QPFs for Typhoon Koppu (2015)

Due to the slow motion and long impact duration of Koppu, three 24 h target periods are selected for this case to examine the QPFs in Figure 8. Starting and ending at 1200 UTC, the first one is over 16–17 October (top row, Figure 8a–f). During this first period, Koppu's center was still over the ocean east of Luzon and thus the rain gauges over land did not record too much rainfall, which peaked at 109.6 mm at Virac (Figure 8f). In the QPF for this target period by the first run (t_0 = 0000 UTC 13 October), on day 4.5 (halfway between days 4 and 5), the model rainfall (Figure 8a) was maximized (along the eyewall) south of the track that had a northward bias. As the lead time decreased, the northward bias also reduced and the TC rainfall area gradually shifted southward (Figure 8b–e). At the short range, the model was producing rainfall along the eastern coast of northern Luzon (Figure 8c–e), where there is a lack of rain gauges (Figure 1b).

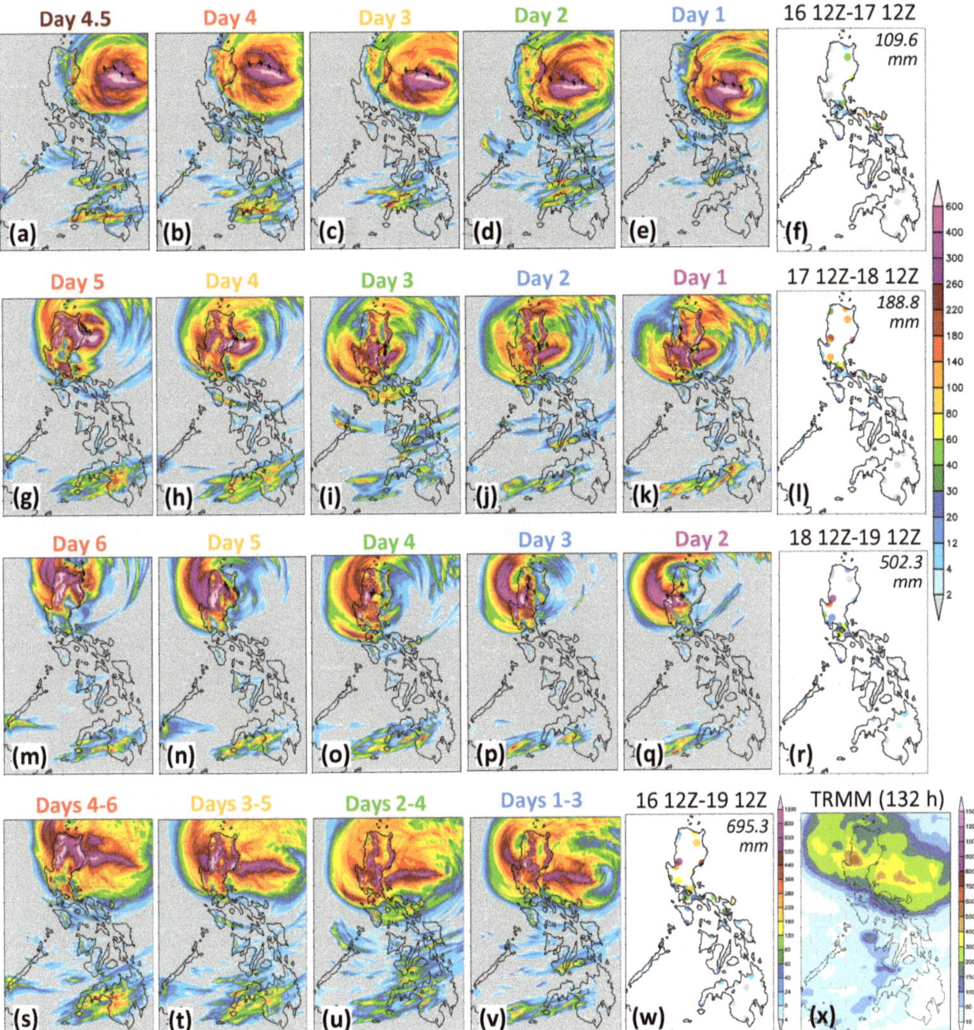

Figure 8. As in Figure 4a–f, except for predictions (ranges as labeled) and gauge observations for the 24 h rainfall distributions over (**a–f**) 1200 UTC 16-1200 UTC 17 October, (**g–l**) 1200 UTC 17-1200 UTC 18 October, and (**m–r**) 1200 UTC 18-1200 UTC 19 October 2015 during TY Koppu (2015), respectively. (**s–x**) As in (**a–f**), except at (**s–v**) four different ranges and (**w**) from the gauge data for 72 h QPFs over 1200 UTC 16–1200 UTC 19 October. (**x**) TRMM total rainfall (mm) for the entire typhoon event (132 h starting from 1800 UTC 15 October). The same color scale is used for (**a–r**) and doubled for (**s–v**). Only (**x**) has its own scale.

During the next 24 h target period from 1200 UTC 17 to 1200 UTC 18 October, Koppu continued to approach, made landfall, penetrated the middle section of Luzon, and then exited the land to move offshore (Figure 1a). Therefore, rainfall was the most along the middle section of Luzon, with a peak amount of 188.8 mm again at Baguio (second highest at Casiguran), and it decreased toward both the north and south (Figure 8l). From the longest (day 5) to the shortest range (day 1), the model rainfall area gradually shifted south as the northward track bias reduced (Figure 8g–k), with ample rainfall in central Luzon (mainly along the two mountain ranges) at lead times within day 4. However, within day 3, the model QPFs also appear to have too much rainfall in southern Luzon, in the area north of Manila (Figure 8i–k), most likely because of the reduced translation speed associated with the track errors during landfall as mentioned earlier.

During the third (and last) 24 h period from 1200 UTC 18 to 1200 UTC 19 October, Koppu was moving northward slowly offshore of northwestern Luzon, so that the intense and asymmetric rainband seen in Figure 7c produced heavy rainfall at Baguio, with a peak 24 h amount reaching 502.3 mm, and also at Dagupan (176 mm) facing the Lingayen Gulf (Figures 2b and 8r). This maximum of over 500 mm was much higher than those combined in the 48 h earlier. At ranges within day 4, the model is seen to be able to produce comparable amounts of rainfall (reaching 600 mm) in the nearby region (Figure 8o–q), because the asymmetric, northward extending rainband seen in Figure 7f could be captured reasonably well. The precision in heavy-rainfall location also improved at shorter lead times. On day 5 (t_0 = 1200 UTC 14 October), the QPF also appeared reasonably good in quality, but the rainfall amounts seemed over-predicted (Figure 8n). Starting at 1200 UTC 13 October, the QPF on day 6 did not extend its heavy rainfall area down to the Baguio area and was not as ideal (Figure 8m), since Koppu made landfall at the northeastern corner of Luzon in this run at such a long range (Figure 6a). In [85], a single forecast run initialized at 1200 UTC 17 October (same as the last t_0 in our ensemble) using the Typhoon Weather Research and Forecasting (TWRF) model of the Central Weather Bureau (CWB) was examined for Koppu, but the heavy rainfall missed the Baguio region because the track was slightly too far south. In addition, the model QPFs was only examined through visual comparison with the observations, and no objective measure was used.

When the accumulation period is the combined 72 h from 1200 UTC 16 to 1200 UTC 19 October 2015, again the observed rainfall was the most in central Luzon, with a peak total of 695.3 mm at Baguio (Figure 8w) while the TRMM data indicated a maximum less than 600 mm in 132 h (Figure 8x). The predicted rainfall distributions for the 72 h period by the four runs initialized at 1200 UTC on each day from 13–16 October, thus at ranges from days 4–6 to days 1–3 (longer to shorter), are depicted in Figure 8s–v, respectively. One can see that, with the gradual reduction in track errors with time, the rainfall in the middle section of Luzon near Baguio as well as along the two mountain ranges of CCM and SMM became more concentrated (Figure 8s–v). Additionally, the rainfall extending northward from the Lingayen Gulf and caused by the primary rainband on 18–19 October became more distinct and clear with time. Overall, the comparison in Figure 8 indicates that the rainfall of Koppu was predicted reasonably well within the range of about 3–4 days, especially for rainfall toward the last day of the 72 h period, when the primary rainband could be captured at nearly the right location despite some biases in the TC track.

4.3. Categorical Skill Scores and Rainfall Similarity

The results of 24 h and 72 h QPFs for Koppu in categorical skill scores, verified against the rain gauges, are shown in Figure 9. For the first 24 h period starting from 1200 UTC 16 October, meaningful results in Figure 9a (with a possibility of TS > 0) are confined to 100 mm (per 24 h) and below, since the observed peak amount was only 109.6 mm (Figure 8f). As seen, QPFs for this period have TS values about 0.3–0.65 at 0.05 mm and 0.15–0.5 at both 10 and 50 mm (Figure 9a). At the threshold of 100 mm, none of the hindcasts initialized before and at 1200 UTC 16 October produced a TS above zero. In terms of the BS, the data points at 0.05 mm are between about 0.7 and 1.3, and between about 0.4 and 1.8 at 10 mm (Figure 9a), and thus are quite reasonable. However, at 50 mm their values are larger at about 1–3.7 and suggested some over-prediction, most likely near the northeastern coast of Luzon (see first row of Figure 8).

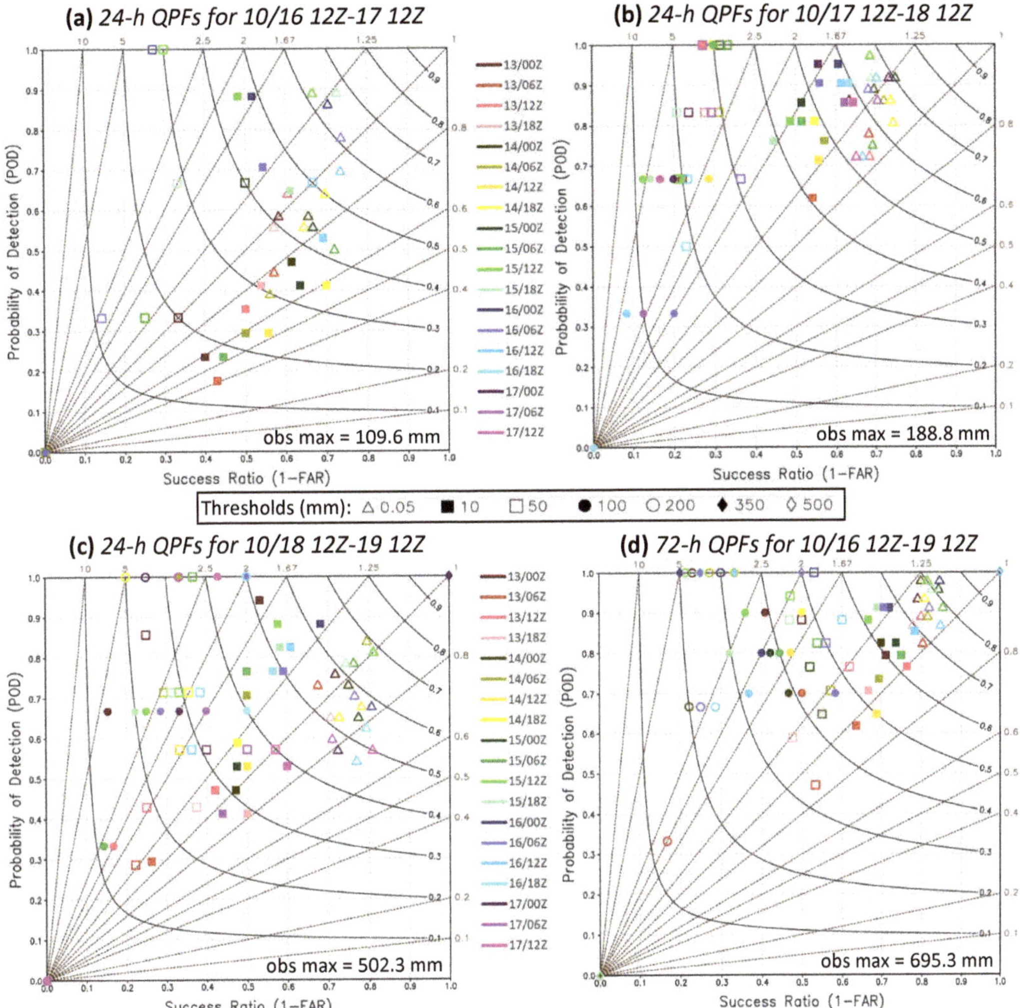

Figure 9. As in Figure 5a–c, except for 24 h QPFs targeted for (**a**) 1200 UTC 16–1200 UTC 17, (**b**) 1200 UTC 17–1200 UTC 18, and (**c**) 1200 UTC 18–1200 UTC 19 October 2015, and (**d**) 72 h QPFs for the combined period (for 1200 UTC 16–1200 UTC 19 October) during Koppu at six selected thresholds from 0.05 to 350 mm (symbol), plus 500 mm in (**d**).

For the second 24 h period starting from 1200 UTC 17 October, the observed peak rainfall increased to 188.8 mm as the storm approached and made landfall, but still meaningful results are restricted to below the threshold of 200 mm. For this period (Figure 9b), the TSs are roughly over 0.5–0.7 at 0.05 mm, 0.4–0.6 at 10 mm, 0.2–0.33 at 50 mm, and 0.1–0.3 at 100 mm, respectively. While the TSs improved compared to those for the first 24 h period, the over-prediction seemed to be more serious as the BSs of all points are above 1 and up to about 5 in Figure 9b. In addition, the BSs tend to be larger toward the higher thresholds. This increased over-forecasting on day 2 when Koppu made landfall and the rainfall over Luzon increased is similar to the situation shown in Figure 5a,b for Mangkhut.

For the third and final 24 h period starting from 1200 UTC 18 October (Figure 9c), the observed peak rainfall amount was the most among the three days and reached 502.3 mm, so the results in categorical statistics can be examined at higher thresholds. At the lowest threshold of 0.05 mm (per 24 h), the TS values were about 0.45–0.7 and BSs were about 0.7 to 1.1, thus suggesting good performance in rainfall areas over the Philippines. At 10 and 50 mm, most of the TSs were lowered to about 0.3–0.6 and further to 0.2–0.4, while the BSs increased to about 0.8–1.6 and further to 1–2.5 (Figure 9c). Toward the higher thresholds, the TSs were 0.1–0.5 at 100 mm with BSs roughly between 1.3 and 4.5, and the highest TS also reached 0.5 at 200 mm (t_0 = 1200 UTC 15 and 1800 UTC 16 October), where the BSs were about 2 to 5. Thus, over-prediction also occurred toward the higher thresholds, but overall it appeared somewhat less serious compared to the second 24 h period. Finally, eight experiments, of which the first was initialized at 0600 UTC 14 and the last at 0000 UTC 17 October, registered a TS = POD = SR = BS = 1.0 at 350 mm (overlapping points). As eight out of 19 runs achieved this and the range was 102–126 h for the earliest member (beyond day 5), this result was quite good and very encouraging (Figure 9c).

For the three-day total rainfall during Koppu from 1200 UTC 16 to 1200 UTC 19 October 2015, the categorical statistics of QPFs from the time-lagged ensemble are depicted in Figure 9d. With an observed peak amount of 693.5 mm at Baguio (Figure 8w), many runs yielded TS values of about 0.7–0.83 at 0.05 mm (per 72 h), 0.5–0.7 at 10 mm, 0.35–0.55 at 50 mm, 0.3–0.45 at 100 mm, and 0.2–0.33 at 200 mm, respectively (Figure 9d). At the higher threshold of 350 mm, all runs also produced TSs of 0.2–0.35 except for three earlier ones with t_0 before 0600 UTC 14 October. Even at 500 mm, eight out of the last nine members starting from 1200 UTC 14 October also yielded TSs of either 0.5 (three runs) or 1.0 (five runs), except the one at 1800 UTC 15 October. Again, the TS values tend to rise when the period for accumulation is lengthened, because now the events (rainfall reaching specific thresholds) are more likely to occur (in both the observation and prediction) to increase the chance of hits, and some timing errors in rainfall are also tolerated. Similar to the results of 24 h periods, the BS values of the 72 h QPFs indicated some over-forecasting since the majority of the points are above BS = 1 rather than below (Figure 9d). For those with BS < 1, they were all produced by runs at and before 1200 UTC 14 October, so that larger northward track biases existed. Again, the BS values tended to increase with rainfall thresholds in Figure 9d, from about 1–1.2 at 0.05 mm (per 72 h) to roughly 3–5 at 350 mm and 2–4 at 500 mm. Correspondingly, for such points with 0 < TS < 1 at high thresholds of 350 and 500 mm (16 occasions), they all have POD = 1 but SR < 1 (except for one). Overall, Figure 9d indicates an improved performance in QPFs for Koppu toward the high thresholds up to 500 mm (per 72 h). This is encouraging because even for the last run (t_0 = 1200 UTC 16 October), most of the rainfall produced by the northward-extending rainband near Baguio and the Lingayen Gulf took place on day 3, i.e., 48–72 h into the hindcast experiment. Therefore, the CReSS model at Δx = 2.5 km is able to reproduce realistic rainfall for TY Koppu (2015) at lead times beyond 48 h, allowing for timely preparation by the authorities before the main rainfall period of the TC.

For TY Koppu (2015), the results of SSS for the 72 h rainfall (from 1200 UTC 16 to 1200 UTC 19 October) are shown in Figure 10a. For the entire Philippine Archipelago, the SSS was less ideal between 0600 UTC 13 and 0000 UTC 14 October, but mostly at least around 0.7 afterwards (red curve), again with some run-to-run variations. It is noteworthy that the earliest run, with t_0 at 0000 UTC 13 October, also registered an SSS of 0.66 and comparable to some of the later runs executed from 14–16 October. This result is encouraging considering that the target period is in the range of 84–156 h. For the small domain of Luzon, the SSS values (blue curve) are mostly also slightly better, by ~0.05 at most. Among all runs, the highest SSS is close to 0.8 by the last run with t_0 at 1200 UTC 16 October (and thus could fully cover the 72 h target period), whereas the lowest one is about 0.4 (with t_0 at 0600 UTC 13 October). Compared to those for Mangkhut (see Figure 5d), the SSS values for Koppu in Figure 10a appear better to some extent, in agreement with the better results in Figure 9.

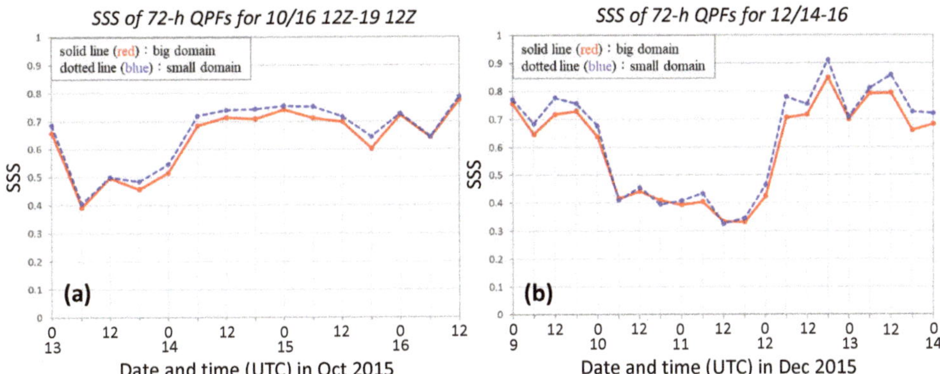

Figure 10. As in Figure 5d, except for SSS of 72 h QPFs for the period of (**a**) 1200 UTC 16–1200 UTC 19 October during TY Koppu (2015) and (**b**) 14–16 December (in UTC) during TY Melor (2015), as functions of initial time for the big and small domains (see Figure 1b).

5. Time-Lagged Ensemble for Typhoon Melor (2015)

5.1. Track Forecast and Examples of Rainfall Structure

The third and final typhoon is Melor (2015), for which a total of 28 hindcast runs were carried out every 6 h from 0000 UTC 9 to 1800 UTC 15 December. As shown in Figure 1a, TY Melor moved westward through the central part of the Philippines, and across the northern end of Samar, the far end of southeastern Luzon, and south of Luzon on 14 December, then the northeastern shore of Mindoro during early 15 December, and finally moved into the ocean west of Manila. In the hindcasts made during the first three days of 9–11 December, the model TCs also all developed a northward bias that tended to be the most serious and about 300–350 km too north around 13 December (Figure 11a). While these TCs would turn south again to make landfall in the central Philippines, the timing was too late for some of them, so the track errors remained quite large (Figure 11b). From 12 December, on the other hand, the tracks started to improve considerably and all those executed at and after 0600 UTC 12 December had track errors within about 75 km at or near the first landfall at Samar, with a lead time up to about 45 h (Figure 11b). The moving speeds of TCs in these later runs were also roughly correct. With time, however, several of them still had some northward bias and the predicted the TC moved toward the Manila area instead of Mindoro (such as those initialized at 0000 UTC 12 and 0600–1800 UTC 13 December), while a few others exhibited slight southward bias near Mindoro (Figure 11). Thus, using the NCEP GFS gross analyses and forecasts as IC/BCs, the majority of the earlier hindcasts exhibited a northward bias for all three typhoons in our study. These runs were initialized at least about 2 days before landfall for Melor (Figure 11), 2.5 days for Koppu (Figure 6), and ≥3 days before landfall for Mangkhut (Figure 2). Similar characteristics using NCEP data are also found in some typhoons in Taiwan [58,59].

In Figure 12, the rainfall structure of Melor in the model run at t_0 = 1800 UTC 12 December is compared with the TRMM retrievals at two different times. This particular run had small track errors (≤75 km) throughout the landfall period. Before landfall at Samar, at around 1900 UTC 13 December, the model is shown to capture the correct eye size of Melor and the rainfall in the northern and eastern quadrants of the storm (Figure 12a,b). Similarly, shortly after landfall when Melor's center was right between Luzon and Mindoro at 1000 UTC 15 December, the intense rainfall near northwestern Mindoro was also nicely captured by the model (Figure 12c,d), although the model rainfall across Luzon was seemingly not as widespread as in the TRMM observation. Overall, the rainfall structures of Melor in the CReSS model are realistic with small track errors (Figure 12), similar to the examples shown earlier for the other two typhoons.

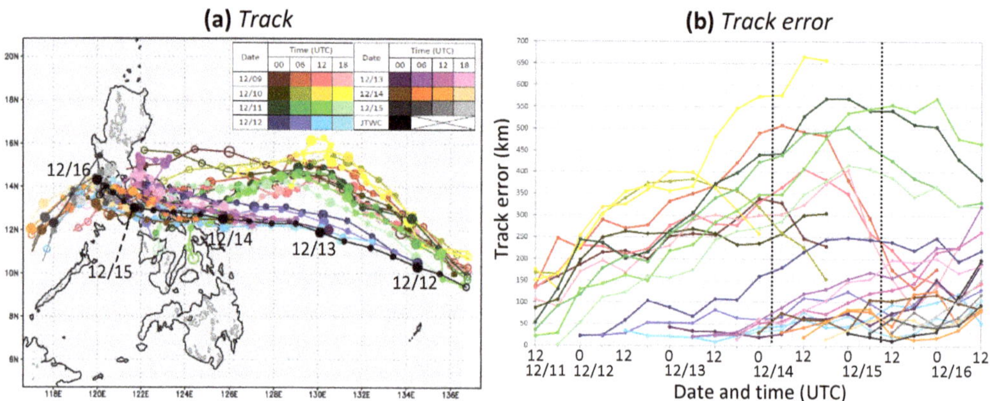

Figure 11. As in Figure 2, but for (**a**) the tracks and (**b**) track errors of TY Melor (2015) by the time-lagged forecasts, against the JTWC best track.

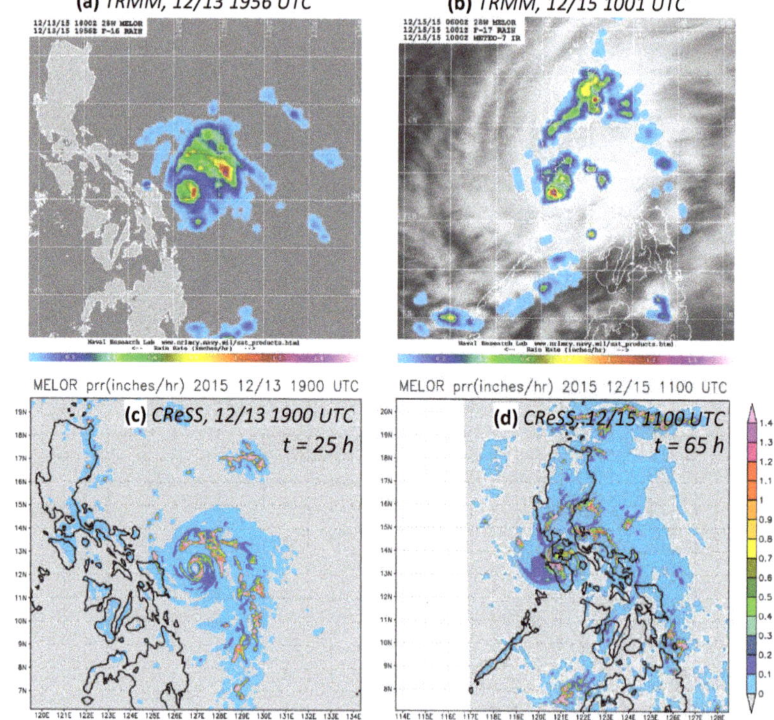

Figure 12. As in Figure 7, except for TRMM rainrate (inch h^{-1}) of TY Melor (2015) at (**a**) 1956 UTC 13 (no cloud imagery from stationary satellite) and (**b**) 1001 UTC 15 December (source: NRL), and model rainrate (inch h^{-1}) valid at (**c**) 1900 UTC 13 and (**d**) 1100 UTC 15 December from the forecast initialized at 1800 UTC 12 December 2015.

5.2. Model QPFs for Typhoon Melor (2015)

As TY Melor (2015) moved across many islands in the central Philippines and also slowed down from 15–16 December, its impact period was relatively long. Therefore, three 24 h target periods covering 14–16 December were chosen as well for this event (rows 1–3 of Figure 13, respectively). On the first day of 14 December, significant rainfall occurred in northern Samar, the southeastern end of Luzon, and in the Manila area, with a peak amount of 169.6 mm (Figure 13f). The model predictions, selected to be all from 0000 UTC of earlier days, show better QPFs over the central Philippines only within the range of day 2 and not before (Figure 13a–e), but none of the runs captured the significant but localized rainfall near Manila. Due to the northward track bias, the rainfall in the central Philippines was missed by earlier runs at ranges beyond day 3 (Figure 13a,b). Similarly, the rainfall on day 3 (t_0 = 0000 UTC 12 December, Figure 13c) also just missed the northern part of Samar.

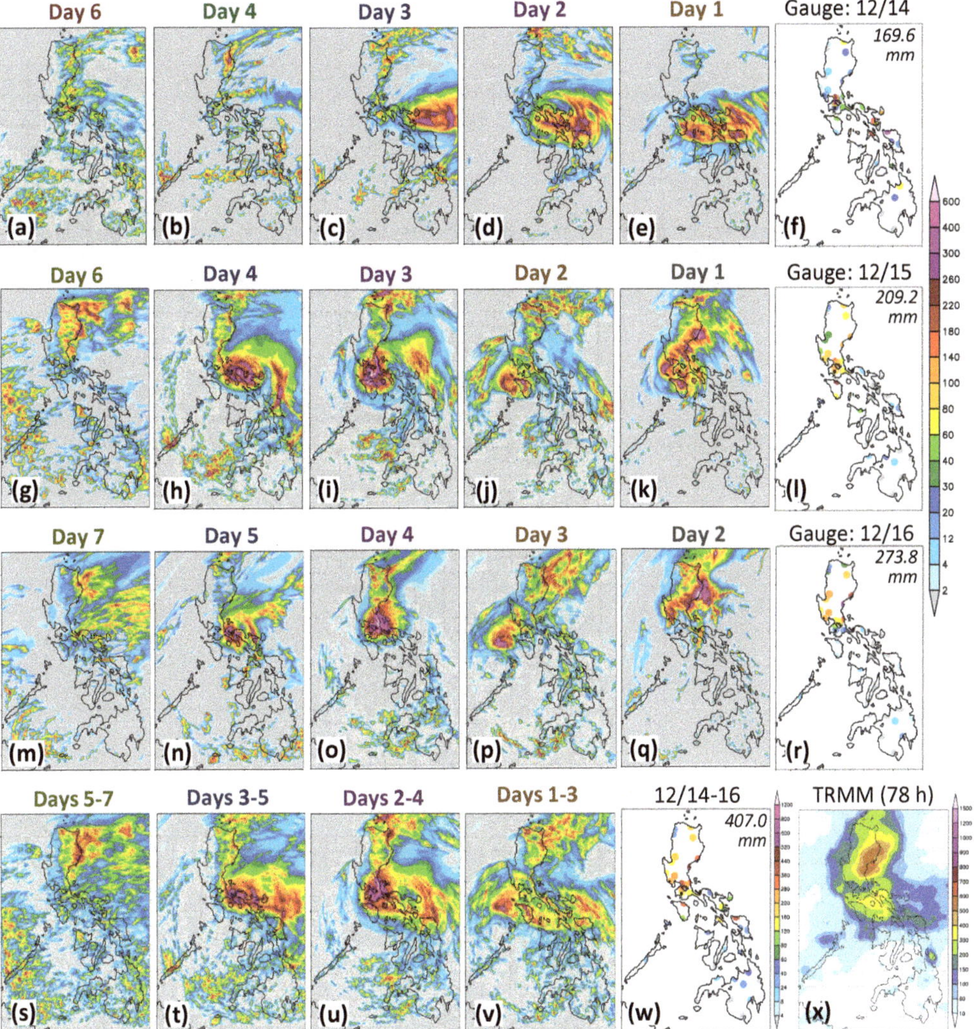

Figure 13. As in Figure 8, except for predictions (ranges as labeled) and gauge observations for the 24 h daily rainfall distributions on (**a–f**) 14, (**g–l**) 15, and (**m–r**) 16 December, respectively, (**s–w**) 72 h QPFs for 14–16 December 2015, and (**x**) TRMM total rainfall (mm) for the entire typhoon event (78 h starting from 1800 UTC 13 December) during TY Melor.

On the second day of 15 December, the rain over the eastern part of the central Philippines had subsided, while northern Mindoro and southwestern Luzon (near Manila) had more rain instead, with a peak of 209.2 mm at Calapan to the north of the High Rolling Mountains (Figures 2b and 13l). The handful of hindcast experiments selected (Figure 13g–k) showed fairly different rainfall distributions among them, but all captured the rainfall in Mindoro within the range of 3 days despite their differences (Figure 13i–k). On day 4 (t_0 = 0000 UTC 12 December), this run was still producing rainfall over the central Philippines linked to a slower TC motion (Figure 13h). In the run on day 6 (t_0 on 10 December), the storm was weakening and still east of Luzon without making landfall, thus the model rainfall was mainly along the northeastern and northern coast of Luzon (Figure 13g).

On the third day of 16 December, Melor was lingering offshore to the west of the Manila area, and most rainfall was observed along the eastern shore of Luzon with a peak daily value of 273.8 mm at Baler (Figures 1b and 13r). The hindcasts at ranges from day 7 to day 2 again show very different rainfall distributions as somewhat expected due to the track differences (Figure 13m–q). At the shortest range selected to show, the rainfall along the coast of eastern Luzon was captured on day 2 (Figure 13q), but to a lesser and lesser degree at longer ranges on days 3–4 (Figure 13o,p) where the differences in TC track also seemed apparent. The day-5 hindcast for 16 December (Figure 13n) is the same run to produce day 4 QPF for 15 December (Figure 13h), and the rain was still in the central Philippines due to slow TC motion. Likewise, the day 7 QPF for 16 December is an extension from day 6 QPF for 15 December (Figure 13g), and thus the rainfall decreased as the TC continued to diminish off the coast of eastern Luzon (Figure 13m). Therefore, while the QPFs for fixed target dates show different characteristics for Melor in Figure 13, they are largely dictated by the TC track in the model.

For the combined three-day period of 14–16 December, the observed rainfall from Melor was the most in eastern Luzon, the Manila area, northern Samar, and northern Mindoro, with a peak amount reaching 407 mm at Baler (Figure 13w). Compared to rain-gauge observations, the TRMM retrieval indicated more rainfall, reaching 500–600 mm in eastern Luzon and almost 300 mm in northern Mindoro, but only about 200 mm or less in Manila and northern Samar (Figure 13x). The predicted 72 h rainfall within the short range of days 1–3 (Figure 13v) was reasonable and close to gauge observations at Samar, Mindoro, and eastern Luzon, but not enough near Manila. On days 2–4, the hindcast from 0000 UTC 13 December (Figure 13u) made good 72 h QPFs, with much rainfall (~500 mm) over Manila and southern Luzon, and was perhaps the most ideal among the four runs selected for visual comparison. At the longer range of days 3–5, due to the slow TC motion in this run from 12 December as mentioned, the rainfall in the central Philippines had not extended westward into the Manila area and Mindoro (Figure 13t). At an even longer range, again, the 72 h rainfall only reached eastern Luzon associated with the erroneous track (Figure 13s). In summary, the QPFs were reasonable and captured most of the heavy rainfall regions along the track of Melor through the archipelagos in the central Philippines up to the range of days 2–4, but a rainy scenario in southern Luzon was delayed at longer lead times due to the track errors linked to slow TC motion in the model.

5.3. Categorical Skill Scores and Rainfall Similarity

The results of categorical scores of 24 h and 72 h QPFs for Melor are shown in Figure 14. On the first day of 14 December, the observed peak rainfall was 169.6 mm, and many of the 24 h QPFs for this day exhibited TS values of about 0.4–0.7 at the threshold of 0.05 mm (per 24 h), 0.3–0.7 at 10 mm, and 0.05–0.4 at 50 mm (Figure 14a), respectively. At the highest threshold of 100 mm (per 24 h) where the TSs could be above zero, the values were over a wide range from about 0.1 all the way up to 0.8. However, all eight such QPFs (with TS > 0 at 100 mm) in Figure 14a were from model runs at and after 0600 UTC 12 December, i.e., those with reduced northward track bias (cf. Figure 11a) and inside the range of 66 h. For this day, the BS values are mostly between 0.4 and 2.0 and quite reasonable, although again some under-prediction tended to occur at low thresholds with slight over-prediction toward the higher thresholds.

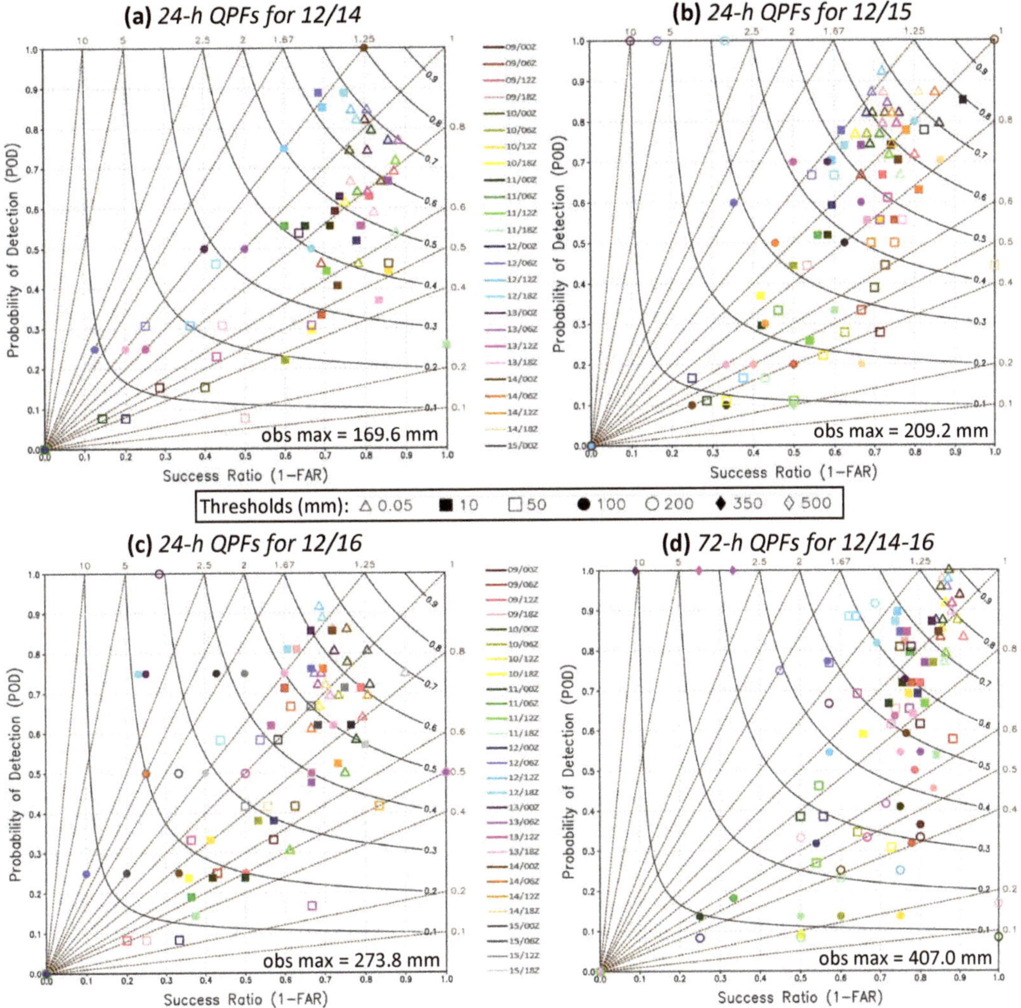

Figure 14. As in Figure 9, except for 24 h QPFs targeted for (**a**) 14, (**b**) 15, and (**c**) 16 December 2015 (in UTC), and (**d**) 72 h QPFs for the combined period (of 14–16 December) during Melor at six selected thresholds from 0.05 to 350 mm, plus 500 mm in (**d**).

The peak 24 h daily rainfall increased slightly to 209.2 mm (in northern Mindoro) on 15 December (cf. Figure 13l). The QPFs for this day had TS values of roughly 0.5–0.75 at 0.05 mm (per 24 h) but over wide ranges of about 0.2–0.8 at 10 mm and 0.1–0.65 at both 50 and 100 mm, respectively (Figure 14b). At the threshold of 200 mm, which is very close to the observed maximum, a few runs registered a TS above zero, including those initialized at 0600 (TS = 0.17) and 1800 UTC on 12 (TS = 0.33) and at 0000 UTC on 14 December (TS = 1). Many of the hindcasts also produced higher TS values for 15 December compared to those for the day before (Figure 14a,b). Overall, the BS values from the QPFs for 15 December indicated slight under-prediction, as there were more points with BS below 1 than above, except at the higher thresholds of 200 and 350 mm (Figure 14b).

For the third day of 16 December, when the observed peak amount increased again to 273.8 mm, the TS values were about 0.5–0.7 at 0.05 mm (per 24 h), 0.15–0.65 at 10 mm, 0.05–0.5 at 50 mm, and 0.1–0.5 at 100 mm, respectively (Figure 14c). At 200 mm, again, a few hindcasts were able to produce hits and TS values above 0, including those initialized at times from 1800 UTC 12 to 1800 UTC 13 December (TS = 0.28–0.5), and at 0000 (TS = 0.25) and 1200 UTC 15 December (TS = 0.2). Distributed quite evenly between about 0.25 and 3.5, the BS values suggested some under-prediction at 50 mm and some over-prediction at 100 mm, but little preference elsewhere (Figure 14c).

For the 3-day total rainfall over 14–16 December during Melor, the observed peak amount reached 407 mm at Baler and the categorical score results are shown in Figure 14d. When the accumulation period is lengthened to 72 h, the QPFs yielded TS values about 0.7–0.9 at the lowest threshold of 0.05 mm (per 72 h), 0.55–0.75 at 10 mm, 0.2–0.65 at 50 mm, and 0.1–0.6 at 100 mm, respectively. At 200 mm, all but three runs yielded TS > 0. The highest value was 0.65 by the run with t_0 at 1800 UTC 12 December, while 10 runs produced TSs of 0.2–0.44, and seven others (between 1800 UTC 9 and 0000 UTC 12 December) had TSs below 0.2 (Figure 14d). At 350 mm, which is not much below 407 mm, the three hindcasts initialized from 0000–1200 UTC 13 December also produced non-zero TSs of 0.09–0.33. In terms of BS, many data points indicated slight under-forecasting for Melor, more serious with BS around 0.2–0.6 at the thresholds of 100 and 200 mm. Over-forecasting only occurred at 350 mm, but at such a high threshold that the statistics were based on very few data points (Figure 14d). Thus, over the accumulation period for the entire duration of the three typhoons, Melor was the only case for which the time-lagged ensemble produced mostly under-prediction in QPFs. For both Mangkhut (Figure 5c) and Koppu (Figure 9d), over-prediction occurred, especially toward the higher thresholds.

The values of SSS from the lagged members for the total 3-day rainfall from Melor (14–16 December) are shown in Figure 10b. One can see that higher scores of ≥0.65 were achieved by the first five and last eight runs, while the middle eight runs executed mainly from 10–11 December had worse SSS values near and below 0.4 (Figure 10b), evidently due to the northward TC track biases in those runs (see Figure 11). Therefore, the first several runs also had good SSS values comparable to the runs at shorter ranges after 0000 UTC 12 December, and thus indicate the potential to produce decent QPFs at longer lead times. For the big domain, the best SSS (=0.85) was made by the run at 1800 UTC 12 December, consistent with the best TS of 0.65 at 200 mm shown in Figure 14d above, while a few other runs also had SSS near or above 0.8. Over the small domain, which for Melor is slightly larger and extends down to include Mindoro (see Figure 1b), the SSS values also tend to be higher (except for some runs with larger track errors), and the highest value is SSS = 0.91 and very impressive (Figure 10b). Overall, the SSS values of the three typhoons are in agreement with the verifications using a categorical matrix.

6. Discussion

In this section, the QPF performance for the three typhoons is further discussed. Since there is a lack of studies utilizing categorical statistics or other objective methods to verify model QPFs for typhoons in the Philippines in the open literature, in this section our results are compared with those for selected typhoons in Taiwan in earlier studies at the same thresholds and similar ranges. Such a comparison is made in Table 3, sorted by the observed peak rainfall amounts. While the typhoons in Taiwan are chosen somewhat arbitrarily and the best TSs are used, one can see that the TSs of the QPFs at the short range (within 72 h) exhibit a clear tendency of higher scores in larger rainfall events as demonstrated before [41–45], although this phenomenon is certainly not exclusive because the predictability varies from case to case. Owing to this dependency property, it would not be fair to compare the TSs for very rainy TCs in Taiwan, such as those ranked near the top [41,45,64], with those for the three typhoons in the Philippines. As such, when the TC events in Taiwan with a peak 24 h amount of about 500–600 mm or less are used for comparison, perhaps below Kong-Rey (2013), one can see that the TSs for the typhoon cases at the short range [41,45,55] in the two regions are roughly comparable (Table 3, top half). At forecast ranges beyond 72 h (up to 8 days), the situation is similar and the scores are comparable. As shown in Section 4.3, quite a number of runs were able to predict a rainfall amount of $\geq$350 mm at Baguio (and not at any other site) from Koppu over the third 24 h period selected (Figure 9c), and thus had a perfect TS of 1.0, even at ranges beyond 72 h (the earliest one with t_0 at 0600 UTC 14 October). One should also be aware that such high scores are more likely to occur in the Philippines due to the much fewer total verification points available ($N = 56$), such that very few points can reach the high thresholds, compared to Taiwan where the gauge sites operated by the CWB are around 450.

Table 3. Comparison of TS values of 24 h QPFs by the CReSS model between selected typhoons in Taiwan and the three TCs in the Philippines in this work, at six rainfall thresholds from 50 to 750 mm. Results are grouped into those within the short range (when the target period is inside 72 h from t_0, top half) or beyond (>72 h, bottom half), and sorted by the observed peak rainfall amount (mm) from high to low. For each typhoon, the best TS values are listed from all available runs in the ensemble (if applicable). A "-" indicates threshold exceeding the peak amount, and the letter "F" in the source column represents a figure in this study.

Category/ Typhoon	Observed Peak Amount (mm)	TS at Threshold Value (mm)						Source
		50	100	200	350	500	750	
Taiwan, $\leq$72 h								
Morakot (2009)	1663	0.81	0.80	0.80	0.62	0.50	0.42	[64]
Fanapi (2010)	1110	0.80	0.67	0.61	0.45	0.30	0.18	[41]
Saola (2012)	889	0.89	0.81	0.67	0.41	0.22	0.05	[45]
Soudelor (2015)	842	0.86	0.78	0.50	0.50	0.35	0.36	[45]
Fung-Wong (2014)	797	0.76	0.70	0.32	0.33	0.33	0.00	[45]
Kong-Rey (2013)	706	0.79	0.65	0.41	0.32	0.00	-	[45]
Tembin (2012)	621	0.67	0.50	0.34	0.00	0.00	-	[45]
Matmo (2014)	556	0.68	0.61	0.61	0.40	0.20	-	[45]
Nanmadol (2011)	488	0.64	0.66	0.40	0.20	-	-	[45]
Talim (2012)	415	0.72	0.58	0.38	0.38	-	-	[55]
Meranti (2011)	298	0.42	0.12	0.00	-	-	-	[41]
Jelawat (2012)	216	0.56	0.30	0.50	-	-	-	[55]
Philippines, $\leq$72 h								
Mangkhut (2018)	536	0.33	0.75	1.00	0.00	0.00	-	F5b
Koppu (2015)	502	0.40	0.43	0.50	1.00	0.00	-	F9c
Melor (2015)	274	0.50	0.43	0.25	-	-	-	F14c
Taiwan, >72 h								
Saola (2012)	884 [#]	0.79	0.73	0.70	0.50	0.20	0.00	[55]
Tembin (2012)	621	0.17	0.16	0.03	0.00	0.00	-	[55]
Kong-Rey (2013)	426 [#]	0.63	0.48	0.26	0.00	-	-	[55]
Talim (2012)	415	0.70	0.30	0.21	0.00	-	-	[55]
Jelawat (2012)	216	0.45	0.38	0.50	-	-	-	[55]
Philippines, >72 h								
Mangkhut (2018)	536	0.36	0.20	0.00	0.00	0.00	-	F5b
Koppu (2015)	502	0.37	0.43	0.50	1.00	0.00	-	F9c
Melor (2015)	274	0.47	0.50	0.50	-	-	-	F14c

[#] Different peak rainfall amount in the same typhoon is because a different 24 h target period was used for verification in a different study.

For comparison, the best TSs for the three typhoons in the Philippines for the entire event using accumulation length of 48 or 72 h are shown in Table 4, where the runs with a t_0 up to 48 h ahead of the starting time of the accumulation period are considered as at short range, such that the first 24 h of the target period is also within 72 h, to make a fair comparison with Table 3. As one can see, when the accumulation period is lengthened and the peak rainfall increased, the TSs also tend to improve and the non-zero values may reach a higher threshold, at 500 mm for Koppu, for example. For entire TC rainfall events, it is encouraging to see that the cloud-resolving models can produce fairly good TS values at high thresholds $\geq$200 mm prior to the event, even at lead times beyond the short range.

Table 4. As in Table 3, but summarizing the best TSs of QPFs over accumulation periods of 48 or 72 h for entire typhoon events in the Philippines. Results are grouped into runs with a t_0 within ($\leq$) 48 h prior to the starting time of accumulation (i.e., first 24 h of target period within 72 h, top) and those beyond (>) 48 h (bottom), respectively.

Range/ Typhoon	Obs. Peak Amount (mm)	Length of Accum.	TS at Threshold Value (mm)					Source
			50	100	200	350	500	
t_0 within 48 h								
Mangkhut (2018)	786	48 h	0.54	0.41	0.46	0.00	0.00	F5c
Koppu (2015)	695	72 h	0.56	0.47	0.33	0.33	1.00	F9d
Melor (2015)	407	72 h	0.64	0.60	0.65	0.33	-	F14d
t_0 beyond 48 h								
Mangkhut (2018)	786	48 h	0.50	0.28	0.29	0.00	0.00	F5c
Koppu (2015)	695	72 h	0.52	0.47	0.27	0.33	0.33	F9d
Melor (2015)	407	72 h	0.55	0.42	0.31	0.00	-	F14d

In Table 5, the SSS values for selected typhoons in Taiwan available in earlier studies [59,60,83] are compared with those for the three TCs in the Philippines as well, since SSS can reflect the overall quality of the QPF quantitatively. For some TCs in Taiwan such as Soulik and Soudelor [59], the SSS can be consistently high (say, $\geq$0.85) inside the short range when the track errors are small enough. Such high SSS values are likely because the rainfall in Taiwan can reach higher amounts and be more concentrated due to its higher topography and smaller size, thus the larger denominator in the second RHS term in Equation (5). However, when track errors are too large in these selected and rainy cases, the SSS values can also be very low and near zero, when the abundant rainfall in reality is almost completely missed in the model (i.e., when predicted TCs move too far from Taiwan). Therefore, although the SSS values for the three TCs in the Philippines might not reach very high values like some of the cases in Taiwan did within the short range, a near complete miss (with a very low SSS) also seems less likely to happen. Beyond the short range, however, there was a fair chance (about 33%) for the QPFs to be of decent quality (with SSS reaching 0.65) for both Koppu and Melor, although the rainfall of Mangkhut was less predictable at such a longer range, obviously linked to the northward track biases. This likelihood of 33% is in fact slightly higher than the probabilities for some of the more rainy TCs in Taiwan, when t_0 is more than 48 h prior to the starting time of the target accumulation period (Table 5), and therefore is certainly quite encouraging.

Table 5. Similar to Table 3, but comparing the SSS values of QPFs over accumulation periods for entire events of selected typhoons in Taiwan and the Philippines, for runs with a t_0 within ($\leq$) or beyond (>) 48 h prior to the starting time of accumulation (as used in Table 4). In each lead-time category, the total number of runs available, the range of SSS (computed over the entire Philippines), and the number of runs with an SSS $\geq$ 0.65 (percentage in parentheses) are listed.

Region/ Typhoon	Obs. Peak Amount (mm)	Length of Accum.	t_0 within 48 h			t_0 Beyond 48 h			Source
			Runs	SSS	$\geq$0.65	Runs	SSS	$\geq$0.65	
Taiwan									
Morakot (2009)	2635	48 h	9	0.33–0.94	5 (56%)	18	0.00–0.23	0 (0%)	[60]
Saola (2012)	1298	48 h	9	0.60–0.93	5 (56%)	16	0.06–0.89	9 (56%)	[59]
Soulik (2013)	1054	24 h	9	0.87–0.94	9 (100%)	18	0.00–0.83	4 (22%)	[59]
Soudelor (2015)	1045	24 h	9	0.86–0.92	9 (100%)	19	0.03–0.87	5 (26%)	[59]
Usagi (2013)	758	36 h	5	0.88–0.91	5 (100%)	17	0.76–0.91	17 (100%)	[83]
Dujuan (2015)	716	32 h	4	0.62–0.71	2 (50%)	9	0.01–0.63	0 (0%)	[83]
Nepartak (2016)	643	56 h	4	0.58–0.69	1 (25%)	15	0.13–0.82	1 (7%)	[83]
Fitow (2013)	268	27 h	4	0.67–0.85	4 (100%)	5	0.70–0.90	5 (100%)	[83]
Chan-Hom (2015)	166	18 h	4	0.50–0.73	2 (50%)	13	0.17–0.76	1 (8%)	[83]
Philippines									
Mangkhut (2018)	786	48 h	9	0.49–0.73	3 (33%)	8	0.34–0.54	0 (0%)	F5d
Koppu (2015)	695	72 h	9	0.60–0.78	7 (78%)	6	0.39–0.69	2 (33%)	F10a
Melor (2015)	407	72 h	9	0.42–0.85	8 (89%)	12	0.33–0.76	4 (33%)	F10b

Overall, Tables 3–5 indicate that the quality of the QPFs by the 2.5 km CReSS using the time-lagged approach for the three typhoons of Mangkhut, Koppu, and Melor are fairly good and comparable to that for TCs in Taiwan. Targeted for the most-rainy 24 h in each event, the highest TS can reach 1.0 at 200 mm for Mangkhut, also 1.0 at 350 mm for Koppu (by many runs, observed peak = 502 mm), and 0.25 at 200 mm for Melor (observed peak = 274 mm). For the entire event over 48 or 72 h, TS = 1 can even be achieved at 500 mm for Koppu (in 33% of all runs, and TS = 0.33–0.5 in another 27% of runs; observed peak = 695 mm), while TS = 0.33 can be reached at 350 mm for Melor (observed peak = 407 mm). Beyond the short range, some earlier predictions with a longer lead time can also produce QPFs of good quality to provide a useful rainfall scenario for early preparation. Among such runs, 33% of them produced SSS $\geq$ 0.65 for both Koppu and Melor, but not for Mangkhut due to larger track biases. In terms of predictability, the rainfall of Koppu is the highest, followed by Melor, while that of Mangkhut appears to be the lowest. For Mangkhut, which was the least predictable, nevertheless, all 11 later runs still produced non-zero TSs at 200 mm for the entire event (48 h), and one run with t_0 at 1200 UTC 11 September did yield a TS of 0.33 at 250 mm (not plotted in Figure 5c), at the range of 60–108 h.

7. Conclusions

In this study, the 2.5 km CReSS is applied to three typhoons in the Philippines: Mangkhut (2018), Koppu (2015), and Melor (2015), using a time-lagged strategy in order to maximize the computational resource to produce rainfall forecasts at high resolution. Following the order, the three typhoons made landfall in northern Luzon, central Luzon, and the middle part of the Philippine Archipelago through southern Luzon and Mindoro, respectively, and the track predictions and QPFs for each are verified using rain-gauge data (56 stations) and categorical statistics at seven thresholds (0.05, 10, 50, 100, 200, 350, and 500 mm) herein. The QPF results are compared with those for selected TCs in Taiwan, since there is a lack of such objective verifications in the Philippines in the literature. The similarity skill score (SSS) that measures the overall similarity between the predicted and observed rainfall patterns is also used. The major findings can be summarized as the following.

Among the three TCs, the rainfall of Koppu (2015) that penetrated central Luzon was the most predictable, as the observed rainband and heavy rainfall near Baguio could be captured by many of the lagged runs in spite of some track errors. For the most-rainy 24 h (1200 UTC 18 to 1200 UTC 19 October, 502.3 mm at Baguio), 8 out of 19 total runs produced a perfect TS of 1.0 at 350 mm, including two (out of five) within the short range (target period inside 72 h from t_0) and six (out of 14) at longer lead times, among which the earliest run was at 0600 UTC 14 October (at t = 102–126 h). For the entire 72 h event (1200 UTC 16 to 1200 UTC 19 October, 695.3 mm at Baguio), eight of the last nine runs since 1200 UTC 14 yielded TS of 0.5 (three runs) or 1.0 (five runs) at 500 mm, and one run at longer lead times had TS = 0.33 at 500 mm. Using SSS $\geq$ 0.65 as an indication of good overall quality of QPFs, then, seven out of nine runs (78%) since 1200 UTC 14 October, and two out of six runs (33%) initialized before that time, reached this criterion. Overall, the ability of the system to capture heavy rainfall at high thresholds for Koppu is rather impressive, although some tendency of over-prediction existed.

The predictability of Melor (2015), which struck the central portion of the Philippine Archipelago, was in the middle of the three TCs. For the most-rainy day of 16 December (peak amount was 273.8 mm at Baler), TS values of 0.2–0.5 at 200 mm were produced in 7 of the 13 runs initialized between 1800 UTC 12 and 1800 UTC 15 December, after the reduction in some northward track biases that led to larger rainfall errors (in location and timing) in some lagged members in the middle. Consequently, for the 72 h total TC rainfall from Melor (14–16 December, peak amount was 407 mm at Baler), three runs executed from 0000–1200 UTC 13 December produced TSs of 0.09–0.33 at 350 mm, and all eight runs from 0600 UTC 12 (at shorter lead times) and the four earliest runs (before 0000 UTC 10 December, at longer lead times) yielded TS of 0.17–0.65 at 200 mm, but only 0.08 at most in between. In terms of SSS, eight out of last nine runs (89%) from 0000 UTC 12 December had an SSS $\geq$ 0.65, while four out of 12 (33%) at longer lead times did the same. The BS values indicated little preference in either over- or under-prediction for Melor.

For Mangkhut (2018) that made landfall in the northern part of Luzon and produced the highest peak rainfall among the three TCs, its rainfall seemingly had the lowest predictability. The earlier runs exhibited northward track bias, which prevented good QPFs beyond the short range. For the most-rainy 24 h (15 September, peak rainfall was 535.6 mm at Baguio), none of these runs had a TS above 0.2 at 100 mm. However, as the track error reduced with time, all seven runs within the short range had TSs of 0.29–0.75 at 100 mm, and two had TS = 0.5 or 1.0 at 200 mm. For the two-day total rainfall from Mangkhut (14–15 September, peak amount = 785.5 mm at Baguio), six out of nine runs from 0000 UTC 12 September produced a TS $\geq$ 0.2 at 200 mm, while three out of eight runs at longer lead times achieved the same. While three short-range runs and two longer-range ones also had TSs of 0.2–0.33 at 250 mm (not plotted), no run was able to produce hits at 350 mm (and thus TS = 0). Thus, the TS values for Mangkhut were slightly lower than those attainable for the other two TCs, although many runs inside the short range were able to produce rainfall amounts near 600 mm near Baguio (but not at the site). For the SSS, only 33% of runs from 0000 UTC 12 September produced an SSS $\geq$ 0.65, while none at the longer lead times could do the same. The overall lower SSS values are likely linked to the smaller area of significant rainfall in this case, especially in earlier runs with a TC too far north.

The comparison of results from categorical measures and the SSS with those for selected TCs in Taiwan, overall, indicates that the quality of the QPFs by the 2.5 km CReSS using the time-lagged approach for the three typhoons in the Philippines studied here was fairly good and comparable, especially for Koppu. Due to the limitation on length, only the tracks and QPFs using rain-gauge observations are verified in this study (Part I). In a follow-up study (Part II), verification of TC intensity, heavy-rainfall probability, and QPFs using recent satellite products will be reported to complement the present work.

Author Contributions: Conceptualization, C.-C.W.; Formal analysis, C.-C.W. and C.-H.T.; Funding acquisition, C.-C.W. and B.J.-D.J.; Investigation, C.-C.W. and C.-H.T.; Data curation, C.-H.T. and S.J.D.; Methodology, C.-C.W. and C.-H.T.; Project administration, C.-C.W. and B.J.-D.J.; Software, C.-H.T.; Supervision, C.-C.W.; Visualization, C.-H.T.; Writing—original draft, C.-C.W.; Writing—review and editing, all authors. All authors have read and agreed to the published version of the manuscript.

Funding: This study was jointly supported by the Ministry of Science and Technology (MOST) of Taiwan, under grants MOST-109-2923-M-002-008, MOST-110-2923-M-002-006, MOST-110-2111-M-003-004, and MOST-110-2625-M-003-001, and by the Department of Science and Technology—Philippine Council for Industry, Energy and Emerging Technology Research and Development (DOST—PCIEERD) of the Philippines, under grant PMIS ID No. 3566.

Institutional Review Board Statement: Not applicable.

Informed Consent Statement: Not applicable.

Data Availability Statement: The CReSS model and its user's guide are open to researchers and available at http://www.rain.hyarc.nagoya-u.ac.jp/~tsuboki/cress_html/index_cress_eng.html (accessed on 1 February 2019). The NCEP GFS analysis/forecast data are available at http://rda.ucar.edu/datasets/ds335.0/#!description (accessed on 13 October 2015).

Acknowledgments: The authors thank the anonymous reviewers for their valuable comments and suggestions to improve the manuscript. The GFS analyses and forecasts used to drive the CReSS hindcasts are produced and made available by the NCEP. The PAGASA, JTWC, and NRL are acknowledged for providing the rain-gauge data, best track data, and plots of TRMM rainrates used in this study.

Conflicts of Interest: The authors declare no conflict of interest.

References

1. UNU-EHS. *World Risk Report 2016*; United Nations University-Institute for Environment and Human Security, Bündnis Entwicklung Hilft: Berlin, Germany; United Nations University-EHS: Bonn, Germany, 2016; 69p, ISBN 978-3-946785-02-6. Available online: https://weltrisikobericht.de/wp-content/uploads/2016/08/WorldRiskReport2016.pdf (accessed on 20 June 2022).
2. Guha-Sapir, D.; Hoyois, P.; Wallemacq, P.; Below, R. *Annual Disaster Statistical Review 2016*; The Numbers and Trends; Centre for Research on the Epidemiology of Disasters (CRED), Institute of Health and Society (IRSS), and Université catholique de Louvain: Brussels, Belgium, 2017; 80p. Available online: https://www.emdat.be/sites/default/files/adsr_2016.pdf (accessed on 20 June 2022).
3. Jaganmohan, M. *Countries with the Highest Disaster Risk Worldwide in 2021*; Statista: Hamburg, Germany, 2022. Available online: https://www.statista.com/statistics/1270469/disaster-risk-index-most-affected-countries/ (accessed on 20 June 2022).
4. Wannewitz, S.; Hagenlocher, M.; Garschagen, M. Development and validation of a sub-national multi-hazard risk index for the Philippines. *GI_Forum* **2016**, *1*, 133–140. [CrossRef]
5. Brown, S. The Philippines is the most storm-exposed country on Earth. *Time*, 11 November 2013.
6. Czajkowski, J.; Villarini, G.; Michel-Kerjan, E.; Smith, J.A. Determining tropical cyclone inland flooding loss on a large scale through a new flood peak ratio-based methodology. *Environ. Res. Lett.* **2013**, *8*, 044056. [CrossRef]
7. Opiso, E.M.; Puno, G.R.; Alburo, J.L.P.; Detalla, A.L. Landslide susceptibility mapping using GIS and FR method along the Cagayan de Oro-Bukidnon-Davao City route corridor, Philippines. *KSCE J. Civ. Eng.* **2016**, *20*, 2506–2512. [CrossRef]
8. Cabrera, J.S.; Lee, H.S. Flood risk assessment for Davao Oriental in the Philippines using geographic information system-based multi-criteria analysis and the maximum entropy model. *J. Flood Risk Manag.* **2020**, *13*, e12607. [CrossRef]
9. NDRRMC (National Disaster Risk Reduction and Management Council). Effects of Typhoon "Yolanda" (Haiyan). SitRep. 108. 2014 NDRRMC, Quezon City, Philippines; 67p. Available online: https://www.ndrrmc.gov.ph/attachments/article/1329/Effects_of_Typhoon_YOLANDA_(HAIYAN)_SitRep_No_108_03APR2014.pdf (accessed on 7 April 2020).
10. Soria, J.L.A.; Switzer, A.D.; Villanoy, C.L.; Fritz, H.M.; Bilgera, P.H.T.; Cabrera, O.C.; Siringan, F.P.; Maria, Y.Y.S.; Ramos, R.D.; Fernandez, I.Q. Repeat storm surge disasters of Typhoon Haiyan and its 1897 predecessor in the Philippines. *Bull. Am. Meteorol. Soc.* **2016**, *97*, 31–48. [CrossRef]
11. Mori, N.; Kato, M.; Kim, S.; Mase, H.; Shibutani, Y.; Takemi, T.; Tsuboki, K.; Yasuda, T. Local amplification of storm surge by Super Typhoon Haiyan in Leyte Gulf. *Geophys. Res. Lett.* **2014**, *41*, 5106–5113. [CrossRef]
12. Ren, D. *Storm-Trigger Landslides in Warmer Climates*; Springer: Cham, Switzerland, 2015; 365p, ISBN 978-3-319-08517-3. [CrossRef]
13. Yonson, R.; Noy, I.; Gaillard, J.C. The measurement of disaster risk: An example from tropical cyclones in the Philippines. *Rev. Dev. Econ.* **2018**, *22*, 736–765. [CrossRef]
14. Desquitado, A.M.S.; Perez, M.R.R.; Puchero, R.S.R.; Macalalad, E.P. A climatological study of typhoons over the Philippine Area of Responsibility from 1989–2018. *E3S Web Conf.* **2020**, *200*, 02001. [CrossRef]

15. Wang, C.-C.; Lee, C.-Y.; Jou, B.J.-D.; Celebre, C.P.; David, S.; Tsuboki, K. High-resolution time-lagged ensemble prediction for landfall intensity of Super Typhoon Haiyan (2013) using a cloud-resolving model. *Weather Clim. Extrem.* **2022**, *37*, 100473. [CrossRef]

16. Emanuel, K. Increasing destructiveness of tropical cyclones over the past 30 years. *Nature* **2005**, *436*, 686–688. [CrossRef]

17. Kubota, H.; Chan, J.C.L. Interdecadal variability of tropical cyclone landfall in the Philippines. *Geophys. Res. Lett.* **2009**, *36*, L12802. [CrossRef]

18. Takagi, H.; Esteban, M. Statistics of tropical cyclone landfalls in the Philippines: Unusual characteristics of 2013 Typhoon Haiyan. *Nat. Hazards* **2016**, *80*, 211–222. [CrossRef]

19. Cinco, T.A.; de Guzman, R.G.; Ortiz, A.M.D.; Delfino, R.J.P.; Lasco, R.D.; Hilario, F.D.; Juanillo, E.L.; Barba, R.; Ares, E.D. Observed trends and impacts of tropical cyclones in the Philippines. *Int. J. Climatol.* **2016**, *36*, 4638–4650. [CrossRef]

20. Hsu, P.-C.; Chen, K.-C.; Tsou, C.-H.; Hsu, H.-H.; Hong, C.-C.; Liang, H.-C.; Tu, C.-Y.; Kitoh, A. Future changes in the frequency and destructiveness of landfalling tropical cyclones over East Asia projected by high-resolution AGCMs. *Earth's Future* **2021**, *9*, e2020EF001888. [CrossRef]

21. Golding, B.W. Quantitative precipitation forecasting in the UK. *J. Hydrol.* **2000**, *239*, 286–305. [CrossRef]

22. Fritsch, J.M.; Carbone, R.E. Improving quantitative precipitation forecasts in the warm season. A USWRP research and development strategy. *Bull. Am. Meteorol. Soc.* **2004**, *85*, 955–965. [CrossRef]

23. Cuo, L.; Pagano, T.C.; Wang, Q.J. A review of quantitative precipitation forecasts and their use in short- to medium-range streamflow forecasting. *J. Hydormeteorol.* **2011**, *12*, 713–728. [CrossRef]

24. Kneis, D.; Abona, C.; Bronsterta, A. Heistermanna, Verification of short-term runoff forecasts for a small Philippine basin (Marikina). *Hydrol. Sci. J.* **2017**, *62*, 205–216. [CrossRef]

25. MacLeod, D.; Easton-Calabria, E.; de Perez, E.C.; Jaime, C. Verification of forecasts for extreme rainfall, tropical cyclones, flood and storm surge over Myanmar and the Philippines. *Weather Clim. Extrem.* **2021**, *33*, 100325. [CrossRef]

26. Wang, S.-T. *Track, Intensity, Structure, Wind and Precipitation Characteristics of Typhoons Affecting Taiwan (in Chinese)*; Research Rep. 80-73, NSC 80-04140-P052-02B; National Science Council (NSC): Taipei, Taiwan, 1989; 285p.

27. Chang, C.-P.; Yeh, T.-C.; Chen, J.-M. Effects of terrain on the surface structure of typhoons over Taiwan. *Mon. Weather Rev.* **1993**, *121*, 734–752. [CrossRef]

28. Cheung, K.K.W.; Huang, L.-R.; Lee, C.-S. Characteristics of rainfall during tropical cyclone periods in Taiwan. *Nat. Hazards Earth Syst. Sci.* **2008**, *8*, 1463–1474. [CrossRef]

29. Su, S.-H.; Kuo, H.-C.; Hsu, L.-H.; Yang, Y.-T. Temporal and spatial characteristics of typhoon extreme rainfall in Taiwan. *J. Meteorol. Soc. Jpn.* **2012**, *90*, 721–736. [CrossRef]

30. Lee, C.-S.; Huang, L.-R.; Shen, H.-S.; Wang, S.-T. A climatology model for forecasting typhoon rainfall in Taiwan. *Nat. Hazards* **2006**, *37*, 87–105. [CrossRef]

31. Lee, C.-S.; Huang, L.-R.; Chen, D.Y.-C. The modification of the typhoon rainfall climatology model in Taiwan. *Nat. Hazards Earth Syst. Sci.* **2013**, *13*, 65–74. [CrossRef]

32. Hong, J.-S.; Fong, C.-T.; Hsiao, L.-F.; Yu, Y.-C.; Tzeng, C.-Y. Ensemble typhoon quantitative precipitation forecasts model in Taiwan. *Weather Forecast.* **2015**, *30*, 217–237. [CrossRef]

33. Lorenz, E.N. Deterministic nonperiodic flow. *J. Atmos. Sci.* **1963**, *20*, 130–141. [CrossRef]

34. Epstein, E.S. Stochastic dynamic prediction. *Tellus* **1969**, *21*, 739–759. [CrossRef]

35. Hsiao, L.-F.; Yang, M.-J.; Lee, C.-S.; Kuo, H.-C.; Shih, D.-S.; Tsai, C.-C.; Wang, C.-J.; Chang, L.-Y.; Chen, D.Y.-C.; Feng, L.; et al. Ensemble forecasting of typhoon rainfall and floods over a mountainous watershed in Taiwan. *J. Hydrol.* **2013**, *506*, 55–68. [CrossRef]

36. Yang, T.-H.; Yang, S.-C.; Ho, J.-Y.; Lin, G.-F.; Hwang, G.-D.; Lee, C.-S. Flash flood warnings using the ensemble precipitation forecasting technique: A case study on forecasting floods in Taiwan caused by typhoons. *J. Hydrol.* **2015**, *520*, 367–378. [CrossRef]

37. Teng, H.-F.; Done, J.M.; Lee, C.-S.; Kuo, Y.-H. Dependence of probabilistic quantitative precipitation forecast performance on typhoon characteristics and forecast track error in Taiwan. *Weather Forecast.* **2020**, *35*, 585–607. [CrossRef]

38. Gentry, M.S.; Lackmann, G.M. Sensitivity of simulated tropical cyclone structure and intensity to horizontal resolution. *Mon. Weather Rev.* **2010**, *138*, 688–704. [CrossRef]

39. Tsuboki, K.; Sakakibara, A. Large-scale parallel computing of cloud resolving storm simulator. In *High Performance Computing*; Zima, H.P., Joe, K., Sato, M., Seo, Y., Shimasaki, M., Eds.; Springer: Berlin/Heidelberg, Germany, 2002; pp. 243–259.

40. Tsuboki, K.; Sakakibara, A. Numerical Prediction of High-Impact Weather Systems. In *The Textbook for the Seventeenth IHP Training Course in 2007*; Hydrospheric Atmospheric Research Center, Nagoya University, and UNESCO: Nagoya, Japan, 2007; p. 273.

41. Wang, C.-C. The more rain, the better the model performs—The dependency of quantitative precipitation forecast skill on rainfall amount for typhoons in Taiwan. *Mon. Weather Rev.* **2015**, *143*, 1723–1748. [CrossRef]

42. Wang, C.-C. Corrigendum. *Mon. Weather Rev.* **2016**, *144*, 3031–3033. [CrossRef]

43. Wang, C.-C. Paper of notes: The more rain from typhoons, the better the models perform. *Bull. Am. Meteorol. Soc.* **2016**, *97*, 16–17.

44. Wang, C.-C.; Chang, C.-S.; Wang, Y.-W.; Huang, C.-C.; Wang, S.-C.; Chen, Y.-S.; Tsuboki, K.; Huang, S.-Y.; Chen, S.-H.; Chuang, P.-Y.; et al. Evaluating quantitative precipitation forecasts using the 2.5 km CReSS model for typhoons in Taiwan: An update through the 2015 season. *Atmosphere* **2021**, *12*, 1501. [CrossRef]

45. Wang, C.-C.; Paul, S.; Huang, S.-Y.; Wang, Y.-W.; Tsuboki, K.; Lee, D.-I.; Lee, J.-S. Typhoon quantitative precipitation forecasts by the 2.5 km CReSS model in Taiwan: Examples and role of topography. *Atmosphere* **2022**, *13*, 623. [CrossRef]
46. Fang, X.; Kuo, Y.-H. Improving ensemble-based quantitative precipitation forecasts for topography-enhanced typhoon heavy rainfall over Taiwan with a modified probability-matching technique. *Mon. Weather Rev.* **2013**, *141*, 3908–3932. [CrossRef]
47. Toth, Z.; Kalnay, E. Ensemble forecasting at NMC: The generation of perturbations. *Bull. Am. Meteorol. Soc.* **1993**, *74*, 2317–2330. [CrossRef]
48. Molteni, F.; Buizza, R.; Palmer, T.N.; Petroliagis, T. The ECMWF Ensemble Prediction System: Methodology and validation. *Q. J. R. Meteorol. Soc.* **1996**, *122*, 73–119. [CrossRef]
49. Roebber, P.J.; Schultz, D.M.; Colle, B.A.; Stensrud, D.J. Toward improved prediction: High-resolution and ensemble modeling systems in operations. *Weather Forecast.* **2004**, *19*, 936–949. [CrossRef]
50. Hoffman, R.N.; Kalnay, E. Lagged average forecasting, an alternative to Monte Carlo forecasting. *Tellus A* **1983**, *35*, 100–118. [CrossRef]
51. Mittermaier, M.P. Improving short-range high-resolution model precipitation forecast skill using time-lagged ensembles. *Q. J. R. Meteorol. Soc.* **2007**, *133*, 1487–1500. [CrossRef]
52. Lu, C.; Yuan, H.; Schwartz, B.E.; Benjamin, S.G. Short-range numerical weather prediction using time-lagged ensembles. *Weather Forecast.* **2007**, *22*, 580–595. [CrossRef]
53. Yuan, H.; McGinley, J.A.; Schultz, P.J.; Anderson, C.J.; Lu, C. Short-range precipitation forecasts from time-lagged multimodel ensembles during the HMT-West-2006 campaign. *J. Hydrometeorol.* **2008**, *9*, 477–491. [CrossRef]
54. Trilaksono, N.J.; Otsuka, S.; Yoden, S. A time-lagged ensemble simulation on the modulation of precipitation over West Java in January–February 2007. *Mon. Weather Rev.* **2012**, *140*, 601–616. [CrossRef]
55. Wang, C.-C.; Huang, S.-Y.; Chen, S.-H.; Chang, C.-S.; Tsuboki, K. Cloud-resolving typhoon rainfall ensemble forecasts for Taiwan with large domain and extended range through time-lagged approach. *Weather Forecast.* **2016**, *31*, 151–172. [CrossRef]
56. Wang, C.-C.; Huang, S.-Y.; Chen, S.-H.; Chang, C.-S.; Tsuboki, K. Paper of notes: Cloud-resolving, time-lagged typhoon rainfall ensemble forecasts. *Bull. Am. Meteorol. Soc.* **2016**, *97*, 1128–1129.
57. Elsberry, R.L.; Tsai, H.-C.; Chin, W.-C.; Marchok, T.P. Predicting rapid intensification events following tropical cyclone formation in the western North Pacific based on ECMWF ensemble warm core evolutions. *Atmosphere* **2021**, *12*, 847. [CrossRef]
58. Wang, C.-C.; Chen, S.-H. High-resolution time-lagged ensemble quantitative precipitation forecasts (QPFs) for typhoons in Taiwan using the Cloud-Resolving Storm Simulator (CReSS). In Proceedings of the 4th WMO Workshop on Monsoon Heavy Rainfall (MHR-4): Science and Prediction of Monsoon Heavy Rainfall, Shenzhen, China, 16–18 April 2019.
59. Wang, C.-C.; Chen, S.-H.; Chen, Y.-H.; Tsuboki, K. Cloud-resolving time-lagged rainfall ensemble forecasts for typhoons in Taiwan: Examples of Saola (2012), Soulik (2013), and Soudelor (2015). In Proceedings of the International Conference on Heavy Rainfall and Tropical Cyclone in East Asia (Virtual Meeting), Online, 1–3 March 2022.
60. Hsu, H.-H.; Kuo, H.-C.; Jou, J.-D.; Chen, T.-C.; Lin, P.-H.; Yeh, T.-C.; Wu, C.-C. *Scientific Report on Typhoon Morakot (2009)*; National Science Council: Taipei, Taiwan, 2010; 192p. (In Chinese)
61. Wang, C.-C.; Kuo, H.-C.; Yeh, T.-C.; Chung, C.-H.; Chen, Y.-H.; Huang, S.-Y.; Wang, Y.-W.; Liu, C.-H. High-resolution quantitative precipitation forecasts and simulations by the Cloud-Resolving Storm Simulator (CReSS) for Typhoon Morakot (2009). *J. Hydrol.* **2013**, *506*, 26–41. [CrossRef]
62. Wang, C.-C.; Chen, S.-H.; Tsuboki, K.; Huang, S.-Y.; Chang, C.-S. Application of time-lagged ensemble quantitative precipitation forecasts for Typhoon Morakot (2009) in Taiwan by a cloud-resolving model. *Atmosphere* **2022**, *13*, 585. [CrossRef]
63. Zhang, F.; Weng, Y.; Kuo, Y.-H.; Whitaker, J.S.; Xie, B. Predicting Typhoon Morakot's catastrophic rainfall with a convection-permitting mesoscale ensemble system. *Weather Forecast.* **2010**, *25*, 1816–1825. [CrossRef]
64. Fang, X.; Kuo, Y.-H.; Wang, A. The impact of Taiwan topography on the predictability of Typhoon Morakot's record-breaking rainfall: A high-resolution ensemble simulation. *Weather Forecast.* **2011**, *26*, 613–633. [CrossRef]
65. Hendricks, E.A.; Jin, Y.; Moskaitis, J.R.; Doyle, J.D.; Peng, M.S.; Wu, C.-C.; Kuo, H.-C. Numerical simulations of Typhoon Morakot (2009) using a multiply nested tropical cyclone prediction model. *Weather Forecast.* **2016**, *31*, 627–645. [CrossRef]
66. Racoma, B.A.B.; David, C.P.C.; Crisologo, I.A.; Bagtasa, G. The change in rainfall from tropical cyclones due to orographic effect of the Sierra Madre Mountain Range in Luzon, Philippines. *Philipp. J. Sci.* **2016**, *145*, 313–326.
67. Lin, Y.-L.; Farley, R.D.; Orville, H.D. Bulk parameterization of the snow field in a cloud model. *J. Clim. Appl. Meteorol. Climatol.* **1983**, *22*, 1065–1092. [CrossRef]
68. Cotton, W.R.; Tripoli, G.J.; Rauber, R.M.; Mulvihill, E.A. Numerical simulation of the effects of varying ice crystal nucleation rates and aggregation processes on orographic snowfall. *J. Clim. Appl. Meteorol. Climatol.* **1986**, *25*, 1658–1680. [CrossRef]
69. Murakami, M. Numerical modeling of dynamical and microphysical evolution of an isolated convective cloud—The 19 July 1981 CCOPE cloud. *J. Meteorol. Soc. Jpn.* **1990**, *68*, 107–128. [CrossRef]
70. Ikawa, M.; Saito, K. Description of a nonhydrostatic model developed at the Forecast Research Department of the MRI. *MRI Technol. Rep.* **1991**, *28*, 238.
71. Murakami, M.; Clark, T.L.; Hall, W.D. Numerical simulations of convective snow clouds over the Sea of Japan: Two-dimensional simulation of mixed layer development and convective snow cloud formation. *J. Meteorol. Soc. Jpn.* **1994**, *72*, 43–62. [CrossRef]
72. Deardorff, J.W. Stratocumulus-capped mixed layers derived from a three-dimensional model. *Bound.-Layer Meteorol.* **1980**, *18*, 495–527. [CrossRef]

73. Louis, J.F.; Tiedtke, M.; Geleyn, J.F. A short history of the operational PBL parameterization at ECMWF. In *Proceedings of the Workshop on Planetary Boundary Layer Parameterization, 25–27 November 1981*; ECMWF: Reading, UK, 1982; pp. 59–79.

74. Kondo, J. Heat balance of the China Sea during the air mass transformation experiment. *J. Meteorol. Soc. Jpn.* **1976**, *54*, 382–398. [CrossRef]

75. Segami, A.; Kurihara, K.; Nakamura, H.; Ueno, M.; Takano, I.; Tatsumi, Y. 1989: Operational mesoscale weather prediction with Japan Spectral Model. *J. Meteorol. Soc. Jpn.* **1989**, *67*, 907–924. [CrossRef]

76. Kanamitsu, M. Description of the NMC global data assimilation and forecast system. *Weather Forecast.* **1989**, *4*, 335–342. [CrossRef]

77. Kleist, D.T.; Parrish, D.F.; Derber, J.C.; Treadon, R.; Wu, W.S.; Lord, S. 2009: Introduction of the GSI into the NCEP global data assimilation system. *Weather Forecast.* **2009**, *24*, 1691–1705. [CrossRef]

78. Huffman, G.J.; Bolvin, D.T.; Nelkin, R.J.; Wolff, D.B.; Adler, R.F.; Gu, G.; Hong, Y.; Bowman, K.P.; Stocker, E.F. The TRMM Multi-satellite Precipitation Analysis: Quasi-global, multiyear, combined-sensor precipitation estimates at fine scale. *J. Hydrometeorol.* **2007**, *8*, 38–55. [CrossRef]

79. Schaefer, J.T. The critical success index as an indicator of warning skill. *Weather Forecast.* **1990**, *5*, 570–575. [CrossRef]

80. Wilks, D.S. *Statistical Methods in the Atmospheric Sciences*; Academic Press: Cambridge, MA, USA, 1995; p. 467.

81. Wang, C.-C. On the calculation and correction of equitable threat score for model quantitative precipitation forecasts for small verification areas: The example of Taiwan. *Weather Forecast.* **2014**, *29*, 788–798. [CrossRef]

82. Roebber, P.J. Visualizing multiple measures of forecast quality. *Weather Forecast.* **2009**, *24*, 601–608. [CrossRef]

83. Chen, S.-H.; Wang, C.-C. Developing objective guidance for the quality of quantitative precipitation forecasts of westward-moving typhoons affecting Taiwan through machine learning. *Atmos. Sci.* **2022**. *accepted* (In Chinese with English abstract).

84. Roberts, N.M.; Lean, H.W. Scale-selective verification of rainfall accumulations from high-resolution forecasts of convective events. *Mon. Weather Rev.* **2008**, *136*, 78–97. [CrossRef]

85. Yeh, T.-C.; Cayanan, E.O.; Hsiao, L.-F.; Chen, D.-S.; Tsai, Y.-T. A study of the rainfall and structural changes of Typhoon Koppu (2015) over northern Philippines. *Terr. Atmos. Ocean. Sci.* **2021**, *32*, 619–632. [CrossRef]

Article

An ISOMAP Analysis of Sea Surface Temperature for the Classification and Detection of El Niño & La Niña Events

John Chien-Han Tseng

Central Weather Bureau, Taipei 100, Taiwan; jchtsenghome@gmail.com

Abstract: Isometric feature mapping (ISOMAP) is a nonlinear dimensionality reduction method used for extracting features from spatiotemporal data. The traditional principal component analysis (PCA), a linear dimensionality reduction method, measures the distance between two data points based on the Euclidean distance (line segment), which cannot reflect the actual distance between the data points in a nonlinear space. By contrast, the ISOMAP measures the distance between two data points based on the geodesic distance, which more closely reflects the actual distance by the view of tracing along the local linearity in the original nonlinear structure. Thus, ISOMAP-reconstructed data points can reflect the features of real structures and can be classified more accurately than traditional PCA-reconstructed data points. Moreover, these ISOMAP-reconstructed data points can be used for cluster analysis by emphasizing the differences among the points more than those by the traditional PCA. In this study, sea surface temperature (SST) data points reconstructed using the traditional PCA and ISOMAP were compared. The classification based on these reconstructed SST points was tested using the Niño 3.4 index, which labels El Niño, La Niña, or normal events. The mean differences from the ISOMAP data points were larger than those from the traditional PCA data points. The ISOMAP not only helped differentiate the points in two different events but also provided better difference measurement of the points belonging to the same class (e.g., 82/83 and 97/98 El Niño events). On examining the evolution of the leading three temporal eigen components of the SST PCA, or especially the SST ISOMAP, we found that the trajectories were similar to the Lorenz 63 model on a phase space figure. This implies that NWP perturbations can be traced using the ISOMAP to measure growing unstable behaviors. Spatial eigenmodes (empirical orthogonal function) between the traditional PCA and ISOMAP were also determined and compared herein.

Keywords: El Niño; EOF; ISOMAP; La Niña; Niño 3.4; PCA; SST

Citation: Tseng, J.C.-H. An ISOMAP Analysis of Sea Surface Temperature for the Classification and Detection of El Niño & La Niña Events. *Atmosphere* **2022**, *13*, 919. https://doi.org/10.3390/atmos13060919

Academic Editors: Bo-Wen Shen, Roger A. Pielke Sr. and Xubin Zeng

Received: 27 April 2022
Accepted: 31 May 2022
Published: 6 June 2022

1. Introduction

Classification or clustering is performed to understand differences among events (e.g., numerical data, colors, and object shapes or figures). The success of the classification method depends on the effective clarification for describing, measuring, and recognizing these differences. For example, to analyze the high dimensional data takes time and computing cost to extract information. Traditionally, the analyzed high dimensional data are dimensional reduction to low dimension 2D or 3D points and the difference can be measured simply by the distance in the low-dimensional space. Certainly, a meaningful low-dimensional space that can appropriately extract features from data must be identified.

Principal component analysis (PCA) is a traditional linear dimensionality reduction method. Leading PCA components extract the main variances of the original data, which are called explained variances. Fewer the leading PCA components and more the explained variances, the better the PCA results. For example, if the original data dimensions are 10,000, then three leading PCA components can be used to explain 80% original variances and the PCA results are very good. By contrast, the PCA results are worse when an excess of 100 leading PCA components are used to explain only 50% original variances. Therefore, many PCA modifications and other alternative techniques [1–3] attempted to address this imperfect interpretation in real data analysis. For example, the modifications including: rotation PCA [4], probabilistic PCA [5–7], Bayesian PCA [5,6], and kernel PCA [8], are available. Other alternative techniques are the independent component analysis (ICA) and independent subspace analysis (ISA), which found the hidden possible factors behind of the physical phenomena based on source signals rather than prominent variances as PCA. The framework of ICA or ISA was built on non-Gaussian distributions and the assumption of composed linearly of source signals [9–11]. The ICA or ISA looked for components or subspaces which were the most statistically independent as possible under the view of non-Gaussian probability distributions. Better classification results can be obtained after retrieving low-dimensional PCA or ICA components. Low-dimensional data points from the PCA or ICA can be classified more easily than the original high-dimensional data points.

Tenebaum et al. [12] proposed iso metric feature mapping (ISOMAP) to solve the classification problem and obtain well-distributed low-dimensional data points. They pointed out that the traditional PCA considers the data linearly; for example, time evolution is resolved by the linear evolution of the original data arrangement. The geopotential height can be imagined to evolve in a month by daily data. The 30-times data being considered is constrained by the linear time variation. The Geopotential height does not evolve linearly in 1 month. However, despite this, the covariance matrix of the Geopotential height is counted linearly in the PCA. When these linear considerations are used to determine nonlinear variation, the data points cannot be discriminated and are sticky together by the view of low dimensional principal axes. Classifying concentrated or sticky together data points is difficult, and it leads to classification failure. The linear PCA separated points cannot represent the actual distances between the data points, leading to imperfect and false classification. In the ISOMAP, the original nonlinear relations in the data are built by establishing the nearest neighbors. The ISOMAP maintains the linearity of small domains but reflects nonlinear variations in the larger domain. In other words, this is called manifold consideration or manifold learning. That means the small local domain of the manifold is the homeomorphism as the Euclidean space. Dimensionality reduction through the ISOMAP reflects the real and nonlinear variations between the data points; in brief, data points can be pushed away more than that with the traditional linear PCA. Thus, classification can be effectively performed after ISOMAP dimensionality reduction.

In this study, we used sea surface temperature (SST) data to perform traditional El Niño classification. Classification results obtained through the traditional PCA and ISOMAP were compared. We presented the leading eigen components constructed using the PCA/ISOMAP as the space coordinate values over time, which depicted trajectories. On examining the data point trajectories after the ISOMAP, we found that the evolution exhibited the Lorenz 63 model on a phase space figure. The SST data points switched to different El Niño, normal, and La Niña events and spiral trajectories were never repeated. The ISOMAP SST point trajectories indicated that El Niño cycles like the chaotic behaviors as the phase space plot showing. Based on this indication, we believe that ISOMAP analysis can be a diagnostic tool for tracing the circulation evolution and evaluating ensemble perturbations, ensemble forecast spread, or even the numerical weather prediction (NWP) score between forecasts and analysis.

Some studies have reported that there are several types of El Niño/La Niña events (e.g., warm pool, dateline, Central Pacific, Eastern Pacific, and El Niño Modoki [pseudo El Niño] types) [13–15] and have differentiated these events by the geographical positions in which the SST anomaly centers/patterns are located. They combined PCA, correlation, regression-PCA methods, and different indices (Niño 1 + 2, Niño 3, Niño 4, and Niño 3.4) for an improved recognition of these different types of El Niño/La Niña events. These different events have varied connections with different weather/climate patterns. The improved recognition allows a more accurate forecast and greater resilience of their corresponding extreme weather events. However, the aforementioned methods are not straightforward and their results do not allow simple data visualization. We found the SST points constructed using the ISOMAP to reflect the differences in the types of El Niño/La Niña events in a simpler manner because these differences could be measured directly on the basis of the distance between the constructed SST points.

The remainder of this paper is organized as follows. In Section 2, we explain the definition of Niño 3.4 index for different SST events, the SST data, and the concept of ISOMAP. In Section 3, we show PCA/ISOMAP-reconstructed points and classification results. In the final section, we summarise major findings and the future applications regarding ISOMAP.

2. Data and Methods

The SST data used herein were obtained from version 5 of the NOAA NCDC ERSST (Extended Reconstructed global Sea Surface Temperature) data set, based on COADS data, collected from January 1980 to December 2021. To distinguish the El Niño, normal, or La Niña events are based on the Niño 3.4 (170° W–120° W, 5° S–5° N) index from NOAA's Climate Prediction Center. El Niño events were defined in Niño 3.4 region when the moving 3-month average SST anomaly exceeded 0.5 °C for at least 5 months. By contrast, La Niña events/anti-El Niño events were defined when the average SST anomaly was lower than 0.5 °C for 5 months. Determining the relationship between this index and the Pacific Ocean domain (120° E–60° W, 30° S–30° N) SST pattern would be interesting. Most studies on El Niño have focused on the large-scale circulation patterns and not on the Niño 3.4 index area.

The concept of ISOMAP is shown in Figure 1. The real data points are located in the warp surface, as shown in the arc curve in Figure 1. During PCA calculation, we assume that the variation (temporal or spatial) is linear and the relation between data points is like a short straight line. The PCA relation can be considered a type of Euclidean distance. However, the real distance between point c and point d is greater than the Euclidean distance. Thus, PCA always fails to present the real situation. To use a linear tool such as linear algebra eigen solutions, we must rearrange the relation, the distance, according to the longer straight line in Figure 1. The distance between point c and point d on the geodesic line truly reflects the distance in the real warp surface, the curve, in Figure 1. In brief, we consider the neighbor's situation and we do not take the 'shortcut' between the given data points.

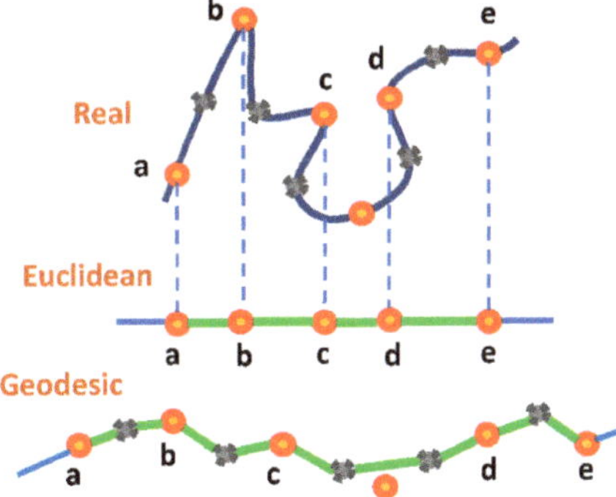

Figure 1. The concept of ISOMAP. The given data points are marked by a–e. Their neighbors are gray star points. The green segments are the measured distances of these data points.

Tenebaum et al. [12] proposed to plot the nearest neighbor graph that is used to reflect the distance on the warp surface. This means the distance between point c and point d is not calculated using the Euclidean distance, the shortcut straight distance method, but by including the other three points. Under geodesic framework, after considering the neighbors, the shortest distance is decided again and one red circle point is bypass to the distance counting. The geodesic distance is calculated on the basis of this neighbor graph, and then, this distance relation is used to form the covariance matrix that can be solved by the traditional PCA or multi-dimensional scaling (MDS) method [16]. The ISOMAP (Algorithm 1) is shown below [17–19].

Algorithm 1: ISOMAP

Step 1. construct the matrix of squared pairwise similarities $\mathbf{D}$, $\mathbf{D}_{ij} = \left\| y_i - y_j \right\|^2$, the distance matrix, measured on temporal dimension. The i, j are temporal indexes.
Step 2. build the weighted graph based on the $\mathbf{D}$ according to how many neighbors of each point
Step 3. estimate the geodesic distances $\mathbf{D}_G$ by finding the shortest paths on the weighted graph (Dijkstra's algorithm [20])
Step 4. define $\mathbf{B} = -1/2\,\mathbf{J}\,\mathbf{D}_G\,\mathbf{J}^\mathbf{T}$, where $\mathbf{J} = \mathbf{I} - 1/N$, $\mathbf{I}$ is the identity matrix, and N is the number of data points
Step 5. solve eigen problem $\mathbf{BP} = \mathbf{P}\Sigma$
Step 6. computing the leading principal vectors by $\mathbf{X} = \mathbf{P}\Sigma^{1/2}$

To measure the ISOMAP low dimensionality sufficient criterium is the residual variance [12], which is defined as follows:

$$1 - R^2(\mathbf{D}_M\,,\,\mathbf{D}_G).$$

$\mathbf{D}_G$ is the geodesic distance matrix used in steps 3 and 4 in Algorithm 1 for solving the eigen problem, $\mathbf{D}_M$ is the geodesic distance matrix reconstructed from the ISOMAP, the low-dimensional eigen (principal) components (leading modes of step 6 in Algorithm 1), and R is the coefficient of correlation. The larger the correlation between $\mathbf{D}_M$ and $\mathbf{D}_G$ is, the lower the residual variance is, and low-dimensional components can be used to approximate the original high-dimensional structure [12,18]. In this paper, we constructed the distance matrix based on the temporal dimension. We assumed that the time variation of the SST is not linear. The detailed arrangements of the matrices used are given in the Appendix A.

The classifier used in this study is the smooth support vector machine (SSVM) that replaces the plus function in the non-smooth SVM by a smooth function [21]. All test results presented in the next paragraphs were obtained from 20-times 5-fold cross-validation. This means that 80% data were randomly selected as the training set and the remaining 20% data formed the testing set in each validation. Then, the average training errors/testing errors of this 20-times validation were calculated.

3. Classification

Three leading temporal eigenvectors from the PCA and ISOMAP were used for presenting points reconstructed after dimensionality reduction (step 6 in the Algorithm 1). The leading 20 eigenvectors were used for testing the classification results. Figure 2a presents the 3D structure of the PCA-reconstructed points, and Figure 2b shows the 2D structure of the PCA-reconstructed points. The El Niño events were labeled red, normal events were marked yellow, and La Niña events were labeled blue. In Figure 2, most El Niño, normal, and La Niña events were already well separated. This means that meteorologists are correctly using the Niño 3.4 index to define the El Niño event. Inevitably, some points were stuck together, which probably led to the classifier failure. Under this situation, if the sufficient distance between the PCA data points can be obtained, the better is for the classification. As mentioned in the previous section, the solution can be the ISOMAP that efficiently separates the points by measuring the geodesic distance, thus reflecting the actual space distance variation between the data points.

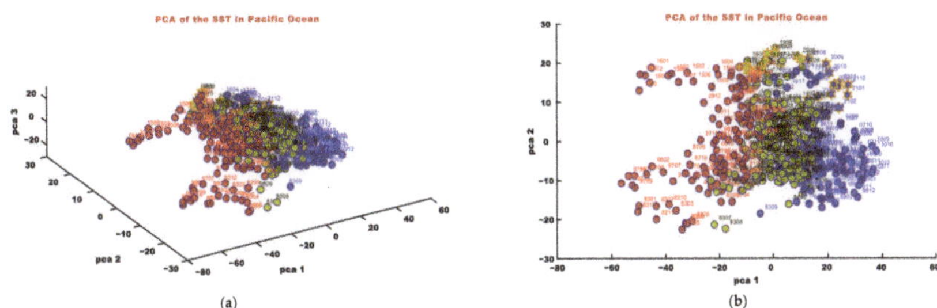

Figure 2. (**a**) 3D structure composed of the three leading PCA components. (**b**) 2D structure composed of the two leading PCA components.

During the application of the ISOMAP, the number of neighbors should be determined first, which was done by evaluating the residual variance. In this study, the number of neighbors for 41 years of SST data was 44. Similar to the PCA calculation, the leading three eigen components of the ISOMAP are presented in Figure 3a. The ISOMAP points were indeed more separated than the PCA points. Some events were significantly different from others even if they belonged to the same class (El Niño: 82/83, 97/98, 15/16; La Niña: 84/85, 88/89, 98/99). The 3D structure of the ISOMAP-reconstructed data points exhibited more space variations, with the data points being well separated, and allowed grouping of the data points into different clusters. Figure 3b is the 2D structure of Figure 3a and shows well-separated data points compared with those in Figure 2b (PCA calculation).

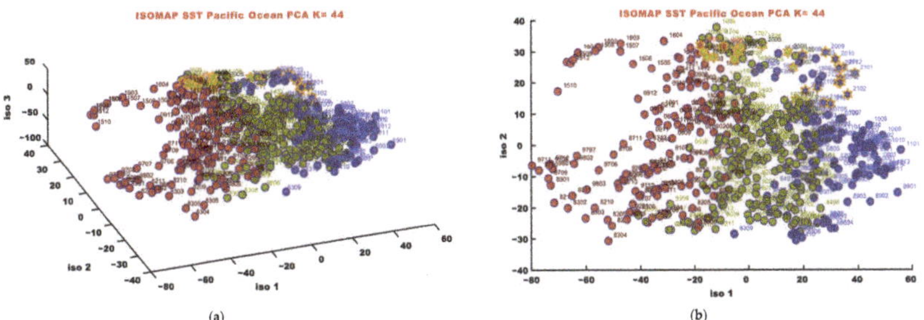

Figure 3. (**a**) 3D structure composed of the three leading ISOMAP components. (**b**) 2D structure composed of the two leading ISOMAP components.

The data points for the recent 31 months were marked by a star sign outside the circle of points. If the center of Figures 2 and 3 represented the climatology mean, the recent 31-month trajectory was far from the center. This indicates that the recent SST data evolved differently from the previous SST data. Using the past data to describe the recent 31-month variation would be difficult. To highlight this situation, we faded the data points before the recent 31-month period but made those for the recent 31 months clear (Figure 4). We found the trajectory to be like number 8, the circular shape. On using animation to demonstrate these 41-year SST data points, we could find the trajectories in a circular motion, similar to the Lorenz 63 model [22] on a phase space figure. The SST points swung between the El Niño, normal, and La Niña events and would not repeat their passages. These trajectory behaviors corresponded to those reported in previous studies [23,24], which indicated that the change in ENSO events is a low-order chaotic system.

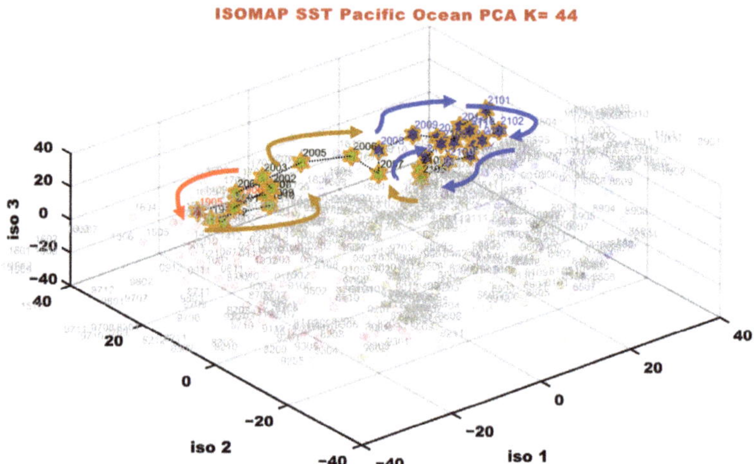

Figure 4. 3D structure composed of the three leading ISOMAP components but only has the recent 31-month data points highlighted. The curves marked by arrow s depict the time evolution (May 2019 to December 2021) of these points. The trajectory shows the end of last El Niño (red arrows), circling in normal events (yellow arrows), dramatically changing to be La Niña (blue arrows) with a bigger circle route then back to two-month normal events, and to be La Niña again.

One spatial eigenmode and its corresponding temporal eigenmode acted as a pair in the PCA or ISOMAP calculation. The original data matrix could simply be multiplied with the temporal mode to obtain the spatial mode and vice versa. After the examination of the temporal modes, the leading three spatial eigenmodes, the empirical orthogonal function (EOF), were presented in Figures 5 and 6 (from the PCA and ISOMAP respectively). The number in the top legend of all panels of Figures 5 and 6 is the ratio of the explained variance. With the three leading eigenmodes, the explained variances of the ISOMAP were not as good as those of the PCA. The first eigenmodes of both methods were similar, but the second and third eigenmodes were slightly different. These small differences in the three leading eigenmodes led to obvious differences in the distribution of the corresponding temporal points. In fact, no obvious differences were observed between the PCA and ISOMAP in the first 20 leading spatial eigenmodes (not shown). Slight rearrangement by using the ISOMAP can help separate the data points, allowing their grouping into different clusters.

The residual variances mentioned in the previous section reflect the similarity between the original covariance and low-dimensional covariance. Residual variances of the ISOMAP with the different nearest neighbor numbers are presented in Figure 7. Residual variances became smaller as the neighbor number increased. However, the lower residual variance values did not mean that the nearest neighbor number in ISOMAP was considerably better. It depends on what kind of purpose the ISOMAP needs to achieve. If we want to do classification, we cannot choose the largest nearest neighbor number. Basically, when we connected all the points together in the ISOMAP, the matrix used to solve eigenmodes was similar to the covariance matrix in the PCA. The method connected all points is actually the MDS. When the distance is measured in Euclidean space, the MDS is identical to the PCA [18]. The PCA-reconstructed points in Figure 2 were not distributed well for classification. This is also the reason for not considering all neighbors to count the ISOMAP because the temporal data points are all connected to each other, and thus, the advantage of the ISOMAP is lost. On the other hand, the explained variances from the ISOMAP were not good, but their residual variances were sufficiently lower to reconstruct low-dimensional structures approximate to the high-dimensional structures. The advantage of ISOMAP gave the reasonable distance estimations between the disconnected points through the Dijkstra's algorithm, which decided the shortest path through the neighbors and visited all the points. In brief, the disconnected two points were connected by their neighbors, while preferably no single point is isolated. It also pointed out that try to avoid using too small nearest neighbor number to prevent some points from being isolated.

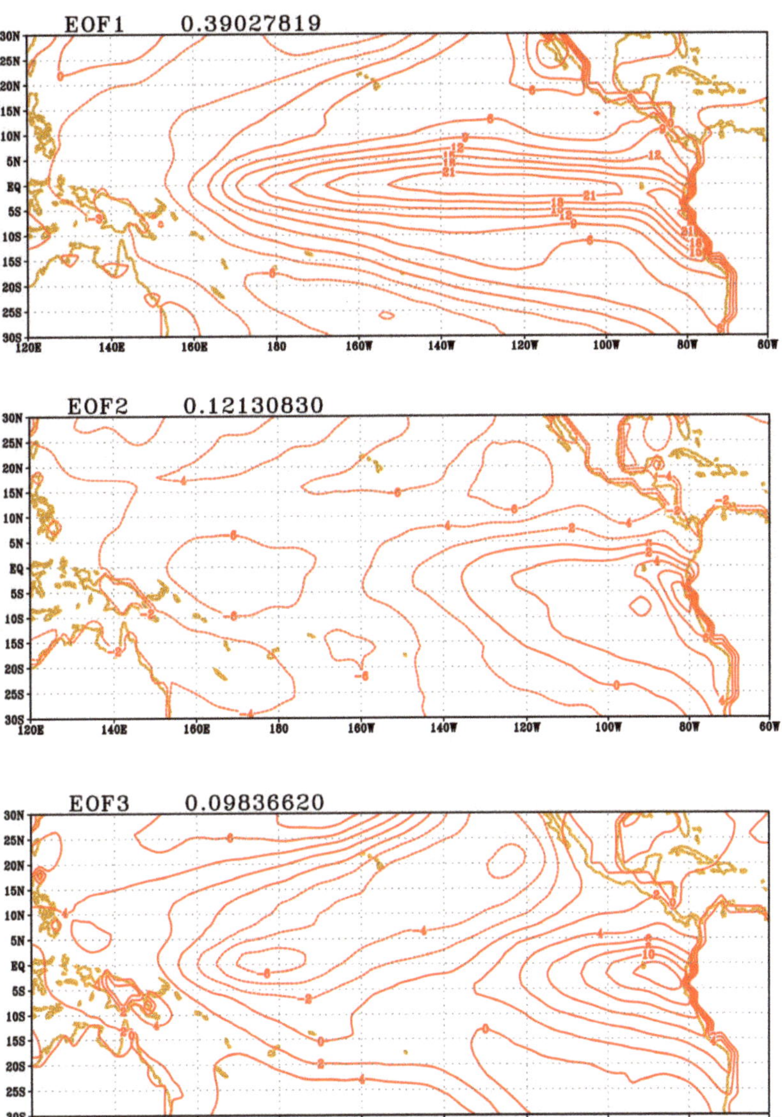

Figure 5. The first three spatial eigen modes (EOFs) of the PCA. The number in the top legend of every figure is the ratio of the explained variance.

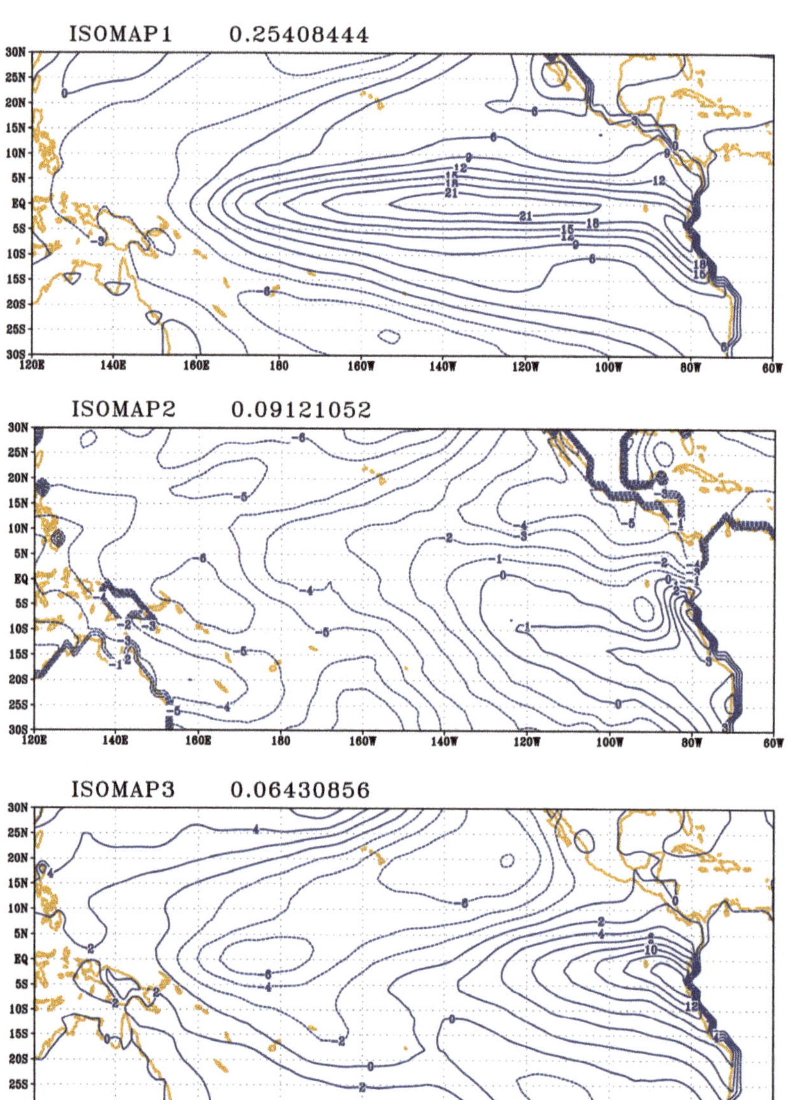

Figure 6. The first three spatial eigen modes (EOFs) of the ISOMAP. The number in the top legend of every figure is the ratio of the explained variance.

The first 20 leading components were considered as data points for classification, and SSVM results are presented in Table 1 (the PCA) and Table 2 (the ISOMAP). We performed 20-times 5-fold cross-validation to obtain these results. The ISOMAP results were slightly better than the PCA results. We defined two classes problem as El Niño/non El Niño or La Niña/non La Niña, because the simple SVM was the 2-class classifier. To classify El Niño/non El Niño was little easier than to classify La Niña/non La Niña. The clues were already in Figures 2 and 3, because the blue La Niña points were closer each other and more difficult to distinguish. No special reason existed for taking the 20 dimensionalities to form data points for the classification. We tested from 1–20 dimensionalities for classification, and the accuracy was approximately 90% with 2–20 dimensionalities, whereas it was approximately 84% with 1 dimensionality. Removal of the normal events from the testing would have led to considerably better classification results and 99% accuracy. All these test results support that the Niño 3.4 index is a good index for defining ENSO events.

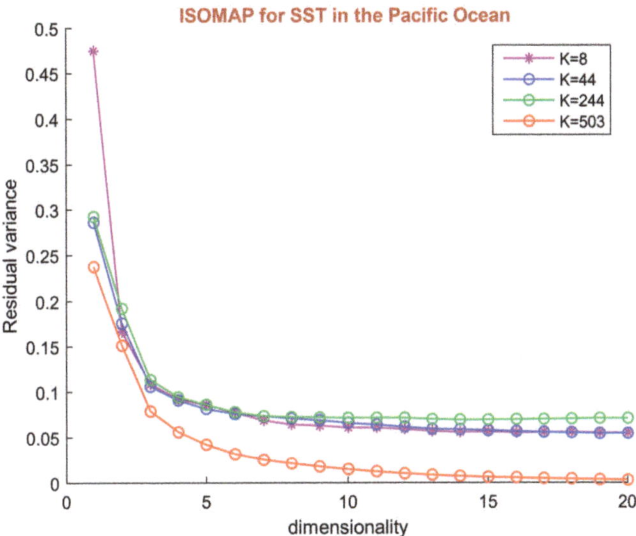

Figure 7. ISOMAP residual variances. The different nearest neighbor numbers in ISOMAP calculation are shown in different colors.

Table 1. The training errors and the testing errors from the 20-dimensionality PCA with SVM.

PCA with SVM	Training Error	Testing Error
El Niño and non El Niño	0.0525	0.1115
La Niña and non La Niña	0.0525	0.1001

Table 2. The training errors and the testing errors from the 20-dimensionality ISOMAP with SVM.

ISOMAP (44 *) with SVM	Training Error	Testing Error
El Niño and non El Niño	0.0385	0.0922
La Niña and non La Niña	0.0545	0.0801

* 44 is the number of neighbors.

4. Conclusions and Discussion

The ISOMAP could help identify extreme El Niño events and easily measure the differences between any two events from the reconstructed space and data points. The distances far or close in the ISOMAP-reconstructed space provide the measurement of the similarities between data points and were more accurate than the distances in the traditional PCA-reconstructed space. The ISOMAP residual variances provided the reference values to decide whether the number of lower dimensionalities was sufficient for the classification. The ISOMAP results could be easily used to perform clustering. Moreover, the ISOMAP allowed grouping of some events or clarified why two events belonging to the El Niño class were different. Although studies have indicated that no two El Niño events are identical, measuring the extent of differences between two events in the same class was possible because of the ISOMAP tool. Meanwhile, we are also proceeding the test that if it is possible to define or predict the ENSO event through SSVM with ISOMAP-reconstructed points instead of the Niño 3.4 index.

Besides solving the El Niño problem, the ISOMAP method can also be used to perform composite analysis when similar cases are to be selected among other meteorological problems. The ISOMAP can be used as a tool for diagnosing different atmospheric circulations. Of course, the ISOMAP can be a score measurement for evaluating the NWP outputs and observation. For example, the SST NWP model output can be projected on the leading components of the ISOMAP SST observational data and some forecasts with a high probability to be true can be analyzed. Moreover, the same procedure can be used to trace the NWP perturbations and detect the growing unstable behaviors. The ISOMAP trajectory behaviors showing in Figure 4 are only similar to Lorenz 63 model trajectory in shapes. It is worth doing more researches in ISOMAP trajectory behavior to check the profound meaning of sensitivity to initial conditions in Lorenz 63 model or other NWP models.

In this study, the number of nearest neighbors was 44 for the ISOMAP calculation. In fact, we tested the neighbor number from 8 to 60, and the number 44 was selected as it could provide the best SSVM classification results. We used the monthly SST data and do not know whether the selected neighbor number is related to the El Niño period of 2–10 years. With a neighbor number greater than 480, the classification results are similar to those of the PCA method. Actually, if we use Euclidean space to measure the distance and take all points connected (any one point connecting to others), the ISOMAP is degenerated to MDS which is identical to the PCA [18]. The reconstructed data points are closer to each other (same as Figure 2), and the advantage of using the ISOMAP is lost.

The traditional PCA is sensitive to the counting domain chosen. If we are concerned about the ENSO, the SST in Pacific Ocean, we should perform a PCA calculation in the tropical Pacific region. If the global domain SST is used to calculate the PCA, global EOF structures that are difficult to explain are obtained. However, the ISOMAP can maintain the local structures even it is not wise to take this kind of calculation. Because the PCA or ISOMAP extracts the counting domain maximum variance patterns out, the bigger domain contains the irrelevant pattern with Pacific Ocean SST pattern. The ISOMAP builds the nearest neighbors first and it has the limitation of the irrelevant pattern.

During this study, the SST ISOMAP-reconstructed point returned to the cluster of the last La Niña event since August 2021. At that time, according to the Niño 3.4 index, La Niña was defined as 3-month running mean SST anomalies lower than 0.5 °C for at least 5 consecutive months. Thus, NOAA declared the recent La Niña event until January 2022. Are there any methods to determine the ISOMAP point trajectory in the moment of August 2021 toward the La Niña event or back to the normal year? The ISOMAP-reconstructed points from the NWP ensemble results can probably predict the SST after the moment of August 2021 would be a La Niña event.

Funding: This research is grateful for support from the Ministry of Science and Technology, Taiwan, ROC (grant # MOST 110-2634-F-008-008) and Central Weather Bureau, Taiwan, ROC.

Institutional Review Board Statement: Not applicable.

Informed Consent Statement: Not applicable.

Data Availability Statement: The SST data are from IRI web, and the link is https://iridl.ldeo.columbia.edu/SOURCES/.NOAA/.NCDC/.ERSST/ (accessed on 27 April 2022).

Acknowledgments: The author thanks three anonymous reviewers for their constructive and helpful comments. The author appreciates reviewers for pointing out some mistakes in the original manuscript. During this research, the author thanks Bo-Wen Shen for his guidance and encouragement.

Conflicts of Interest: The authors declare no conflict of interest.

Appendix A

The data matrix is $\mathbf{Y}_{KN}$. It can be formed by the SST anomaly, that is, the SST value subtracting its climatological mean. Moreover, M is the dimension of the space, and N is the dimension of time. During the PCA calculation, we solved the eigen problem of

$\mathbf{Y}^T\mathbf{Y}$ or $\mathbf{Y}\mathbf{Y}^T$. In meteorology, temporal eigen components are called principal components, and spatial eigen components are called empirical orthogonal functions. $\mathbf{Y}^T\mathbf{Y}$ or $\mathbf{Y}\mathbf{Y}^T$ is the covariance matrix of the data $\mathbf{Y}$. However, we can count the distance between any two temporal points (pairwise) and take the sum of all spatial points to build the square distance matrix $\mathbf{D}_{NN}$. This distance matrix can also be used for counting the eigen problem. This method is often called multidimensional scaling.

The detailed arrangements are as follows:

(1) The covariance matrix for the PCA is

$$\mathbf{Y}^T\mathbf{Y} = \mathbf{C}_{NN} = \begin{pmatrix} \sum_{k=1}^{K} y^T_{k,\,t=1} y_{k,\,t=1} & \sum_{k=1}^{K} y^T_{k,\,t=1} y_{k,\,t=2} & \cdots & \sum_{k=1}^{K} y^T_{k,\,t=1} y_{k,\,t=N} \\ \sum_{k=1}^{K} y^T_{k,\,t=2} y_{k,\,t=1} & \cdots & \cdots & \cdots \\ \cdots & \cdots & \cdots & \cdots \\ \cdots & \cdots & \cdots & \sum_{k=1}^{K} y^T_{k,\,t=N} y_{k,\,t=N} \end{pmatrix}$$

(2) The distance matrix for MDS is

$$\mathbf{D} = \begin{pmatrix} \sum_{k=1}^{K} \left(y_{k,\,t=1} - y_{k,\,t=1}\right)^2 & \sum_{k=1}^{K} \left(y_{k,\,t=1} - y_{k,\,t=2}\right)^2 & \cdots & \sum_{k=1}^{K} \left(y_{k,\,t=1} - y_{k,\,t=N}\right)^2 \\ \sum_{k=1}^{K} \left(y_{k,\,t=2} - y_{k,\,t=1}\right)^2 & \cdots & \cdots & \cdots \\ \cdots & \cdots & \cdots & \cdots \\ \cdots & \cdots & \cdots & \sum_{k=1}^{K} \left(y_{k,\,t=N} - y_{k,\,t=N}\right)^2 \end{pmatrix}$$

We can choose the number of neighbors (e.g., 44 neighbors in this study) required to plot the weighted graph and construct the geodesic distance matrix. This geodesic distance matrix can be implemented using Dijkstra's algorithm [20], one of the shortest path algorithms. This means that the direct distance (shortcut) between number 1 and number 45 is replaced by the shortest path through the 44 neighbors to neighbor 45. The new geodesic distance matrix is called $\mathbf{D}_G$ in this article. The ISOMAP method first establishes the weighted graph and constructs the geodesic distance matrix and then solves the MDS problem.

The data point coordinate is calculated as the PCA/ISOMAP temporal principal component multiplied by its square root of the eigenvalue (step 6 in Algorithm 1). We employ subscript M to represent the number of eigen components used. For example, we take the first three leading components to form the distance matrix $\mathbf{D}_M$ as follows:

$$\mathbf{D}_M = \begin{pmatrix} d_{M(t=1,\,t=1)} & \cdots & \cdots & d_{M(t=1,\,t=N)} \\ \cdots & \cdots & \cdots & \cdots \\ \cdots & \cdots & \cdots & \cdots \\ d_{M(t=N,\,t=1)} & \cdots & \cdots & d_{M(t=N,\,t=N)} \end{pmatrix}$$

The element of $\mathbf{D}_M$, the distance between any two-time interval steps is defined as follows:

$$d_{M(t,\,t)} = \sum_{i=1}^{M} \left(\sigma_i^{1/2} pca_{i\,(t,\,t)} - \sigma_i^{1/2} pca_{i\,(t,\,t)}\right)^2$$

where *pca* is the temporal eigenmode from the PCA/ISOMAP (steps 4 and 5 in the Algorithm 1), and σ is the corresponding eigenvalue.

The coefficient of correlation $R^2(\mathbf{D}_M, \mathbf{D}_G)$ is calculated using all the matrix elements of $\mathbf{D}_M$ and $\mathbf{D}_G$ as follows:

$$R^2 = \frac{\left(\sum (d_M - \bar{d}_M)(d_G - \bar{d}_G)\right)^2}{\sum (d_M - \bar{d}_M)^2 \sum (d_G - \bar{d}_G)^2}$$

Then, the residual variance is defined as

$$1 - R^2(\mathbf{D}_M, \mathbf{D}_G)$$

We can count $M = 1, 2, \ldots, 20$ consecutively to obtain residual variances of the leading 20 eigen components. In PCA's residual variance calculation, we use the $\mathbf{D}$ instead of $\mathbf{D}_G$.

References

1. Alpaydin, E. *Introduction to Machine Learning*, 3rd ed.; MIT Press: Cambridge, MA, USA, 2014; p. 640.
2. Bishop, C.M. *Pattern recognition and machine learning*; Springer-Verlag Press: New York, NY, USA, 2006; p. 738.
3. Hsieh, W.W. *Machine learning methods in the environmental sciences: Neural networks and kernels*; Cambridge university press: New York, NY, USA, 2009; p. 349.
4. Richman, M.B. Rotation of principal components. *J. Climatol.* **1986**, *6*, 293–335. [CrossRef]
5. Tipping, M.E.; Bishop, C.M. Probabilistic principal component analysis. In *Technical Report, NCRG/97/010*; Neural Computing Research Group, Aston University: Birmingham, UK, 1997; p. 13.
6. Tipping, M.E.; Bishop, C.M. Mixtures of probabilistic principal component analyzers. *Neural Comput.* **1999**, *11*, 443–482. [CrossRef] [PubMed]
7. Roweis, S. EM algorithms for PCA and SPCA. In *Advances in Neural Information Processing Systems*; Jordan, M.I., Kearns, M.J., Solla, S.A., Eds.; MIT Press: Cambridge, MA, USA, 1998; Volume 10, pp. 626–632, 1107.
8. Schölkopf, B.; Smola, A.; Müller, K.-R. Nonlinear component analysis as a kernel eigenvalue problem. *Neural Comput.* **1998**, *10*, 1299–1319. [CrossRef]
9. Hyvärinen, A.; Oja, E.E. Independent component analysis: Algorithms and applications. *Neural Netw.* **2000**, *13*, 411–430. [CrossRef]
10. Hannachi, A.; Unkel, S.; Trendafilov, N.T.; Jolliffe, I.T. Independent component analysis of climate data: A new look at EOF rotation. *J. Clim.* **2009**, *22*, 2797–2812. [CrossRef]
11. Pires, C.; Hannachi, A. Independent subspace analysis of the sea surface temperature variability: Non-Gaussian sources and sensitivity to sampling and dimensionality. *Complexity* **2017**, *2017*, 1–23. [CrossRef]
12. Tenebaum, J.B.; Silva, V.D.; Langford, J.C. A global geometric framework for nonlinear dimensionality reduction. *Science* **2000**, *290*, 2319–2323. [CrossRef]
13. Kao, H.-Y.; Yu, J.-Y. Contrasting Eastern-Pacific and Central-Pacific types of ENSO. *J. Clim.* **2009**, *22*, 615–632. [CrossRef]
14. Kug, J.-S.; Jin, F.-F.; An, S.-I. Two types of El Niño and warm pool El Niño. *J. Clim.* **2009**, *22*, 1499–1515. [CrossRef]
15. Ashok, K.; Behera, S.K.; Weng, H.; Yamagata, T. El Niño Modoki and its possible teleconnection. *J. Geophys. Res.* **2007**, *112*, C11007. [CrossRef]
16. Cox, T.F.; Cox, M.A.A. *Multidimensional Scaling*; CRC press: Boca Raton, FL, USA, 2001; p. 328.
17. Tripathy, B.K.; Sundareswaren, A.; Ghela, S. *Unsupervised learning approaches for dimensionality reduction and data visualization*; CRC Press: Boca Raton, FL, USA, 2021; p. 160.
18. Marsland, S. *Machine Learning: An Algorithmic Perspective*, 2nd ed.; CRC Press: Boca Raton, FL, USA, 2014; p. 457.
19. Hannachi, A.; Turner, A.G. ISOMAP nonlinear dimensionality reduction and bimodality of Asian monsoon convection. *Geophys. Res. Lett.* **2013**, *40*, 1653–1658. [CrossRef]
20. Neapolitan, R.E. *Foundations of Algorithms*, 5th ed.; Jones & Bartlett Learning Press: Burlington, MA, USA, 2015; p. 676.
21. Lee, Y.-J.; Mangasarian, O.L. 2000: SSVM: A smooth support vector machine for classification. *Comput. Optim. Appl.* **2000**, *20*, 5–22. [CrossRef]
22. Lorenz, E.N. Deterministic nonperiod flow. *J. Atmos. Sci.* **1963**, *20*, 130–141. [CrossRef]
23. Tziperman, E.; Cane, M.A.; Zebiak, S.E. Irregularity and locking to the seasonal cycle in an ENSO prediction model as explained by the quasi-periodicity route to chaos. *J. Atmos. Sci.* **1995**, *52*, 293–306. [CrossRef]
24. Wang, B.; Fang, Z. Chaotic oscillations of the tropical climate: A dynamic theory for ENSO. *J. Atmos. Sci.* **1996**, *53*, 2786–2802. [CrossRef]

Review

Challenges and Progress in Computational Geophysical Fluid Dynamics in Recent Decades

Wen-Yih Sun [1,2,3]

1   Department of Earth, Atmospheric, and Planetary Sciences, Purdue University, West Lafayette, IN 47907, USA; wysun@purdue.edu
2   Department of Atmospheric Sciences, National Central University, Zhongli 320, Taiwan
3   Institute for Space-Earth Environmental Research, Nagoya University, Nagoya 464-8601, Japan

Abstract: Here we present the numerical methods, applications, and comparisons with observations and previous studies. It includes numerical analyses of shallow water equations, Sun's scheme, and nonlinear model simulations of a dam break, solitary Rossby wave, and hydraulic jump without smoothing. We reproduce the longitude and transverse cloud bands in the Equator; two-day mesoscale waves in Brazil; Ekman spirals in the atmosphere and oceans, and a resonance instability at 30° from the linearized equations. The Purdue Regional Climate Model (PRCM) reproduces the explosive severe winter storms in the Western USA; lee-vortices in Taiwan; deformation of the cold front by mountains in Taiwan; flooding and drought in the USA; flooding in Asia; and the Southeast Asia monsoons. The model can correct the small-scale errors if the synoptic systems are correct. Usually, large-scale systems are more important than small-scale disturbances, and the predictability of NWP is better than the simplified dynamics models. We discuss the difference between Boussinesq fluid and the compressible fluid. The Bernoulli function in compressible atmosphere conserving the total energy, is better than the convective available potential energy (CAPE) or the Froude number, because storms can develop without CAPE, and downslope wind can form against a positive buoyancy. We also present a new terrain following coordinate, a turbulence-diffusion model in the convective boundary layer (CBL), and a new backward-integration model including turbulence mixing in the atmosphere.

Keywords: numerical weather prediction (NWP); Courant–Friedrichs–Lewy (CFL) criterion; Purdue regional climate model (PRCM); convective available potential energy (CAPE); shallow water equations (SWE); Bernoulli function; chaos; dynamics system; forward–backward scheme; leapfrog scheme

Citation: Sun, W.-Y. Challenges and Progress in Computational Geophysical Fluid Dynamics in Recent Decades. *Atmosphere* **2023**, *14*, 1324. https://doi.org/10.3390/atmos14091324

Academic Editors: Bo-Wen Shen, Roger A. Pielke Sr. and Xubin Zeng

Received: 17 May 2023
Revised: 27 July 2023
Accepted: 28 July 2023
Published: 22 August 2023

Contents:

1. Introduction

The foundation of fluid dynamics includes conservations of mass, momentum, and energy, as well as the thermodynamic equation of state. Fluids are assumed to obey the continuum assumption. At infinitesimal space, it contains infinite numbers of molecules and the property of the fluid (i.e., temperature, density, velocity) inside is uniform and well-defined. Fluid dynamics consists of a set of nonlinear differential equations to present the interactions among multi-length and temporal scales under different mechanical and thermodynamic environments. The computational geofluid dynamics (CGFD) has been applied to study problems ranging from small turbulence, pollution, storm, weather, to global climate, oceanic circulations, weather on other planets, and convection in the Sun. The Navier–Stokes equations (NS) do not have a well-defined size or analytical solutions. Usually, the partial differential equations are converted to the finite difference or spectrum equations at discrete point/mode, which are difficult to keep the original property of the equations. The numerical equations should be consistent and converge to the original differential equations, but it is difficult to prove that the numerical solution is unique and converges to the true solution, except a few special cases. It is also difficult to calculate the sub-grid scale turbulences/variables which cannot be resolved. Furthermore, those equations have frequently been simplified and solved by numerical methods with the help of high-performance computers. The results depend upon the detailed formulation of the forcings, the numerical schemes applied, and the initial and boundary conditions.

Bjerknes (1904) [1] formulated the primitive equations in numerical weather prediction (NWP) and climate modeling. Richardson (1922) [2] took six weeks to produce a six-hour forecast at two points in Central Europe based on the equations proposed by Bjerknes. Charney et al. (1950) [3] used a computer to simulate a barotropic model. Phillips (1956) [4] developed the first successful weather/climate model. Now, the weekly weather forecasting generated from the NWP models are shown in TV and newspapers everywhere. People are concerned about the accuracy of the forecast, especially for severe weather and how long the forecasting is still valuable.

Poincare (1882, 1890, 1892) [5–7], discovered that a small change in one state of a deterministic nonlinear system can result in large differences in a later state in his research of the three-body problem, became the first person to discover a chaotic deterministic system which laid the foundations of modern chaos theory. He also introduced phase space (Poincare 1881), Poincare section (map) [7,8], and other concepts to work on dynamic systems. Lorenz (1963) [9] had rediscovered the chaotic behavior and attractors of a nonlinear system when he worked on simplified Benard–Rayleigh convection. Both Poincare and Lorenz emphasized the chaos could come from the uncertainty of initial state. A predictability limit of two weeks (Lorenz 1969) [10] has been suggested and highly cited. Meanwhile, Smagorinsky (1969) [11] documented that "In general even at 21 days, the perturbed map is still far from randomly related to the control—another way of saying that the deterministic limit has not yet been reached". Arakawa (1966) [12] believed that the predictability limit is not necessarily a fixed number. Pagel et al. (1997) [13] found that regional models are not sensitive to the small variation of initial conditions. Wiin-Nelson (1998) [14] suggested that the practical limit is determined by the uncertainty of the initial state due to the accuracy and distributions of observations and the uncertainty in describing the forcing of a given system. Charney and Shukla (1981) [15], and Hoskins (2012) [16] pointed out that the improvement of the initial data, modeling, parameterization, and better understanding of the large-scale systems (blockings, equatorial waves, etc.), as well as the influence of soil and ocean, which has much longer predictable timescale, increase our predictability significantly. The idea of a predictability limit of two weeks is based on a highly simplified model, which is sensitive to the parameters and model chosen (Hand and Finch 2008) [17]. The dissipation in the atmosphere and oceans also makes the systems differ from the conservative systems discussed by Poincare and Lorenz. The sensitivity of the numerical model was also found in Beltrami (1987) [18] and Wiin-Nielsen (1998) [14]. Recently, Shen (2020) [19], and Shen et al. (2022a, b) [20,21] added more modes in Lorenz models and obtained both point attractors and chaotic attractors due to the aggregated negative feedback of small-scale convective process. Poincare (1892) [7] noted that to study the evolution of a physical system over time, one must model based on a choice of laws of physics and the necessary and sufficient parameters that characterize the system. The equations of the motions in the atmosphere and oceans are much more complicated than those studied by Poincare or Lorenz. It may be impossible to predict the long-term behavior in the chaotic weather/climate systems, but the improvement of the NWP; intensive research and test on prediction systems in many institutes/organizations; and our experiences working on various models indicate that we will be able to predict the change in weather/climate near the saddle points and extend the short-term forecast from a couple weeks to a season or longer.

Even if the dynamic system is stable, there may be numerical instabilities resulting from the method of applying (Greenwood 2003) [22], Sun (2010, 2011) [23,24]. Instead of working on the initial condition, we used the ECMWF reanalysis as the initial and lateral boundary conditions for our real data simulations. We have focused on the equations and numerical modeling, which correspond to the parameters and the mathematics models discussed by Poincare, Lorenz, and Wiin-Nielsen, etc. We also used the observations data to verify the numerical results and improve the numerical model. We are aware of uncertainty and difficulty in CGFD, and the nature is always more complicated than those presented by CGFD. But CGFD is still the best tool to study and predict the motions in the atmosphere and oceans.

This article is limited to the author's learning and working experiences with the teachers, former students, and colleagues, which only covers a small portion of CGFD. Although majority of the material comes from the previous publications, it may be different from the traditional texts: (Mesinger and Arakawa 1976 [25], Gill 1982 [26], Haltiner and Williams 1983 [27], Pielke Sr. 2002 [28], Lin 2007 [29], Holton and Hakim 2012 [30], Versteeg and Malalasekera, 1995 [31], Vallis 2013 [32], etc.) or the conventional theories. This paper tries to link dynamics instability and weather/climate research. Each section includes the basic equations, approach, and results, although the detailed derivations are incomplete. The details should be referred to the original papers. The reader with numerical analysis, computational fluid dynamics, dynamic meteorology or oceanography should be able to follow. Comments, suggestions, and corrections will be appreciated. We hope readers will discover the beauty, uncertainty, and challenge in CGFD and linkage between nature and sciences.

2. Shallow Water Equations and Numerical Schemes

The shallow water equations (SWE) are simple but capable of capturing the major characteristics of the large-scale motions in the atmosphere and oceans, because of a small ratio of the vertical to the horizontal length scale and in the hydrostatic equilibrium. They have been applied to simulate the Rossby wave, Kevin wave, Poincare wave, geostrophic adjustment, and surface gravity waves, etc. (Pedlosky 1979 [33], Gill 1982 [26], Vallis 2013 [32]), as well as the flow in lakes and rivers. It is noted that the barotropic model is a non-divergence shallow water system. Lorenz (1980) [34] added diffusion in the continuity equation to represent thermal dissipation, and derive a low-order primitive equation by dropping the time derivatives in the divergence equations to study attractor and quasi-geostrophic flow in SWE. His results showed that gravity wave decays with time and the quasi-geostrophic mode remains as expected. It is likely that Kelvin waves may be affected too. The dispersive relation of an inertial surface wave is

$$\sigma^2 = f^2 + (k^2 + l^2)gH$$

where σ is frequency, f is the Coriolis parameter, k and l are the wave number, H is the mean water depth. Another important equation is the conservation of potential vorticity PV:

$$\frac{D}{Dt}\left(\frac{\varsigma + f}{h}\right) = 0$$

where ς is the relative vorticity, and h is the depth of water. The corresponding PV in the compressible atmosphere:

$$\frac{D}{Dt}\left(\frac{\omega_a \cdot \nabla \theta}{\rho}\right) = 0$$

where ω_a is the absolute vorticity, θ is potential temperature, and ρ is atmosphere density.

2.1. Instability of Forward–Backward (FB) and Leapfrog (LP) Schemes (Sun 2010) [23]

The 1D linearized shallow water equations are

$$\frac{\partial h}{\partial t} = -H\frac{\partial u}{\partial x} + \delta v\frac{\partial^2 h}{\partial x^2} \tag{1}$$

$$\frac{\partial u}{\partial t} = -g\frac{\partial h}{\partial x} + v\frac{\partial^2 u}{\partial x^2} \tag{2}$$

where H is the mean depth, g is gravity, h and u are depth perturbation and velocity, v is viscosity. The control parameter $\delta = 0$ or 1. In the staggered C-grids, the finite difference equations in the forward–backward (FB) scheme become:

$$\frac{(h_p^{n+1}-h_p^n)}{\Delta t} = -H\frac{u_{p+1/2}^n-u_{p-1/2}^n}{\Delta x} + \delta v\frac{h_{p+1}^n-2h_p^n+h_{p-1}^n}{\Delta x^2}$$

$$\frac{(u_q^{n+1}-u_q^n)}{\Delta t} = -g\frac{h_{q+1/2}^{n+1}-h_{q-1/2}^{n+1}}{\Delta x} + v\frac{u_{q+1}^n-2u_q^n+u_{q-1}^n}{\Delta x^2} \tag{3}$$

where $q = p \pm 1/2$, p is the grid for h, and Δx is the spatial interval. A wave-type (normal mode) solutions at the nth time step are:

$$h_p^n = \hat{h}^n \exp[ikp\Delta x] = \hat{h}^0 \lambda^n \exp[ikp\Delta x]$$

$$u_q^n = \hat{u}^n \exp[ikq\Delta x] = \hat{u}^0 \lambda^n \exp[ikq\Delta x] \tag{4}$$

where $i = \sqrt{-1}$, $\hat{h}^n$ and $\hat{u}^n$ are the amplitude of h and u at the nth time step; k is the wave number. We obtain the difference equations

$$\begin{pmatrix} \hat{h}^{n+1} \\ \hat{u}^{n+1} \end{pmatrix} = \begin{pmatrix} 1-\delta S^2 & -i\Delta t HX \\ -ig\Delta t X(1-\delta S^2) & 1-S^2-R^2 \end{pmatrix}\begin{pmatrix} \hat{h}^n \\ \hat{u}^n \end{pmatrix} \tag{5}$$

and the eigenvalue

$$\lambda = \left[\left\{2-(1+\delta)S^2-R^2\right\} \pm \sqrt{[2-(1+\delta)S^2-R^2]^2-4(1-S^2)(1-\delta S^2)}\right]/2 \tag{6}$$

where $X = \frac{\sin[k\Delta x/2]}{\Delta x/2}$, $R^2 = gH\Delta t^2 X^2 = (CX\Delta t)^2 = 4C_o^2\sin^2(k\Delta x/2)$, $S^2 = v\Delta t X^2$, $C = \sqrt{gH}$, and Courant number $C_0 = \frac{C\Delta x}{\Delta t}$.

For simplicity, we set $g = 1$, $H = 1$, and $\Delta x = 1$.

The difference equations for the Leapfrog (LF) scheme are:

$$\frac{(h_p^{n+1}-h_p^{n-1})}{2\Delta t} = -H\frac{u_{p+1/2}^n-u_{p-1/2}^n}{\Delta x} + \delta v\frac{h_{p+1}^{n-1}-2h_p^{n-1}+h_{p-1}^{n-1}}{\Delta x^2}$$

$$\frac{(u_q^{n+1}-u_q^{n-1})}{2\Delta t} = -g\frac{h_{q+1/2}^n-h_{q-1/2}^n}{\Delta x} + v\frac{u_{p+1}^{n-1}-2u_p^{n-1}+u_{p-1}^{n-1}}{\Delta x^2} \tag{7}$$

and the corresponding eigenvalue

$$\lambda^2 = \left\{1-(1+\delta)S^2-2R^2\right\} \pm \sqrt{\left\{1-(1+\delta)S^2-2R^2\right\}^2-(1-2\delta S^2)(1-2S^2)} \tag{8}$$

Without viscosity (i.e., $\nu = 0$), the FB is neutrally stable, $|\lambda| = 1$, when the Courant number $Co = \frac{C\Delta t}{\Delta x} < 1$; and the LF is also neutrally stable when $Co < 0.5$ (Mesinger and Arakawa 1976 [25], Haltiner and Williams, 1980 [27]). At $Co = 1$, the FB has a repeated eigenvalue. $\lambda_{1,2} = \left[\{2 - R^2\} \pm \sqrt{[2 - R^2]^2 - 4} \right]/2 = -1$ for $2\Delta x$ waves. Equation (5) becomes

$$\mathbf{x}^{n+1} = \begin{pmatrix} \hat{h}^{n+1} \\ \hat{u}^{n+1} \end{pmatrix} = \begin{pmatrix} 1 & -2i \\ -2i & -3 \end{pmatrix} \begin{pmatrix} \hat{h}^n \\ \hat{u}^n \end{pmatrix} = \mathbf{A}\mathbf{x}^n \tag{9}$$

An eigenvector corresponding to $\lambda_1 = -1$ is $\mathbf{x}_1 = \begin{pmatrix} 1 \\ -i \end{pmatrix}$. The generated eigenvector $\mathbf{x}_2$ can be found from

$$\mathbf{A}\mathbf{x}_2 = \lambda_1 \mathbf{x}_2 + \mathbf{x}_1 \tag{10}$$

and

$$\mathbf{x}_2 = \begin{pmatrix} 0.5 \\ 0 \end{pmatrix} \tag{11}$$

If we define a matrix $\mathbf{P} = (\mathbf{x}_1, \mathbf{x}_2) = \begin{pmatrix} 1 & 0.5 \\ -i & 0 \end{pmatrix}$, then, we can have

$$[\mathbf{A}]_B = \mathbf{P}^{-1}\mathbf{A}\mathbf{P} = \begin{pmatrix} 0 & i \\ 2 & -2i \end{pmatrix} \begin{pmatrix} 1 & -2i \\ -2i & -3 \end{pmatrix} \begin{pmatrix} 1 & 0.5 \\ -i & 0 \end{pmatrix} = \begin{pmatrix} -1 & 1 \\ 0 & -1 \end{pmatrix} = \begin{pmatrix} \lambda & 1 \\ 0 & \lambda \end{pmatrix} \tag{12}$$

Which is in Jordan block form (Nobel 1969) [35], and since

$$\begin{pmatrix} \lambda & 1 \\ 0 & \lambda \end{pmatrix}^n = \begin{pmatrix} \lambda^n & n\lambda^{n-1} \\ 0 & \lambda^n \end{pmatrix} \tag{13}$$

We obtain

$$\mathbf{A}^n = \mathbf{P}([\mathbf{A}]_B)^n \mathbf{P}^{-1} = \begin{pmatrix} 1 & 0.5 \\ -i & 0 \end{pmatrix} \begin{pmatrix} \lambda^n & n\lambda^{n-1} \\ 0 & \lambda^n \end{pmatrix} \begin{pmatrix} 0 & i \\ 2 & -2i \end{pmatrix} \tag{14}$$

$$= \begin{pmatrix} 1 & 0.5 \\ -i & 0 \end{pmatrix} \begin{pmatrix} (-1)^n & n(-1)^{n-1} \\ 0 & (-1)^n \end{pmatrix} \begin{pmatrix} 0 & i \\ 2 & -2i \end{pmatrix}$$

Red dashed line in Figure 1 shows the magnitude of $2\Delta x$ wave linearly increases with time and with $2\Delta t$ oscillation, which can be controlled by the fourth-order Shuman smoothing (blue solid line) (Sun 2010) [23]. When $Co = 0.5$, (i.e., $R = 1$), and $S^2 = 0$, (i.e., $\nu = 0$), the LF scheme has the repeated eigenvalues too, $\lambda_1 = \lambda_2 = -i$ and $\lambda_3 = \lambda_4 = i$. Hence, it becomes weakly unstable, the same as the FB scheme.

Sun (1982) [36] applied viscosity to the burger equation to suppress instability. But in SWE, viscous terms create an unstable $2\Delta x$ and $3\Delta x$ waves at $Co = 1$, and an unstable $2\Delta x$ wave at $Co = 0.8$ in the FB. The growth rate becomes less than one for longer waves and the maximum damping occurs at $4\Delta x$ or $3\Delta x$ waves with $\delta = 1$ (Figure 2). The LF also becomes unstable at $R^2 = 1$, (i.e., $2\Delta x$-wave at $Co = 0.5$). The amplification factor for $2\Delta x$-wave equals 2 for $\nu = 0.25$ (i.e., $S^2 = 0.5$), as shown in Figure 3, because viscosity enhances the 2-Δt oscillation for the short waves (wavelength $\leq 3\Delta x$). Instability also shows up for $2\Delta x$-wave when $\nu > 0.13$ at $Co = 0.4$ in LF. Viscosity has the largest damping at $4\Delta x$ waves when $Co = 1$ in FB and $Co = 0.5$ in LF; and at $3\Delta x$ waves when $Co = 0.8$ in FB and $Co = 0.4$ in LF. The instability becomes weaker with $\delta = 0$.

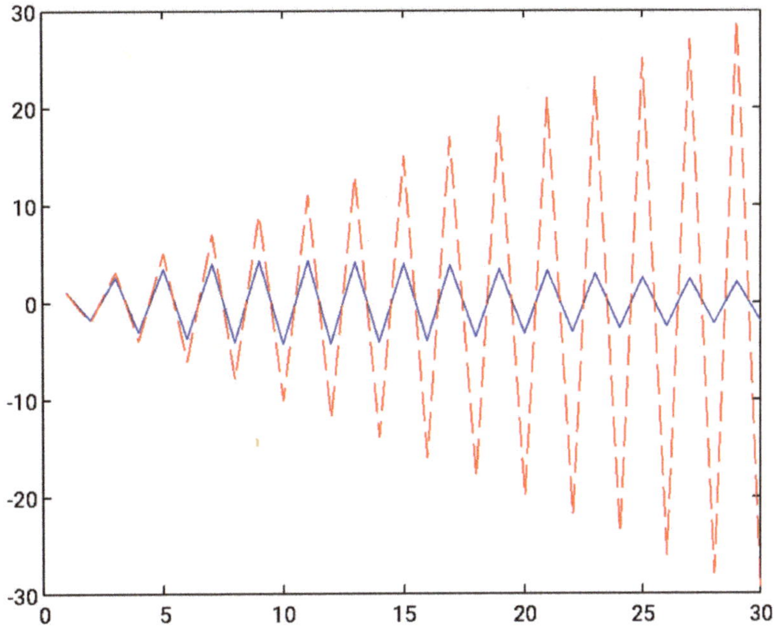

Figure 1. Magnitude of $n(-1)^{n-1}$: h or u without smoothing, (red dashed line), and $n(-0.91)^{n-1}$ with smoothing (blue solid line) as function of nth time step, where $\lambda' = Sh * \lambda = 0.91$ for $\eta = 0.15$. (Sun 2010) [23].

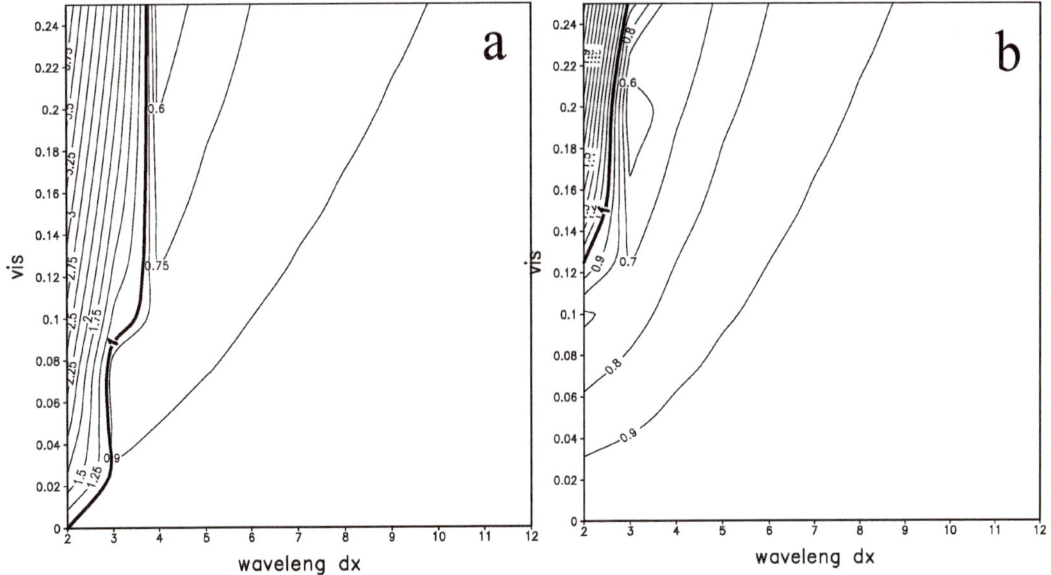

Figure 2. Growth rate for forward–backward scheme as function of wavelength (Δx) and viscosity (ν) at: (**a**) Courant number $Co = 1$; (**b**) $Co = 0.8$. (Sun 2010) [23].

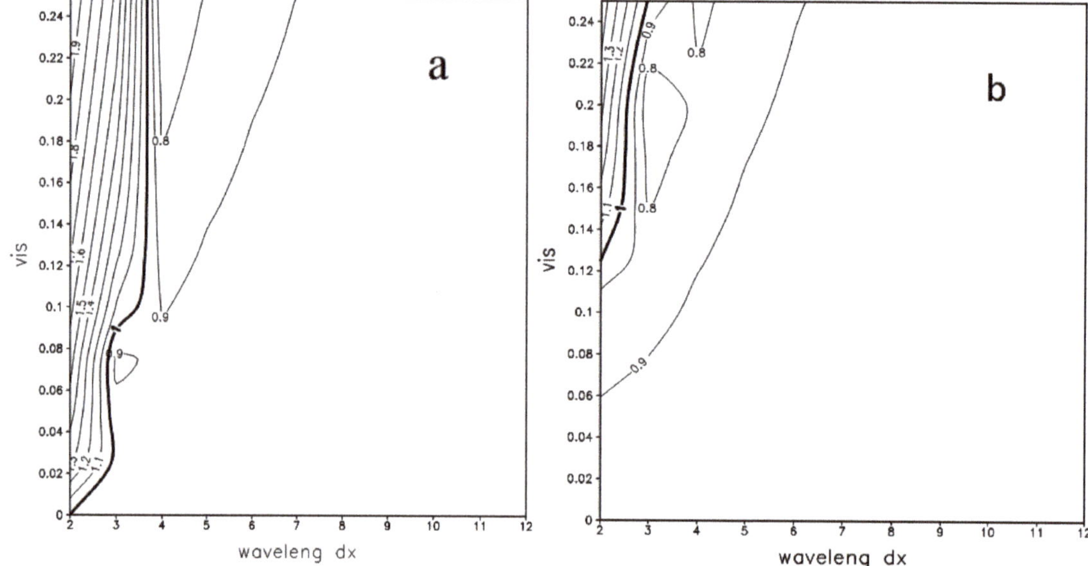

Figure 3. Growth rate for Leapfrog scheme as function of wavelength (Δx) and viscosity (ν) at: (**a**) Courant number $Co = 0.5$; (**b**) $Co = 0.4$. (Sun 2010) [23].

2.2. Modified Leapfrog Scheme for Shallow Water Equations

Sun and Sun (2011) [37] proposed to replace $h^n_{q+1/2}$ in (7) by $\overline{h}^n_{q+1/2} = \frac{h^{n+1}_{q+1/2} + 2h^n_{q+1/2} + h^{n-1}_{q+1/2}}{4}$. The finite different equations become

$$\frac{(h^{n+1}_p - h^{n-1}_p)}{2\Delta t} = -H\frac{u^n_{p+1/2} - u^n_{p-1/2}}{\Delta x} \tag{15}$$

$$\frac{(u^{n+1}_q - u^{n-1}_q)}{2\Delta t} = -g\frac{\overline{h}^n_{q+1/2} - \overline{h}^n_{q-1/2}}{\Delta x} \tag{16}$$

After solving (15), we obtain $h^{n+1}_{q+1/2}$ in (16). No iteration is required. It is a second-order accuracy in both space and time. The eigenvalues are:

$$\lambda = \left\lfloor \left\{2 - R^2\right\} \pm \sqrt{[2 - R^2]^2 - 4} \right\rfloor / 2 \tag{17}$$

$|\lambda| = 1$, if $Co < 1$ which is twice allowed in the original LF scheme. The eigenvalue of (17) is identical to (6) of the FB scheme without viscosity ($S = 0$) (Sun 1984a, 2010) [23,38]. This simple method enables the LF users to save one half of the computing time.

2.3. New Scheme for Shallow Water Equations, Sun (2011) [24]

Typically, finite difference schemes produce oscillations behind the shock in simulating the dam break. Therefore, more sophisticated methods have been proposed (Roe 1981 [39], Toro 1999 [40]), including characteristic based method (Wang and Yeh 2005 [41], Oh 2007 [42], etc.) and various Riemann solvers (MacCormack 1969 [43], etc.). Unfortunately, it is impossible to diagonalize the matrices on both x- and y-directions of characteristic equations simultaneously, those methods require complicated algorithms with some uncertainty (Oh 2007) [42]. Sun (2011) [24] proposed a new method by using the time average on the right-hand sides of (3):

$$\frac{(h_p^{n+1}-h_p^n)}{\Delta t} = -H\frac{u_{p+1/2}^{n+1/2}-u_{p-1/2}^{n+1/2}}{\Delta x} = -H\left[\frac{\partial\left(u+\frac{\Delta t}{2}\frac{\partial u}{\partial t}\right)}{\partial x}\right]_p^n = -H\left[\frac{\partial\left(u-\frac{\Delta t}{2}\left[g\frac{\partial h}{\partial x}\right]\right)}{\partial x}\right]_p^n$$
$$= -H\left[\frac{\partial u}{\partial x}\right]_p^n + \frac{gH}{2}\left[\frac{\partial^2 h}{\partial x^2}\right]_p^n \tag{18}$$

$$h_p^{n+1} = h_p^n + \frac{H\Delta t}{\Delta x}\left(u_{p-1/2}^n - u_{p+1/2}^n\right) + \frac{gH\Delta t^2}{2\Delta x^2}\left(h_{p-1}^n - 2h_p^n + h_{p+1}^n\right) \tag{19}$$

$$\frac{(u_q^{n+1}-u_q^n)}{\Delta t} = -g\frac{h_{q+1/2}^{n+1/2}-h_{q-1/2}^{n+1/2}}{\Delta x} = -\frac{g}{2}\left[\frac{\partial\left(h^n+h^{n+1}\right)}{\partial x}\right]_q \tag{20}$$

$$u_q^{n+1} = u_q^n + \frac{\Delta t g}{2\Delta x}\left[h_{q-1/2}^n - h_{q+1/2}^n + h_{q-1/2}^{n+1} - h_{q+1/2}^{n+1}\right] \tag{21}$$

The last term in (19) is like the diffusion term in the Lax–Wendroff scheme (Mesinger and Arakawa 1976) [25]. Using the normal mode solutions of (4), we obtain the eigenvalues

$$\lambda = 1 - 2C_o^2\sin^2(k\Delta x/2) \pm 2C_o\sin(k\Delta x/2)\sqrt{\left(C_o^2\sin^2(k\Delta x/2)-1\right)} \tag{22}$$

where $C_o = \sqrt{gH}\Delta t/\Delta x$ is the Courant number. (22) is identical to (17) or (6) with S = 0. Equation (16) shows that $|\lambda| = 1$ if $C_o < 1$. Same as the FB discussed previously (Sun 2010) [23]. The phase angle ϕ and the phase speed $\hat{C}$ of the numerical scheme are

$$\lambda = |\lambda|\exp(i\phi) = |\lambda|\exp(ikC'\Delta t) \tag{23}$$

and

$$C_{ph} = \varphi/(k\Delta t) = \sin^{-1}\left[\frac{\sqrt{4-\left[2-C_o^2\sin^2(k\Delta x/2)\right]^2}}{2}\right]/(k\Delta t) \tag{24}$$

The phase speed C_{ph} is the same as the FB (Sun 1984a; 2010) [23,38]. It is noted that the diffusion term in (19) does not cause damping in the solutions because the amplification factor $|\lambda| = 1$ when $C_o < 1$. On the other hand, the diffusion term in the Lax–Wendroff scheme can severely damp the short waves (Mesinger and Arakawa, 1976) [25]. The ratio of the computed phase speed to the analytical phase speed C_{ph}/C is presented in Table 1.

Table 1. Ratio of computed phase speed to analytic phase speed C_{ph}/C as function of wavelength (λ) and Courant number *Co* for 2nd-order scheme according to (24) (Sun and Sun 2011) [37].

Co \ λ	$2\Delta x$	$3\Delta x$	$4\Delta x$	$5\Delta x$	$6\Delta x$	$7\Delta x$	$8\Delta x$	$9\Delta x$	$10\Delta x$
0.250	0.643	0.834	0.905	0.939	0.957	0.969	0.976	0.981	0.985
0.500	0.667	0.855	0.920	0.950	0.965	0.975	0.981	0.985	0.988
0.750	0.613	0.900	0.949	0.969	0.979	0.985	0.988	0.991	0.993
1.00	0.000	0.500	1.00	1.00	1.00	1.00	1.00	1.00	1.00

2.4. Nonlinear Shallow Water Simulations

The flux form of the nonlinear SWE can be written as:

$$\frac{\partial h}{\partial t} = -\frac{\partial\psi}{\partial x} - \frac{\partial\phi}{\partial y} \tag{25}$$

$$\frac{\partial\psi}{\partial t} + \frac{\partial(u\psi)}{\partial x} + \frac{\partial(v\psi)}{\partial y} - f\phi = -gh\frac{\partial(h+h_s)}{\partial x} - k\psi \tag{26}$$

$$\frac{\partial \varphi}{\partial t} + \frac{\partial(u\varphi)}{\partial x} + \frac{\partial(v\varphi)}{\partial y} + f\psi = -gh\frac{\partial(h+h_s)}{\partial y} - k\varphi \tag{27}$$

where $\psi = hu$ and $\phi = hv$, and h_s is bottom elevation, k is friction

2.4.1. Simulation of Dam Break: Water Depth Is 10 m within a Radius of 11 m, and 1 m Outside

It is also assumed $f = h_s = k = 0$. Following the procedure of Section 2.3, we can write the difference–differential equations of (25)–(27):

$$
\begin{aligned}
\frac{h^{n+1}_{p,q} - h^{n}_{p,q}}{\Delta t} &= -\left(\frac{\partial \psi^{n+1/2}}{\partial x}\right)_{p,q} - \left(\frac{\partial \phi^{n+1/2}}{\partial y}\right)_{p,q} \\
&= -\left[\frac{\partial\left(\psi + \frac{\Delta t}{2}\frac{\partial \psi}{\partial t}\right)}{\partial x}\right]^{n}_{p,q} - \left[\frac{\partial\left(\phi + \frac{\Delta t}{2}\frac{\partial \phi}{\partial t}\right)}{\partial y}\right]^{n}_{p,q} \\
&= -\left(\frac{\partial \psi^{n}}{\partial x}\right)_{p,q} - \left(\frac{\partial \phi^{n}}{\partial y}\right)_{p,q} \\
&\quad + \left[\frac{\partial\left(\frac{\Delta t}{2}\left[\frac{\partial(u\psi)}{\partial x} + \frac{\partial(v\psi)}{\partial y} + \frac{g}{2}\frac{\partial h^2}{\partial x}\right]\right)}{\partial x}\right]^{n}_{p,q} + \left[\frac{\partial\left(\frac{\Delta t}{2}\left[\frac{\partial(u\phi)}{\partial x} + \frac{\partial(v\phi)}{\partial y} + \frac{g}{2}\frac{\partial h^2}{\partial y}\right]\right)}{\partial y}\right]^{n}_{p,q}
\end{aligned}
\tag{28}
$$

$$
\begin{aligned}
\frac{\psi^{n+1}_{p*,q} - \psi^{n}_{p*,q}}{\Delta t} &= -\left[\frac{\partial(u\psi)^{n+1/2}}{\partial x}\right]_{p*,q} - \left[\frac{\partial(v\psi)^{n+1/2}}{\partial y}\right]_{p*,q} - \frac{g}{2}\left[\frac{\partial\left(h^{n+1/2}\right)^2}{\partial x}\right]_{p*,q} \\
&= -\left[\frac{\partial\left(\{u\psi\} - \frac{\Delta t}{2}\left\{u\left[\frac{\partial(u\psi)}{\partial x} + \frac{\partial(v\psi)}{\partial y} + \frac{g}{2}\frac{\partial h^2}{\partial x}\right] + \psi\left[u\frac{\partial u}{\partial x} + v\frac{\partial u}{\partial y} + g\frac{\partial h}{\partial x}\right]\right\} + \frac{g}{2}\left(\frac{h^n+h^{n+1}}{2}\right)^2\right)}{\partial x}\right]^{n}_{p*,q} \\
&\quad - \left[\frac{\partial\left(\{v\psi\} - \frac{\Delta t}{2}\left\{v\left[\frac{\partial(u\psi)}{\partial x} + \frac{\partial(v\psi)}{\partial y} + \frac{g}{2}\frac{\partial h^2}{\partial x}\right] + \psi\left[u\frac{\partial v}{\partial x} + v\frac{\partial v}{\partial y} + g\frac{\partial h}{\partial y}\right]\right\}\right)}{\partial y}\right]^{n}_{p*,q}
\end{aligned}
\tag{29}
$$

and

$$
\begin{aligned}
\frac{\phi^{n+1}_{p,q*} - \phi^{n}_{p,q*}}{\Delta t} &= -\left[\frac{\partial(u\phi)^{n+1/2}}{\partial x}\right]_{p,q*} - \left[\frac{\partial(v\phi)^{n+1/2}}{\partial y}\right]_{p,q*} - \frac{g}{2}\left[\frac{\partial\left(h^{n+1/2}\right)^2}{\partial y}\right]_{p,q*} \\
&= -\left[\frac{\partial\left(\{u\phi\} - \frac{\Delta t}{2}\left\{u\left[\frac{\partial(u\phi)}{\partial x} + \frac{\partial(v\phi)}{\partial y} + \frac{g}{2}\frac{\partial h^2}{\partial x}\right] + \phi\left[u\frac{\partial u}{\partial x} + v\frac{\partial u}{\partial y} + g\frac{\partial h}{\partial x}\right]\right\}\right)}{\partial x}\right]^{n}_{p,q*} \\
&\quad - \left[\frac{\partial\left(\{v\phi\} - \frac{\Delta t}{2}\left\{v\left[\frac{\partial(u\phi)}{\partial x} + \frac{\partial(v\phi)}{\partial y} + \frac{g}{2}\frac{\partial h^2}{\partial x}\right] + \phi\left[u\frac{\partial v}{\partial x} + v\frac{\partial v}{\partial y} + g\frac{\partial h}{\partial y}\right]\right\}^{n} + \frac{g}{2}\left(\frac{h^n+h^{n+1}}{2}\right)^2\right)}{\partial y}\right]_{p,q*}
\end{aligned}
\tag{30}
$$

where $p^* = p \pm 1/2$, and $q^* = q \pm 1/2$.

This scheme is simple, accurate, and conserves the mass and other properties in an inviscid fluid. Most important is that it can be applied to the nonlinear SWE without any diffusion or numerical smoothing. Hence, it can reproduce the solitary Rossby wave, hydraulic jump, the flow moving over mountains, and vortex-merging, etc. Without smoothing/diffusion, our numerical results are usually different from the previous studies that were generated by the models with smoothing. Hence, the physical interpretations can be different. Solutions are usually diverse among elliptic, hyperbolic, and parabolic partial differential equations.

The domain consists of 400×400 grids with $\Delta x = \Delta y = 1$ m. The fourth-order Shuman smoothing with coefficient $\alpha = 0$, 0.1, or 0.2 is applied to the prognostic variables each time step on both second-order and fourth-order schemes (Sun 2011) [24]. Most previous publications show the results at $t = 0.69$ s. The simulated vertical cross section at $t = 0.69$ s for both the second-order ($\alpha = 0$, and 0.2) and fourth-order schemes ($\alpha = 0.2$) are shown in Figure 4. The differences among the second-, third- and fourth-order schemes are insignificant. The results show that the water level h remains at 10 m near the center; the shock reaches at $r = 18$ m from the center with a crest of $h \approx 3.7$ m at $r = 16$ m; and a band of low water ($h \approx 3.2$ m) forms around $r = 13$ m. They are consistent with the second-order Godunov-type numerical model (Fujihara and Borthwick 2000) [44]; the unstructured triangular meshes and Roe's flux function solver (Anastasiou and Chan 1997) [45]; and the characteristic Lagrangian method (Oh 2007) [42], etc.

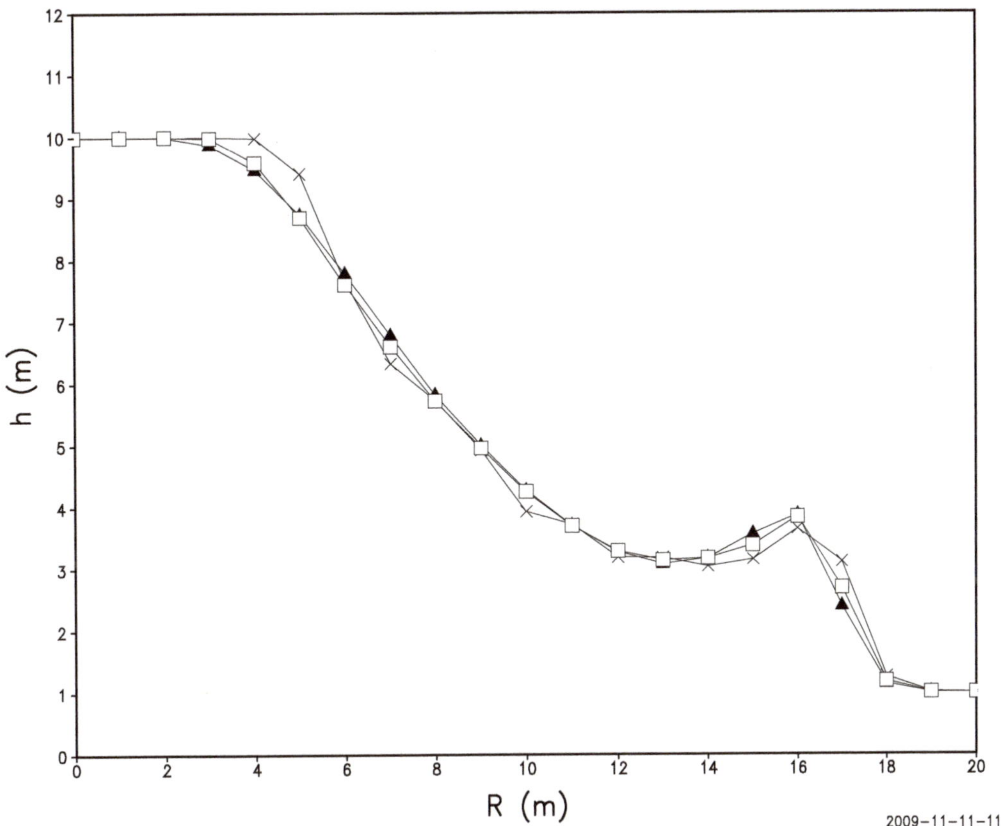

Figure 4. Vertical cross section of water depth h at distance R from the center at t = 0.69 s from 2nd-order scheme with Shuman smoothing coefficient $\alpha = 0$ (×) and $\alpha = 0.2$ (▲), and 4th-order scheme with $\alpha = 0.2$ (□). (Sun 2011) [24].

As pointed out by Tseng and Chu (2000) [46], the traditional MacCormack scheme breaks down at $t = 0.22$ s for this case. The current model running to 4 s without smoothing indicates that the diffusion term in (19) may stabilize the system. The water moves away and creates a deep hole near the center by $t = 3$ s. The hole deepens and widens while the shock propagates away. After $t = 22$ s, water starts returning to the center, eventually a high-water column with a sharp gradient forms near the center, while the original shock continuously propagates away, as shown at $t = 25$ s in Figure 5. The simulations agree excellently with Toro (1999) [40].

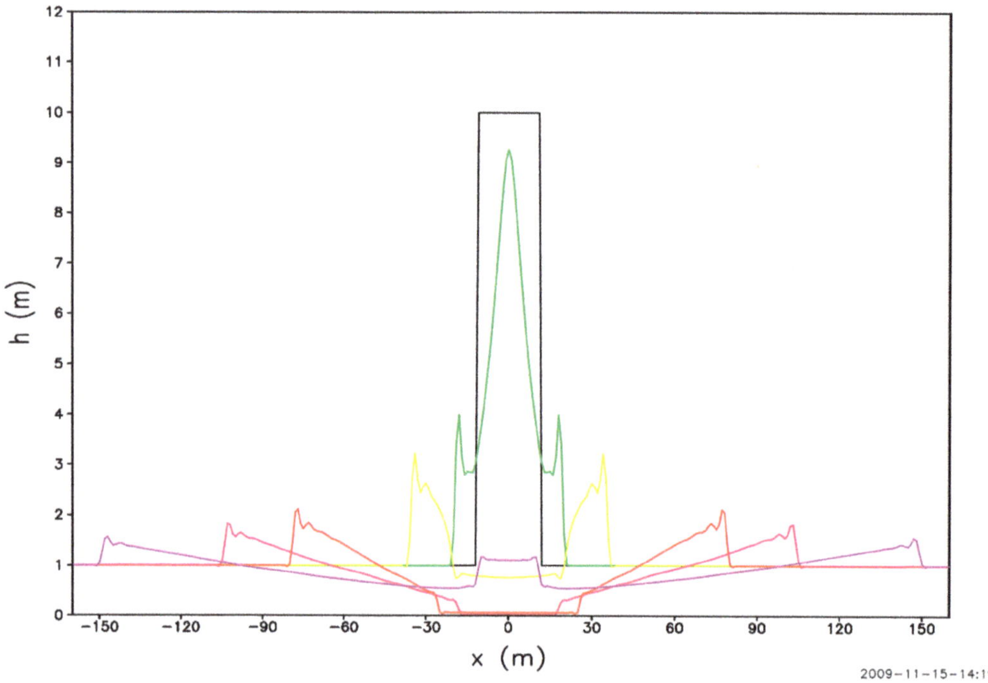

Figure 5. Vertical cross sections of water depth at t = 1, 3, 10, 15 and 25 s from the 2nd-order scheme with smoothing coefficient α = 0.2. Sun (2011) [24].

The total mass is conserved. The total energy slightly decreases with time due to the truncation errors and smoothing applied. The total energy is given by

$$E = \sum_p \sum_q \left\{ \frac{h_{p,q}\left(u_{p,q}^2 + v_{p,q}^2\right)}{2} + \frac{g(h_{p,q})^2}{2} \right\} \Delta x \Delta y \tag{31}$$

The ratio of the total energy to its initial value $Ra(t) = E(t)/E(t=0)$ is about 0.96 at $t = 1$ s and 0.91 at $t = 3$ s. Using the Lagrangian method, Oh (2007) [42] obtains $Ra = 0.85$ at $t = 1.5$ s. The current scheme manages both mass and kinetic energy well, which is a major challenge to dam-break simulations (Liang et al., 2004) [47].

2.4.2. Solitary Rossby Wave (Sun and Sun 2013) [48]

The flux form upstream scheme, the center difference scheme, and the Lax–Wendroff scheme, etc., failed to simulate the Rossby soliton (Chu and Fan, 2010) [49]. The major difference between the Lax–Wendroff scheme and Sun (2011) [24] is that the diffusion term in the Lax–Wendroff scheme can severely damp the short waves, but no damping in Sun scheme as discussed in Section 2.3. Boyd (1980, 1985) [50,51] has shown that an asymptotic solution exists to the inviscid, nonlinear shallow water equations for the Rossby soliton on an equatorial beta plane, which are governed by Korteweg–de Vries (KDV) equation (1895) [52].

$$\frac{\partial u}{\partial t} - 6u\frac{\partial u}{\partial x} + \frac{\partial^3 u}{\partial x^3} = 0: \text{ solution}: u(x,t) = \frac{c}{2}\text{sech}^2\left[\frac{\sqrt{c}}{2}(x - ct)\right]$$

The soliton propagates to the west at a fixed phase speed of linear wave c, without change in shape. After a coordinate transfer $s = x - ct$. The equations become

$$u_0(s, y, t) = \zeta(s, t) \frac{(-9 + 6y^2)}{4} e^{-0.5y^2} \tag{32}$$

$$v_0(s, y, t) = 2y \frac{\partial \zeta(s, t)}{\partial s} e^{-0.5y^2} \tag{33}$$

$$h_0(s, y, t) = \zeta(s, t) \frac{(3 + 6y^2)}{4} e^{-0.5y^2} \tag{34}$$

and the variable $\zeta(s, t)$ satisfies

$$\frac{\partial \zeta}{\partial t} - 1.5366 \zeta \frac{\partial \zeta}{\partial s} - 0.098765 \frac{\partial^3 \zeta}{\partial s^3} = 0 \tag{35}$$

Which is the Korteweg–de Vries (KDV) equation with the exact solution

$$\zeta(s, t) = 0.772 B^2 \text{sech}^2 (B(s + 0.395 B^2 t)) \tag{36}$$

Equations (32)–(36) were also referred as the zeroth-order solution (Boyd 1980, 1985) [50,51]. The initial condition is given by (32)–(36) with $t = 0$.

The model equations including the Coriolis parameter f and the terrain height h_s are shown in (25)–(27) without friction ($k = 0$). Using Sun (2011) [24] scheme, Sun and Sun (2013) [48] simulated solitary waves which are comparable with solutions of Boyd (1980, 1985) [50,51]. The dispersive waves from the linearized equations are also comparable with Boyd's results as shown in Figures 6 and 7. The amplitude difference in soliton may come from the initial condition not being resolved by the grids. The phase lag is due to model phase speed is slightly slower than analytic solution, the lag becomes larger for the short waves in Figure 7.

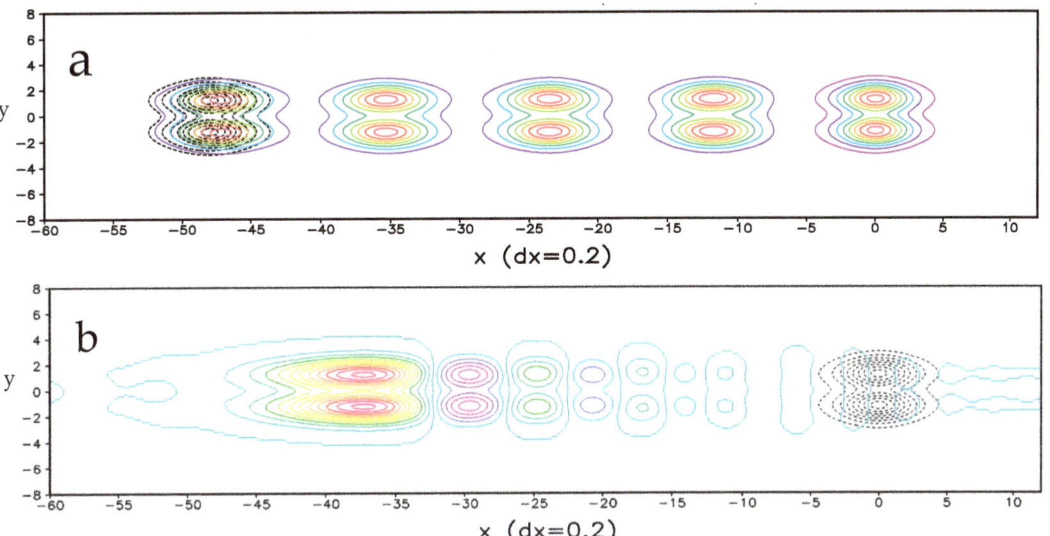

Figure 6. Height of Rossby wave propagated westward (**a**): analytic solution at t = 120 (dashed line) and simulated height at t = 0 (at x = 0) and t = 30, 60, 90, 120 (solid lines), the contours from 1.0 to 1.14 with interval 0.02; (**b**) simulated height (solid line) at t = 120 from linearized wave, contours are from 0.95 to 1.11 with interval of 0.01, and initial condition at t = 0 (dashed line) (Sun and Sun 2013) [48].

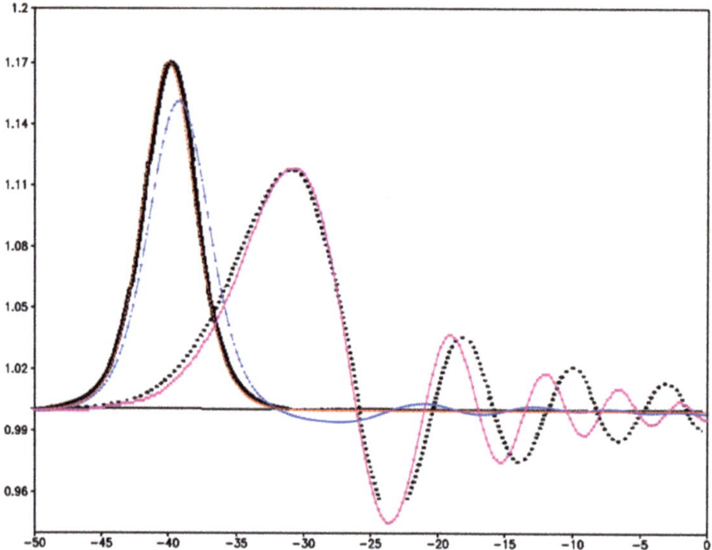

Figure 7. Simulated solution of linearized equations (purple solid line), soliton of nonlinear equations (blue dashed line), and Boyd's solutions from nonlinear equations (bold solid line) and linearized equations (black dotted line) at t = 100. (Sun and Sun 2013) [48].

The results also confirm that solitary waves of large amplitude containing a recirculating fluid trapped within the moving disturbance (Figure 8). However, the simulated recirculation for B = 0.6 is more complicated than Boyd's, which may indicate the pattern may not be well represented by Boyd's asymptotic series. The simulations also show that terrain can slow down the speed of the soliton but do not significantly change the general pattern of the waves farther downstream (Sun and Sun 2013) [48]. Solitary wave solutions are also related to the non-dissipative equations of the Lorenz 1963 model [9]:

$$dX/dt = \sigma(-X+Y), \quad dY/dt = -XZ + rX - Y, \quad \text{and} \quad dZ/dt = XY - bZ$$

with $\sigma = 10$, $r = 28$, and $b = 8/3$.

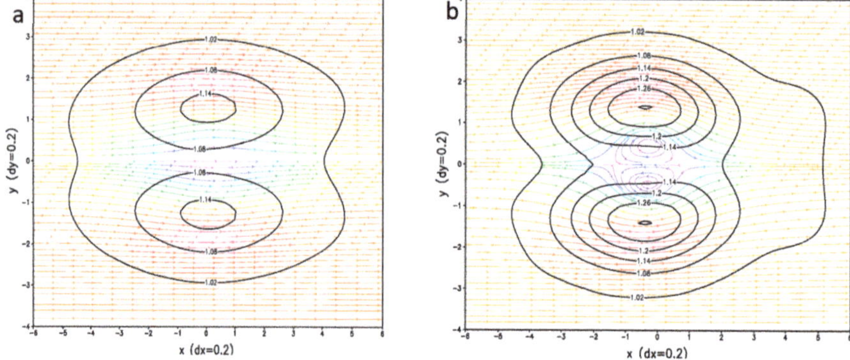

Figure 8. Simulated heights and streamlines in coordinate moving with wave from sum of zeroth-and first-order solutions as initial conditions with $\Delta x = \Delta y = 0.2$ and smooth coefficient = 0.1: (**a**) at t = 121.5 for B = 0.395; and (**b**) at t = 100.9389 for B = 0.6. (Sun and Sun 2013) [48].

Combining those equations into a single equation of Z, Shen (2020) [19] obtained:
$\frac{d^2 Z}{d\varsigma^2} + 3Z^2 - 4rZ = 0$, which is identical to the integral of Korteweg–de Vries equation
(i.e., $\frac{d^2 f}{d\varsigma^2} + 3f^2 - cf = 0$). It will be interested to see the modification in Figure 1 of Shen
(2020) [19] if he keeps more terms in his model.

2.4.3. Hydraulic Jumps

The inviscid, one-dimensional SWE without f is

$$\frac{\partial u}{\partial t} + \frac{\partial}{\partial x}\left[\frac{u^2}{2} + g(h + h_s)\right] = 0 \tag{37}$$

And $Be = \left(\frac{u^2}{2}\right) + g(h_s + h)$ is the Bernoulli function, and $h_f = h_s + h$. The jump is
defined where h and u are discontinuous. The flux from becomes

$$\frac{\partial hu}{\partial t} + \frac{\partial}{\partial x}\left[hu^2 + \frac{gh^2}{2}\right] + gh\frac{\partial h_s}{\partial x} = 0 \tag{38}$$

Let us define the potential of mass flux $J = J_1 + J_2$ where

$$J_1 = hu^2 + \frac{gh^2}{2} \text{ and } J_2 = gh\frac{\partial h_s}{\partial x} \tag{39}$$

The equation of total energy density per unit length E becomes:

$$\frac{\partial}{\partial t}[E] = \frac{\partial}{\partial t}\left[h\left(\frac{u^2}{2}\right) + gh\left(h_s + \frac{h}{2}\right)\right] = -\frac{\partial}{\partial x}\left[uh\left(\frac{u^2}{2}\right) + uhg(h_s + h)\right]$$
$$= -\frac{\partial}{\partial x}[uh \cdot Be] \tag{40}$$

It is assumed that mass flux remains constant; $\partial h/\partial x = 0$ at the inflow; and open
boundary condition at the outflow, they are more realistic than the periodic boundary
conditions used by Houghton and Kasahara (1968) [53] (will be referred as HK) and others.

Sun (2018) [54] has presented six different cases, only Case B in (Sun 2018) [54] will be
discussed here. The terrain is:

$$h_s = h_c\left[1 - \left(\frac{x - x_c}{x_L}\right)^2\right] \text{ for } 0 \leq |x - x_c| \leq x_L \text{ and } h_s = 0 \text{ otherwise.}$$

where $h_0 = 20$ cm, $h_c = 10$ cm, $\Delta x = 1.0$ cm, $x_L = 40\,\Delta x$, $x_c = 600$ or 400 cm (600 or 400 grid),
$g = 980$ cm s^{-2}, and the domain $L = 1200\,\Delta x$, and $u_0 = 42$ cm s^{-1}. Combining the conservation of mass flux and the steady Bernoulli function, HK obtained:

$$\frac{F_o^2}{2}U^3 + U(M - \frac{F_o^2}{2} - 1) + 1 = 0 \tag{41}$$

where $U = \frac{u}{u_0}$, $M = h_s/h_0$, and the Froude number $F_0 = \frac{u_0}{c_0} = \frac{u_0}{\sqrt{gh_0}}$, where u_0 and h_0 are
velocity and depth at inflow. There exist three real roots of (41) when

$$M \leq M_* \equiv \frac{F_o^2}{2} - \frac{3}{2}F_o^{2/3} + 1 \tag{42}$$

Figure 9a,b show the time sequence of the simulated h_f and u at t = 0 (black), 2.4
(red), 4.8 (green), 7.2 (dark blue), and 9.6 (purple). At beginning, the shock waves
propagate both directions. The simulated $\Phi c = h_c/h_0$ and $F_c = u_c/c_0$ at crest at =4.8 s
agree with asymptotic solution of HK. There are no real solutions of Equation (41),
because $M_* = \frac{F_o^2}{2} - \frac{3}{2}F_o^{2/3} + 1 = 0.3728$.

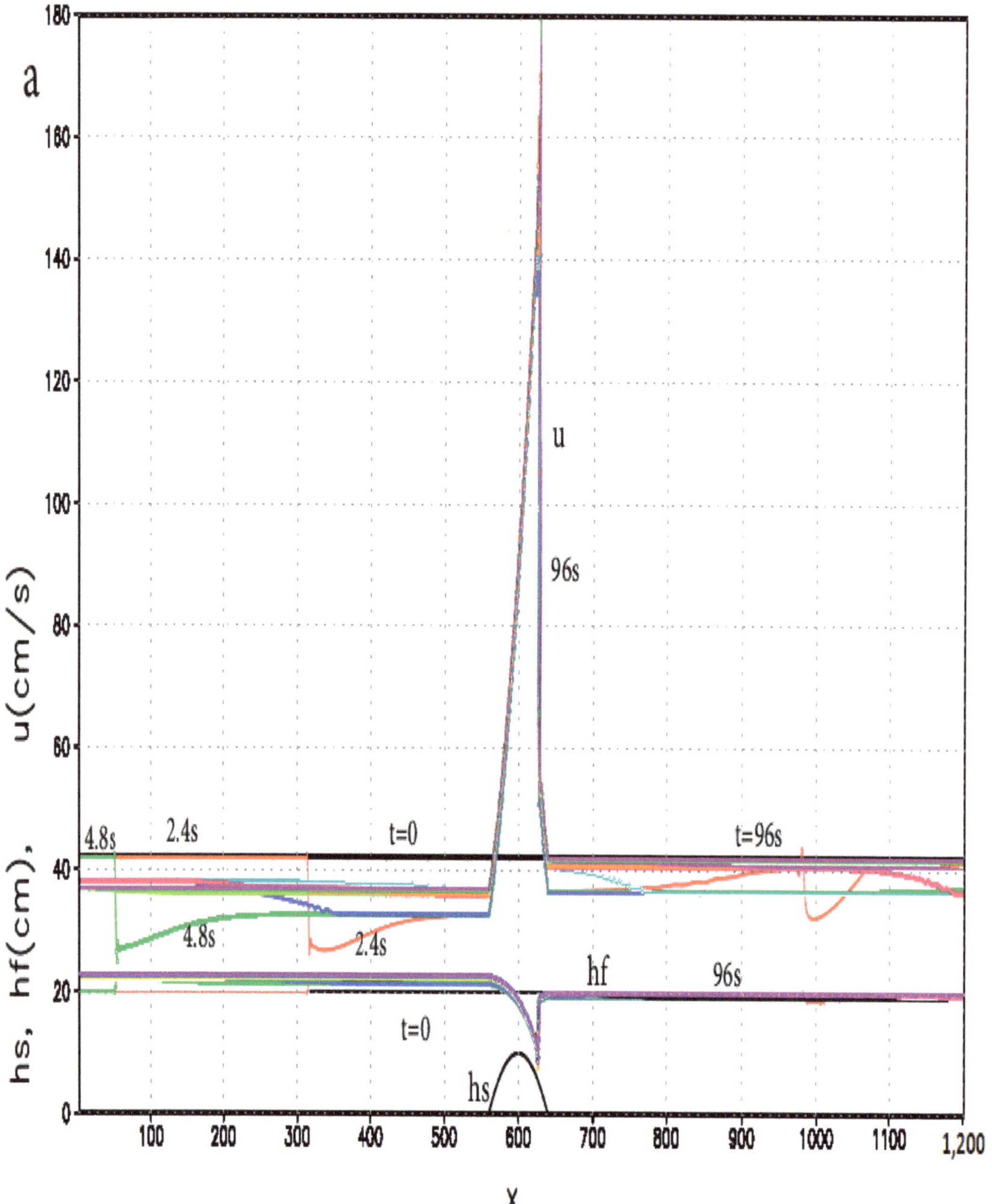

Figure 9. *Cont.*

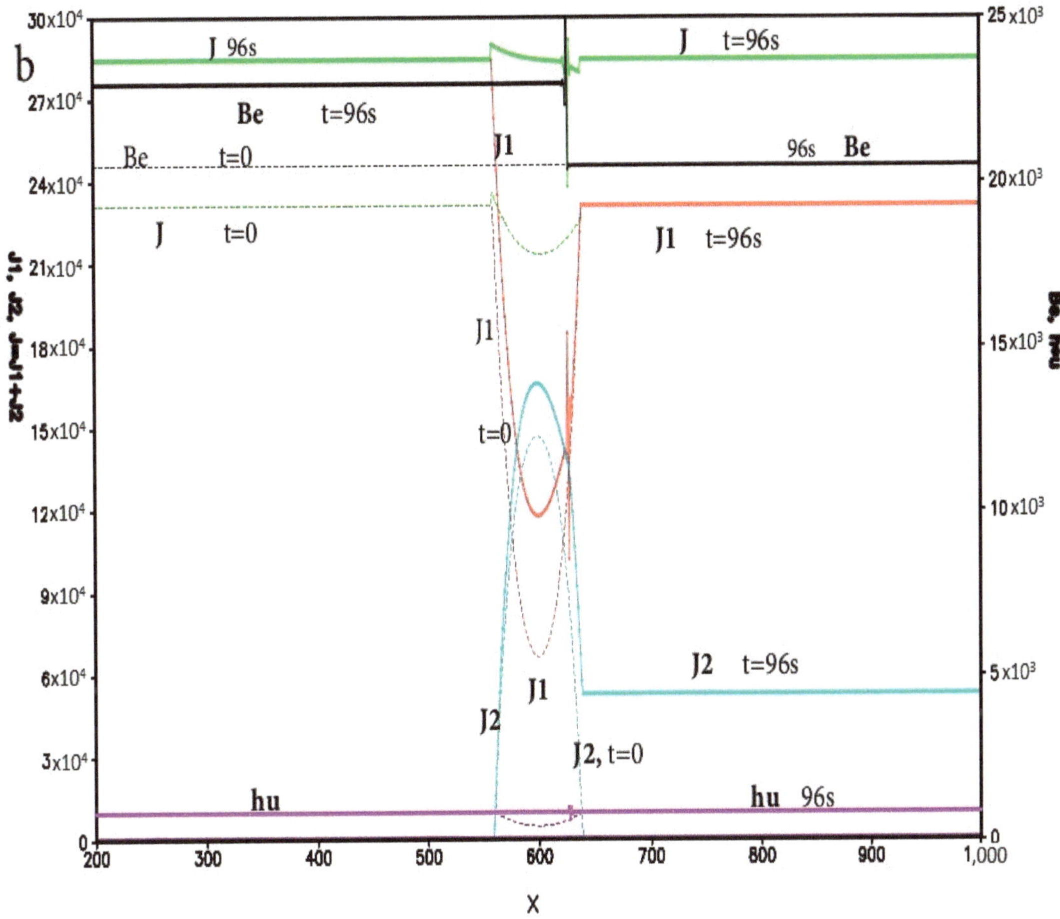

Figure 9. (**a**): Time sequences of water surface height h_f and velocity u (velocity at inflow $u_0 = 42$ cm s^{-1} and Froude number at inflow $F_0 = 0.3$), red line at t = 2.4 s, green at t = 4.8 s, dark blue at t = 7.2 s, light blue at t = 9.6 s, purple at t = 96 s; (**b**): Left: mass flux without terrain J_1 (red), mass flux with terrain J_2 (blue), and J (green) = $J_1 + J_2$; Right: h × u (purple), Bernoulli function Be (black); dashed lines at t = 0, solid lines at t = 96 s. (Sun 2018) [54].

Therefore, the flow with singularity originally has changed to a critical flow by decreasing velocity and increasing flow height at the inflow. At the crest, the simulated velocity U_c^s is identical to U_{*}^s, which has a physical solution (i.e., a real number instead of a complex number) of Equation (41). Equation (37) is also held at the crest, but not at the jump at x = 627 with u =175.7 cm s^{-1}, where the velocity calculated from Equation (41) is only 157.58 cm s^{-1}. The Bernoulli function (black line) drops from 22,948 to 20,485 cm^2 s^{-2} across the jump at t = 96 s, but J remains almost the same (284,582 vs. 284,581 cm^3 s^{-2}), except small kinks at the jump and obstacle foots, because it is derived from an inviscid Equation (38) of mass flux in Figure 9.

The shockwave reflected from the inflow boundary changes the flow property from the region II of HK initially to the critical flow at the final steady state, which occurs to all cases having hydraulic jumps. It may imply that the critical flow is a stable state according to Poincare or an attractor to Lorenz.

2.4.4. Vortex Moving over Central Mountains Range in Taiwan (Sun 2016) [55]

Previously, the surface friction was believed secondary and not included in Lin et al. (2005) [56], Huang and Lin (2008) [57]. The surface friction in momentum equations in (26) and (27) is assumed proportional to terrain elevation:

$$k = \alpha \sqrt{h_s} \tag{43}$$

The potential vorticity in the SWE becomes

$$\frac{d}{dt} \ln\left(\frac{\zeta + f}{h}\right) = \frac{1}{\zeta + f}\left(\frac{\partial(ku)}{\partial y} - \frac{\partial(kv)}{\partial x}\right) \tag{44}$$

Equation (44) is used to simulate the flow in the troposphere (with a depth ~10 km), which is topped by an undisturbed stratosphere (Matsuno 1966) [58]. In water, we obtain: the reduced gravity $g' = 4$ m s^{-2}; depth H = 100 m; the horizontal length scale R = 0.2 R*, where R* is the radius of the vortex in the real atmosphere; the Coriolis parameter f = f_{23} ($\phi = 23°$ N) = 5.7 × 10^{-5} s^{-1}; and the peak of mountain in the model $h_m = 10$ m at the grid (100, 85) corresponds to the mountain peak of $h_m^* = 3658$ m with a 4 km horizontal resolution. Table 2a,b show the radius of vortex (R); fluid depth (H), the height of mountain peak (h_m), Rossby number (Ro), Froude number of the mountain ($Fr = \frac{U}{\sqrt{g'h_m}}$) (Sun and Chern 1994 [59], Sun and Sun 2015 [60]), Froude number far away from topography ($Fr_\infty = \frac{U}{\sqrt{g'H}}$), and L_R/R, where L_R the radius of deformation. They are comparable to the corresponding parameters in the atmosphere (Sun 2016) [55] indicated by superscript *. The initial surface height depression of the vortex is

$$h' = -h'_o / \left(1 + \left(\frac{x - x_o}{R}\right)^2 + \left(\frac{y - y_o}{R}\right)^2\right) = -h'_o / \left(1 + \left(\frac{x - i_o\Delta x}{R}\right)^2 + \left(\frac{y - j_o\Delta y}{R}\right)^2\right) \tag{45}$$

and geostrophic wind relation is used to represent the wind around the vortex initially.

Table 2. Parameters in atmosphere and in shallow-water model with (a) R = 24 km and U = 0.8 ms^{-1}, and (b) R = 40 km and U = 1.2 m s^{-1}, (b) Same as Table 2a except R = 40 km and U = 1.2 m s^{-1}. (Sun 2016) [55].

(a)	R (km)	H (m)	$(10^{-5}$ f s$^{-1})$	h_m (m)	vel (m/s)	N or. g'	Ro	L_R	Fr_m	Fr_α
Atm	R* = 120	H* = 10^4	5.783	$h_m^* = 3500$	U* = 4	N = 10^{-2} s^{-1}	U*/f*R* = 0.58	NH*/(f*R*) = 10^2/(f*R*) = 14.41	U*/(Nh$_m$) = U/(35 ms^{-1}) = 0.11	U*/(NH*) = 0.04
Water	R = 24	H = 100	5.783	$h_m = 10$	U = 0.8	g' = 4 m s^{-2}	U/fR = 0.58	$\sqrt{g'H}$/(fR) = 20/(fR) = 14.41	U/$\sqrt{g'h_m}$ = 0.2 U*/6.3 = U/(32 ms^{-1}) = 0.12	U/$\sqrt{g'H}$ = 0.04

(b)	R (km)	H (m)	$(10^{-5}$ f s$^{-1})$	h_m (m)	vel (m/s)	N or. g'	Ro	L_R	Fr_m	Fr_α
Atm	R* = 200	10^4	5.783	$h_m^* = 3500$	U* = 6	N = 10^{-2} s^{-1}	U*/f*R* = 0.52	NH*/(f*R*) = 8.65	U*/(Nh$_m$) = 0.17	U*/(NH*) = 0.06
Water	R = 40	H* = 100	5.783	$h_m = 10$	U = 1.2	g' = 4 m s^{-2}	U/fR = 0.53	$\sqrt{g'H}$/(fR) = 8.65	U/$\sqrt{g'h_m}$ = 0.19	U/$\sqrt{g'H}$ = 0.06

2.4.5. Inviscid Vortex with $\alpha = 0$ and U = −1.2 m s^{-1} (U* = −6 m s^{-1})

Figure 10a shows the trajectories of vorticities A_1, A_2, and A_3, initially located at grids LN(280, 170), LC(280, 150), and LS(280,130), respectively. The amplitudes of initial surface depression (h'_o) are 2, 3, and 4 m; the initial radii R are 16, 24, and 40 km (R* = 80, 120, and 200 km). The movement of vortices is independent of size or intensity. The streamlines, and wind speed (in red) in Figure 10b show the maximum wind of 27 m s^{-1} at r = 60 km in the atmosphere for A2 when the vortex is located over the Central Mountain Range (CMR) at 96,000 s (26.7 h). Figure 10c–e show

$$\frac{D\Pi}{Dt} \equiv \frac{D(\text{PV})}{Dt} = \frac{d}{dt}\left(\frac{\zeta+f}{h}\right) = \frac{\partial}{\partial t}\left(\frac{\zeta+f}{h}\right) + \mathbf{v}\cdot\nabla\left(\frac{\zeta+f}{h}\right) = 0.$$

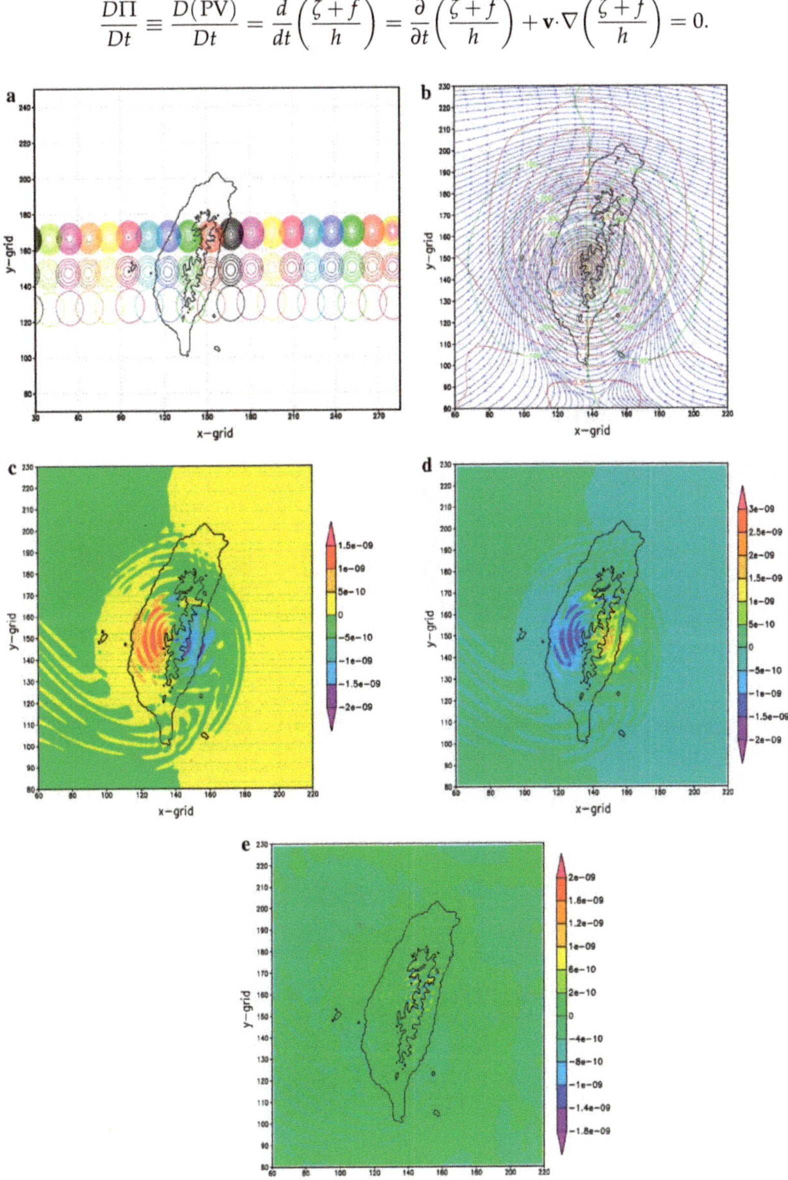

Figure 10. (**a**) Trajectories for A1 (contours from 1.2×10^{-5} to 2.2×10^{-5} m^{-1}s^{-1}), A2 (contours from 1.2×10^{-5} to 1.5×10^{-5} m^{-1}s^{-1}), and A3 (contour is 6×10^{-6} m^{-1} s^{-1}) with $\alpha = 0$, black line is hs = 5 m; (**b**) streamline, wind V (red) and y-component mass flux, hv (green); PV budget for: (**c**) local rate of change, (**d**) advection, and (**e**) total derivative o for A2 at t = 9.6×10^4 s (Sun, 2016) [55].

It confirms that the PV is conserved when an inviscid vortex passes over the mountains.

2.4.6. Trajectories of PV with Surface Friction

The trajectories of the vortex with $h'_o = 3$ m and $R = 24$ km ($R^* = 120$ km in the real atmosphere) are shown in Figure 11; in which the solid lines represent the trajectories with $U = -1.2$ m s^{-1} ($U^* = -6$ m s^{-1}); dash-dot-lines with $U = -0.8$ m s^{-1} ($U^* = -4$ m s^{-1}); the line with * for $U = -1.0$ m s^{-1} ($U^* = -5$ m s^{-1}); the line with Δ for $U = -1.6$ m s^{-1} ($U^* = -8$ m s^{-1}), except the black solid line for F4 (with $U = -1.6$ m s^{-1} ($U^* = -8$ m s^{-1}), $\alpha = 1.5 \times 1.0^{-4}$ m$^{-1/2}$s^{-1}, starting at LN); the lines with o including the beta ($\beta = df/dy$) effect for B4 and B5; the yellow line with black dots for $\alpha = 0$ (no friction) for A2 and B4; magenta for $\alpha = 1 \times 10^{-5}$ m$^{-1/2}$ s^{-1}; cyan for $\alpha = 3 \times 10^{-5}$ m$^{-1/2}$s^{-1}; green for $\alpha = 5 \times 10^{-5}$ m$^{-1/2}$s^{-1}; blue for $\alpha = 8 \times 10^{-5}$ m$^{-1/2}$s^{-1}; red for $\alpha = 1 \times 10^{-4}$ m$^{-1/2}$s^{-1}; and black for $\alpha = 1.5 \times 1.0^{-4}$ m$^{-1/2}$s^{-1}.

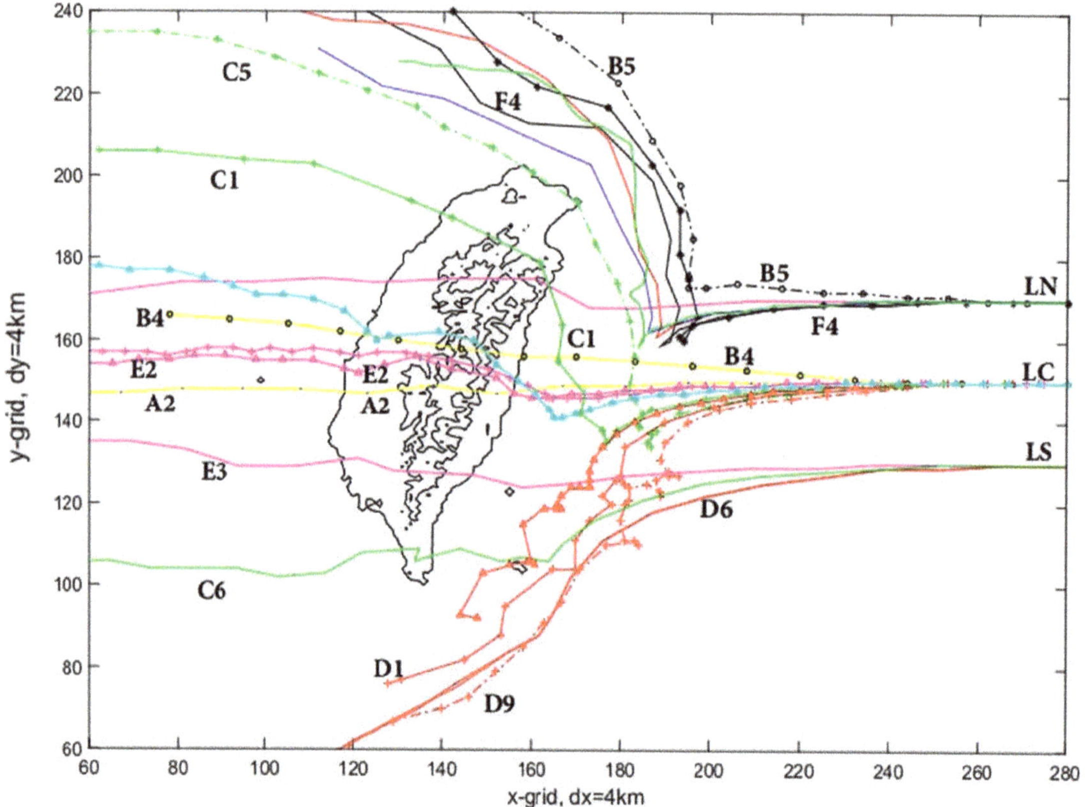

Figure 11. Trajectories of PV with initial surface depression $h'_o = 3$ m, initial radius $R = 24$ km (Sun, 2016) [55].

Since there is no surface friction over the ocean, at beginning vortices move westward. The vortices with beta effect ($\beta = df/dy$), deflect slightly northward initially, because a higher value of f is carried southward by cyclonic circulation ahead of the vortex.

With moderate and large surface friction, when a vortex approaching Taiwan, surface friction creates a strong northerly channel flow westside of the vortex. A strong southerly flow also forms eastside of the vortex because the slowdown vortex is pushed by the easterly flow from behind. Meanwhile, the vortex deforms. The asymmetric vortex and the accompanying two high-wind zones rotate cyclonically while the vortex moves westward. At first, the northerly channel flow is stronger than the southerly flow on the east of the vortex, causing the vortex to deflect southward. The zone of high PV and strong wind continues rotating around the vortex center. The movement of the center depends on the summation of the velocity vector on the vortex. It can continue moving westward or deflect southward or northward. It can also retreat and form a loop. The mechanism is different from the tradition theory of channel flow or the inner core asymmetric flow coming from mid-troposphere (Wu et al., 2015) [61]. If a large vortex moves around the northern coast, it can induce a strong southwesterly flow to create flooding on the Western Taiwan. On the other hand, the secondary circulation is weak when the vortex directly passed over the CMR. Although SWE may explain the effects of mountains and surface friction, the dynamics of a real typhoon are far more complicated than the SWE.

2.4.7. Vortices Merge (Sun and Oh 2022) [62]

The domain consists of 450×450 grids, with $\Delta x = \Delta y = 0.025$ m. Following Waugh (1992) [63] and Oh (2007) [42], we set $g = 1$ m s^{-2}, and the initial surface perturbation is

$$h' = h_0 / \left[\left(\frac{x - x_o}{r_o} \right)^2 + \left(\frac{y - y_o}{r_o} \right)^2 + 1 \right]^{1.5}$$

where $y_o = 0$, $x_o = \pm 0.5 \times d_o$ ($\pm 40\ \Delta x$), $r_o. = 1.2$ m ($48\ \Delta x$), and geostrophic wind at $t = 0$.

Case A ($H = 1$ m, $f = 1$ ms^{-2}, $d_o = 2$ m and $h_o = -0.2$ m at $t = 0$)

If we use the length scale $L = d_o/2$ and the maximum velocity U, we obtain the Rossby number, $Ro = U/fL = 0.16$, and radius of deformation $Ra = \sqrt{gH}/f = 1$ m at $t = 0$. Figure 12a shows the initial PV (thick black), surface heigh h (thin white), streamlines (thin black), and speed (shaded color). Figure 12b reveals that the strong wind zones (red zone) advect downstream at $t = 20$ s. The PV rotates slower at point B, but moves faster at point A; hence, development of the filaments is due to differential velocity across the PV contours. The streamlines show that the vorticity of filaments does not advect or ejected from the vortex cores as proposed by Kimura and Herring (2001) [64], Cerretelli and Williamson (2003) [65], and Melander et al., 1987 [66], etc. Our simulations show that the filaments are shattered with little effect to vortex merging in the inner core.

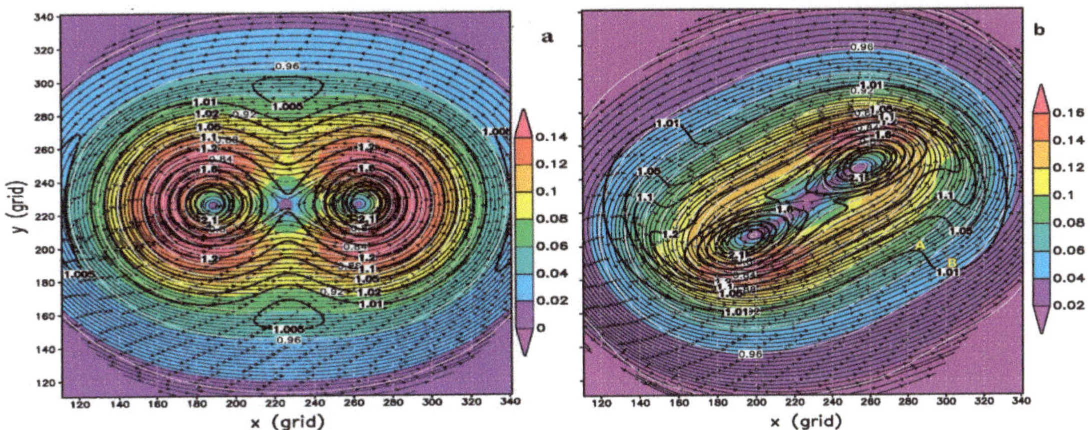

Figure 12. *Cont.*

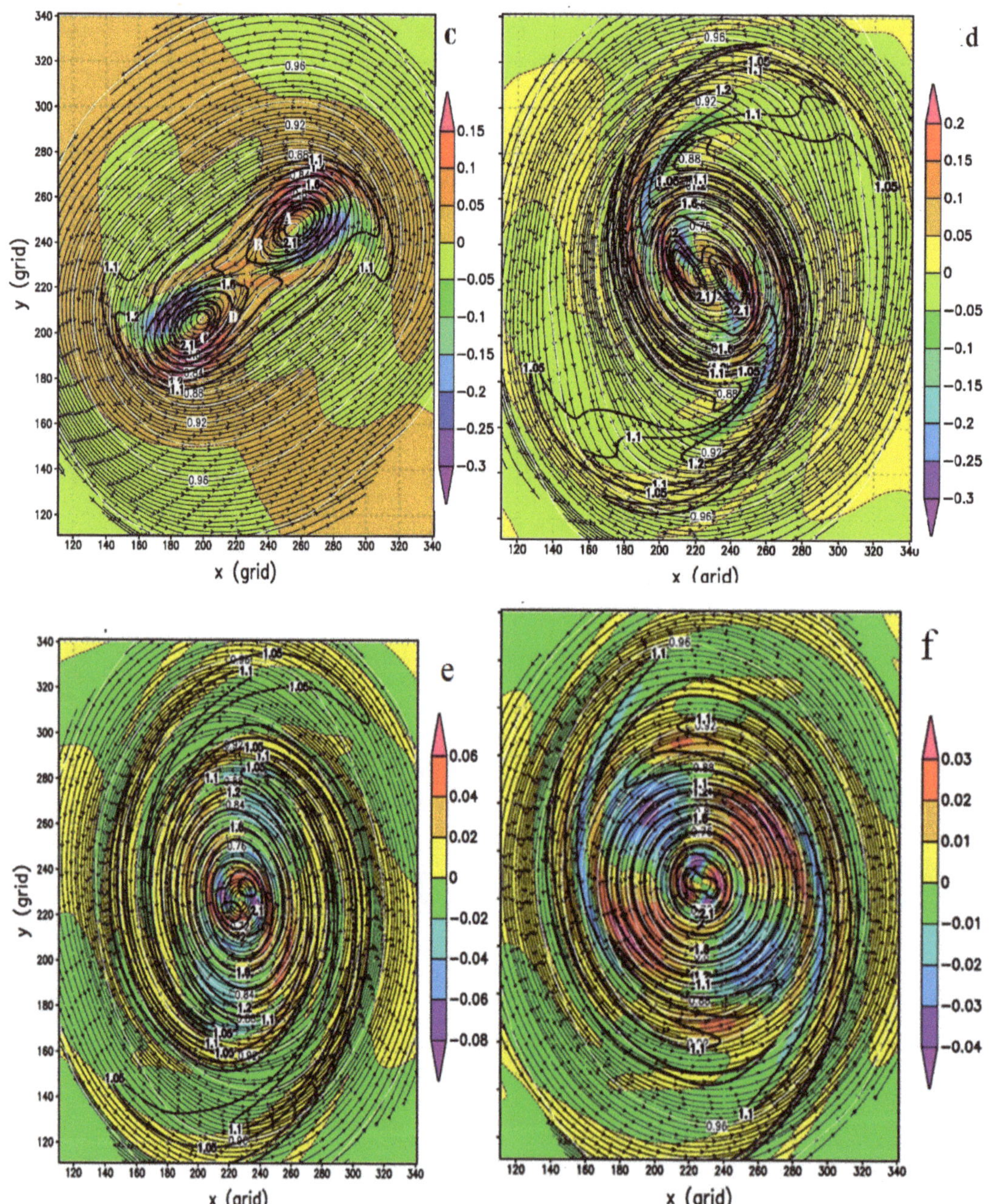

Figure 12. (**a**) PV (thick black), h (white), streamline (thin back), and speed (shaded color) at t = 0; (**b**) Same as (**a**) but at t = 20 s (Sun and Oh 2022) [62]. (**c**) $\Pi \equiv$ PV (thick black), h (white), streamline (thin back), and $\partial\Pi/\partial t$ (shaded color) at t = 20 s; (**d**) Same as (**c**) but at t = 60 s; (**e**) Same as (**c**) but at t = 100 s, (**f**) Same as (**c**) but at t = 120 s (Sun and Oh 2022) [62].

The asymmetric structure of quadrants does exist in the simulated PV. Advection generates positive $\partial\prod/\partial t$ on the lee of the PV ridge, where contours become more convex and the angle between PV contours and streamlines increases. Hence, positive PV can be effectively transported along two parallel but oppositive paths, from A to B or from C to D shown in Figure 12c. Hence, the vortices may move closer while rotating around each other. The vortices never merge when negative $\partial\prod/\partial t$ prevails between two vortices. Hence, the advance of positive $\partial\prod/\partial t$ provides a simple yet eloquent mechanism for vortex merging, which is superior to the following mechanisms: the axisymmetrization principle (Melander et al., 1987 [66], Melander et al., 1988 [67]); the angle between vortex major axis and vortex center line (Huang 2005) [68]; interactions between vorticity and rate of strain (Moore and Saffman, 1975 [69]; Brandt and Nomura 2007 [70]); or production of palinstrophy in Kimura and Herring (2001) [64] and Cerretelli and Williamson (2003) [65], etc.

The contours of h and PV at t = 60 s in Figure 12 are in good agreement with the Lagrangian simulations (Oh 2007) [42]. The maximum speed, 0.18 m s^{-1}, occurs along the inner core boundary. The velocity at the vortex centers, 0.09 m s^{-1}, is identical to the angular velocity, ω = v/r = 0.09 m s^{-1} /0.36 m = 0.25 = 14° s^{-1}. They move along the constant h-contour and rotate around the center of system, like the observed cold front moving along the eastern coastal mountain ranges of Taiwan and China with the observed wind U instead of a Kelvin wave of $U + \sqrt{gh}$ (Sun and Chern 2006) [71]. As time increases, filaments stretch longer and wrap around the shrinking inner core at 100 s (Figure 12e). The shape of inner core gradually changes from elliptic to circular. The filaments become long, thin strips around the center (Figure 12f at t = 120 s). The vorticities are still distinguishable at the end of integration, t = 130 s and total kinetic energy, mass, and PV remain as initial values (Sun and Oh 2022) [62].

Two-dimensional, inviscid turbulence has both kinetic energy and vorticity square (enstrophy) in the inertial range satisfy, $E(k) \sim \varepsilon^{2/3}k^{-5/3}$ and $Z(k) \sim \eta^{2/3}k^{-3}$, where ε is the rate of cascade of kinetic energy per unit mass, η is the rate of cascade of mean-square vorticity. The backward energy cascade from higher to lower wavenumbers k. The -3 range gives an upward vorticity flow (Fjortoft 1953 [72], Kraichnan [73]). They are different from the energy cascade from lower wave number to higher wave number in 3D flow (Kolmogorov 1941 [74], Obukhov 1941 [75]). Vortex tubes/filaments can effectively transfer the energy from the resolvable scale to small and/or subgrid scales. Meanwhile, the unstable disturbance can grow from a small size to large one. The weather/climate models still require rigorous turbulence parameterizations to handle the cascade of energy and enstrophy.

3. Linearized Equations

In fluid dynamics, the dispersive relationship of waves is derived from the linearized equations, including acoustic waves, gravity waves, inertia internal gravity waves, Poincare waves, Kelvin waves, and Rossby waves, etc. Stability analysis of Benard–Rayleigh convection, Eady baroclinic wave (Eady 1949 [76], Pedlosky 1979 [33], Gill 1982 [26], etc.), and numerical schemes are also derived from the linearized equations, as shown in Sections 2.1 and 2.3.

3.1. Coexist of Two Different Types of Cloud Bands (Sun 1978) [77]

Figure 13a shows that both longitudinal and transverse cloud bands coexist to the east of disturbed wave trough near the Equator (Malkus and Riehl 1964) [78]. The rows of longitudinal mode composed of medium or small clouds are parallel to the wind shear with a distance between two rows about 25 km. The large clouds line up at right angles to the wind shear (i.e., transverse mode) with a wavelength of 100 km. Sun (1978) [77] applied the normal mode ($\phi = \Phi(z)\exp(i(kx + ly - \sigma t))$) for perturbation in the linearized equations including a wave-CISK type of latent heat to study those modes. His results show the coexistence of the longitudinal mode consisting of the small–mid size clouds in the lower atmosphere (lest or near 5–6 km) where the buoyancy is greater than effect of stratification, and the transverse model with the cloud tops reaching 9.6 km, in which the total cooling from stratification is slightly larger than buoyancy. Hence, the transverse mode chooses the strongest wind shear direction (Figure 13b). Kuo (1963) [79] pointed out that in a stable stratified fluid, the transverse mode is more unstable than the longitudinal mode, but the reverse is true for an unstable stratified flow.

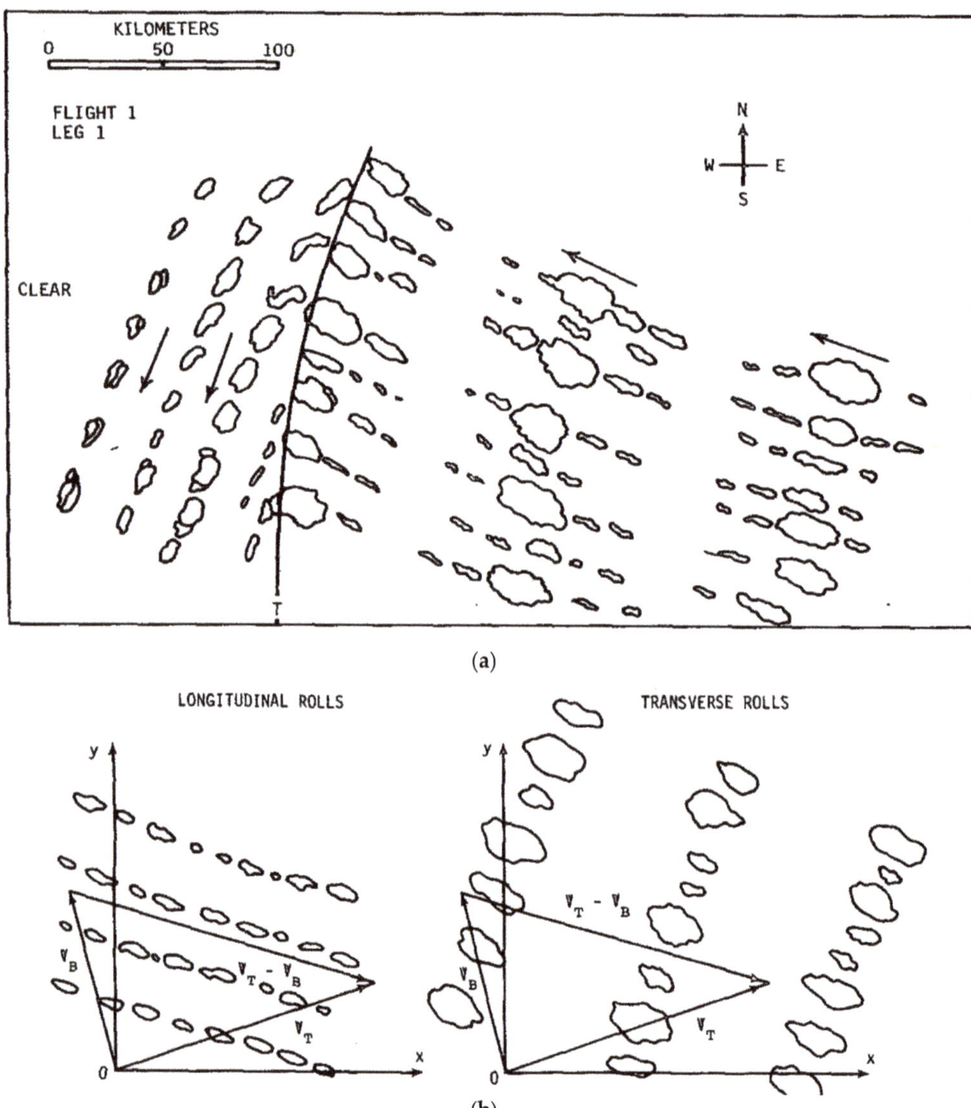

Figure 13. (a): Schematic distribution of cumuliform clouds near a wave trough in the easterlies; the orientation of cloud rows and the spacing between them was measured exactly from the film. The direction of the wind in the lower layer is indicated by an arrow. (Malkus and Riehl, 1964) [78]. **(b)**: Schematic distribution of cumuliform clouds near a wave trough in the easterlies; the orientation of cloud rows and the spacing between them was measured from the film. The wind in the lower layer is indicated by V_B, wind at upper layer is V_T, (Sun 1978) [77].

3.2. Solve Linearized Equation as Initial Value Problems

The normal mode $\phi = \Phi(z)\exp(i(kx + ly - \sigma t))$ or $\phi = \Phi\exp(i(kx + ly + mz - \sigma t))$ cannot be applied to latent heating (positive warming inside cloud and negative cooling outside) or time dependent basic state. Hence, we treated the linearized equations as an initial value problem.

Rainbands and Symmetric Instability (Sun, W. Y., 1984b) [80]

We set the y-axis to the direction of wind shear coinciding with the direction of cloud band when the buoyancy is dominated, as discuss in Section 3.1. Because the variation along y-direction is much smaller than that along x-direction, the linearized equations become:

$$\frac{\partial u'}{\partial t} - fv' + \frac{1}{\rho_0}\frac{\partial p'}{\partial x} = 0, \tag{46}$$

$$\frac{\partial v'}{\partial t} + u'\frac{\partial \tilde{V}}{\partial x} + w'\frac{\partial \tilde{V}}{\partial z} + fu' = 0 \tag{47}$$

$$\frac{\partial w'}{\partial t} + \frac{1}{\rho_0}\frac{\partial p'}{\partial z} + g\frac{\theta'}{\theta_0} = 0 \tag{48}$$

$$\frac{\partial \theta'}{\partial t} + u'\frac{\partial \tilde{\theta}}{\partial x} + w'\frac{\partial \tilde{\theta}}{\partial z} = \frac{\tilde{\theta}Q}{c_p\tilde{T}}\,(\text{heating}) \tag{49}$$

$$\frac{\partial u'}{\partial x} + \frac{\partial w'}{\partial z} = 0 \tag{50}$$

The environment is assumed in thermal wind balance.

$$\frac{g}{\theta_0}\frac{\partial \tilde{\theta}}{\partial x} = f\frac{\partial \tilde{V}}{\partial z} \tag{51}$$

The equations can be combined into a single vorticity equation:

$$\begin{aligned}
\frac{\partial^2 \varsigma}{\partial t^2} &= \frac{\partial^2}{\partial t^2}\left(\frac{\partial w'}{\partial x} - \frac{\partial u'}{\partial z}\right) = \frac{\partial^2}{\partial t^2}\left(\frac{\partial^2 \psi}{\partial z^2} + \frac{\partial^2 \psi}{\partial x^2}\right) \\
&= 2S^2\frac{\partial^2 \psi}{\partial x \partial z} - N^2\frac{\partial^2 \psi}{\partial z^2} - F^2\frac{\partial^2 \psi}{\partial z^2} + \frac{g}{\theta_o}\frac{\tilde{\theta}}{c_p\tilde{T}}\frac{\partial Q}{\partial x}
\end{aligned} \tag{52}$$

where

$$N^2 = \frac{g}{\theta_0}\frac{\partial \tilde{\theta}}{\partial z},\ S^2 = \frac{g}{\theta_0}\frac{\partial \tilde{\theta}}{\partial x} = f\frac{\partial \tilde{V}}{\partial z},\ F^2 = f(f + \frac{\partial \tilde{V}}{\partial x}),\ w = \frac{\partial \psi}{\partial x},\ u = -\frac{\partial \psi}{\partial z} \tag{53}$$

If Q = 0, and assume a normal mode solution:

$$\psi = \Psi\exp(i(kx + mz - \sigma t)) \tag{54}$$

We obtain

$$\sigma^2 = \frac{N^2 k^2 + F^2 m^2 - 2S^2 km}{k^2 + m^2} \tag{55}$$

Equation (55) is identical to inertia internal gravity waves if $\frac{\partial \tilde{V}}{\partial x} = \frac{\partial \tilde{V}}{\partial z} = 0$. Fjortoft (1950) [81], Ooyama (1966) [82], Hoskins (1978) [83] and others found symmetric instability in (55) if

$$Ri = \frac{\frac{g}{\theta_o}\frac{\partial \tilde{\theta}}{\partial z}}{\left(\frac{\partial \tilde{V}}{\partial z}\right)^2} < \frac{f^2}{F^2} = \frac{f^2}{f(f + \frac{\partial \tilde{V}}{\partial x})} \tag{56}$$

Including latent heat by assuming Q > 0 in ascending and Q = 0 in descending motion, Sun (1984b) [80] showed the stream function at t = 2.2743×10^4 s in Figure 14a, and heating rate in Figure 14b. The temperature in the entire cloud layer increases: area inside clouds due to condensation, and cloud-free area by subsidence warming. Cooling occurs right above the cloud layer due to penetrative convection (Sun 1975 [84], Kuo and Sun 1976 [85]). The cooling in the planetary boundary layer (PBL) is not exactly beneath the condensation warming; but slightly toward the cool side because of cold advection coming from the left. It generates a local circulation and induces the rolls to move toward the warm side, as shown in Figure 14c at t = 2.653×10^4 s.

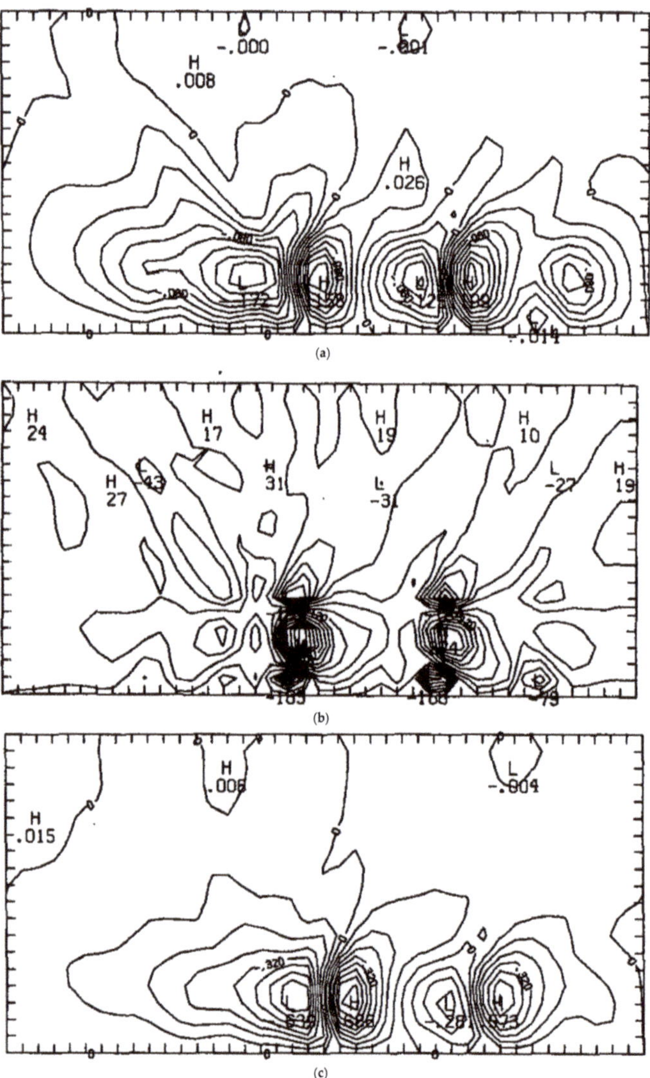

Figure 14. (**a**) Stream function at 2.274×10^4 s for Case D. Contours are from 0.16 to +0.18 with interval 2.0×10^{-2} m^2 s^{-1}; (**b**) Vertical cross section of the heating rate ($\frac{\partial Q}{\partial t}$) at 2.27×10^4 s for Case D. Contours are from -1.800 to 3.9 ($\times 10^{-5}$ K s^{-1}) with interval 3.0×10^{-6} Ks^{-1} and labels scaled by 10^{-7} (**c**). Stream function at 2.653×10^4 s for Case D. Contours are from -0.56 to $+0.64$ with interval of 8.00×10^{-2} m^2 s^{-1}. (Sun 1984b) [80].

Meanwhile, the small bands are suppressed by the subsidence of the larger bands, as observed squall lines in Midwest of the USA on 3 April 1974 (Figure 15). The downward motion also carries the high wind from the upper layer down to PBL, which enhances the low-level convergence and may cause the wide spread of tornados (Agee et al., 1975 [86]). The descending flow carrying warm, dry air may play a significant role in the development of severe storms too.

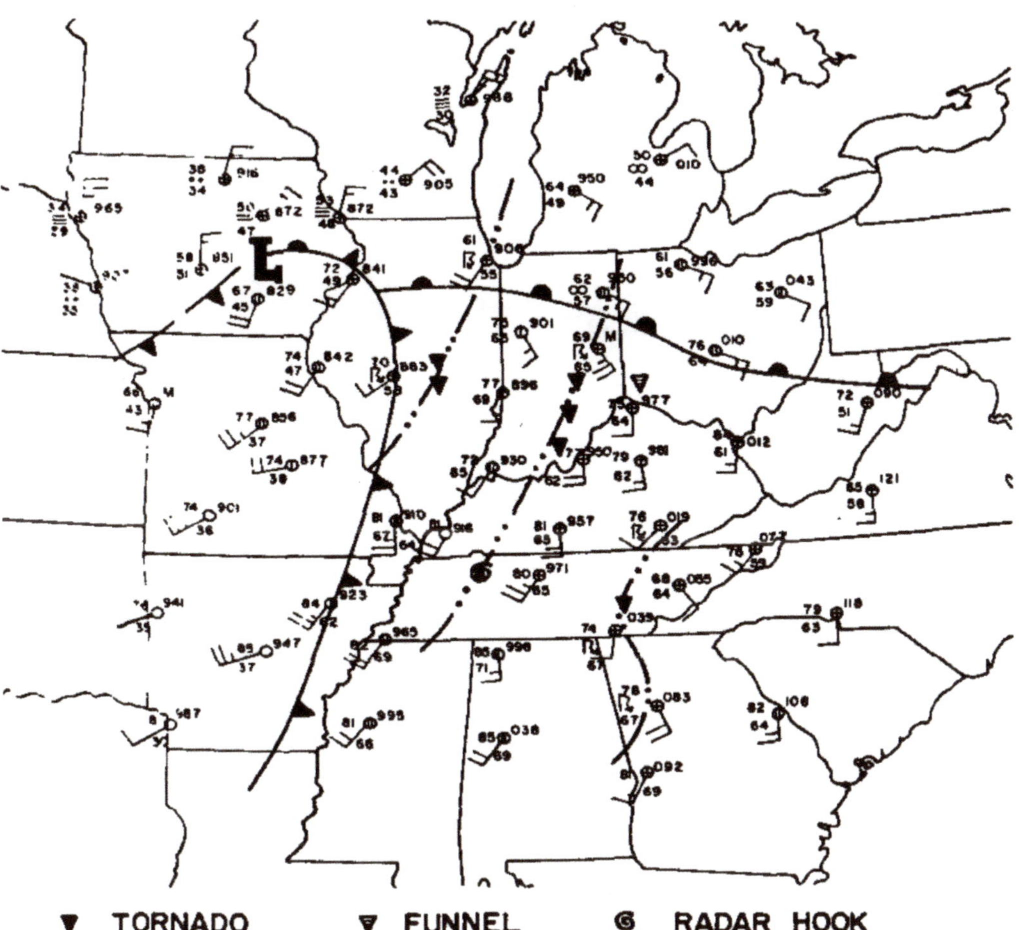

Figure 15. Surface weather depiction at 2000 GMT (1500 EST) 3 April 1974. Three active squall lines were all producing tornadoes in Illinois, Indiana, and Tennessee. (After Agee et al., 1975) [86].

3.3. Cloud Bands over Tropical Continent

Sun and Orlanski (1981a) [87] solved the linearized equations and found the coexisting 1-day and 2-days mesoscale waves (Figure 16a,b) excited by resonance of trapezoidal instability and the land sea breeze. The energy of those inertia internal gravity waves propagates upward (Figure 16b) as simulated by Kuo and Sun (1976) [85]. They spread inland too, but there was no convection over the ocean (Figure 17) because the variation of the sea surface temperature is small. Those mesoscale convections were also confirmed by the nonlinear simulations (Sun and Orlanski, 1981b) [88] and the observed cloudiness in tropical regions (Orlanski and Polinsky 1977) [89]. Including the beta effect, Sun (1995b) [90] showed that the growth rate of instability is unsymmetric with respect to latitude.

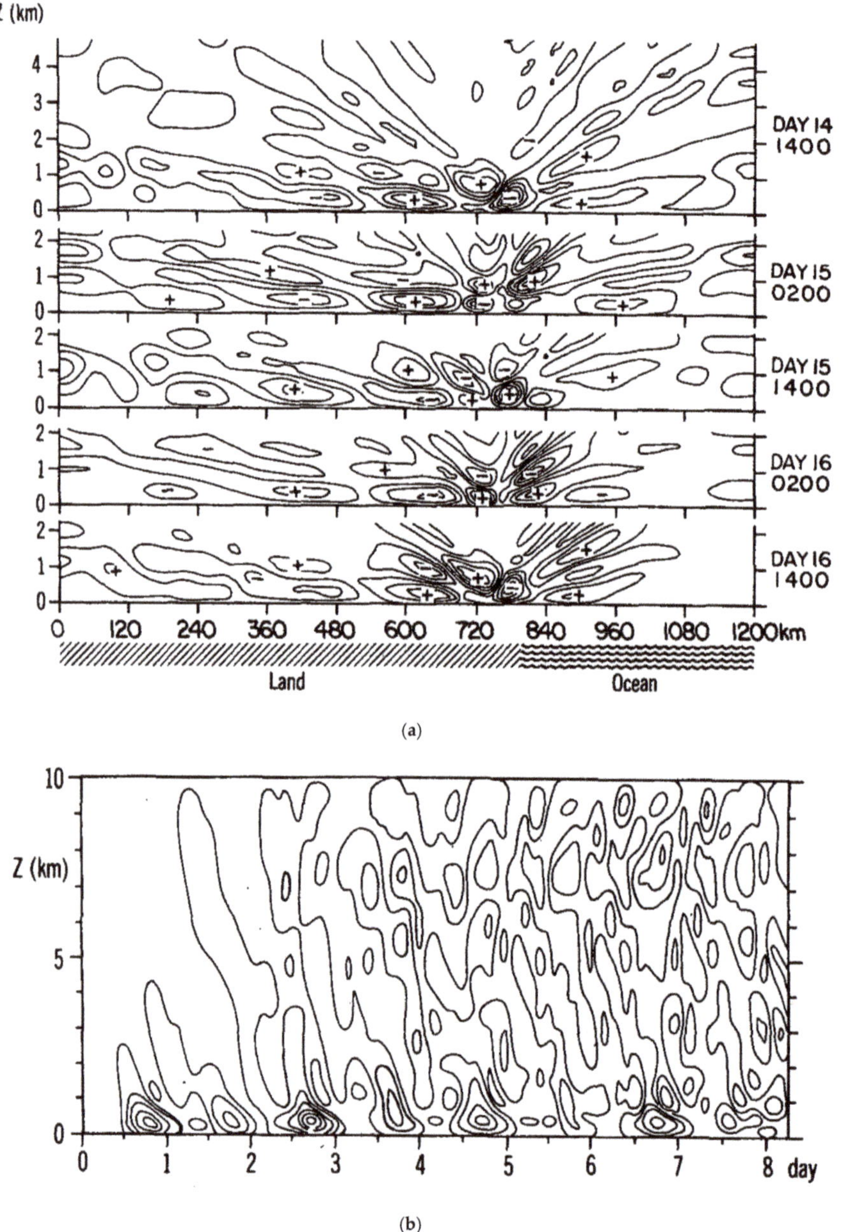

Figure 16. (**a**) A composite picture of the vertical velocity in the lower atmosphere from 1400 LST of day 14 to 1400 LST of day 16 at 12 h intervals for case B. The maximum is 20.1 cm s^{-1} and the minimum -32.4 cm s^{-1} at 1400 LST of day 14; the maximum 26.5 cm s^{-1} and minimum -24.4 cm s^{-1} at 0200 LST of day 15; the maximum 45.6 cm s^{-1} and minimum -21.6 cm s^{-1} at 1400 LST of day 15; the maximum 52.6 cm s^{-1} and minimum -42.4 cm s^{-1} at 0200 LST of day 16; the maximum 41.4 cm s^{-1} and minimum -51.8 cm s^{-1} at 1400 LST of day 16. (**b**) Temporal variation of z velocity at x = 768 km from 00 LST of day 0 to 06 LST of day 8 for case C. The maximum is 3.1 cm s^{-1}, the minimum -7.53 cm s^{-1}, and the interval is 1.06 cm s^{-1} (Sun and Orlanski 1981a) [87].

Figure 17. GOES East satellite picture at 2130 GMT 5 September 1974, showing three cloud bands aligned parallel to the coast of South America. Their horizontal wavelength is of a few hundred kilometers. There is a clear sky off the coast (Sun and Orlanski 1981a) [87].

3.4. Ekman Layers in Atmosphere and Ocean (Sun and Sun 2020) [91]

The equations of 1D Ekman layer are:

$$\frac{\partial u\rho}{\partial t} - fv\rho = -fv_g\rho + \frac{\partial}{\partial z}\left(\mu\frac{\partial u\rho}{\partial z}\right)$$
$$\frac{\partial v\rho}{\partial t} + fu\rho = fu_g\rho + \frac{\partial}{\partial z}\left(\mu\frac{\partial v\rho}{\partial z}\right) \tag{57}$$

They can be combined into a single equation :

$$\frac{\partial W}{\partial t} + ifW = ifW_g + \frac{\partial}{\partial z}\left(\mu\frac{\partial W}{\partial z}\right) \tag{58}$$

where $W = u\rho + iv\rho$, and $W_g = u_g\rho + iv_g\rho$ (u_g and v_g) are geostrophic wind in x and y direction, respectively, and i $=\sqrt{-1}$. It is assumed $\rho = 1000$ kg m^{-3} in the ocean, and $\rho = 1$ kg m^{-3} in the atmosphere. The domain is from -400 m to 4 km with a spatial interval of 2 m in the ocean and 4 m in the atmosphere, respectively. The stress $\rho\mu\frac{\partial W}{\partial z}$ is calculated at the mid-point. The velocity (u, v) is continuous at the interface, z = 0, but the surface stresses can be different across it. Viscosity μ is function of both space and time in both atmosphere and ocean (Figure 18), which consists of 1 day, half-day, and other modes.

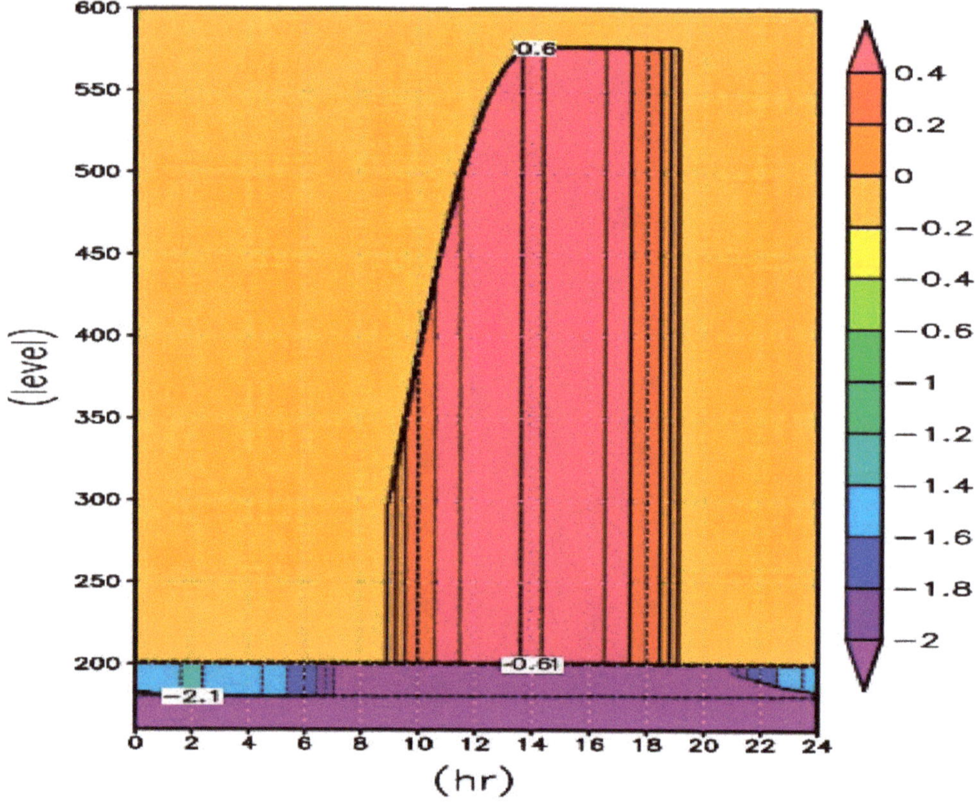

Figure 18. Diurnal variation of $\log_{10}$ (μ) in atmosphere and ocean (Sun and Sun 2020) [91].

3.4.1. Simulation at 30° N with Diurnal Variation in Atmosphere and Ocean (with $\delta z_{0\_max} = 40$ m and $\mu_{0\_max} = 0.04$ m^2 s^{-1}) (Case C of Sun and SUN 2020)

The time sequences at z = 0 (blue, multiplied by 10) and 600 m (red) in Figure 19a show that the parcels rotate clockwise, and the amplitudes grow with time. The blue Δ shows a weak wind at 12L12 (12 p.m. day 12), and blue o reveals a strong wind at 24L12 (midnight), corresponding to a nocturnal low-level jet (LLJ) (Blackadar 1957 [92], Palmer and Newton 1969 [93], etc.) in the atmosphere. Meanwhile a jet develops in the upper oceanic mixed layer during daytime and week current at night (Price et al., 1986 [94]). The growth rate of velocity is about 1.00043/h at z = 600 m in Figure 19b.

Equation (58) is identical to the equation of parametric resonances (Kalashnik 2018 [95]). The velocities at 03L12 (dotted dashed green),12L12 (thick dashed red), 15L12 (dotted dashed magenta), 24L12 (thick dashed black), and hourly velocity oscillations at various heights in the atmosphere (Figure 19c) show that the individual spiral significantly departs from the daily average (thick blue), and the shape is close to the analytic solution (thick green). The velocity grows from a small elliptical shape near surface to larger and more circular circulations at 200 m (blue), and 600 m (red). Large amplitude oscillations also exist at the upper mixed layer and beyond: at 1 km (green); 1.6 km (blue), because of resonance with inertia oscillation of the Coriolis force. Velocities with a large magnitude also develop in the oceanic mixed layer and beyond. Ekman spirals rotate a complete cycle in 24 h (Figure 19d). Figure 19e shows that in the morning, the velocity at surface is parallel to the stress, and the magnitude reaches ~0.1 m s^{-1} as observed at 30° N (Price et al., 1986 [94]). The average velocity (thick blue) at −4 m is ~60° to the right of the averaged wind stress, the averaged surface stress is 0.1 Pa and the angle ψ between the velocity at -2 m and surface stress, is 58°. They agree with 0.08 Pa and $\psi = 60°$, respectively, observed at 30.9° N, 123.5° W (Price et al., 1986 [94]). The simulations also show a compacted spiral at night (Figure 19d,e). The daily averaged spiral of D12 (thick blue in Figure 19e) is between 12L and 24L, but quite different from either of them. It is close to the analytic solution (thick green). The averaged spiral is also comparable with the observations shown in Figure 19f when the mean stress is aligned to the direction of the observed stress. If we re-align the averaged stress as the y-axis, we obtain the hourly stress and mass flux in Figure 19g. The integrated averaged spirals in the atmosphere and ocean are not as flat as suggested by Brown (1970) [96], Price et al. (1986) [94] and others. Polton et al. (2013) [97] believed that observed spiral is not flatten if the geostrophic shear can be removed from the data.

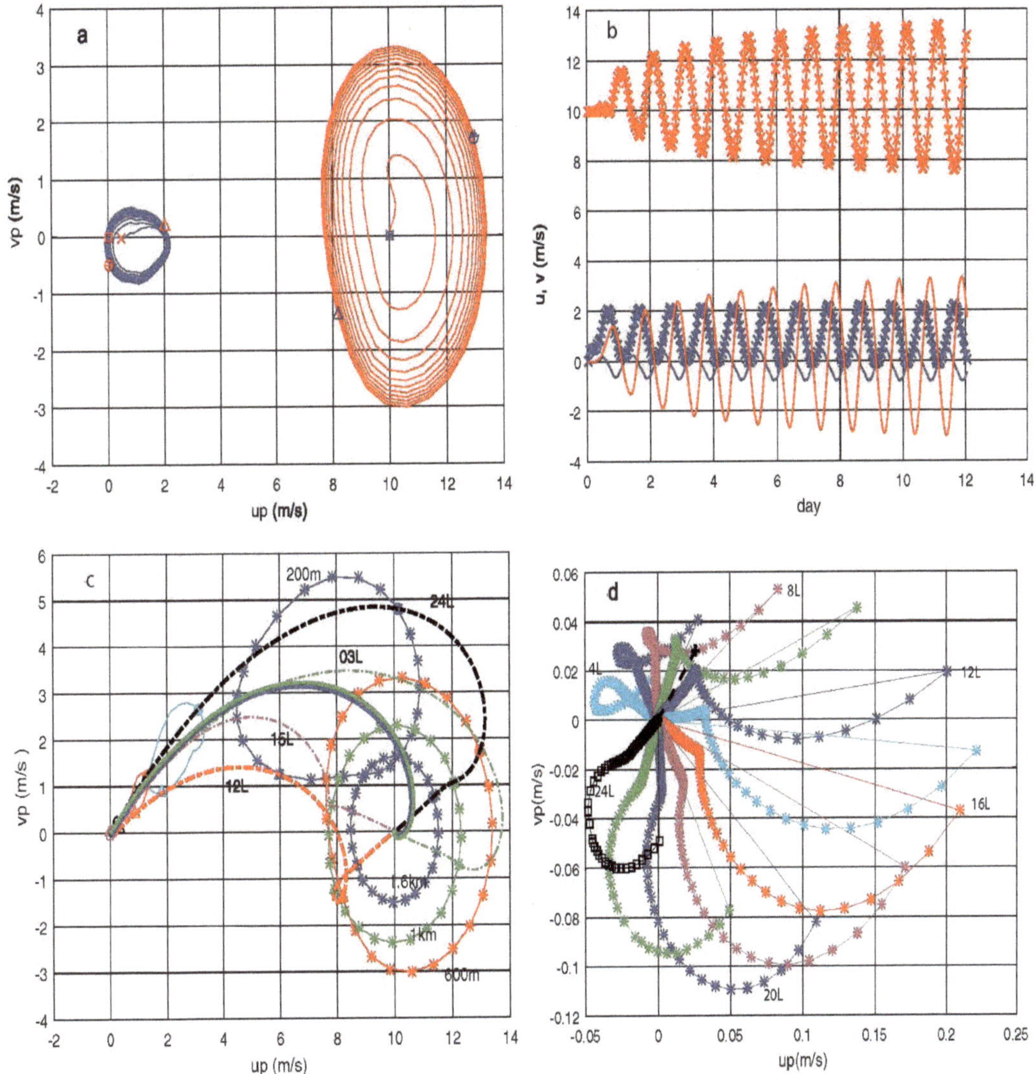

Figure 19. *Cont.*

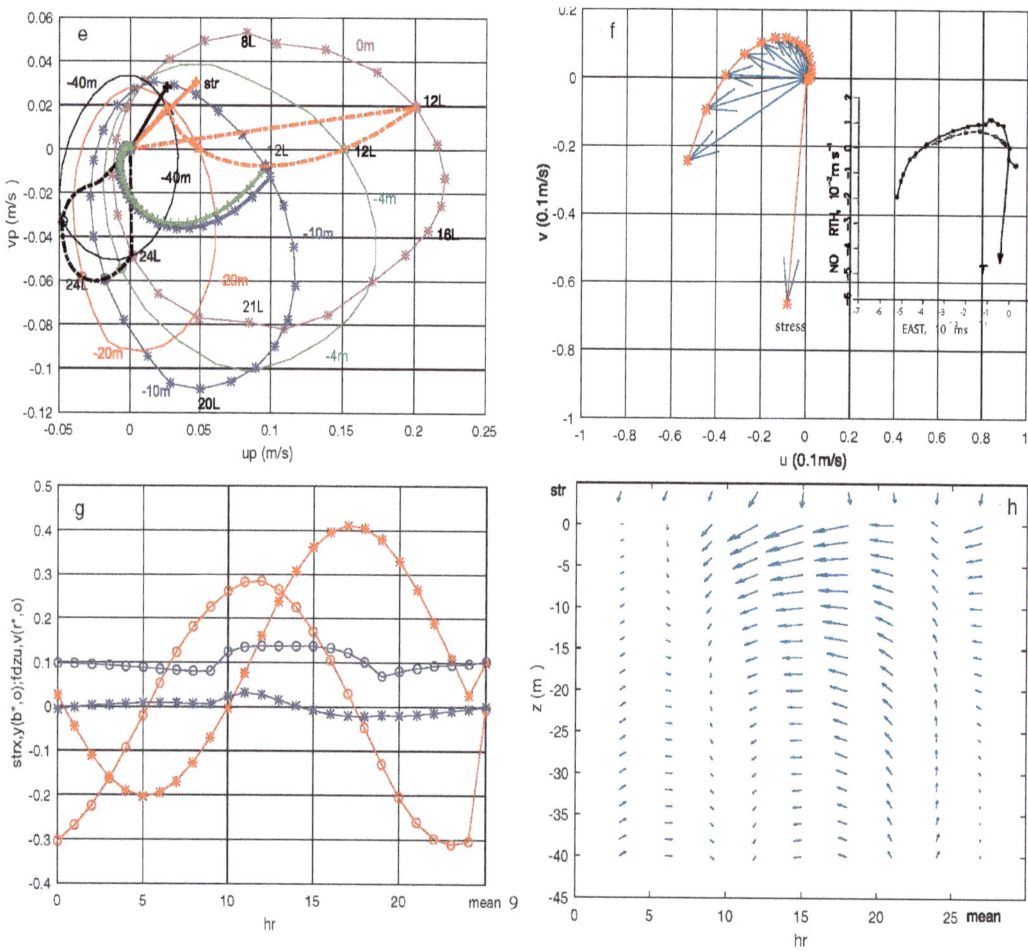

Figure 19. (**a**) Time evolutions of velocity at z = 600 m (red), and z = 0 (bule, multiplied by 100), box indicates at t = 0, × at t = 1 s, o at 12L12, Δ at 24L12. (**b**) Time sequences of u (red line with ×), v (red line) at 600 m, and 10 × u (blue line with ×), 10 × v (blue line) at z = 0 for Case C (30° N). (**c**) Simulated velocities (u, v) in atmosphere on D12: at 03L (thin dotted–dashed green); 12L (thick dashed red); 15 L (thin dotted–dashed magenta); and 24L (dashed black), and hourly velocity at z = 200 m (blue with *), z = 600 m (red with *). (**d**) Oceanic Ekman spiral at every 3 h and 0.4 × stress at 24L12 (dashed black). (**e**) Simulated (u, v) in ocean at 12L (thick dashed red) and 24L (thick dashed black), hourly (u, v) at z = 0 (thin magenta with *), −4 m (thin green), −10 m (thin blue with *), −20 m (thin red), −40 m (thin black); analytic velocity (thick green), daily average (thick blue); and 0.4 × stress at 12L (thick red), at 24L (thick black) on Day 12. (**f**) Simulated daily mean Ekman spiral (blue, red), and observation from Price et al. (1986) at 30° N. Hodograph starting from 2 m down to 70 m with 4 m interval in solid black curve of observed and red line of simulation. (**g**) Hourly surface stress ($\tilde{\tau}_x$ (blue *), $\tilde{\tau}_y$ (blue o) and mass flux $flux(\tilde{u})$ (red *), $flux(\tilde{v})$ (red o) and their daily averages (at 25 h). (**h**) Simulated (u, v) in water on Day 12, the top row shows the surface stress, and the last column is daily mean. (Sun and Sun 2020) [91].

The velocity in Figure 19h shows a weak velocity at night and before sunrise, a strong diurnal jet trapped in the upper mixed layer due to a strong surface stress and weak viscosity in the ocean in the daytime. The Coriolis force becomes dominated as the PBL force decays away from the boundary. The rotation of velocity changes 180° in 12 h (half of inertia period); the small velocities cancel out each other near 30 m ($\delta_{Eo} = 33$ m) and below. Hence, the magnitude decreases from 8.9 cm s^{-1} to 1 cm s^{-1}, but the angle changes is no more than 90 degrees from surface to -32 m. It usually took from weeks to months to collect observed data (Chereskin 1995 [98], Price and Sundermeyer 1999 [99], and others), which cancels out oscillations with the period less than weeks or so. Hence, many scientists proposed a large viscosity for rotation but small value for the magnitude of the mean velocity (Lenn et al., 2009 [100], etc.). Although the magnitude of averaged spiral is small, the parcels rotate with large amplitude near and beyond the Ekman depth in both atmosphere (blue and green lines with * in Figure 19c) and ocean (thin black at -40 m in Figure 19e). Therefore, the change in magnitude and rotation of velocity of the mean Ekman spiral cannot be interpreted by viscosity alone because the Coriolis can be important when the time to take the average is comparable or longer than the inertia period. Furthermore, viscosity should not be a complex number either, as it is used by many scientists.

The resonance of f and PBL forcing explains the frequent development of the nocturnal low-level jet (LLJ), dryline, and severe storms in the lee of the Rocky Mountains around 30° N (Fujita 1958 [101], Wexler 1961 [102], Palmer and Newton 1969 [93], Sun and Ogura 1979 [103], and Sun and Wu 1992 [104]).

3.4.2. Simulation at 60° S with Diurnal Variation in Atmosphere and Ocean (Case G of Sun and Sun 2020 [91])

We set $\delta zo\_$ max = 60 m, $\mu o\_$ max = 0.5 m^2 s^{-1}, and $\delta Ea = 2va\_anl/f = 252$ m in the atmosphere and the depth of Ekman thickness, $\delta Eo = 2vo\_anl/f = 89$ m at 60° S. When the direction of daily mean stress is aligned with the observed stress (Figure 20a,b), the average oceanic spiral in Figure 20b is comparable with the observation at Drake Passage between $z = -26$ and $z = -90$ m (Lenn and Chereskin, 2009 [100]), and Polton et al. (2013) [97].

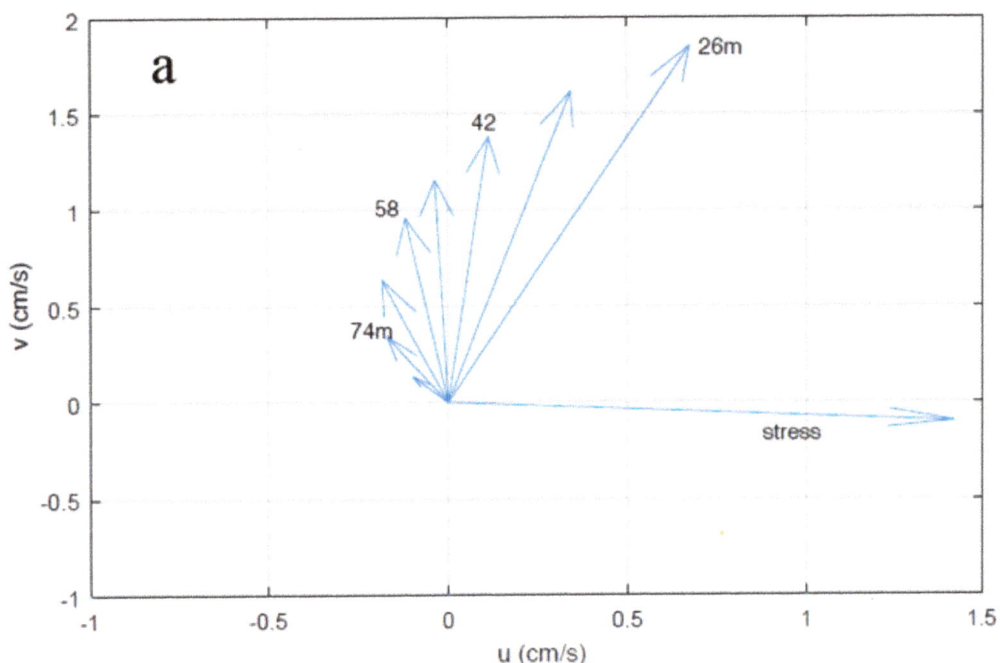

Figure 20. *Cont.*

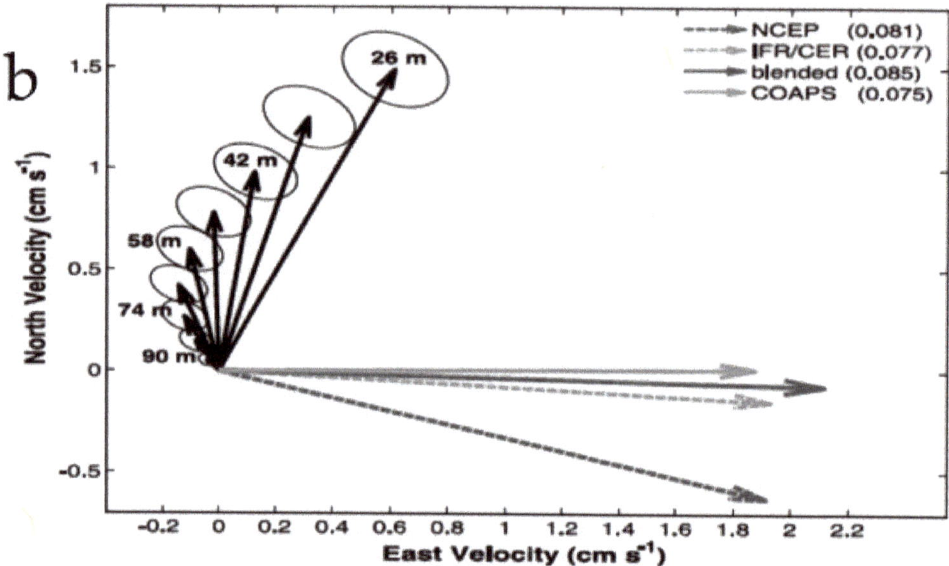

Figure 20. (a) Model simulations, (b) observed surface stress and velocity vectors near 60° S (Lenn and Chereskin 2009) [100], (Sun and Sun 2020) [91].

As discussed before, the time-averaged Ekman spiral is different from the steady state solution of (58). It is unlikely to find a viscosity to fit the observed average profile. Hence, "Diffusion theory is a convection and not a testable scientific theory" (Price and Sundermeyer 1999) [99] is incorrect.

4. Purdue Regional Climate Model (PRCM)—A Hydrostatic Model

The linear equations cannot predict the amplitude of disturbances, unless they are driven by a given force and the equations become inhomogeneous, as in the sea breeze circulation (Sun and Orlanski 1981a [87]), or Ekman layer simulation (Sun and Sun 2020 [91]). Hence, we need the nonlinear atmospheric and oceanic models.

4.1. Basic Equations

The PRCM, a hydrostatic primitive equation model, utilizes the terrain-following coordinate (σ) in the vertical direction:

$$\sigma = \frac{p - p_t}{p_*} = \frac{p - p_t}{p_s - p_t} \tag{59}$$

where p, p_t, and p_s are pressure, pressure at domain top, and surface pressure, respectively.

$$\frac{\partial u}{\partial t} = \mathrm{Adv}(u) + fv - \left[\frac{m}{\rho} \frac{\partial p'}{\partial x} + \frac{m}{p_*} \frac{\partial p'}{\partial \sigma} \left(\frac{\partial \varphi}{\partial x} \right) \right] + \mathrm{Diff}(u) \tag{60}$$

$$\frac{\partial v}{\partial t} = \mathrm{Adv}(v) + fu - \left[\frac{m}{\rho} \frac{\partial p'}{\partial y} + \frac{m}{p_*} \frac{\partial p'}{\partial \sigma} \left(\frac{\partial \varphi}{\partial y} \right) \right] + \mathrm{Diff}(v) \tag{61}$$

where φ (geopotential) $= gz$, $f = 2\Omega \sin \varphi$, and m is a mapping factor.

$$\mathrm{Adv}(\,) = -mu \frac{\partial (\,)}{\partial x} - mv \frac{\partial (\,)}{\partial y} - \dot{\sigma} \frac{\partial (\,)}{\partial \sigma} \tag{62}$$

The ice equivalent potential temperature θ_{ei} is

$$\theta_{ei} = \theta + \left(\frac{\theta}{T}\right)\left(\frac{L_v}{c_p}q_v - \frac{L_f}{c_p}q_i\right) \tag{63}$$

and

$$\frac{\partial \theta_{ei}}{\partial t} = Adv(\theta_{ei}) + \left(\frac{L_v}{c_p}q_v - \frac{L_f}{c_p}q_i\right)\frac{d}{dt}\left(\frac{\theta}{T}\right) + Diff(\theta_{ei}) + Rad(\theta_{ei}) + Conv(\theta_{ei}) \tag{64}$$

The change in the surface pressure is calculated by

$$\frac{\partial p_*}{\partial t} = -\int_0^1 \nabla_\sigma \cdot (p_* V) d\sigma \tag{65}$$

where

$$\nabla_\sigma \cdot (p_* V) = m^2\left[\frac{\partial(u p_*)}{m \partial x}\right] + \left[\frac{\partial(v p_*)}{m \partial y}\right] \tag{66}$$

The vertical velocity in the sigma coordinate is

$$\dot{\sigma} = \frac{-1}{p_*}\int_0^\sigma \nabla_\sigma \cdot (p_* V) d\sigma + \frac{\sigma}{p_*}\int_0^1 \nabla_\sigma \cdot (p_* V) d\sigma \tag{67}$$

The hydrostatic equation is

$$\frac{\partial \varphi}{\partial(\ln\, p)} = -R_d T(1 + 0.61 q_v - q_l - q_i) \tag{68}$$

The total water content q_w, liquid water q_l and ice q_i are given by

$$\frac{\partial q_w}{\partial t} = Adv(q_w) + Diff(q_w) + Conv(q_w) + Micro(q_w) \tag{69}$$

$$\frac{\partial q_l}{\partial t} = Adv(q_l) + Diff(q_l) + Conv(q_l) + Micro(q_l) \tag{70}$$

$$\frac{\partial q_i}{\partial t} = Adv(q_i) + Diff(q_i) + Conv(q_i) + Micro(q_i) \tag{71}$$

The sub-grid scale turbulence kinetic energy, $\overline{E}$, is governed by:

$$\frac{\partial \overline{E}}{\partial t} = -\frac{\partial}{\partial x_i}\left(\overline{u_i}\overline{E}\right) - \overline{u'_i u'_j}\frac{\partial \overline{u_i}}{\partial x_i} + \frac{g}{\theta_0}\overline{w'\theta'_v} - \frac{\partial \overline{u'_i(e' + p'/\rho_0)}}{\partial x_i} - \varepsilon \tag{72}$$

where $e' \equiv \frac{1}{2}\left(u'^2 + v'^2 + w'^2\right)$ and $\overline{E} \equiv \overline{e'}$. The sub-grid fluxes were parameterized by

$$\overline{u'_i u'_j} = -K_m\left(\frac{\partial \overline{u_i}}{\partial x_j} + \frac{\partial \overline{u_j}}{\partial x_j}\right), \tag{73}$$

$$\overline{w'_i \theta'_{ie}} = -K_h\left(\frac{\partial \overline{\theta_e}}{\partial z} - \gamma_c\right), \tag{74}$$

$$\overline{w' q'_w} = -K_h\frac{\partial \overline{q_w}}{\partial z}, \tag{75}$$

$$\overline{w'\theta'_v} = A_1\overline{w'\theta'_{ei}} + B_1\overline{w'q'_w}, \tag{76}$$

$$\overline{w'(e' + p'/\rho_0)} = -3K_m \partial \overline{E}/\partial z, \tag{77}$$

where K_m is the sub-grid scale eddy coefficient for momentum, K_h the sub-grid eddy coefficient for scalar quantities. The coefficients, A and B in (76) for unsaturated air are given by:

$$A_1 = 1 + 0.61 q_w \tag{78}$$

$$B_1 = 0.61\theta - \frac{L_v}{c_p}\left(\frac{\theta}{T}\right)(1 + 0.61 q_w) \tag{79}$$

And for saturated air,

$$A_1 = \frac{R_v T^2(1 + 1.61 q_s - q_w) + 1.61 T L_v q_s}{R_v T^2 + \frac{L_v^2}{c_p} q_s} \tag{80}$$

$$B_1 = -\alpha_i \frac{L_f}{c_p}\frac{\theta}{T} A - \theta \tag{81}$$

Cloud water and cloud ice can coexist in this model. The averaged saturation specific humidity is given by

$$q_s = \alpha_i q_{si} + (1 - \alpha_i) q_{sw} \tag{82}$$

where

$$\alpha_i = \frac{q_i}{q_i + q_\updownarrow} \tag{83}$$

The average latent heat of saturated water vapor can be written as

$$L q_s = \alpha_i L_s q_{si} + (1 - \alpha_i) L_v q_{sw} \tag{84}$$

The cloud water flux is diagnosed from the equivalent potential temperature flux and total water flux by

$$\overline{w' q_\updownarrow'} = A_2 \overline{w'\theta'}_{ei} + B_2 \overline{w'q'}_w \tag{85}$$

where

$$A_2 = \frac{-(1 - \alpha_i)\frac{T}{\theta} L q_s}{R_v T^2 + \left(\frac{L_v}{c_p} + \alpha_i \frac{L_f}{c_p}\right) L q_s} \tag{86}$$

$$B_2 = \frac{(1 - \alpha_i)\left(R_v T^2 + \frac{L_v}{c_p} L q_s\right)}{R_v T^2 + \left(\frac{L_v}{c_p} + \alpha_i \frac{L_f}{c_p}\right) L q_s} \tag{87}$$

Similarly, the cloud ice flux can be expressed as

$$A_3 = \frac{-\alpha_i \frac{T}{\theta} L q_s}{R_v T^2 + \left(\frac{L_v}{c_p} + \alpha_i \frac{L_f}{c_p}\right) L q_s} \tag{88}$$

$$B_3 = \frac{\alpha_i\left(R_v T^2 + \frac{L_v}{c_p} L q_s\right)}{R_v T^2 + \left(\frac{L_v}{c_p} + \alpha_i \frac{L_f}{c_p}\right) L q_s} \tag{89}$$

The covariance of the vertical velocity is parameterized as

$$\overline{w'^2} = \frac{2}{3}\overline{E}, \tag{90}$$

And the eddy viscosities are

$$K_m = 0.10 \updownarrow \overline{E}^{1/2}, \tag{91}$$

$$K_h = \begin{cases} (1 + 2\updownarrow/\Delta z) K_m & \text{if } \updownarrow < \Delta z \\ 3 K_m & \text{otherwise} \end{cases} \tag{92}$$

where ℓ is the mixing length scale of the ensemble average turbulence and Δz is vertical grid space. Following Sun and Ogura (1980) [105], Deardorff (1980) [106], the length scale in the stable atmosphere is assumed:

$$\ell = \begin{cases} 0.6\overline{E}^{1/2}\left(\frac{g}{\theta_o}\frac{\overline{w'\theta_v'}}{K_h}\right)^{-1/2} & \text{if } \ell < \Delta z \\ \Delta z & \text{otherwise.} \end{cases} \tag{93}$$

In an unstable PBL, according to Caughey and Palmer (1979) [107], it is

$$\ell = 0.25\{1.8Z_i[1 - \exp(-4z/Z_i) - 0.003\exp(8z/Z_i)]\}. \tag{94}$$

And the upper limit of $Z_i \leq 530$ m. Dissipation rate ε is given by

$$\varepsilon = 0.41\frac{\overline{E}^{3/2}}{\ell} \tag{95}$$

The model has the prognostic equations for the momentum, surface pressure, ice-equivalent potential temperature θ_{ei}, turbulent kinetic energy (TKE), vapor, cloud, ice, rain, snow, and graupel, etc. (Sun and Chang 1986a [108], Sun 1986 [109], Chern 1994 [110], Chen and Sun 2002 [111]). It consists of multi-layers of snow, soil temperature, and wetness, etc. (Sun and Wu 1992 [104], Bosilovich and Sun 1995 [112], Chern and Sun 1998 [113], Sun and Chern 2005 [114]). It also includes radiation parameterizations (Chou and Suarez 1994 [115], Chou et al., 2001 [116]), and cumulus parameterizations (Kuo 1965, 1974 [117,118], Anthes 1977 [119], Molinari 1982 [120]). The forward–backward scheme is applied in the Arakawa C grid to permit more computational accuracy and efficiency (Sun 1980, 1984a [38,121]). A fourth-order advection scheme (Sun 1993c [122]) or mass-conserved semi-Lagrangian scheme (Sun et al., 1996 [123], Sun and Yeh 1997 [124], and Sun and Sun 2004 [125]) is used to calculate the advection terms. A local reference is applied to calculate the pressure gradient force, which significantly reduces the error of the pressure gradient terms in the σ_p coordinate over steep topography (Sun 1995a [126]). Yang (2004a, b) [127,128] has added the transport of dusts and the atmospheric chemistry module, as shown in Figure 21.

4.2. Inland Sea Breeze and Dryline

Sun and Ogura (1979) [103] proposed the "Inland-sea-breeze" hypothesis that the pre-storm convergence, vertical circulation, and the nocturnal low-level jet along dryline (Figure 22) on the lee of the Rocky Mountains are triggered by a strong horizontal soil moisture and temperature gradients. Sun and Wu (1992) [104] successfully simulated the formation and diurnal movement of the dryline. In the late spring and early summer, on a slope terrain with a strong soil moisture gradient under a vertical wind shear, a dryline can form within 12 h in Oklahoma and Texas. They reproduced a deeper intrusion of the moisture field far above the potential temperature inversion (Figure 23a,b), as observed by Schaefer (1974) [129]. In the daytime, due to the strong mixing with the warm, dry air descending from the Rocky Mountains, the moisture on the west boundary of dryline dilutes quickly and results in an eastward movement of dryline. At night without mixing, the cool, moist air on the eastside thrusts into the deep, warm, dry air on the westside (Figure 23a–d), and it also gradually turns northward due to the Coriolis force and forms a nocturnal LLJ on the slope of the Rocky Mountains (Figure 23c), as proposed by Sun and Ogura (1979) [103]. The southerly LLJ can effectively transport warm, moisture northward. Hence, the vicinity of the dryline becomes a favorable zone for convections. A weak instability caused by the resonance of a diurnal oscillation of the PBL, and inertial oscillation of the Coriolis force may enhance the disturbances around 30° N, as discussed in Section 3.4.1 (Sun and Sun 2020) [91].

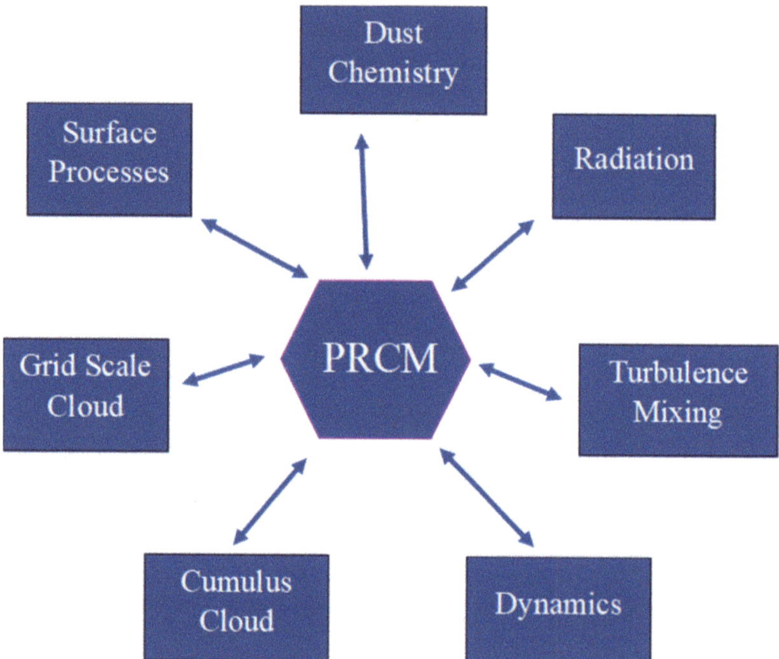

Figure 21. Schematic diagram of PRCM (Purdue Regional Climate Model).

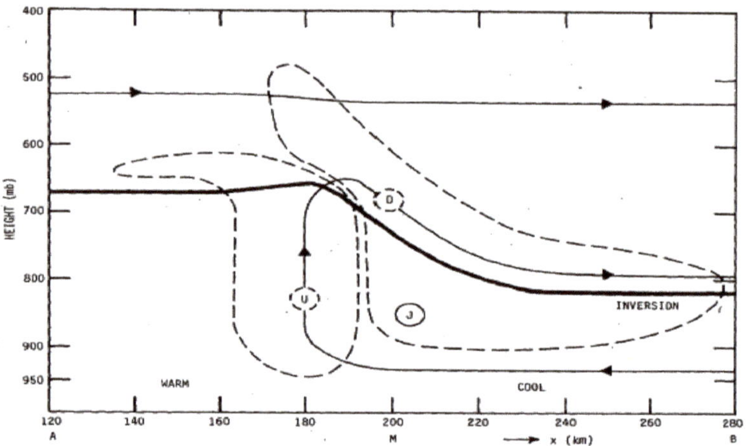

Figure 22. Schematic vertical cross section across the simulated dryline at 1630 LT 8 June 1966. J, U, and D denoted the location of the cores of the jet in the y-direction, the upward motion and downward motion, respectively. The dashed lines are the contour lines of 2 cm s^{-1} for the vertical velocity (Sun and Ogura 1979) [103].

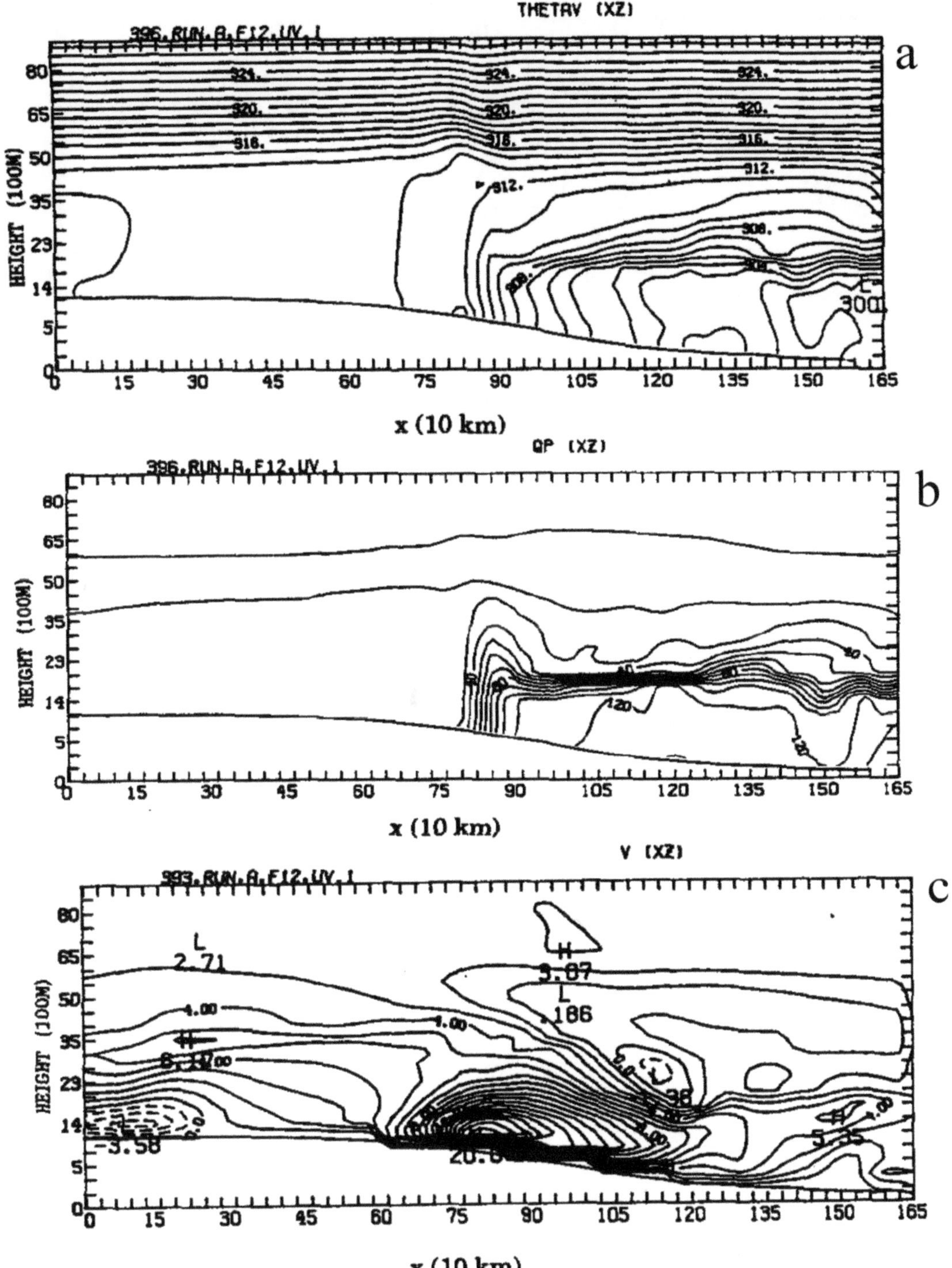

Figure 23. *Cont.*

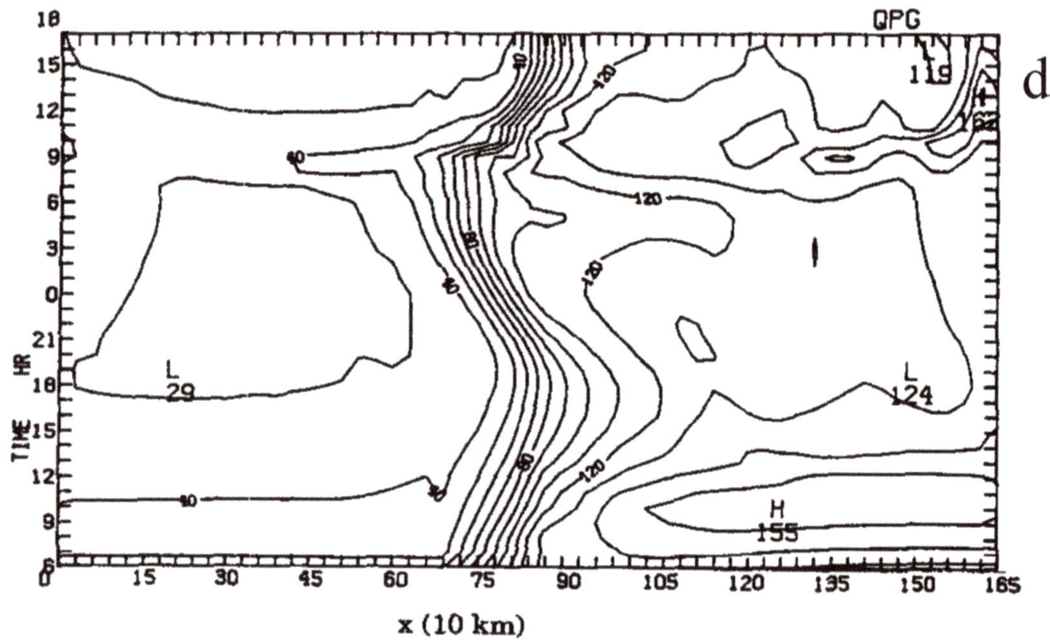

x (10 km)

Figure 23. (a) θ_v; (b) q_w (total water content) at 1800 CST on day 1; (c) southerly wind v and low-lvel-jet (LLJ) at 0200 CST on day 2; (d) time–space variation of q_w at z~10 m, with a contour interval of 1 g Kg^{-1}, labels scale by 10. (Sun and Wu 1992) [104].

4.3. Effect of Mountain

4.3.1. Lee Vortices and Mountains

With detailed physics on the non-slip surface, Sun and Chern (1993) [130] confirmed the importance of vertical stretching on the formation of vortices proposed by Sun et al. (1991) [131]. They also reproduced the observed lee vortices, meso-low, and downslope wind on the east of the CMR under a southwesterly monsoon during the TAMEX-IOP-2 (Taiwan Area Mesoscale Experiment intensive observation period 2). The lee vortex formed near the southeast coast at 2000LST 16 May 1987 (Figure 24a,b) moved to the east of Taiwan at 0200 LST 17 (Figure 25a,b). The air moved across the CMR, then sank adiabatically and formed a warm, meso-low to the south of the lee vortex as observed in Figure 26a,b. Hence, it should not have an artificial meso-β front as shown in Figure 26b by Kuo and Chen (1990) [132]. The PRCM also replicated the observed severe front at 00Z 17 May 1987 (Figure 27). This well-developed front extended from Japan and the Japan Sea through Taiwan into the South China Sea and Vietnam, and had the property of the typical Mei-yu front discussed by Hsu and Sun (1994) [133] and others.

The clouds with super-cooled water observed by aircraft and Denver cyclone observed by radar on 13 February 1990 during Winter Icing and Storms Project's (WISP) Valentine's Day storm (VDS) were also reproduced by PRCM as shown in Figure 28 (Haines et al., 1997 [134]).

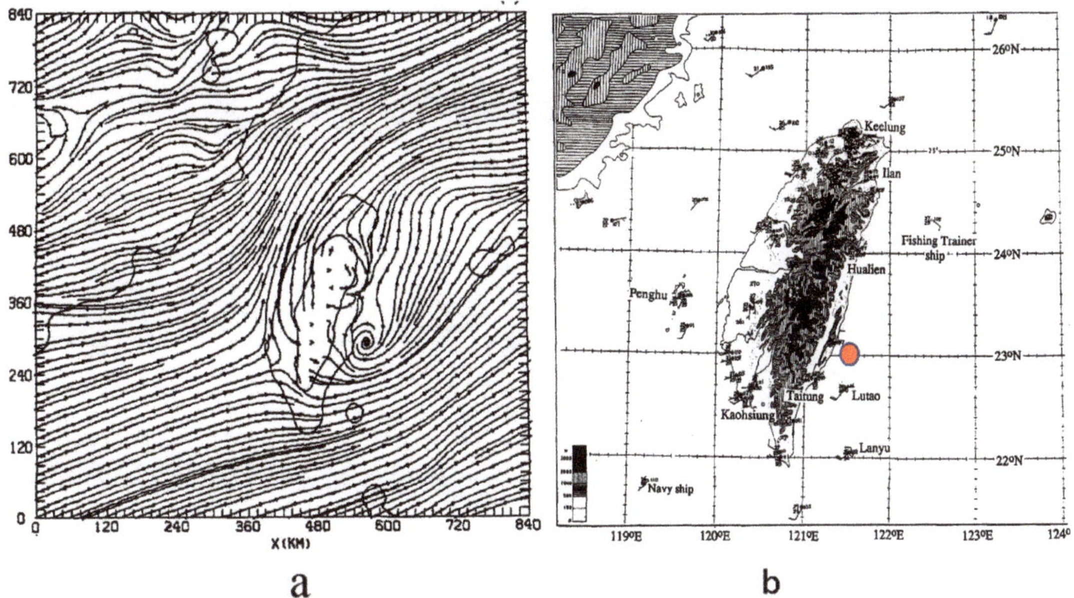

Figure 24. (**a**) Streamlines at z = 1000 m at 2000 LST 16 May (Sun and Chern 1993). (**b**) Observed surface data at 2000 LST 16 May 1987 [130].

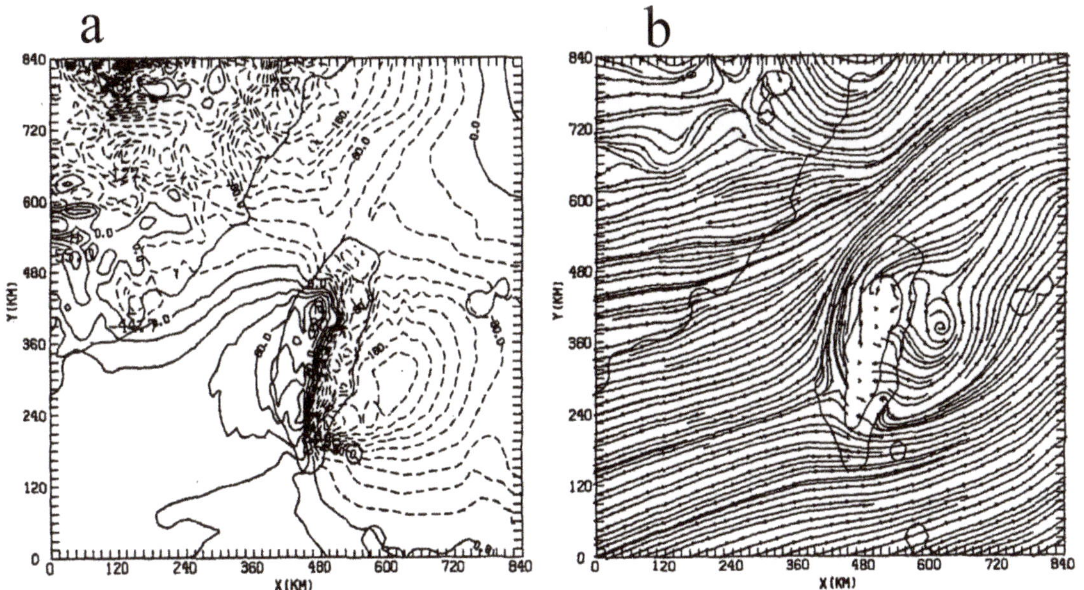

Figure 25. (**a**) Simulated surface pressure deviation from −240 to 120 Pa with interval of 20 Pa. (**b**) Streamline at z = 1000 at 0200 LST 17 May (Sun and Chern 1993) [130].

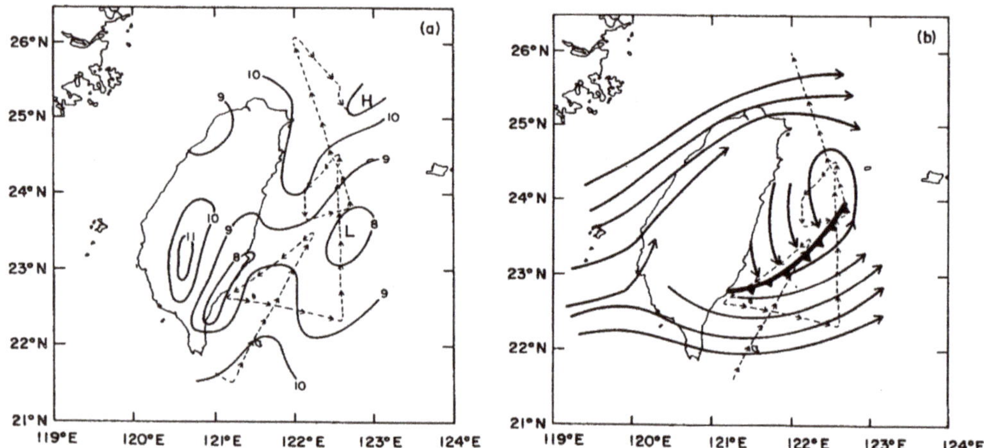

Figure 26. (**a**) Observed sea surface pressure, (**b**) streamline at 900 mb observed by airplane at 0200 LST 17 (1800 UTC 16) May 1989 (Kuo and Chen 1990) [132].

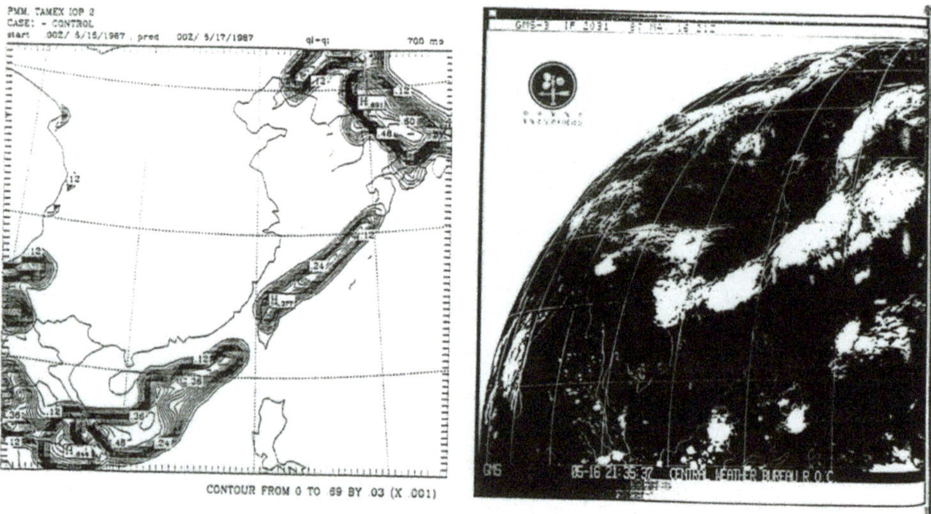

Figure 27. (**Left**) simulated cloud water at 700 mb at 0000 UTC May 17 (after 48 h integration), and (**right**) satellite image at 2137 UTC May 16.

4.3.2. Front Deforms around the Central Mountain Range (CMR) in Taiwan

Sun and Chern (2006) [71] replicated the deformation of a weak cold front by the CMR after 30 h integration. The large-scale patterns are comparable with the ECWMF analysis at 06Z15, 1997 (Figure 29a–c) except that the deformation of the front was missing in the analysis. As the cold air approached Northern Taiwan from the north, a relative high pressure with an anticyclonic circulation built up on the windward side. More oncoming flow was diverted to the northeast of Taiwan than the northwest and caused the difference in the frontal speed between the eastern and western parts of the front. The wind speed also increased as the northeasterly flow was blocked by the mountain range. The speed of the front was comparable to U, the wind behind the front, instead of $(U+\sqrt{g'h})$, the speed of a Kelvin wave.

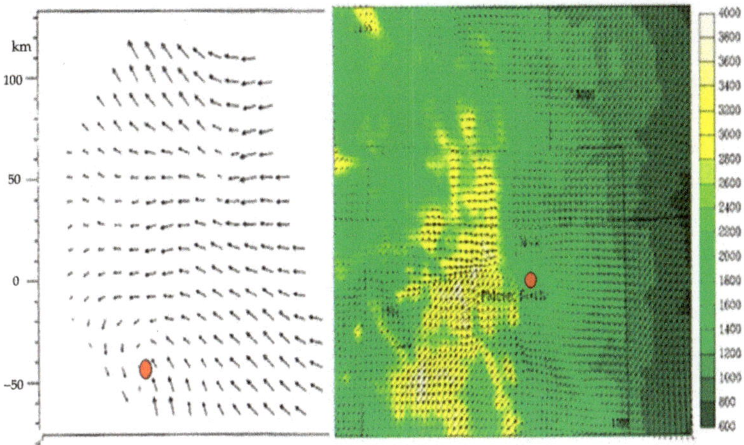

Figure 28. (**Left**) radar observed wind and vortex, and (**Right**) Simulated wind and Denver Cyclone (red) on 13 February 1990 during WISP, color bar indicates elevation height in m (Haines et al., 1997) [134].

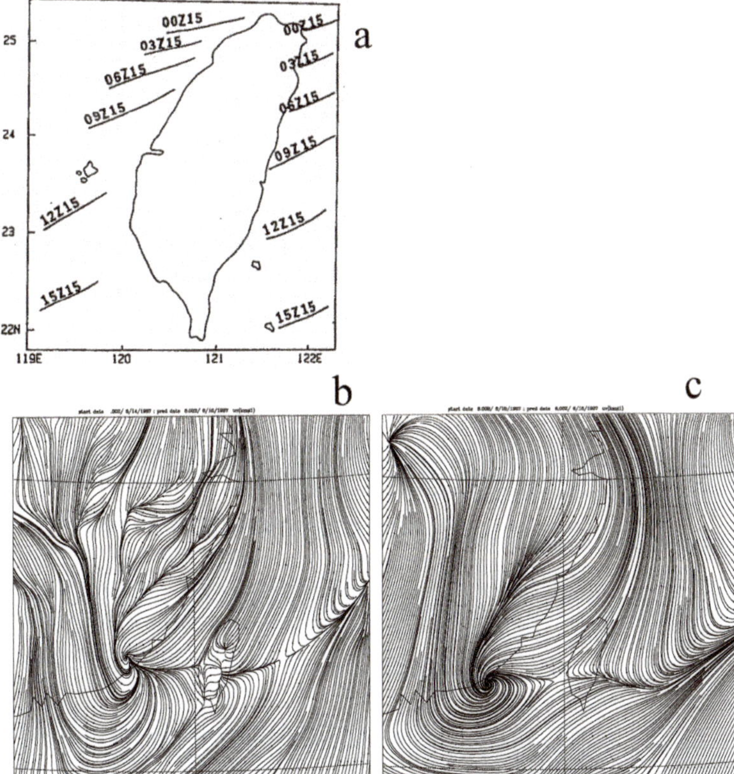

Figure 29. (**a**) Isochrones of leading edge of the cold air from the surface and satellite data. hhZdd is to denote the time and date. (**b**) Simulated streamline at 06Z15, 1997 at $z \approx 25$ m (Sun and Chern 2004). (**c**) ECMWF reanalysis streamline at 06Z15 at $z \approx 25$ m [71].

4.4. Cyclogenesis and Winter Severe Storms in the USA

Chern (1994) [110] has simulated the severe winter storms in the US in March 1985. The unique features of the event are: (1) started from a fair weather, two surface low pressure centers formed near the Great Basin region, one was initiated at 1200 UTC 1 March near the border of Oregon, California, and Nevada; and the other took shape at 0000 UTC 2 March over Eastern Idaho; (2) an explosive development occurred over Nevada with a deepening rate of 18.5 mb in 24 h; (3) the cyclone over Eastern Idaho also exposed with a pressure fall of 13.7 mb in 24 h; (4) a Pacific block with a Rex-type high–low dipole dominated the upstream flow patterns during the entire event; (5) an upper-level short-wave trough grew rapidly from a relatively undisturbed stage and propagated southeastward to the Great Basin region; (6) the upper-level jet stream underwent a dramatic change in wind direction as the upper-level trough developed; and (7) the storm brought heavy snow in excess of two feet over Montana, South Dakota, and southwest of Minnesota.

A control simulation was conducted with full model physics on a 45 km grid to study the rapid cyclogenesis during 1–3 March 1985. The simulation replicated the development and movement of two surface lows in the mountainous areas (Figures 30 and 31). The central pressures of the simulated lows were within 3 mb of the observed values and the predicted cyclone paths were within 135 km of the analysis. The sharp transition in direction of the 300 mb jet stream during the early development of the upper-level ridge/trough system was well captured by the model. The root mean square errors of sea level pressure were 1.90 mb and 3.47 mb for 24 and 48 h predictions, respectively. The correlation coefficients of the change in sea level pressure between the simulation and ECMWF/TOGA analysis were very high, 0.984 for the 24 h prediction and 0.972 for the 48 h prediction (Figures 30 and 31).

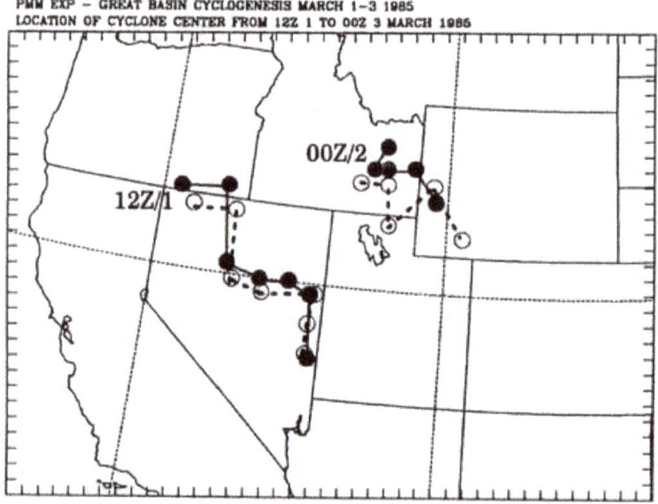

Figure 30. Locations of the cyclones at 6 h interval. Solid circles represent observation from ECMWF/TOGA analysis. Open from PRCM. The cyclones over Nevada (eastern Idaho) first appeared at 1200 UTC1 (0000 UTC 2) March 1985 (Chern 1994) [110].

Six sensitivity tests were performed to identify the impacts of atmospheric radiation, surface fluxes, surface friction, latent heat release, and the effect of topography on the event. Results showed: (a) complex terrain was quite important to the structure, location, and movement of the cyclones. (b) Surface friction and latent heating were more important than the surface fluxes and atmospheric radiation over the entire domain, and (c) latent heating was minor, and it only contributed 5.4 mb increase in surface pressure over South Dakota, to the northeast of the surface low center.

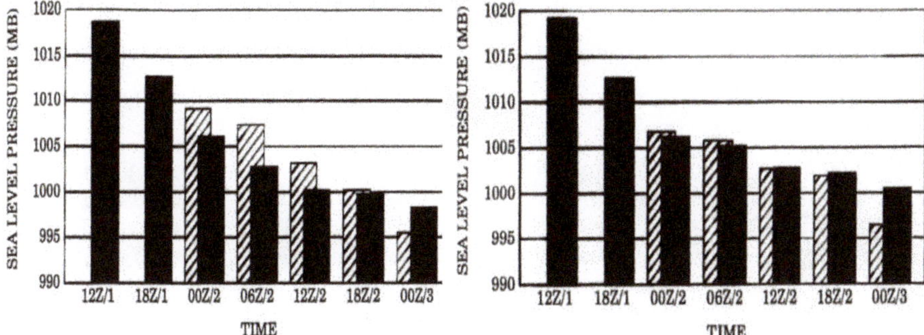

Figure 31. Observed (**left**) and simulated (**right**) surface pressure at low pressure centers in Nevada (filled bar) and Idaho (hatched bar) (Chern 1994) [110].

The development of the Great Basin cyclones was controlled mainly by adiabatic dynamics, and the upper-level forcing. The positive PV anomaly coincided with a tropopause folding through which air originating from the lower stratosphere penetrating deeply into the mid- and lower troposphere as shown in ECMWF and simulated geopotential at 500 mb (Figure 32). The absolute vorticity increased due to the decrease in static stability as the stratospheric high PV air parcels moved into the mid troposphere. The ageostrophic circulations show that the development of the cyclone over Nevada coincided with the rising branch of the indirect circulation in the left quadrant exit region of the upper-level jet streaks upstream from the developing trough. The subsiding branch of the secondary circulation in the warm side of the jet frontal system was important to the tropopause folding. The heavy snowfall and strong upward motion over South Dakota were due to the interactions of the ascending branch of a thermally direct circulation in the entrance of a polar jet streak, the effect of latent heating, and the lifting of air over sloping terrains and a surface front. The system was far more complicated than the idealized baroclinic instability or cyclogenesis presented in the paper (Eady 1949 [76]) or idealized model simulations (Yildirim 1994 [135]).

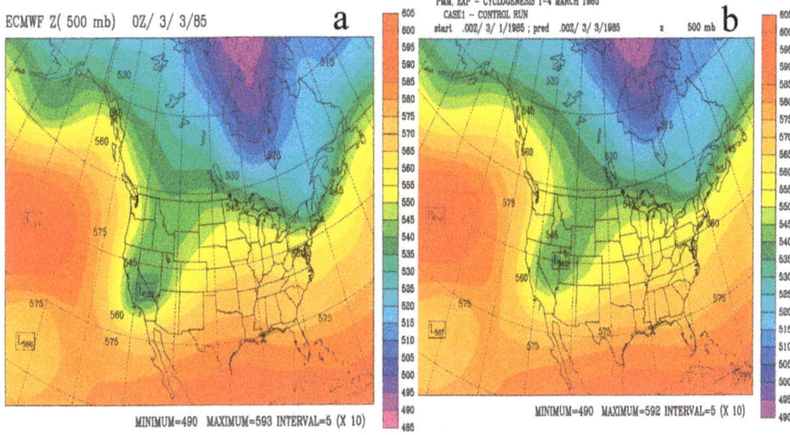

Figure 32. (**a**) ECMWF geopotential Z; (**b**) model simulated Z at 500 m at 00Z/03/1985 after 2 days integration (Chern 1994) [110].

It was a fair weather before 1 March and the NMC model failed to predict the explosive development of the storms, which severely affected the areas from Wisconsin to Nevada longer than a week. Meanwhile, the PRCM simulated the development of storms (unstable mode) well. It indicates that a comprehensive model can predict the development of the system near the saddle point as the dynamic system changes drastically.

5. Regional Climate

5.1. 1993 Flood in the USA

Bosilovich and Sun (1999a, b) [136,137] reproduced the large-scale atmospheric features in the summer of 1993, when heavy precipitation caused a long-lived, catastrophic flood in the Midwestern United States (Figure 33). The upper-level jet stream and trough were over the Northwestern United States, the Great Plains LLJ supported the moisture transport, and heavy precipitation occurred in the midwest. The model also reproduced the observed precipitation pattern: synoptic wave-type in June and mesoscale convective-system (MCS) in July, over the heaviest rainfall area with an excessive ($\geq$20%) simulated precipitation. The sensitivity tests show: the increased surface heating caused by a strong dry anomaly induces a large-scale surface pressure perturbation and weakens the low-level jet and moisture convergence within the flood region. Both wet and dry regional anomalies in the Southern Great Plains cause less precipitation in the flood region. The uniform soil moisture of both anomalies leads to a reduction in differential heating, surface pressure gradient, and the low-level jet. The model also indicates that the long-lived, 1993 catastrophic flood may be related to the large-scale system.

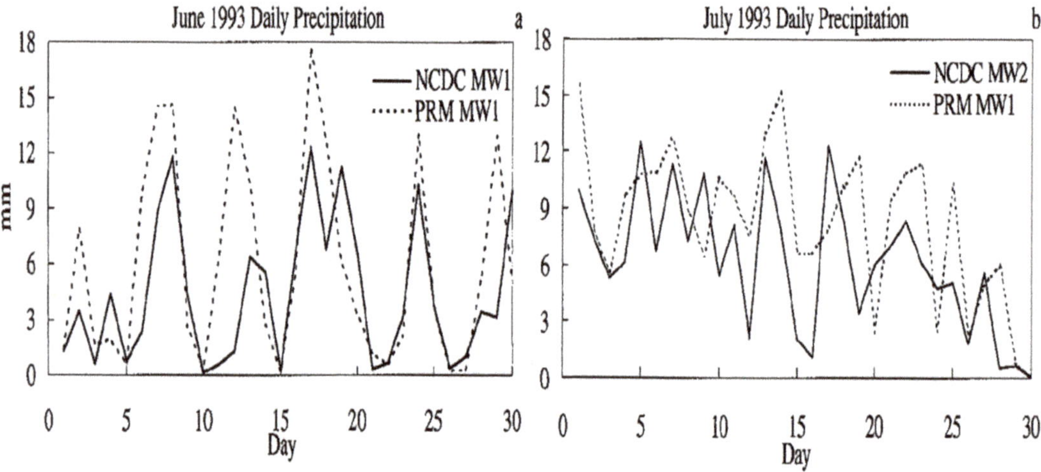

Figure 33. Simulated from PRCM and NCDC observation of daily integrated precipitation over flooding area in the Midwestern US for: (**a**) June associated with synoptic waves and (**b**) July 1993 associated with mesoscale convective system (Bosilovich and Sun 1999a) [136].

5.2. 1988 Drought in the USA

Many scientists have applied numerical models to study the cause(s) of the 1988 drought in the USA shown in Figure 34. However, discrepancies exist among the different hypotheses. Oglesby and Erikson (1989) [138], and Oglesby et al. (1994) [139], using the NCAR-CCM1, suggested that the draught was caused by the persistence of a soil moisture anomaly in the US. However, using the same CCM1, Sun et al. (1997) [140] carried out four experiments, which included those using the climatological SSTs (control case); 1988 SSTs; dry soil moisture anomaly (25% of soil moisture in May in the normal year between 35–50° N in the US); and 1988 SSTs and dry soil moisture anomaly. Our results show that the 1988 SST did not cause the simulated weather pattern over the US to be drier than the control case with the climatological SST. The ensemble mean with the perturbed soil moisture experiment did show less precipitation than the control; however, due to the large variance in the data, the reduction in precipitation was not statistically significant. In the experiment with the soil moisture anomaly and 1988 SST, the most significant differences were obtained, specifically in the reduction of precipitation in the US. This may indicate that the 1988 severe drought may be caused by a combination of dry soil, and a special large scale weather pattern related to the SST, instead of by either SST or dry soil alone.

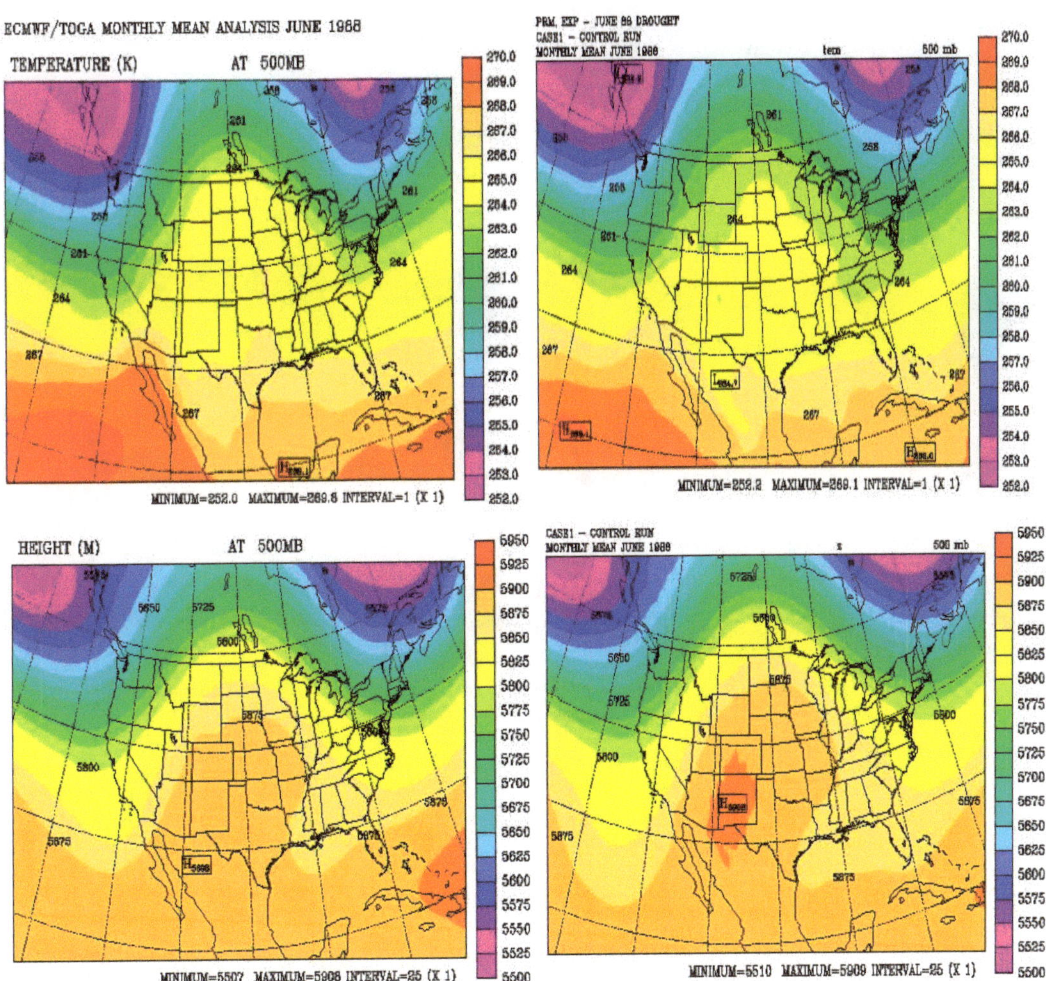

Figure 34. *Cont.*

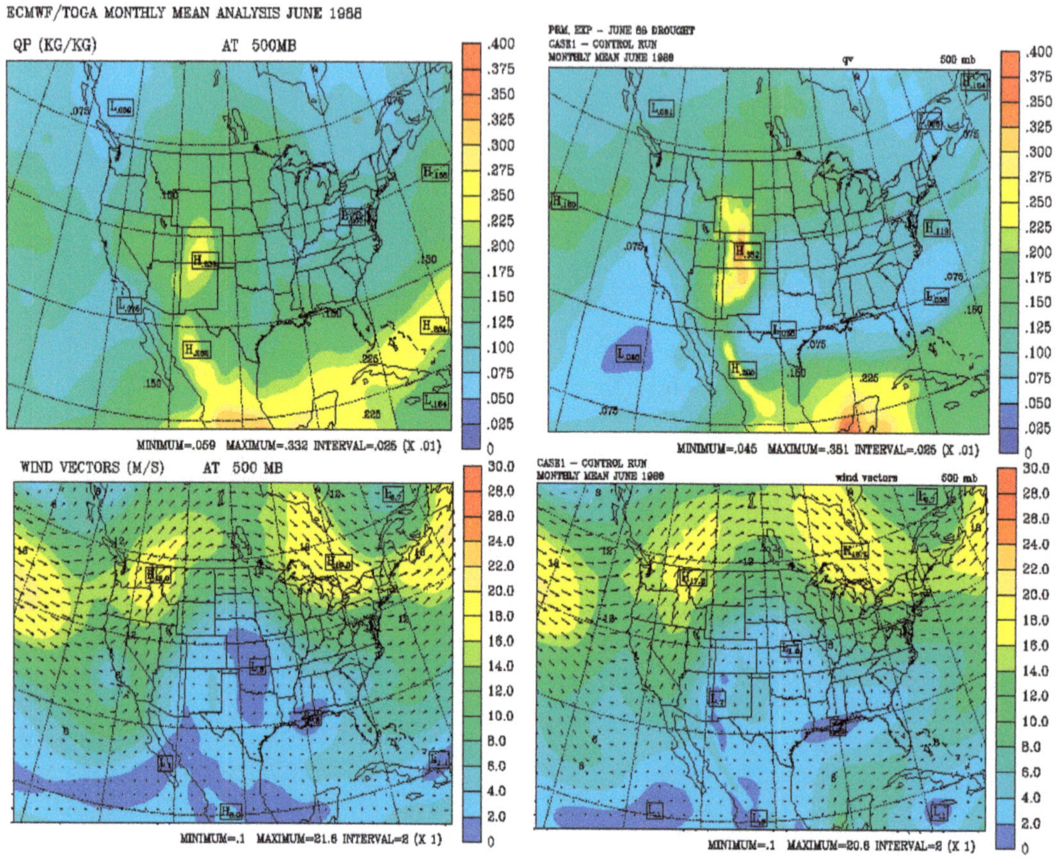

Figure 34. Observed and simulation of temperature, geopotential, moisture, and wind at 500 mb for June 1988 drought in USA (Sun et al., 2004) [141].

Using the PRCM with the observed SST and the ECMWF analysis as initial and lateral boundary conditions, Sun et al. (2004) [141] reproduced a strong warm ridge at 500 hPa in North America over a hot, dry land. The monthly precipitation and soil moisture were far below the normal values in the Midwest and the Gulf States, but more than normal in the Rocky Mountains as observed (Figure 34), corresponding to a stronger N. American monsoon than normal. A sensitivity test shows that the monthly precipitation could significantly increase using the saturated soil moisture as the initial condition. The soil would dry up eventually, because the wet soil does not provide positive feedback with the low-level jet. Hence, the change in the large-scale weather pattern might trigger the dry episode by forming a ridge in North America, which gradually cut down precipitation in the Gulf States and midwestern region. The soil became dry and hot, which further intensified the blocking of the warm ridge in the US. Hence, dry soil had positive feedback for the severe drought in the summer of 1988. But the dry soil alone was not the cause. Soil moisture became much drier in July than June, but precipitation amounts returned to normal after mid-July 1988, which was also well simulated. It indicates that PRCM can reproduce a rapid change from an abnormal status to a normal situation, i.e., a jump from one phase space to another phase space in the dynamics system (Lorenz 1963 [9]).

5.3. Snow, Land Surface, and Regional Climate

5.3.1. One-Dimensional Snow–Soil Model (Sun and Chern, 116 [114])

The land surface module of the PRCM without snow/ice or frozen soil has been described by Bosilovich and Sun (1995, 1998) [112,142], Sun and Bosilovich (1996) [143], and Sun (2001) [144]. Here we present the model with snow/ice. According to Jordan (1991) [145], the heat (enthalpy) equation for a potentially freezing becomes:

$$\left[\frac{\partial(H_{en}\Delta z)}{\partial t}\right]_k = (F_{k+1/2} - F_{k-1/2}) \tag{96}$$

where

$$\begin{aligned} H_{en} &= \{(T - 273.15) * (c_d\rho_d\Delta\zeta_d + c_w\rho_w\Delta\zeta_w + c_i\rho_i\Delta\zeta_i) + L_{ei}\rho_w\Delta\zeta_w\}/\Delta z \\ &= \{(T - 273.15) * (C_d\Delta\zeta_d + C_w\Delta\zeta_w + C_i\Delta\zeta_i) + L_{ei}\rho_w\Delta\zeta_w\}/\Delta z \end{aligned} \tag{97}$$

$L_{ei}(= L_f)$ is latent heat of fusion for ice (3.335×10^5 J/kg); Δz is the thickness of the layer; ρ_d, ρ_w, and ρ_i are density of dry soil, water, and solid ice within the layer; c_d, c_w, and c_i as well as $\Delta\zeta_d/\Delta z$, $\Delta\zeta_w/\Delta z$, and $\Delta\zeta_i/\Delta z$ are the specific heat capacity and component for dry soil, water, and ice, respectively. The mean heat capacity of soil is $C_{soil} = (C_d\Delta\zeta_d + C_w\Delta\zeta_w + C_i\Delta\zeta_i)/\Delta z$. $F_{k\pm1/2}$ is the heat flux at the interface between k and $k \pm 1$ layer

$$F_{k\pm1/2} = \left[-K_{soil}\left(\frac{\partial T_s}{\partial z}\right) + H_{flow}\right]_{k\pm1/2} \tag{98}$$

where the effective thermal conductivity K_{soil} is calculated following Johansen (1975) [146], reported by Farouki (1982) [147]; H_{flow} is the heat transfer due to water passing through layer k. Without the snow coverage, the heat flux at the soil top is

$$F_{1/2} = R_n - H_{s0} - H_{L0} \tag{99}$$

where R_n is net radiation. The downward shortwave radiation $R_s{}^\downarrow$ absorbed at surface is

$$R_s{}^\downarrow = (1 - A)Rs^\downarrow_{sfc-top} \tag{100}$$

where A is surface albedo, and $Rs^\downarrow_{sfc-top}$ is short wave radiation reaching the surface. The net long-wave radiation at surface is

$$R_\uparrow^\downarrow = \varepsilon R_a^\downarrow - \varepsilon\sigma T^4_{sfc} \tag{101}$$

where σ (Stefan − Boltzmann constant) $= 5.67 \times 10^{-8}$ W m^{-2}K^{-4}. The values of A, $Rs^\downarrow_{sfc-top}$, and surface emissivity ε ($\varepsilon = 1$ for soil) are provided by field experiment. Both sensible heat flux H_{s0} and latent heat flux H_{L0} are calculated by similarity equations (Bosilovich and Sun 1995 [112]). When the soil is covered by snow, $F_{k\pm1/2}$, the heat flux inside the snow may include the downward shortwave radiation.

The equation for liquid in a frozen or unfrozen soil is

$$\frac{\partial w}{\partial t} = \frac{\partial}{\partial z}\left(K\frac{\partial\psi}{\partial z}\right) - \frac{\partial K}{\partial z} + S_W \tag{102}$$

where S_w is the source term, including melt of ice/snow, precipitation, evaporation, and runoff within the layer, K is hydraulic conductivity, and ψ is matrix potential. Following Clapp and Hornberger (1978) [148], soil hydraulic property functions can be defined as:

$$\begin{aligned} \psi &= \psi_{ae}\left(\frac{w_s}{w}\right)^b \\ \frac{K}{K_s} &= \left(\frac{w}{w_s}\right)^{3+2b} \end{aligned} \tag{103}$$

The subscript s denotes the value at saturation, and b is an empirical constant which depends upon the texture of the soil.

ψ_{ae} is the air entry tension (Dingman 2002 [149]). Combining (102) and (103), we obtain:

$$\frac{-w}{b\psi}\frac{\partial\psi}{\partial t} = \frac{\partial}{\partial z}\left(K\frac{\partial\psi}{\partial z}\right) - \frac{\partial K}{\partial z} + S_W \tag{104}$$

5.3.2. Equations for Snow Layer

Based on energy and mass balances, the layer's number and thickness of one-dimensional snow model can change with time. The temperature, density, water content, and thickness of each layer are calculated separately. Each snow layer may consist of solid snow and liquid water. Hence, the mass of a snow layer (M_{sn}) is

$$M_{sn} = M_s + M_w \tag{105}$$

where M_s is the mass of solid snow and M_w is the mass of liquid water content. The thickness and density of a snow layer are given by

$$\Delta z_{sn} = \Delta z_s + \Delta z_w, \tag{106}$$

and the mean density is

$$\rho_{sn} = (\rho_s\Delta z_s + \rho_w\Delta z_w)/\Delta z_{sn} \tag{107}$$

where subscripts "sn", "s", "w" are related to snow layer, solid snow, and liquid water, respectively. The temperature (T_{sn}) is derived from equation of enthalpy within snow

$$\left[\frac{\partial(H_{en}\Delta z_{sn})}{\partial t}\right]_k = (F^{sn}_{k+1/2} - F^{sn}_{k-1/2}), \tag{108}$$

where enthalpy is defined as:

$$H_{en} = (C_w + C_s)(T_{sn} - 273.15) + \rho_w\Delta z_w L_{\uparrow i} \tag{109}$$

The heat capacity of a snow layer (C_{sn}) is computed from the heat capacities of solid snow (C_s) and liquid water (C_w), weighted by their respective volume fractions:

$$C_{sn} = (C_s\Delta z_s + C_w\Delta z_w)/\Delta z_{sn}, \tag{110}$$

where $C_w = 4.218 \times 10^6 \text{ J m}^{-3}\text{ K}^{-1}$. According to Verseghy (1991) [150], the heat capacity of solid snow is

$$C_s = 1.9 \times 10^6 \, (\rho_s/\rho_{i0}) \text{ J m}^{-3}\text{ K}^{-1}, \tag{111}$$

where $\rho_{i0} = 920 \text{ kg m}^{-3}$ is the density of pure ice. The heat flux at the snow top is

$$F_{1/2} = R_n - H_{s0} - H_{L0} \tag{112}$$

where R_n is net radiation. The surface energy balance Equations (99)–(101) are also applied to the snow surface. The infrared emissivity is 0.99 for fresh snow and 0.82 for old snow (Oke 1987 [151]). Snow emissivity = 0.95 is used in this study. The snow albedo is calculated by the equation proposed by Bohren and Barkstrom (1974) [152]:

$$A_s = 1.0 - 0.206 \, C_v d_s^{1/2}, \tag{113}$$

where mean snow grain diameter d_s (in mm) is derived from an empirical formula of snow density ρ_s:

$$d_s = G_1 + G_2\rho_s^2 + G_3(10^{-3}\rho_s)^4, \text{ (in mm)} \tag{114}$$

where $C_v = 1.2$, G1 = 0.16 mm, G2 = 0.0, and G3 = 110 mm g^{-4} cm^{12}, and the unit of density of solid snow (ρ_s) is kg m^{-3} (Anderson 1976) [153].

The heat fluxes at the top and bottom of layer k is defined as

$$F^{sn}_{k\pm1/2} = \left[-K_{sn}\left(\frac{\partial T_{sn}}{\partial z}\right) - Rs^{\downarrow}_{sn} - H_{flow} \right]_{k\pm1/2}, \tag{115}$$

where K_{sn} is the effective thermal conductivity; $Rs^{\downarrow}_{sn}$ is the downward solar radiation inside snow; H_{flow} is the heat transfer due to water passing through the layer. The effective thermal conductivity of a snow layer is calculated from volume weighted thermal conductivities of solid snow (K_s) and liquid water (K_w):

$$K_{sn} = (K_s \Delta z_s + K_w \Delta z_w)/\Delta z_{sn}, \tag{116}$$

where the thermal conductivity of liquid water K_w is set to 0.6 Wm^{-1}K^{-1}. The effective conductivity of solid snow represents the heat transport through both conduction and vapor diffusion. The empirical formulation of Yen (1965) [154]

$$K_s = 3.221 \times 10^{-6}\rho_s^2, \tag{117}$$

Equation (117) is used for the effective conductivity of solid snow. Here, the unit of density of solid snow (ρ_s) is kg m^{-3}.

The change in solar radiation inside kth layer is

$$\left\lfloor Rs^{\downarrow}_{sn} \right\rfloor_{k+1/2} = \left\lfloor Rs^{\downarrow}_{sn} \right\rfloor_{k-1/2} \exp(-\beta\Delta z_{sn})_k, \tag{118}$$

where β is the bulk extinction coefficient. Following Jordan (1991) [145], the incident solar radiation is divided into near-infrared and visible components. Infrared radiation is assumed to be totally absorbed within the top layer. The asymptotic bulk extinction coefficient is used for the attenuation of visible radiation, which is computed from the formula proposed by Bohren and Barkstrom (1974) [152], and Anderson (1976) [155].

$$\beta = 100 * \frac{C_v(1.0^{-3}\rho_s)}{\sqrt{\bar{d}_s}}, \text{ in } (\text{m}^{-1}) \tag{119}$$

The unit of d_s is mm in (119). The melting and freezing processes in the model are based on the conservation of mass and heat. Should the calculated new snow temperature from Equations (108) and (109) be greater than the melting point, the excess energy goes toward melting the snow, so that T_{sn} remains at 273.15 K until snow completely melts. Freezing occurs when liquid water exists while snow temperature is below 273.15 K. The changes in density and thickness due to the melting and freezing processes are also considered. The change in mass within each layer is

$$\frac{\partial(\rho_{sn}\Delta z_{sn})}{\partial t} = \left\{ \rho_{precipitation}\text{Pr} - E_{evap} \right\}_{sfc} + \{\text{net water flow}\} \tag{120}$$

where $\left\{ \rho_{precipitation}\text{Pr} - E_{evap} \right\}_{sfc}$ is the net change in the mass at the surface due to evaporation E_{evap} and precipitation $\rho_{precipitation}\text{Pr}$ at surface. If the wet-bulb temperature T_w is less than 0 °C, precipitation is assumed to be snow, otherwise is rain (Anderson 1976) [155]. The density of new snow is calculated by

$$\rho_{ns} = 1000 * [0.09 + 0.0017 * (T_w - 258.16)^{3/2}] \text{ (in kg m}^{-3}) \tag{121}$$

It is noted that 0.09 is used in (121) instead of 0.05 proposed by LaChapelle (1969) [156].

Following Lynch-Stieglitz (1994) [157], we assume that the maximum liquid water holding capacity is 5.5% by height of a compacted snow layer thickness. Should the liquid water content exceed the capacity, the excess water flows to the next lower layer. In the lower layer, the liquid water may freeze, remain in the layer, or flow through to the next lower layer. Liquid water leaving the bottom of the snow either infiltrates into ground or leaves the system as surface runoff. The heat transport due to the water flowing through a snow layer is also considered as shown in Equation (115).

The mechanical compaction is adapted from Pitman et al. (1991) [158], and Lynch-Stieglitz (1994) [5]:

$$\rho_s^{n+1} = \rho_s^n + \left\{ 0.5 \times 10^{-7} \rho_s^n gN_k \times \exp\left[14.643 - \frac{4000}{\min(T, 273.16)} - 0.02\rho_s^n \right] \right\} \Delta t \quad (122)$$

where gN_k is the weight of the snowpack above the midpoint of layer k.

The number of snow layers varies from 0 (no snow) to 6. The depth of each layer changes with time according to the conservation of mass and heat. Furthermore, the thickness of the top and bottom layer is always less than or equal to 6 cm to simulate the drastic change in snow property near the interfaces. The 5-year simulations of snow depth, temperature inside snow and beneath are compared excellently with observations shown in Figure 35, and the predictable time scale of the stiff nonlinear soil–snow model can be longer than a year. The snow–ice model was also coupled with a 1D mixed layer ocean model to simulate the sea ice at the SHEBA Experiment in 1997–1998 (Sun and Chern 2010 [159]).

5.4. Flooding Due to Snow Melt

(Min 2005 [160]), Min and Sun (2015) [161] applied the PRCM with the multilayer of snow–land–surface (SLS) module of Sun and Chern (2005) [114] to study the 1997 spring flooding in Minnesota and the Dakotas due to snowmelt. In the winter of 1996–1997, blizzard from the second half of November through January built up an enormous snowpack; many areas had more than 250 cm of snowfall, 2–3 times the normal annual amount. The frigid conditions were throughout much of the winter and early March in 1997, delaying the onset of snowmelt. Significant melt of the deep snow started with particularly warm conditions at the end of March and early April. Many rivers in South Dakota, Southern Minnesota, and Southern North Dakota were rising well above flood stage. The left panels of Figure 36 show the observed snow depth (top), simulation with a single snow layer (middle, referred as CTL), and with comprehensive snow–soil physics (bottom, referred as EXP 1) in March of 1997. The right panels show the observed wind at 200 hPa (top), the simulation from single snow layer (middle), and from comprehensive snow–soil model (bottom). The sea level pressure simulated with comprehensive snow–soil model (EXP1—solid black line) is also more accurate than the simple snow–soil model (red circle) compared with the ECMWF analysis (green circle) in Figure 37. Including cryosphere model physics lowers the surface temperatures due in part by the initial frozen soil conditions and the reduction of incoming solar radiation at the snow surface because of a higher albedo. These affect the horizontal temperature gradients, and in turn change the location of synoptic weather systems over the cold land region. Hence, an accurate snow–land model is crucial to short term weather forecasting or long-term climate study. Charney and Shukla (1981) [15] showed that the tropical sea surface temperatures or land surface soil moisture may evolve slowly or in predictable ways. Consequently, they can give a positive bias to the subsequent behavior of the atmosphere, and therefore provide the basis for some predictive power.

5.5. Southeast and East Asian Monsoon

Yu et al. (2004a, b) [162,163] applied the PRCM with 60 km resolution and comprehensive physics to study the regional climate in 10 summers (MJJA) during 1991–2000. Compared with ECMWF reanalysis, the simulated mean sea level pressure is excellent with bias = 0.15 hPa, the RMSE = 0.60 hPa, and the pattern correlation is 95%; the mean temperature has bias = 0.47 K, RMSE = 0.72 K, and correlation = 0.99 (Figure 38a,b). The observed and simulated precipitation shows correlation coefficient for mean precipitation = 35% (Figure 38c). Yu et al. suggested that the excessive precipitation may be caused by the misrepresented terrain or the southwestern lateral boundary of the model, and after the systematic error is removed, the errors reduce significantly, and the correlation becomes 0.84 (Table 3). The vorticity at 850 hPa shows excellent intraseasonal variability with the correlation coefficient as high as 0.9 or more (Figure 39).

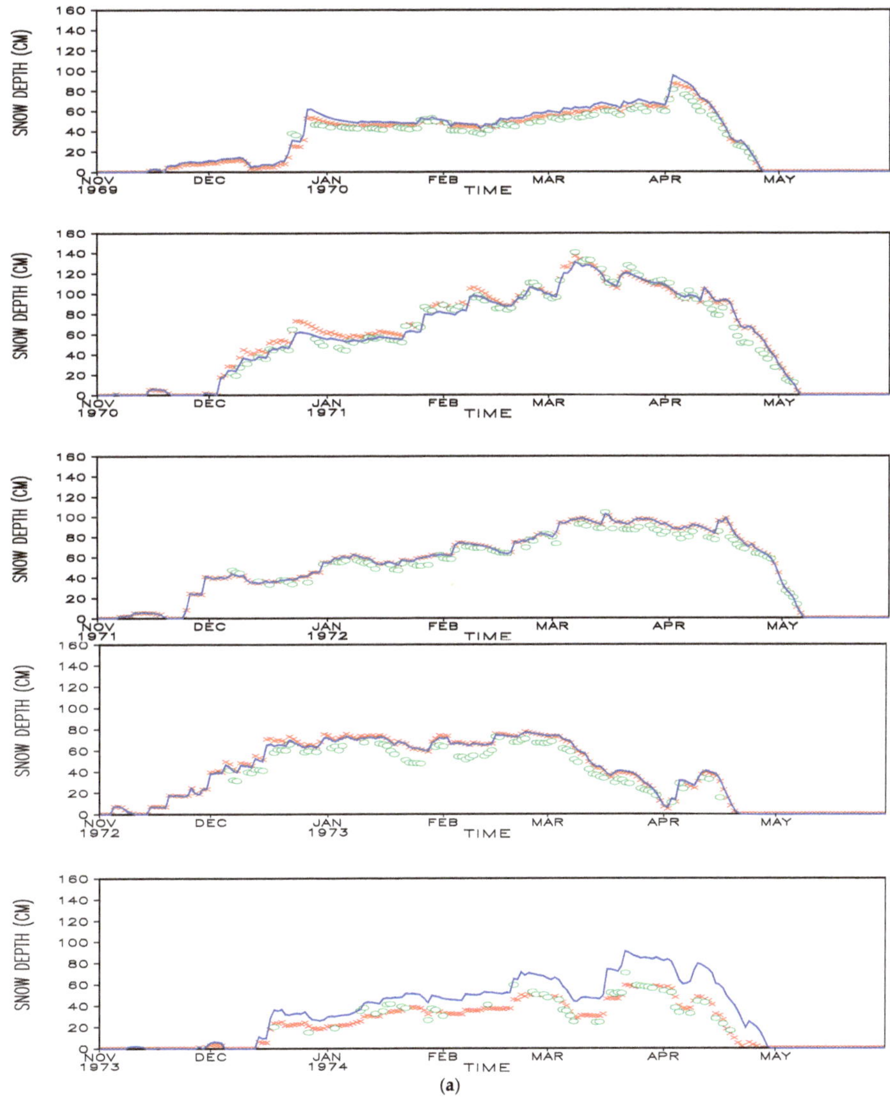

Figure 35. *Cont.*

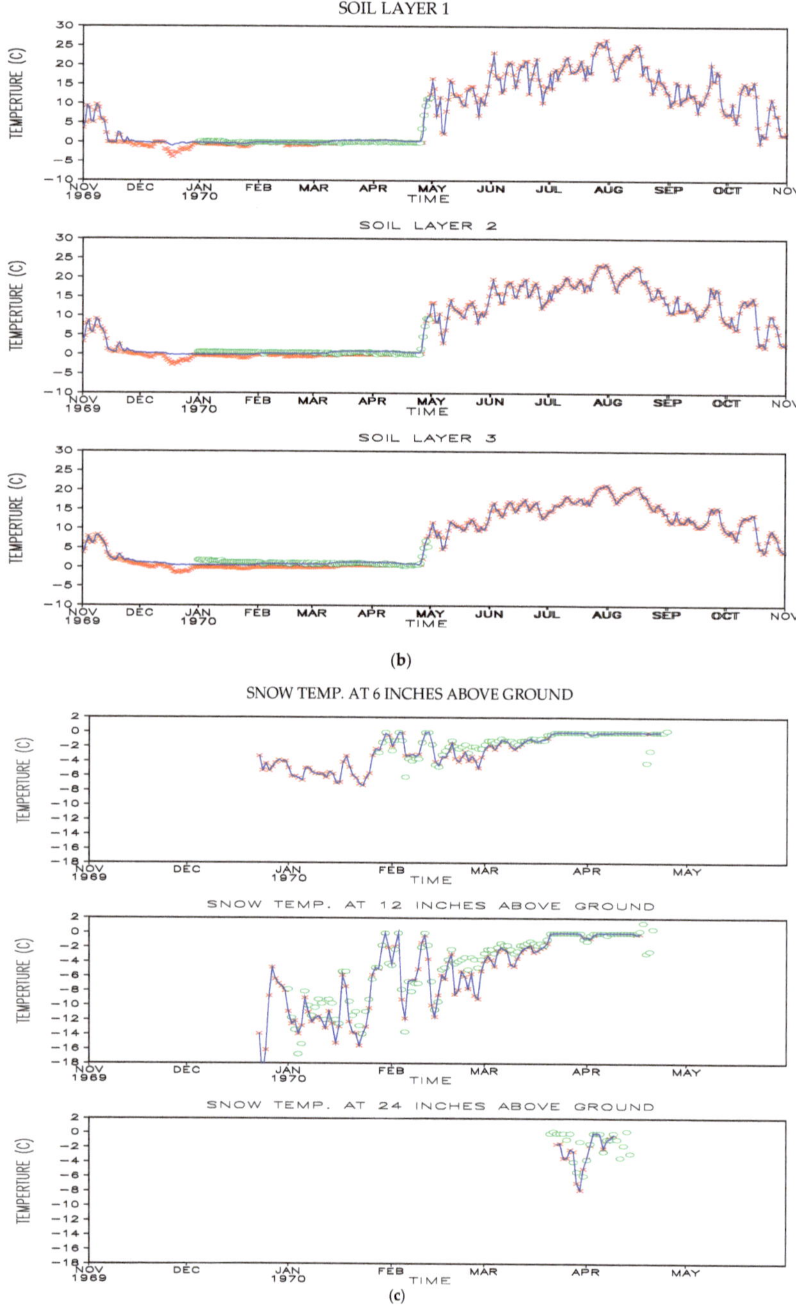

Figure 35. Observed (green circle) and simulated (red x for Case A, blue line for Case B) during November 1970 to May 1974 at Sleepers Watershed. (Case A: Fresh snow density is given by observed mean value of each winter. Case B: Fresh snow density = 150 kg m^{-3}): (**a**) snow depth, (**b**) ground temperature at layer 1–5850 and (**c**) snowpack temperatures at 6, 12, and 24 inches above ground (Sun and Chern 2005) [114].

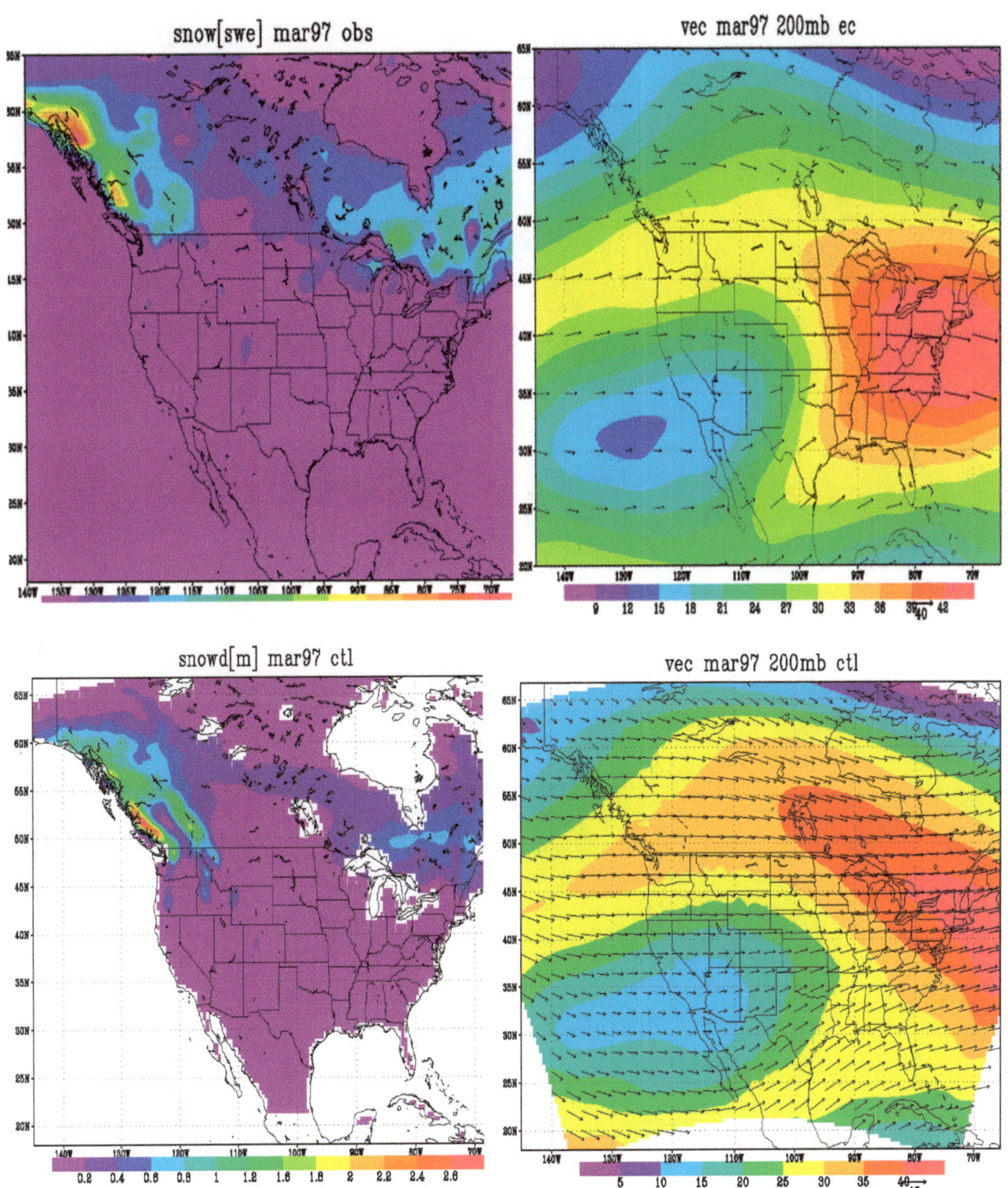

Figure 36. *Cont.*

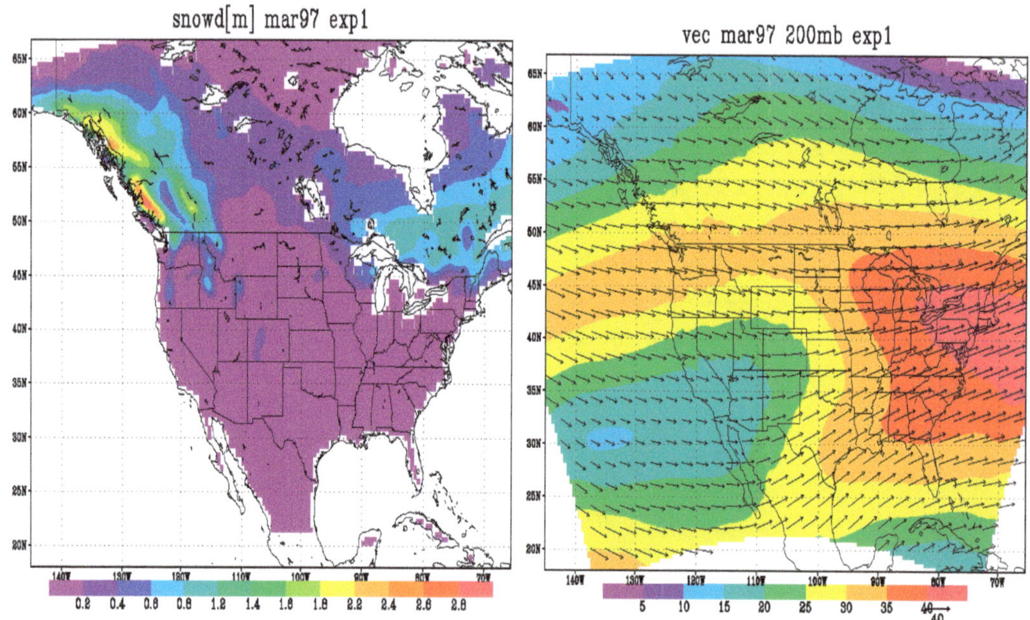

Figure 36. Left panels: the observed snow depth (**top**), simulation with a single snow layer (**middle**, CTL), simulated snow depth with comprehensive snow–soil physics (bottom, EXP 1) in March of 1997. Right panels: observed wind at 200 hPa (**top**), the simulation from single snow layer (**middle**), and simulation from comprehensive snow–soil (**bottom**). (Min and Sun 2015) [161].

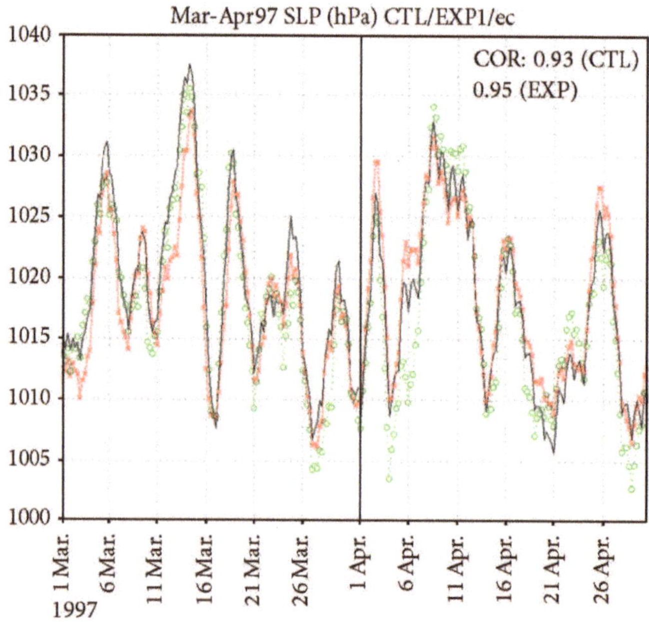

Figure 37. Time series of mean sea level pressure [hPa] from ECMWF (green circle), CTL (red cross), and EXP1 (solid line) simulations in Red River Region in March of 1997 (Min and Sun 2015) [161].

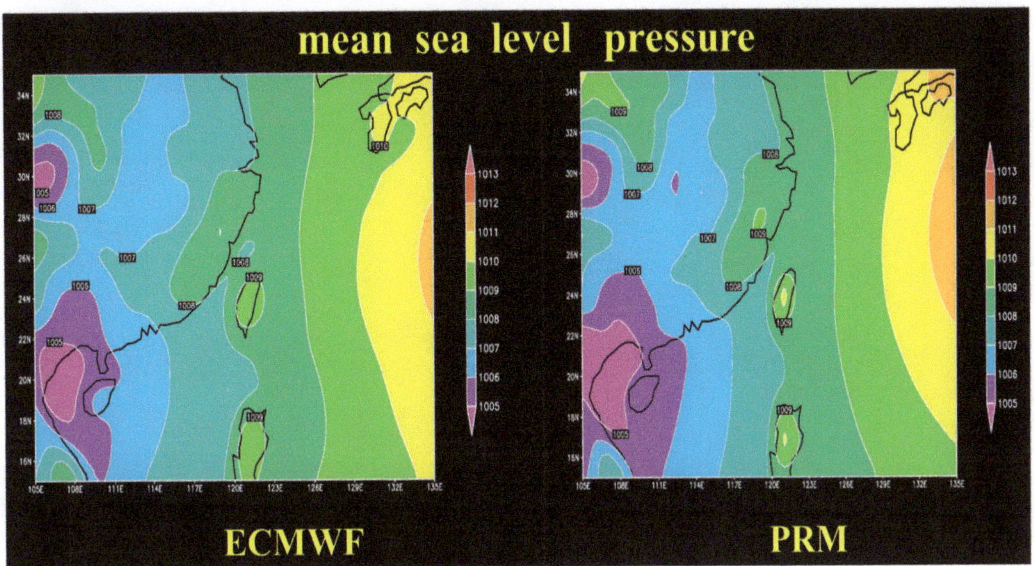

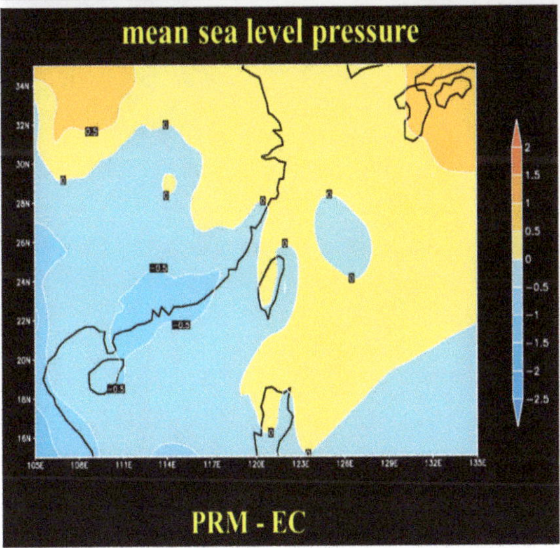

a

Figure 38. *Cont.*

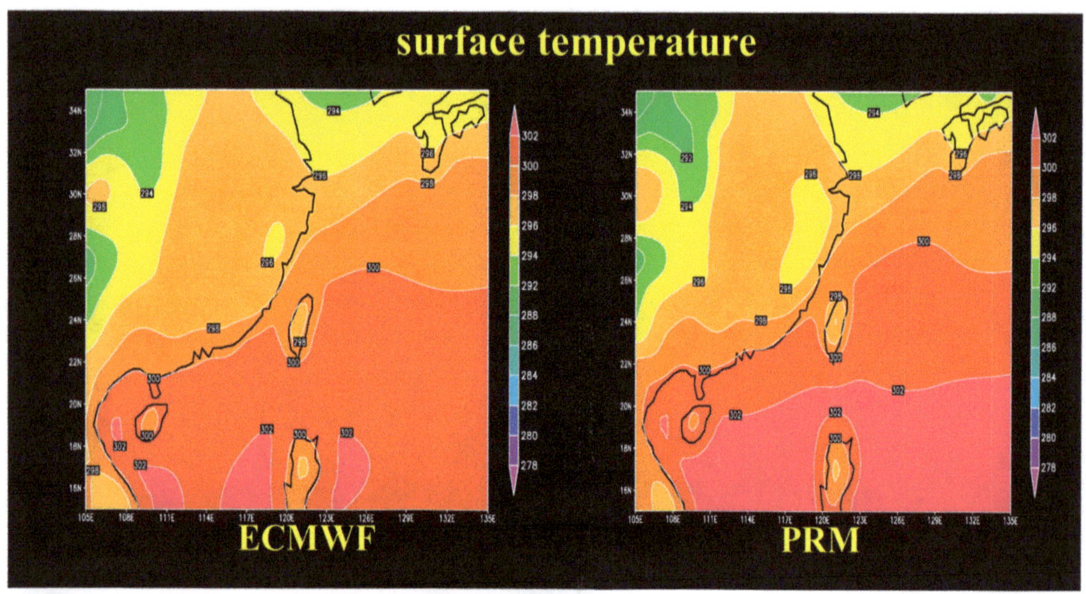

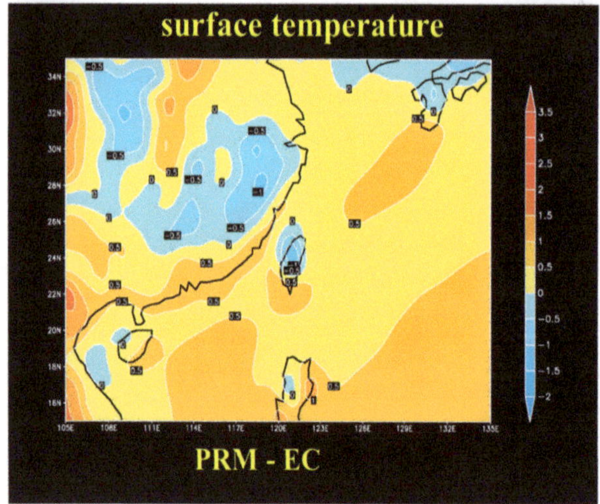

Figure 38. *Cont.*

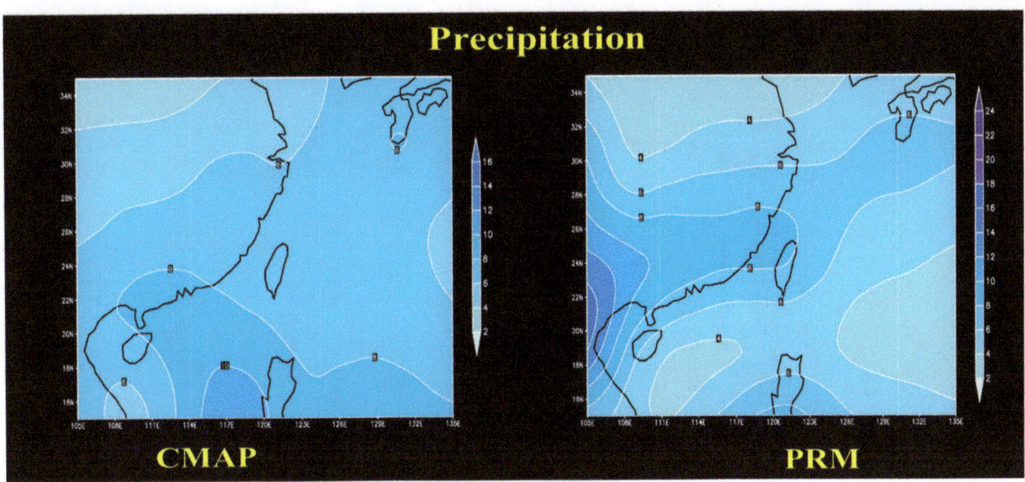

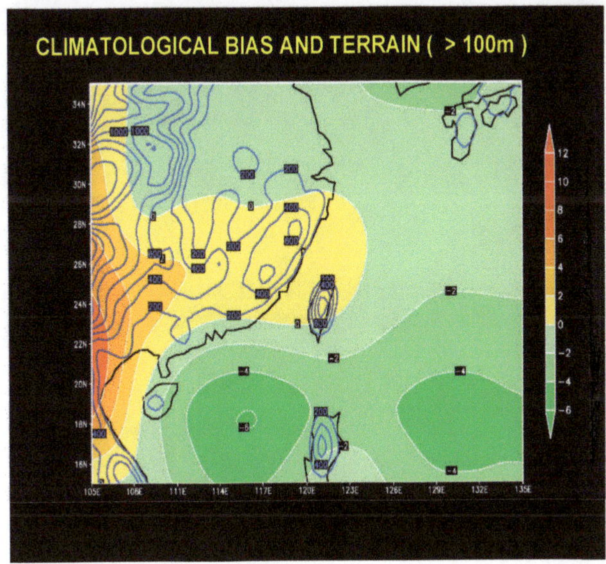

C

Figure 38. (**a**) ECMWF, PRCM in mean sea level pressures and difference, bias = −0.15 Pa, RMSE = 0.60 Pa, pattern correlation coefficient = 0.95 Yu et al. (2004a, b) [162,163]. (**b**) ECMWF, PRCM surface temperature and difference, bias 0.47 K, RMSE = 0.72 K, pattern correlation coefficient 0.99 Yu et al. (2004a, b) [162,163]. (**c**) ECMWF, PRCM precipitation and difference, terrain height indicated by contours in, pattern correlation is 0.35 (Yu et al., 2004a, b) [162,163].

Table 3. Climatological Pattern Correlation of PRCM (Yu et al., 1994b) [162,163].

Climatological Pattern Correlation of PRCM					
	MJJA	**May**	**Jun**	**Jul**	**Aug**
MSLP	0.98	0.96	0.98	0.97	0.96
TEM2M	0.99	0.99	0.99	0.97	0.97
850T	0.87	0.97	0.90	0.73	0.59
200T	0.88	0.83	0.83	0.94	0.94
850Z	0.97	0.96	0.99	0.98	0.97
200Z	0.96	0.99	0.98	0.94	0.77
850Q	0.96	0.98	0.96	0.95	0.91
200Q	0.84	0.93	0.80	0.70	0.71
PRECIP	0.35	0.37	0.52	0.27	0.29
PRECIP					
(bias removed)		0.72	0.89	0.90	0.85

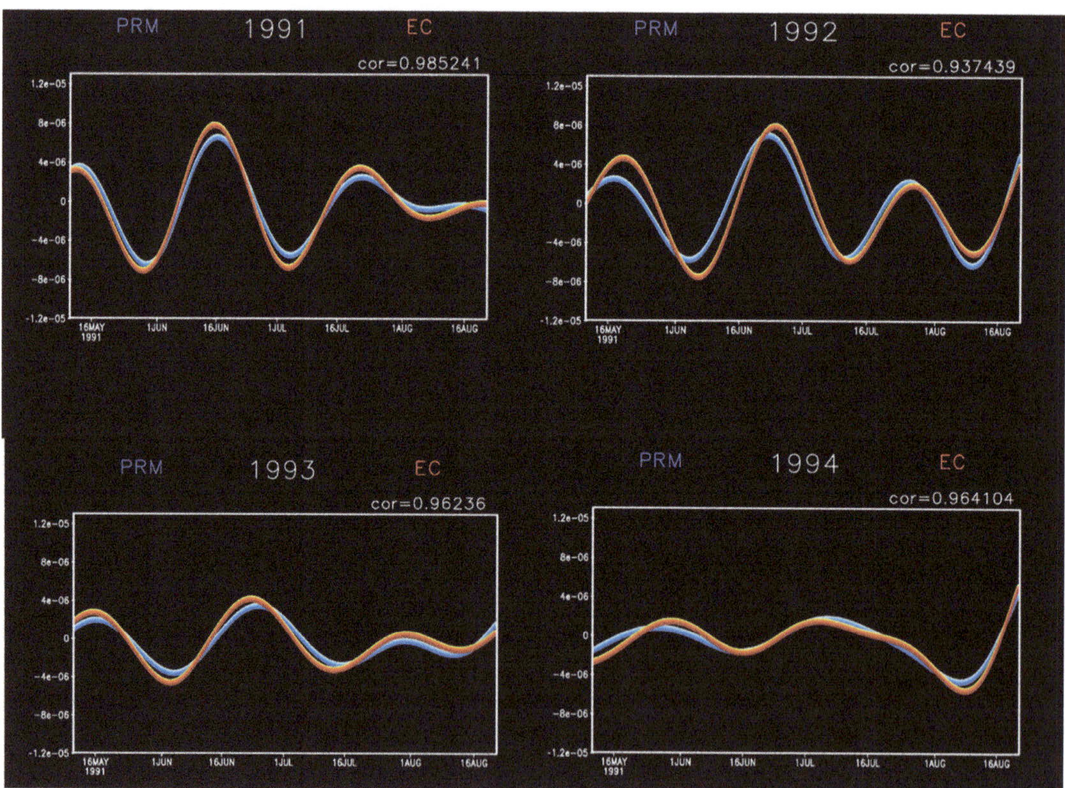

Figure 39. *Cont.*

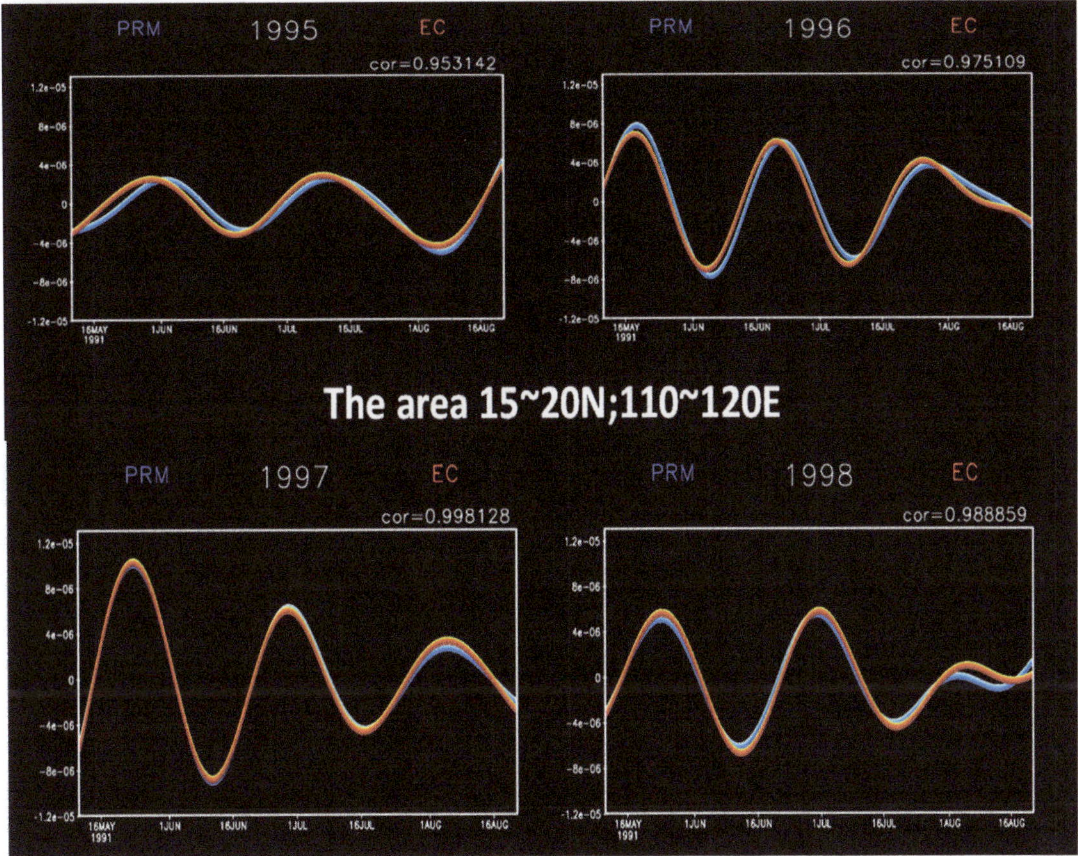

Figure 39. Vorticity at 850 hPa (red for ECMWF, blue for PRCM) and correlation for summers (May, June, July and August) from in 1991–1998 (Yu et al., 2004a, b) [162,163].

The results obtained by Hsu et al. (2004) [164] indicate that the PRCM could simulate the overall characteristics of the 1998 East Asian summer monsoon (EASM) on onset and the daily, seasonal, intraseasonal time scales. On the seasonal time scale, the model tended to produce more precipitation over the land and less precipitation over the ocean. This land–sea contrast in the model was stronger than observed Pacific anticyclone. The over simulated anticyclone by the PRCM, a fundamental problem, was also found in many models.

The northward-propagating rain bands and intraseasonal oscillation events were well reproduced (Figure 40). The model can produce realistic sub-seasonal variability in the inner domain once the proper lateral boundary forcing is provided. It correctly simulated the onset timing and dramatic changes before and after the onset. But the model incorrectly simulated the circulation and precipitation during the onset because it failed to simulate the rapid development of a weak trough in the Northern South China Sea and missed some extreme events that resulted in flooding in China. After the onset period, the model performance became reliable again. This indicates that the model can fix the existing large biases with the proper lateral boundary, and the dynamic system with the strange attractor being locally unstable yet globally stable.

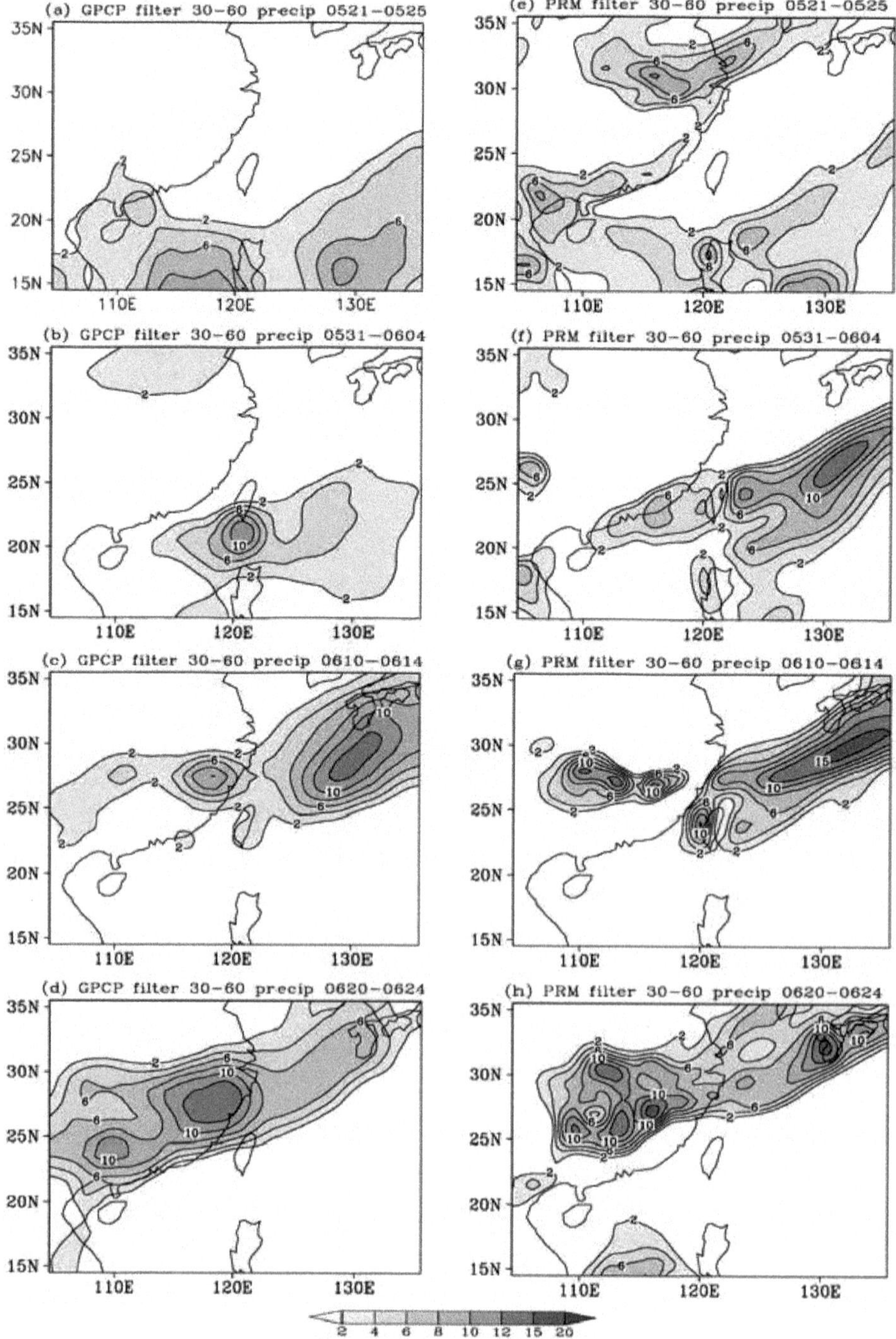

Figure 40. The 30–60-day band-pass filtered GPCP (left) and PRM (right) precipitation for the 5-day periods: (**a,e**) 21–25 May, (**b,f**) 31 May–4 June, (**c,g**) 10–14 June, and (**d,h**) 20–24 June. In 1998. The contour interval is 2 mm/day, and the shading indicates the positive anomalies (Hsu et al., 2004) [164].

Sun et al. (2011) [165] compared the PRCM simulation with ECMWF on heavy precipitation over Korea and China from 30 July to 18 August 1998. They found that heavy rainfall along the Mei-Yu (Changma) front was maintained by: (1) an anomalous 850 hPa subtropical high, which intruded westward and enhanced the LLJ over Southeastern Asia, (2) a stronger baroclinicity around 40° N over Eastern Asia and a low pressure located over Japan, Korean, and Mongolia, which not only enhanced southwesterly wind along the frontal zone but also blocked the northward movement of the rain band, and (3) excessive evaporation from abnormally wet, warm land. The simulated precipitation ended by 18 August when the subtropical high retreated and the low-pressure in Mongolia also moved away from the Asian continent, which was consistent with the ECWMF reanalysis and observed precipitation estimates from GPCP. High correlations were found for both surface and upper-level atmospheric fields when compared to ECMWF reanalysis for the 20-day means, as shown in Table 4. It is noted that the slower propagation of the simulated low pressure in Mongolia compared to the observed data may be responsible for the deepening of the low and excessive precipitation in that area. It is also suspected that the model simulates too strong a convection caused by the modified Kuo-type cumulus parameterization scheme (Molinari and Dudek, 1992 [166]), which may enhance a cyclonic circulation and strengthen the low pressure. This may lead to stronger convergence of moisture and heavier precipitation.

Table 4. 20-day mean statistics pf PRCM compared to reanalysis data of ECMWF except for precipitation which is derived from GPCP.

FIELD	LEVEL	BIAS	RMS	COR
Mean Sea Level Pressure (hPa) Air Qv (kg/kg) Temperature (K) Wind speed (ms^{-1}) Precipitation (mmday^{-1})	SFC	-0.48	1.57	0.95
		2.45×10^{-3}	2.81×10^{-3}	0.96
		0.73	1.75	0.97
		0.78	1.26	0.84
		0.95	1.38	0.73
Height (m)	200 hPa	37	45.1	0.99
	500 hPa	12.5	17.7	0.98
	850 hPa	0.25	10.3	0.97
Temperature (K)	200 hPa	0.74	1.17	0.9
	500 hPa	0.63	0.86	0.98
	850 hPa	0.13	1.22	0.94
Qv (kg/kg)	500 hPa	-1.34×10^{-4}	6.59×10^{-4}	0.83
	700 hPa	3.01×10^{-4}	1.22×10^{-3}	0.82
	850 hPa	1.13×10^{-3}	1.61×10^{-3}	0.92
Wind speed (ms^{-1})	200 hPa	-1.60	3.98	0.91
	500 hPa	-0.72	2.14	0.8
	850 hPa	1.28	2.16	0.83

The well simulations of seasonable variations of 850 vorticity over S. China Sea, the northward propagation of the Monsoon front in East Asia, as well as 1988 drought, 1993 flood, 1997 snow-melted flood in the USA, etc. show that the PRCM can perform the monthly ~ seasonal simulations. It also indicated that the lateral boundary conditions are more important than the initial small-scale disturbances. They are consistent with Paegle et al. (1997) [13] and Hoskins (2012) [16] statement: "Smaller scale phenomena that cannot be represented explicitly may be partially "slave" to the retained scales, like general regions of convection to a front, in which case aspects of their feedback on the retained scales may be well determined by retained scales". Those large-scale systems with slow time variations usually carry more energy, which can transfer to the smaller scales with high frequencies due to nonlinear interactions. The simplified low-order dynamic models contain the nonlinear interaction among the spectral components in a rudimentary form only. Therefore, they do not have cascade processes linking the spectral region of the large-scale forcing with the dissipation range (Wiin-Nielsen 1998 [14]). They also miss the feedback from the smaller waves to the large waves, which can be important in meteorology. Hence, Shen (2019) [167] includes more modes to Lorenz model and found the stable point in addition to the original strange attractors.

5.6. Simulations of Dust Storm

Yang (2004a) [127] has coupled the PRCM and the dust module as illustrated in Figure 41 to simulate the severe dust storms in Asia in April 1998. The dust module consists of: (a) Dust source function derived by Ginoux et al. (2001) [168] based on 1° by 1° terrain and vegetation data set derived from the advanced very high-resolution radiometer (AVHRR) data (DeFries and Townshend 1994 [169]); (b) Particle size, which is a function of the source region's soil properties (Tegen and Fung 1995) [170]). Seven size bins (0.1–0.18 μm, 0.18–0.3 μm, 0.3–0.6 μm, 0.6–1 μm, 1–1.8 μm, 1.8–3 μm, and 3–6 μm, with corresponding effective radii of 0.15, 0.25, 0.4, 0.8, 1.5, 2.5, and 4 μm) are applied in the model; (c) The threshold friction velocity, defined as the horizontal wind velocity required to lift dust particles from the surface, is a function of particle diameter (Marticorena and Bergametti 1995 [171]) and soil wetness; (d) Dust emission, which depends on source function, surface wind speed, threshold velocity, and the fraction of each size class; and (e) Transport and removal processes. Dust aerosols in the PRCM-Dust are transported by advection, dispersion, sub-grid cumulus convection, wet and dry depositions. The dry deposition of dust aerosols is assessed by the gravitational settling for each model vertical layer and surface deposition velocity. The removal of dust aerosols by wet deposition is calculated using the model precipitation rate for both stratified and convective clouds.

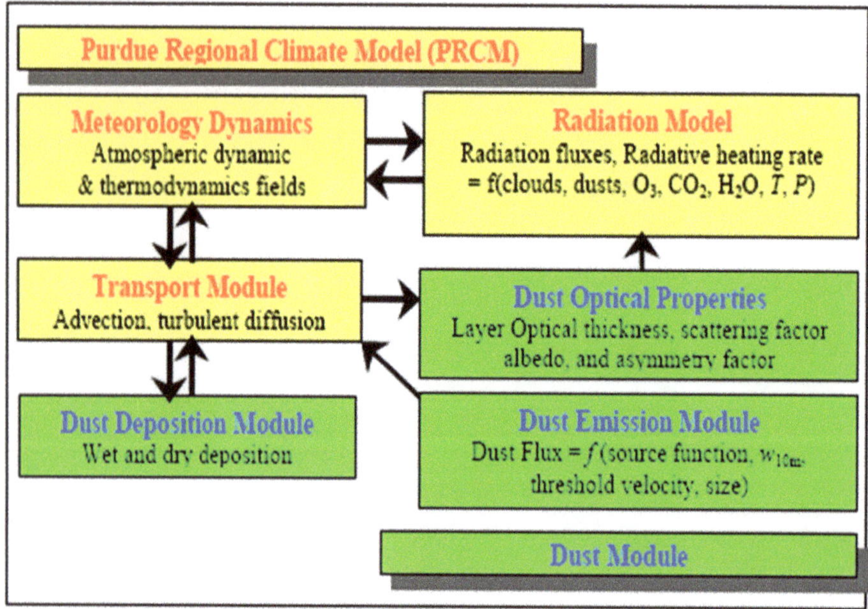

Figure 41. The schematic illustration of the components in the integrated PRCM-Dust model (Yang 2004a) [127].

With a resolution of 60 km × 60 km, Yang (2004a) [127] has integrated the model continuously from 8–24 April 1998 without nudging or restart. The control run is to provide the environment and weather conditions without dusts, and a reference to compare with the case including dusts run (Sun et al., 2013a, b [172,173]). The ECMWF mean geopotential (m) at 850 hPa and wind at 700 hPa during April 8–24 (Figure 42) are well simulated (Figure 43) (Sun et al., 2013a, b) [172,173]. The ending on April 23 is in good agreement (Figure 44). The uplifted dusts reached around 800 hPa or higher over the source region and remained at 3–5 km or higher in the downwind regions. Dusts were transported southeastward with a cutoff low, then moved northeastward before 18 April and then were transported eastward after. They are consistent with the observed trajectory shown in Figure 45. The horizontal distributions of the dust aerosols were consistent with the satellite images, the TOMS Aerosol Index maps, lidars, surface network reports, and the isentropic trajectory derived from the weather map. The radiative forcing of dust aerosols induced a warming in Northern Asia and cooling in Southern Asia. They are similar with KHSS97 (Koepke et al., 1997) [174] and TF95 (Tegen and Fung 1995) [170] data. Figure 46a shows the difference in the simulated surface temperature with and without dust. The difference is +1.81 K in Northeast Asia and −3.03 K in Southeast Asia due to the radiative effects induced by the large amount of dust aerosols in North and Northeast Asia. This is consistent with the results of Myhre and Stordal (2001) [175] as well as the bias of April 1998 from the 10-year mean of the ECMWF reanalysis (Figure 46b). The influence of dust was far beyond the polluted areas and reached all Asia and India. The warming was over the areas with a high concentration of dust in April 1998, because most of the dust stayed above the low clouds or even coexisted with the cirrus clouds. This was also validated by satellite images that showed the dust layer resided on top of low-lying clouds and significantly reduced the cloud albedo, particularly near the ultraviolet (Husar et al., 2001) [176]. The dust layer also trapped longwave radiation from below. The results are different from the fifth generation Penn State/NCAR Mesoscale Model (MM5) and other results. Their simulated mean height of the dusts was lower than observed and resulted in a southward transport of dusts (In and Park, 2002) [177]. It is also noted that many models used nudging, data assimilation to adjust the meteorological field, or restarted the model daily or every other day, and they cannot be used to assess the effect of aerosols on weather or climate.

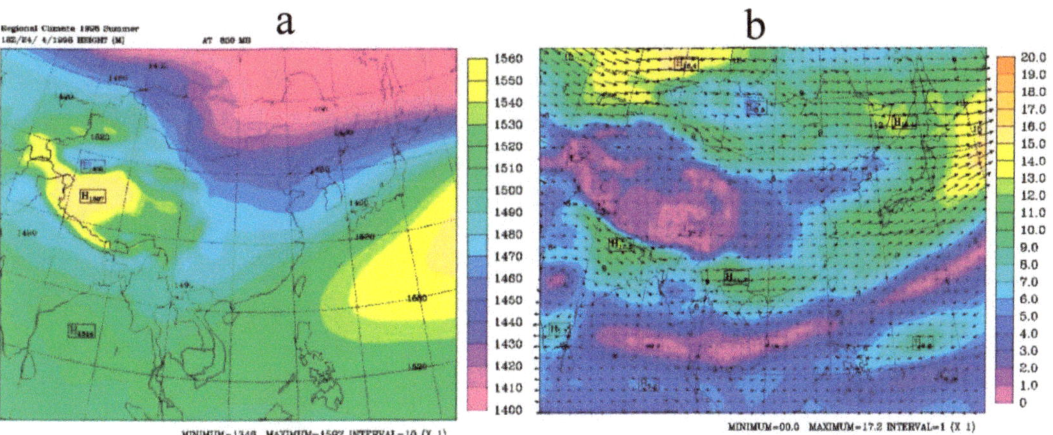

Figure 42. (**a**) ECMWF Mean 850 hPa geopotential (m) during April 8–24; (**b**) wind at 700 hPa (unit: m/s). (Yang 2004a) [127].

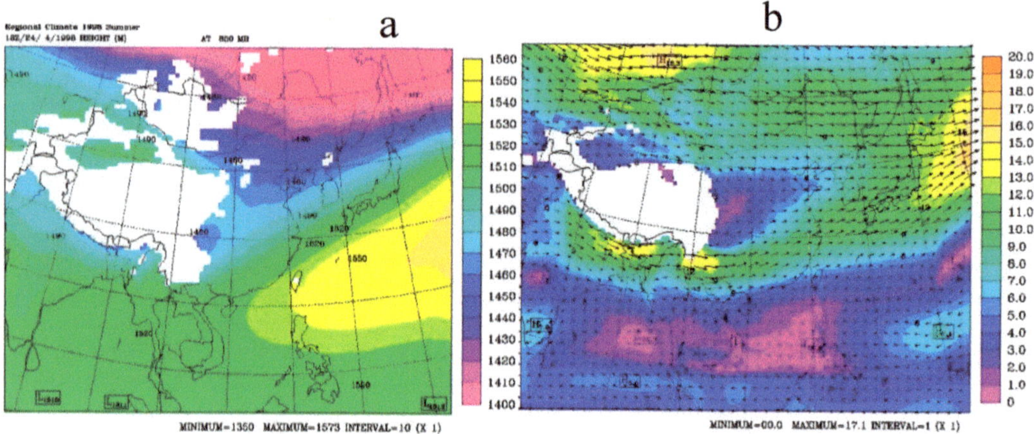

Figure 43. (**a**) Simulated mean 850 hPa geopotential (m) during 8–24 April; (**b**) wind at 700 hPa (unit: m/s). (Sun et al., 2013a, b) [172,173].

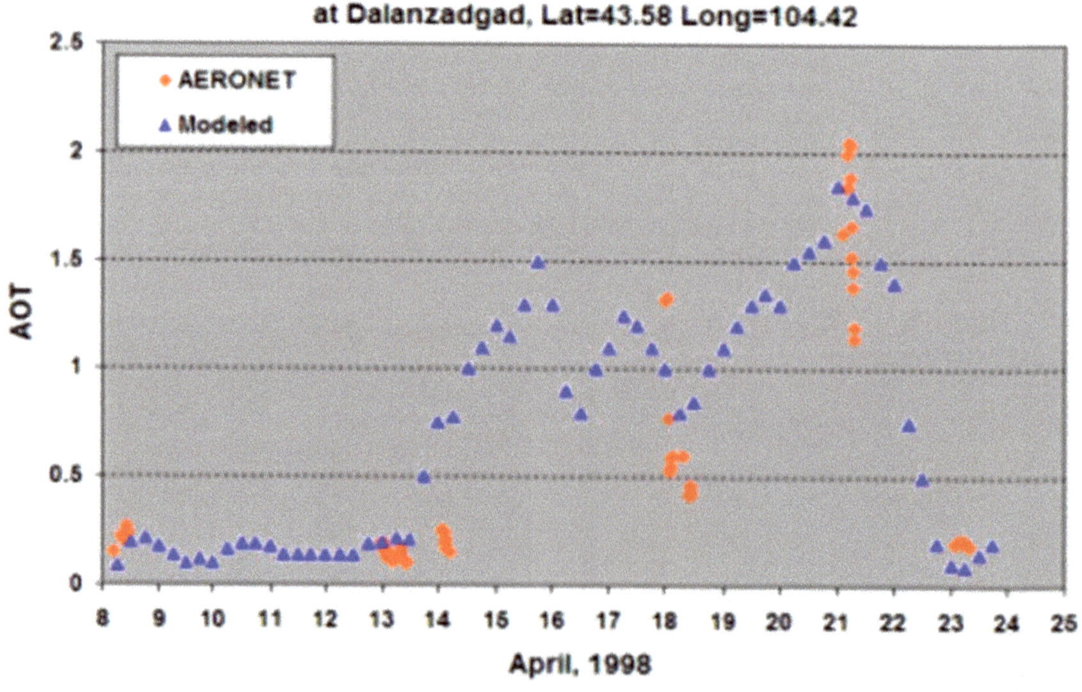

Figure 44. Total column aerosol optical thickness at 500 nm from PRCM-Dust model and AERONET measurement site at Dalanzadgad, Mongolia. (Sun et al., 2013a, b) [172,173].

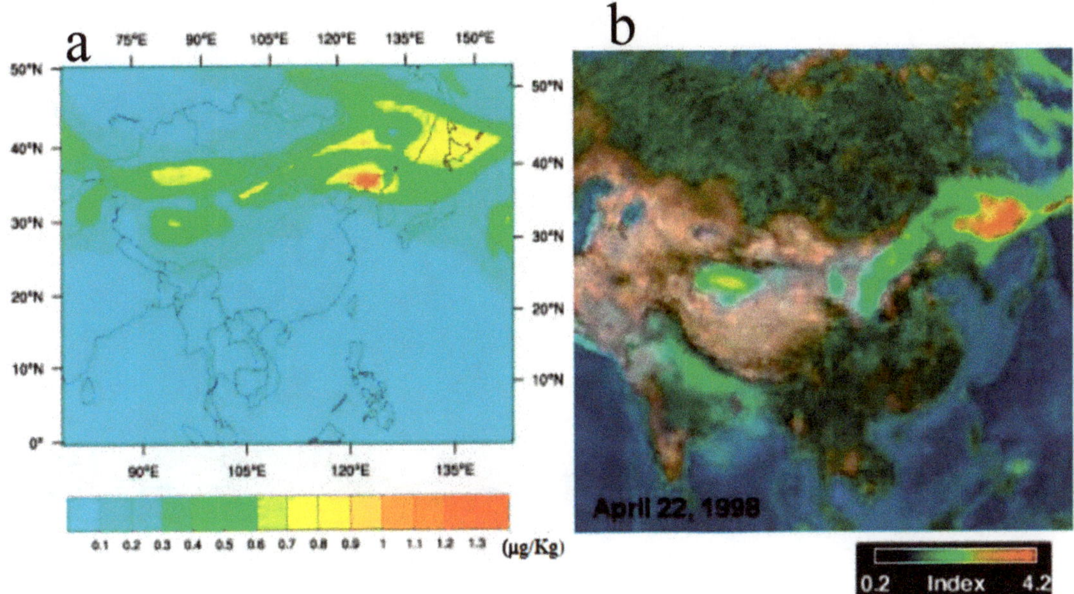

Figure 45. (a) Simulated mixing ratio of dust near 550 hPa and (b) TOM AI maps at 00Z 22 April 1998. (Sun et al., 2013a, b) [172,173].

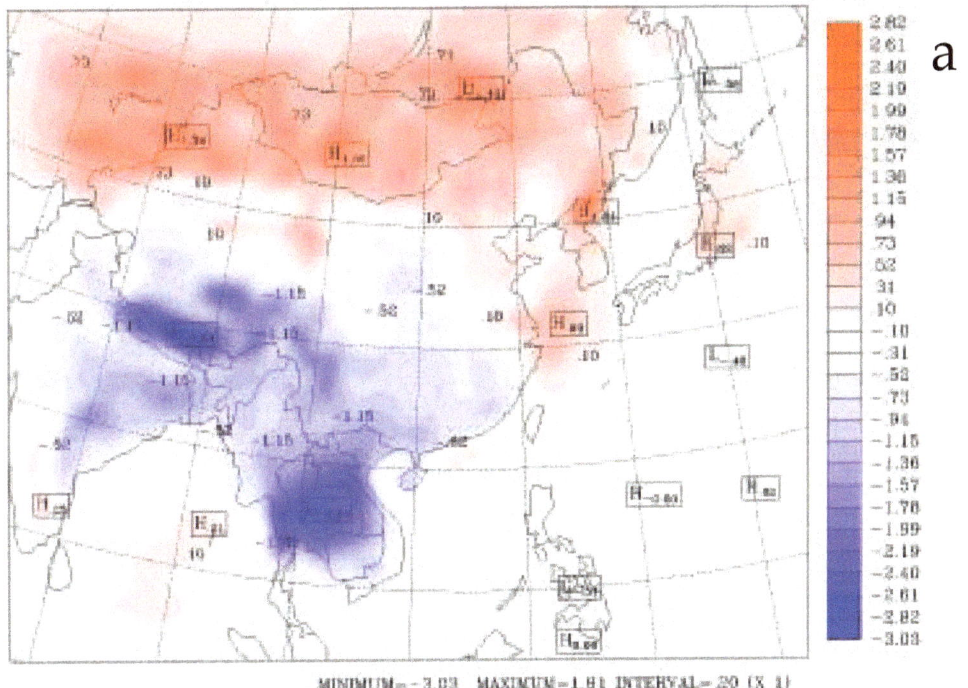

Figure 46. *Cont.*

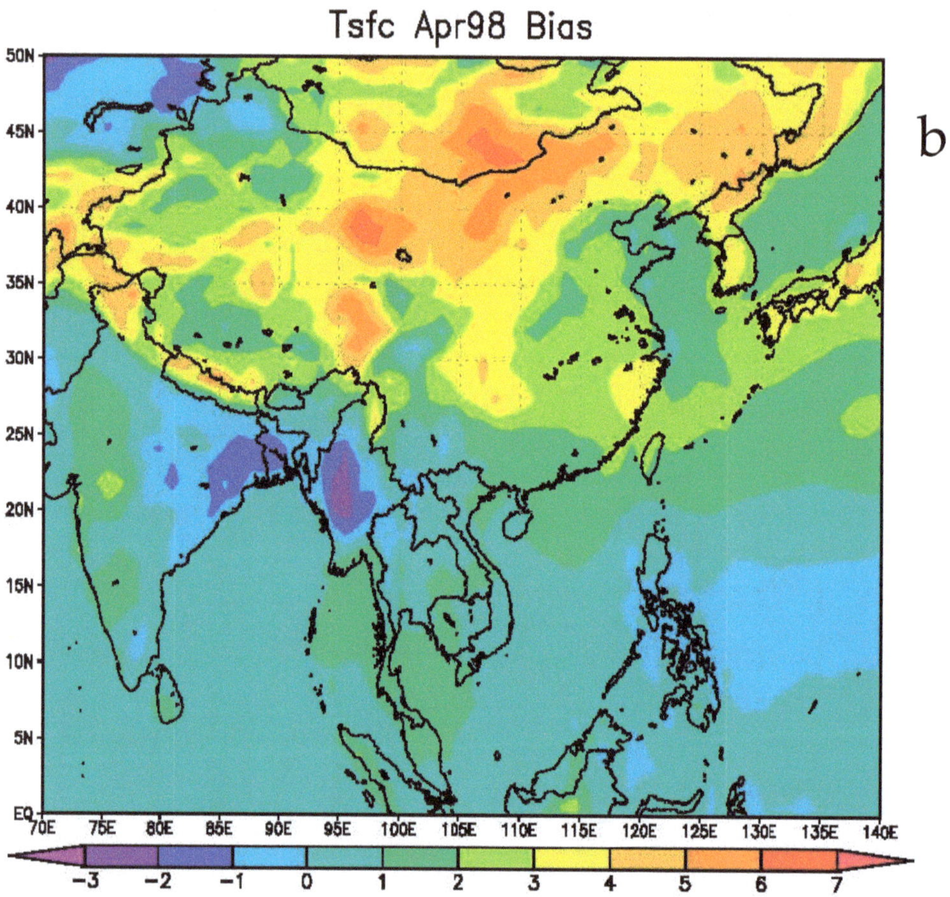

Figure 46. (**a**) PRCM (Dust)–PRCM (without Dust) daily mean surface temperature at 00Z for 8–24 April, (**b**) ECMWF April 1998 monthly surface mean temperature 10-years mean. (Sun et al., 2013a, b) [172,173].

The PRCM was also used by Wu et al. (2003) [178] to study the radiative aerosols released from Chinshan Nuclear Power Plant in Taipei County, Northern Taiwan during winter.

In addition to the initial condition, a regional model requires the lateral boundary conditions from the general circulation models (GCMs) or reanalysis. The PRCM can replicate the weather systems and show the physics and dynamics involved for a month or longer integrations, because the ECMWF reanalysis provides excellent lateral boundaries. If a regional model is applied to do forecasting, the performance depends on the GCMs which provide the lateral boundary (Sun 2006) [179]. In addition to numerical methods, we should improve our understanding of the physical processes among aerosols, clouds, precipitation, radiation, and the turbulent fluxes from sea, land, and mountains to improve the predictability of a model. Meanwhile, if we cannot duplicate the previous and current weather/climate, it is hard to claim that we can predict the climate in the future.

6. Nonhydrostatic Models

6.1. Boussinesq Fluid Versus Compressible Fluid

The vertical momentum equation is:

$$\delta\frac{d\,w}{d\,t} = -\frac{1}{\rho}\frac{\partial\,p}{\partial\,z} - g \tag{123}$$

where $\delta = 0$ for a hydrostatic model, and $\delta = 1$ for a nonhydrostatic model. In Boussinesq fluid, the frequency of an internal gravity wave becomes:

$$\sigma^2 = \frac{k^2 N^2}{\delta k^2 + m^2} \tag{124}$$

It shows that if $k^2 \ll m^2$, nonhydrostatic simulations can be represented by hydrostatic models. The nonhydrostatic models need lots of computing time to solve either the time-consuming Poisson's equation in an anelastic (or incompressible) atmosphere, or acoustic waves with a very small-time interval in the compressible atmosphere. Hence, Hsu and Sun (1991, 1992) [180,181] added the selected diffusion in the hydrostatic model to simulate the hexagon convective clouds over ocean. Laprice (1992) [182] and Yeh et al. (2002) [183] used a hydrostatic pressure as the basis for the vertical coordinate, then incorporated the nonhydrostatic departure from the hydrostatic system as a perturbation. The resulting set of nonlinear equations is solved iteratively. Janjic (2003) [184] used a mass-based vertical coordinate in his nonhydrostatic Meso Model developed at NCEP, and the nonhydrostatic dynamics has been introduced through an add-on module. It is noted that the speed of sound is infinite in an incompressible fluid.

The atmospheric nonhydrostatic models are frequently solved either as a compressible or anelastic fluid. An anelastic fluid that eliminates acoustic waves is required to solve the time-consuming Poisson equation (Kuo and Sun 1976 [85], Sun and Orlanski 1981b [88], and Chen and Sun 2001 [185]). Ogura and Phillips (1962) [186] showed that if the percentage range in potential temperature is small, and the vertical scale is small compared to the depth of an adiabatic atmosphere, the anelastic system reduces to the Boussinesq equations. The Boussinesq approximations include: the density variation is ignored except in the buoyancy term; and the density variation in buoyancy can be represented by

$$d\ln\rho = -d\ln T, \text{ or } d\ln\rho = -d\ln\theta. \tag{125}$$

The equation of state of the air is

$$p = \rho RT, \text{ or } d\ln p = d\ln\rho + d\ln T \tag{126}$$

If $|\,d\ln p\,| \ll |d\ln\rho|$, we have

$$d\ln\rho = -d\ln T, \text{ or } d\ln\rho = -d\ln\theta, \tag{127}$$

which is one of the Boussinesq assumptions. But in an adiabatic process, we also have

$$(c_v/c_p)d\ln p = d\ln\rho. \tag{128}$$

Hence, the Boussinesq assumption is inconsistent when it is applied to the atmosphere. But Boussinesq approximations simplify the equations and become popular in meteorology and oceanography. The traditional parcel method, the Froude number in mountain dynamics, and CAPE are based on the Boussinesq fluid, their limitations will be discussed in Sections 6.5–6.7. The dispersive relation of the inertia internal gravity wave is

$$\sigma^2 = \frac{(k^2 + l^2)N^2 + f^2 m^2}{(k^2 + l^2 + m^2)}. \tag{129}$$

Equation (129) is also derived from the Boussinesq fluid. Property of anelastic fluid ($\frac{\partial \rho}{\partial t} = -\nabla\cdot(\rho\mathbf{v}) = 0$) is similar to the incompressible fluid ($\frac{d\rho}{\rho dt} = -\nabla\cdot\mathbf{v} = 0$). But ρ obtained from the Poisson equation can change due to the change in temperature and velocity field, which may disrupt the conservation of mass. Hence, the compressible fluid may be a better choice. The frequency in an irrotational, compressible atmosphere is

$$\sigma^4 - \sigma^2 C^2 \left[\frac{gS_0}{C^2} + m^2 + k^2 + \mathcal{I}^2 + \Gamma_0^2 \right] + \left(k^2 + \mathcal{I}^2 \right) gS_0 C^2 = 0 \tag{130}$$

where $\Gamma_0 = -\frac{g}{C^2} - \frac{R_0}{2} = S_0 + \frac{R_0}{2}$, and subscript 0 indicates the environment value. Equation (130) contains two acoustic waves and two internal gravity waves, the latter is corresponding to (124).

The compressible nonhydrostatic model requires a small-step to handle acoustic waves according to the Courant–Friedrichs–Lewy (CFL) criterion. MacDonald et al. (2000) [187] proposed a quasi-nonhydrostatic model by multiplying α (typically the square of the vertical to horizontal aspect ratio) to the hydrostatic terms in the vertical momentum equation to decrease both the frequency and amplitude of gravity waves. It enables us to calculate the vertical equations explicitly in a longer time step but slow down the hydrostatic adjustment process. Klemp and Wilhemson (1978) [188] proposed a horizontal explicit and a vertical implicit scheme (HE-VI). Sun and Hsu (2005) [189] used a parameter on the mass divergence term in the continuity equation to damp the high frequency acoustic waves in the NTU/Purdue nonhydrostatic model (Hsu and Sun 2001 [190], Hsu et al., 2004 [191], Sun et al., 2009 [192], and Hsieh et al., 2006, 2010 [193,194], etc.). Using the Cloud Resolving Storm Simulator (CReSS, Tsuboki and Sakakibara 2007 [195]), Sun et al. (2012) [196] proved the method can significantly increase the time interval without affecting the gravity waves or thermal convections. Sun et al. (2012) [196] also demonstrated that the horizontal explicit and a vertical implicit scheme (HE-VI) can introduce big errors if the horizontal length scale is comparable with the vertical length scale.

6.2. Modified Forward–Backward Scheme with Smoothing (MFBS)

The non-hydrostatic equations are:

$$\frac{\partial u_i}{\partial t} + u_j \frac{\partial u_i}{\partial x_j} = -\frac{1}{\rho}\frac{\partial p}{\partial x_i} + g_i + Fr_{ri} = \frac{F_i}{\rho} \tag{131}$$

The buoyancy

$$g_i = \begin{cases} -g & \text{if } i = 3 \\ 0 & \text{otherwise} \end{cases} \tag{132}$$

$$\frac{\partial \theta}{\partial t} + u_j \frac{\partial \theta}{\partial x_j} = \frac{1}{c_p}\frac{\theta}{T}\frac{dq}{dt} + D_\theta \tag{133}$$

$$\delta \frac{\partial \rho}{\partial t} + \frac{\partial(\rho u_j)}{\partial x_j} = 0 \tag{134}$$

$$p = \rho R T \tag{135}$$

$$\theta = T \left(\frac{P_O}{P} \right)^{\frac{R}{c_P}} \tag{136}$$

The 2D linearized equations in (x, z) directions become:

$$\frac{\partial u'}{\partial t} + \frac{1}{\rho_o}\frac{\partial p'}{\partial x} = 0 \tag{137}$$

$$\frac{\partial w'}{\partial t} + \frac{1}{\rho_o}\frac{\partial p'}{\partial z} + g\frac{\rho'}{\rho_o} = 0 \tag{138}$$

$$\frac{\partial \theta'}{\partial t} + w'\frac{\partial \theta_0}{\partial z} = 0 \tag{139}$$

$$\delta\frac{\partial \rho'}{\partial t} + \rho_0\left(\frac{\partial u'}{\partial x} + \frac{\partial w'}{\partial z}\right) + w'\frac{\partial \rho_0}{\partial z} = 0 \tag{140}$$

$$\rho' = \frac{p'}{C^2} - \frac{\rho_0 \theta'}{\theta_0} \tag{141}$$

where primes are deviations from background (with subscript "0"), which are function of height only, and $C = \sqrt{(c_p/c_v)RT_0}$. The background wind is zero. $\delta > 1$ was introduced in the continuity equation to filter the high frequency acoustic waves. If we define: $\hat{p} = p'/\sqrt{\rho_0}$, $\hat{\theta} = \sqrt{\rho_0}\theta'/\theta_0$, $\hat{u} = u'\sqrt{\rho_0}$, $\hat{w} = w'\sqrt{\rho_0}$, $S_0 = \frac{\partial \theta_0}{\theta_0 \partial z}$, and $R_0 = \frac{\partial \rho_0}{\rho_0 \partial z}$, we have:

$$\frac{\partial \hat{u}}{\partial t} + \frac{\partial \hat{p}}{\partial x} = 0 \tag{142}$$

$$\frac{\partial \hat{w}}{\partial t} + \frac{\partial \hat{p}}{\partial z} + \left(\frac{g}{C^2} + \frac{R_0}{2}\right)\hat{p} - g\hat{\theta} = 0 \tag{143}$$

$$\frac{\partial \hat{\theta}}{\partial t} + \hat{w}S_0 = 0 \tag{144}$$

$$\frac{\partial \hat{p}}{\partial t} = -\frac{C^2}{\delta}\left(\frac{\partial \hat{u}}{\partial x} + \frac{\partial \hat{w}}{\partial z} + (S_0 + \frac{R_0}{2})\hat{w}\right) \tag{145}$$

Sun et al. (2013c) [197] showed that the $2\Delta x$ and/or $2\Delta z$ waves require the shortest Δt to satisfy the CFL criterion. To relax the time step limitation, Sun et al. (2013c) [197] apply smoothing on the right-hand side of (145). Thus, equation (145) is replaced by

$$\frac{\partial \hat{p}}{\partial t} = \frac{C^2}{\delta}\overline{f}^{xz} \tag{146}$$

where

$$f = -\left[\frac{\partial \hat{u}}{\partial x} + \frac{\partial \hat{w}}{\partial z} + (S_0 + \frac{R_0}{2})\hat{w}\right]$$

and

$$\overline{f}^{xz} = \overline{\left(\overline{f}^x\right)}^z = \overline{\left(\frac{f_{p-1,q} + 2f_{p,q} + f_{p+1,q}}{4}\right)}^z = \cos^2\left(\frac{k\Delta x}{2}\right)\cos^2\left(\frac{m\Delta z}{2}\right)f$$

where k and m are the wave numbers in the x- and z-directions, respectively. Equation (145) becomes

$$\begin{aligned}\frac{\partial \hat{p}}{\partial t} &= \frac{C^2}{\delta}\overline{f}^{xz} = -\frac{C^2\cos^2\left(\frac{k\Delta x}{2}\right)\cos^2\left(\frac{m\Delta z}{2}\right)}{\delta}\left(\frac{\partial \hat{u}}{\partial x} + \frac{\partial \hat{w}}{\partial z} + (S_0 + \frac{R_0}{2})\hat{w}\right) \\ &= -\frac{C^2}{\tilde{\delta}}\left(\frac{\partial \hat{u}}{\partial x} + \frac{\partial \hat{w}}{\partial z} + (S_0 + \frac{R_0}{2})\hat{w}\right)\end{aligned} \tag{147}$$

and

$$\tilde{\delta} \equiv \frac{\delta}{\cos^2\left(\frac{k\Delta x}{2}\right)\cos^2\left(\frac{m\Delta z}{2}\right)} \tag{148}$$

Equation (148) indicates that $2\Delta x$ and $2\Delta z$ waves are removed from $\frac{\partial \hat{p}}{\partial t}$. The solutions of (142)–(146) can be assumed:

$$\begin{bmatrix}\hat{\theta} \\ \hat{p} \\ \hat{u} \\ \hat{w}\end{bmatrix}(p\Delta x, q\Delta z, n\Delta t) = \tilde{\lambda}^n \begin{bmatrix}\Theta \\ P \\ U \\ W\end{bmatrix} e^{i(kp\Delta x + mq\Delta z)} \tag{149}$$

where $\tilde{\lambda}$ is the complex amplification factor. A forward in time scheme is applied to the momentum fields and a backward to the pressure and temperature, they become:

$$\left(\frac{\tilde{\lambda}-1}{\Delta t}\right)U + iXP = 0 \tag{150}$$

$$\left(\frac{\tilde{\lambda}-1}{\Delta t}\right)W + (iZ - \Gamma_0)P - g\Theta = 0 \tag{151}$$

$$\left(\frac{\tilde{\lambda}-1}{\Delta t}\right)\frac{\tilde{\delta}P}{C^2} + i\tilde{\lambda}XU + \tilde{\lambda}(iZ + \Gamma_0)W = 0 \tag{152}$$

$$\left(\frac{\tilde{\lambda}-1}{\Delta t}\right)\Theta + \tilde{\lambda}S_0 W = 0 \tag{153}$$

where $\Gamma_0 = -\frac{g}{C^2} - \frac{R_0}{2} = S_0 + \frac{R_0}{2}$, $R_0 = \frac{\partial\rho_0}{\rho_0\partial z}$, and $S_0 = \frac{\partial\theta_0}{\theta_0\partial z}$. The eigenvalue for the modified forward–backward scheme with smoothing (MFBS) becomes

$$\frac{\tilde{\delta}}{C^2}\left(\frac{\tilde{\lambda}-1}{\Delta t}\right)^4 + \left(\frac{\tilde{\lambda}-1}{\Delta t}\right)^2\tilde{\lambda}\left[X^2 + Z^2 + \Gamma_0^2 + \frac{\tilde{\delta}gS_0}{C^2}\right] + \tilde{\lambda}^2 gS_0 X^2 = 0 \tag{154}$$

or

$$\frac{\delta}{C^2\cos^2(\frac{k\Delta x}{2})\cos^2(\frac{m\Delta z}{2})}\left(\frac{\tilde{\lambda}-1}{\Delta t}\right)^4 + $$
$$\left(\frac{\tilde{\lambda}-1}{\Delta t}\right)^2\tilde{\lambda}\left[X^2 + Z^2 + \Gamma_0^2 + \frac{\delta gS_0}{C^2\cos^2(\frac{k\Delta x}{2})\cos^2(\frac{m\Delta z}{2})}\right] + \tilde{\lambda}^2 gS_0 X^2 = 0 \tag{155}$$

Which consists of two acoustic waves and two gravity waves, as discussed in Equation (130). Let us define:

$$\Delta\tilde{t}_\delta = 2\sqrt{\tilde{\delta}/\left[C^2\left(X^2 + Z^2 + \Gamma_0^2 + \frac{\tilde{\delta}gS_0}{C^2}\right)\right]} \tag{156}$$

The amplification factor $\left|\tilde{\lambda}\right|$ is then unity if $\Delta t < \Delta\tilde{t}_\delta$. In that case, the frequency $\tilde{\omega}_\delta$ is defined by

$$\tilde{\lambda} = \tilde{\lambda}_r + i\tilde{\lambda}_i = \left|\tilde{\lambda}\right|\exp(-i\tilde{\omega}_\delta\Delta t) = \cos(\tilde{\omega}_\delta\Delta t) - i\sin(\tilde{\omega}_\delta\Delta t) \tag{157}$$

and

$$\tilde{\omega}_\delta = -\tan^{-1}(\tilde{\lambda}_i/\tilde{\lambda}_r)/\Delta t \tag{158}$$

Without smoothing, the values of $\tilde{\lambda}$, $\tilde{\omega}_\delta$, and $\Delta\tilde{t}_\delta$ of Equations (153)–(156) are equal to λ_δ, ω_δ, and Δt_δ of MFB, the solution without smoothing (Sun et al., 2012) [196]. The detailed amplification factor $\left|\tilde{\lambda}\right|$ and phase speed can be found in Sun et al. (2013c) [197].

6.3. Kelvin–Helmholtz Wave with $\Delta x = 10$ m and $\Delta z = 5$ m from CReSS

Figure 47 shows the simulated potential temperatures θ for Kelvin–Helmholtz instability at t = 320 s from the traditional FB with Δts = 0.01 s; modified FB without smoothing (MFB) with δ = 16, Δts = 0.04 s; modified FB with smoothing (MFBS) with δ = 4, Δts = 0.04 s; (MFBS) with δ = 16, Δts = 0.08 s; and HE-VI with Δts = 0.02 s. The five simulations nearly coincide, and their accuracies are comparable. However, the Δts = 0.08 s for (MFBS) is twice that for (MFB) with the same δ, and four times HE-VI. Sun et al. (2012) [196] also showed that even Co being less than 1, HE-VI result can depart from the FB, and the HE-VI model becomes unstable when $\Delta ts \geq 0.04$ s.

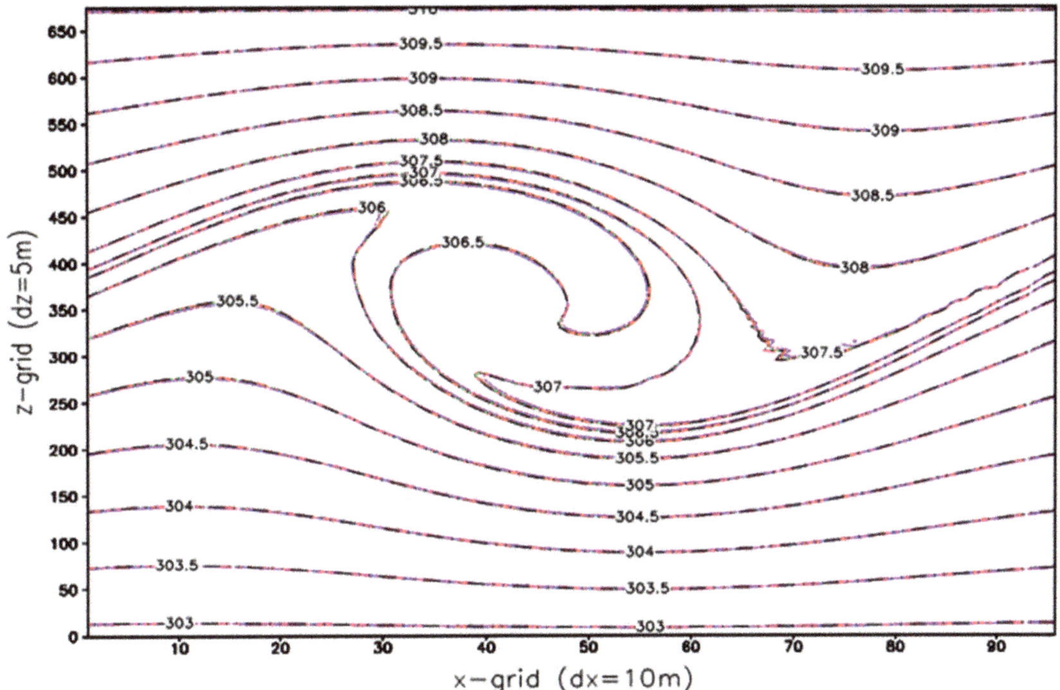

Figure 47. Simulations Kelvin–Helmholtz instability at t = 320 s from: potential temperature from FB: θ_{FB} (multiplication of density $\delta = 1$, $\Delta ts = 0.01$ s); modified FB: MFB ($\delta = 16$, $\Delta ts = 0.04$ s); modified FB with smoothing: MFBS ($\delta = 4$, $\Delta ts = 0.04$ s); modified FB with smoothing: MFBS ($\delta = 16$, $\Delta ts = 0.08$ s), and horizontal explicit–vertical implicit: HE-VI ($\Delta ts = 0.02$ s). (Sun et al., 2013c) [197].

6.4. Equations of NTU/Purdue Nonhydrostatic Model

The vertical coordinate of the NTU/Purdue nonhydrostatic model is defined as:

$$\sigma = \frac{p_0(z) - p_0(z_{top})}{p_0(z_{surface}) - p_0(z_{top})} \tag{159}$$

where p_0, pressure of a reference atmosphere, is strictly a function of height. Following Kasahara (1974) [198], we convert (131)–(134) to the equations in (x, y, σ) coordinate. The cloud physics, PBL, and turbulence parameterizations are the same as those in the hydrostatic model (Sun 1993a [199]). The advection terms were solved based on Sun (1993c) [122], or the semi-Lagrangian schemes (Sun et al., 1996 [123], Sun and Yeh 1997 [124], Sun 2023 [200]).

6.5. Numerical Scheme for Advection Equations

Sun (1993c) [122] combined Crowley fourth-order scheme and Gadd scheme to reduce their phase error, by integrating Crowley scheme and Gadd scheme alternately with time step, because the ratio of phase speed to real phase is slightly less than 1 for Growly scheme, but slightly larger than one for Gadd scheme. The initial square step function and simulations after 800 time steps with the Courant = 0.2 for Crowley, Gadd, and Sun's modified schemes are shown in Figure 48a–d (Sun 1993c) [122], Sun's is better than the original ones. However, phase error and dispersion still exist. Therefore, the semi-Lagrangian schemes is also used to solve the advection equation.

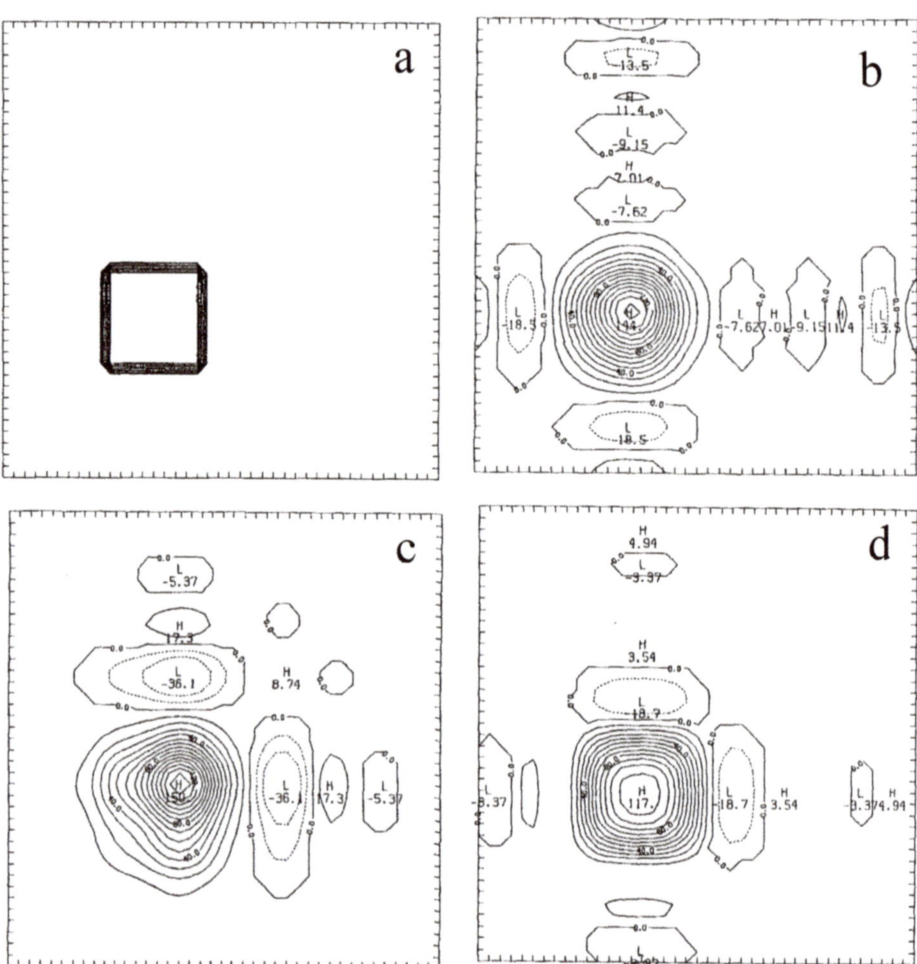

Figure 48. (**a**) Initial step function, and 800 time step integrations with Co = 0.2 for: (**b**) Crowley scheme, (**c**) Gadd scheme, and (**d**) Sun's modified scheme (Sun, 1993c) [122].

In Figure 49, the fluid moves from $O_i(\xi, \eta)$ to $P_i(X, Y)$ (indicated by red circle) within Δt in the regular coordinates. We connect those red circles to form the Lagrangian curve ($\eta = y_j$), which intercepts with the vertical line $X = X_k$ at blue diamond point Q. Because X and Y are well functions of ξ and η, $F(X_k)$ or $Y(X_k)$ can be derived from the coordinates of the neighboring reds or blues by the Lagrangian polynomials (Sun 2023) [200]:

$$
\begin{aligned}
F(X) \;=\; &F_{i-2,j}\frac{(X-X_{i-1,j})(X-X_{i,j})(X-X_{i+1,j})}{(X_{i-2,j}-X_{i-1,j})(X_{i-2,j}-X_{i,j})(X_{i-2,j}-X_{i+1,j})}\\
+&F_{i-1,j}\frac{(X-X_{i-2,j})(X-X_{i,j})(X-X_{i+1,j})}{(X_{i-1,j}-X_{i-2,j})(X_{i-1,j}-X_{i,j})(X_{i-2,j}-X_{i+1,j})}\\
+&F_{i,j}\frac{(X-X_{i-2,j})(X-X_{i-1,j})(X-X_{i+1,j})}{(X_{i,j}-X_{i-2,j})(X_{i,j}-X_{i-1,j})(X_{i,j}-X_{i+1,j})}\\
+&F_{i+1,j}\frac{(X-X_{i-2,j})(X-X_{i-1,j})(X-X_{i,j})}{(X_{i+1,j}-X_{i-2,j})(X_{i+1,j}-X_{i-1,j})(X_{i+1,j}-X_{i,j})}
\end{aligned}
\tag{160}
$$

$$
\begin{aligned}
Y(X) \;=\; &Y_{i-2,j}\frac{(X-X_{i-1,j})(X-X_{i,j})(X-X_{i+1,j})}{(X_{i-2,j}-X_{i-1,j})(X_{i-2,j}-X_{i,j})(X_{i-2,j}-X_{i+1,j})}\\[4pt]
+&Y_{i-1,j}\frac{(X-X_{i-2,j})(X-X_{i,j})(X-X_{i+1,j})}{(X_{i-1,j}-X_{i-2,j})(X_{i-1,j}-X_{i,j})(X_{i-2,j}-X_{i+1,j})}\\[4pt]
+&Y_{i,j}\frac{(X-X_{i-2,j})(X-X_{i-1,j})(X-X_{i+1,j})}{(X_{i,j}-X_{i-2,j})(X_{i,j}-X_{i-1,j})(X_{i,j}-X_{i+1,j})}\\[4pt]
+&Y_{i+1,j}\frac{(X-X_{i-2,j})(X-X_{i-1,j})(X-X_{i,j})}{(X_{i+1,j}-X_{i-2,j})(X_{i+1,j}-X_{i-1,j})(X_{i+1,j}-X_{i,j})}
\end{aligned}
\tag{161}
$$

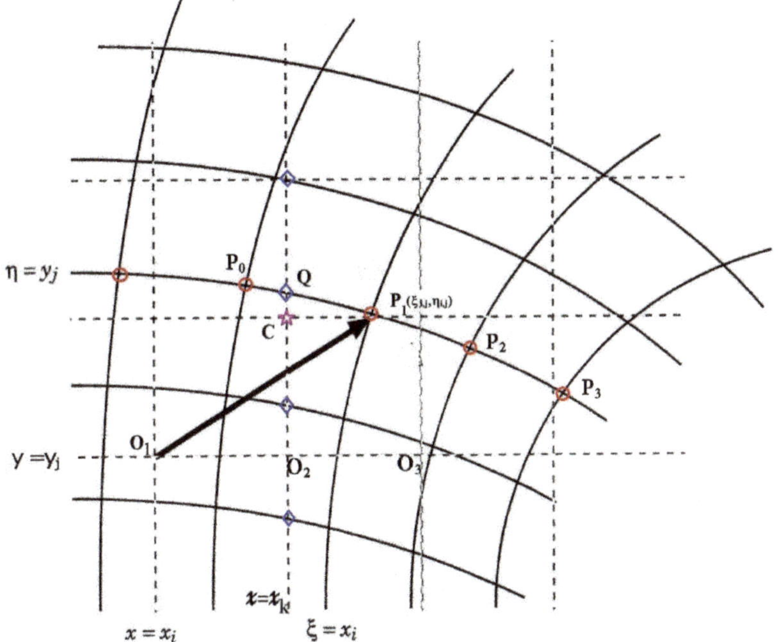

Figure 49. A particle O in Euler grids (dashed lines) moves to P in Lagrangian grids (full lines), and the corresponding Eulerian coordinate lines x = x_i and y = y_j are transformed to the Lagrangian coordinate lines ξ_{ij} and η_{ij}, respectively. the fluid moves from $O_i(\xi,\eta)$ to $P_i(X,Y)$ (indicated by red circle) within Δt in the regular coordinates. We connect those red circles to form the Lagrangian curve (η = yj), which intercepts with the vertical line X = X_k at blue diamond point Q. (Sun 2023) [200].

After obtaining F and Y of all intercepts (blue diamonds) on vertical line at X = X_k, we apply Equation (161) again but on the vertical direction by replacing X by Y to obtain the value F at the Euler grid C. The same procedure is applied to the entire domain. It is noted that following the same procedure, the vertical/(horizontal) Lagrangian curves can be interpolated from the fluids initially coming from the vertical lines, x_i/(horizontal lines y_j). Interpolation from the Lagrangian curves projected by fluids coming from either a horizontal or vertical curve to the regular grids is referred as "Economic Internet Interpolation". The interpolation using both vertical and horizontal lines is called "Complete Internet Interpolation" (Sun and Yeh 1997 [124]). They showed that complete interpolation is slightly more accurate than economic interpolation, but the complete version requires twice computing resources than the economic one. It is also noted that Equations (160) and (161) used in this model can be replaced by the fifth or higher order polynomials (Sun 2023 [200]).

The scheme has been applied to simulate an idealized cyclogenesis of Doswell (1984) [201], which has a tangential velocity of the circular vortex:

$$
V(r) = A\,\mathrm{sech}^2(r)\,\tanh(r),
\tag{162}
$$

where r is the radius of the vortex; and A = 2.598 so that the maximum value of V equals 1. The initial condition is

$$F(x, y, 0) = -\tanh[(y - yc)/\delta], \tag{163}$$

where δ is the characteristic width of the frontal zone, i.e., one half is $F = 1$, the other half $F = -1$ initially. The analytic solution is

$$F(x, y, t) = -\tanh\left[\frac{(y - y_c)}{\delta}\cos(\omega t) - \frac{(x - x_c)}{\delta}\sin(\omega t)\right] \tag{164}$$

where (xc, yc) is the center of rotation and $\omega = V/r$ is the angular velocity. The domain is 10 units long with resolutions of 129 × 129 grid points. An integration time of five units is chosen so that the analytic solution is still resolvable on the low-resolution grid. The numerical simulation for an idealized cyclogenesis after 16-time steps (with δ = 0.05, Co = 4, t = 5, and Error = 0.076) shown in Figure 50a, is in good agreement with the analytic solution shown in Figure 50b (Sun 2023) [200]. The semi-Lagrangian scheme also avoids the phase error or nonlinear instability which may exist in the finite difference or finite volume schemes (Mesinger and Arakawa 1976) [25].

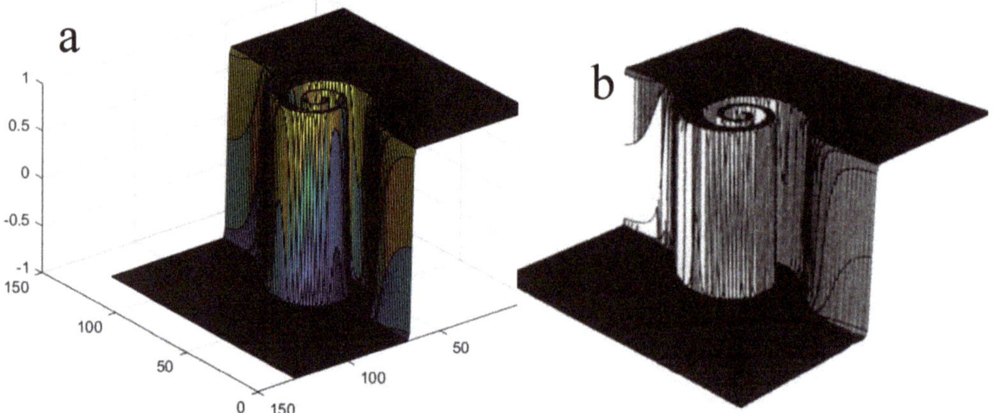

Figure 50. The idealized cyclogenesis test on 129 × 129 grids, with δ = 0.05. (**a**) Numerical simulation after 16 time steps with Co = 4, error = 0.076, (**b**) analytic solution. (Sun 2023) [200].

6.6. Nonhydrostatic Model—Parcel Method, Froude Number, Bernoulli, Downslope Wind, and Waves (Sun and Sun 2015, 2019) [60,202]

The parcel method, which has been cited in many texts, assumes: (a) the air parcel rises or sinks adiabatically; (b) the pressure of the parcel is always the same as the ambient pressure at the same level; and (c) the atmosphere is in hydrostatic equilibrium. Unfortunately, (b) and (c) cannot hold at the same time (Sun and Sun 2015, 2019) [60,202].

Queney (1948) [203] and Queney et al. (1960) [204] first derived the mountain waves and downslope wind based on linearized equations. Long (1953a, b) [205,206], Smith (1985) [207] and others proposed hydraulic jump theory to explain the wave breaking and downslope wind. Clark and Peltier (1984) [208], Peltier and Clark (1979) [209] proposed the amplification of resonant lee waves as well as the trapping and subsequent amplification of internal wave beneath the critical layer. More discussions can be found in Lin (2007) [29]. Both theories propose that the high winds occur in the layer below the dividing streamline in hydraulic jump theory, or the critical layer in the resonant amplification. But simulated large amplitude waves develop far beyond the critical layer (where the wind reverses) as shown in Figure 51, because the density decreases with height. It indicates that energy may not be trapped between the critical layer and ground surface. Long's model states that the increase in kinetic energy (KE) comes from the decrease in potential energy (PE). The simulations show an increase in both KE and PE when the air parcel climbs up the mountain. Then, air descends against buoyancy and produces a warm, low-pressure center on the lee due to adiabatic warming, which enhances the downslope wind and generates a strong updraft.

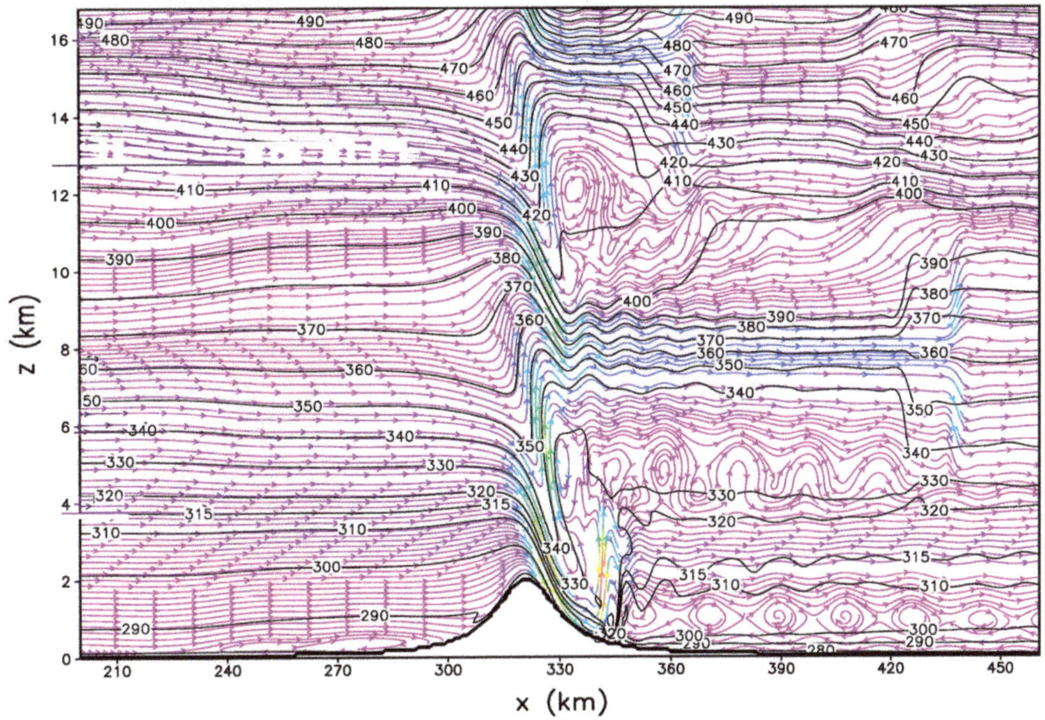

Figure 51. Simulated streamline and θ (dashed black lines) at t = 6 h integration for case with width α = 10 km, and height h_m = 2 km, mean wind U = 20 m s^{-1}, and temperature lapse rate β = 3.5 K km^{-1} for z < 12 km, and β = 0.7 K km^{-1} for z > 12 km at inflow. Vertical velocity is multiple by 10 (Sun and Sun 2015) [60].

Sun and Sun (2015) [60] proposed a hypothesis based on conservations of Bernoulli function that the change in kinetic energy can also come from enthalpy (EN) in compressible fluid. The momentum equation can be written as:

$$\left(\frac{\partial \mathbf{V}}{\partial t}\right) + \mathbf{V} \cdot \nabla \mathbf{V} + 2\Omega \times \mathbf{V} = -\frac{\nabla p}{\rho} + \mathbf{g} - \mathbf{fr} \tag{165}$$

where $\mathbf{V} = (u, w)$. If we define $dr = (dx, dz) = Vdt$ and integrate (165) from i to f:

$$\int_i^f \left(\frac{\partial \mathbf{V}}{\partial t}\right) \cdot d\mathbf{r} + \int_i^f \nabla \left(\frac{\mathbf{V} \cdot \mathbf{V}}{2} + (2\Omega + \nabla \times \mathbf{V}) \times \mathbf{V} + \frac{\nabla p}{\rho} + gz\right) \cdot d\mathbf{r} = -\int_i^f \mathbf{fr} \cdot d\mathbf{r} \tag{166}$$

Because $(\nabla \times \mathbf{V}) \times \mathbf{V}$ is orthogonal to dr, $(2\Omega + \nabla \times \mathbf{V}) \times \mathbf{V} \cdot dr = 0$ and $c_p dT - \frac{dp}{\rho} = 0$ in an adiabatic process, Bernoulli function remains as a constant, for a steady, inviscid flow:

$$\left(\frac{\mathbf{V} \cdot \mathbf{V}}{2}\right)_f + gz_f + c_p T_f = \left(\frac{\mathbf{V} \cdot \mathbf{V}}{2}\right)_i + gz_i + c_p T_i = B. \tag{167}$$

or

$$d\left[\left(\frac{\mathbf{V}\cdot\mathbf{V}}{2}\right) + c_p T + gz\right] = d(KE) + d(EN) + gdz(PE) = 0 \tag{168}$$

or

$$d\left(\frac{\mathbf{V}\cdot\mathbf{V}}{2}\right) + c_p\left[\frac{dT}{dz} + \frac{g}{c_p}\right]dz = 0 \tag{169}$$

where EN (specific enthalpy) $= c_p T$ and $d\,EN = c_p dT$. Bernoulli function is the total energy, i.e., the summation of the kinetic energy (KE), enthalpy (EN), and potential energy (PE).

If we define the hydrostatic–adiabatic process as an air parcel moves adiabatically from z_i to z_f and follows the hydrostatic equation (i.e., $dw/dt = 0$), i.e., according to the parcel method:

$$dp = -\rho gdz \text{ and } c_p dT - \alpha dp = 0 \tag{170}$$

We obtain the hydrostatic–adiabatic lapse rate β_{ha}:

$$\beta_{ha} = -\frac{dT}{dz} = \frac{g}{c_p} \tag{171}$$

β_{ha} is traditionally called "dry adiabatic lapse rate". Let us define $p = p_h + p''$, p'' is nonhydrostatic pressure (or dynamic pressure), and hydrostatic pressure satisfy $\frac{\partial p_h}{\rho \partial z} \equiv -g$. From (136) for a dry adiabatic process, we obtain:

$$d\ln\theta = d\ln T - \frac{R}{c_p}d\ln p = 0 \tag{172}$$

$$\frac{\partial\theta}{\theta\partial z} = \frac{\partial T}{T\partial z} - \frac{R}{c_p}\frac{\partial(p_h + p'')}{p\partial z} = \frac{1}{T}\left[\left(\frac{\partial T}{\partial z} + \frac{g}{c_p}\right) - \frac{1}{c_p}\frac{\partial p''}{\rho\partial z}\right] = 0 \tag{173}$$

and

$$\frac{dw}{dt} = -\frac{\partial(p_h + p'')}{\rho\partial z} - g = \begin{cases} -\frac{\partial p''}{\rho\partial z} = 0 & \text{If } -\frac{dT}{dz} = \frac{g}{c_p} \\ -\frac{\partial p''}{\rho\partial z} \neq 0 & \text{If } -\frac{dT}{dz} \neq \frac{g}{c_p} \end{cases} \tag{174}$$

If the parcel moves adiabatically with the hydrostatic–adiabatic lapse rate, the dynamic pressure is disregarded and w remains constant, that is consistent with $\left(\frac{\mathbf{V}\cdot\mathbf{V}}{2}\right)_f = \left(\frac{\mathbf{V}\cdot\mathbf{V}}{2}\right)_i$ in (169).

Sun and Sun (2015) [60] also proved that conventional Froude number in mountain meteorology

$$Fr = \frac{U_i}{Nh} \tag{175}$$

Equation (175) is based on both hydrostatic and adiabatic approximations. The Froude number can also be derived from a Boussinesq fluid ($\rho'/\rho = -\theta'/\theta$) having a hydraulic jump over an obstacle of height h:

$$Fr = \frac{U_i}{\sqrt{g'h}} = \frac{U_i}{\sqrt{g\frac{\theta'}{\theta}h}} = \frac{U_i}{\sqrt{gh\frac{\left(\theta_i + \frac{d\theta}{dz}h\right) - \theta_i}{\theta}}} = \frac{U_i}{\sqrt{\frac{g}{\theta}\frac{d\theta}{dz}h}}\frac{U_i}{Nh} \tag{176}$$

Since an air parcel cannot follow the hydrostatic adiabatic lapse rate, the Froude number is invalid in the atmosphere, but it can be applied in the incompressible water.

Figure 52a shows the simulated x-component wind (shaded color), potential temperature θ (dashed black lines), Bernoulli function B (thick write line), streamlines (thin black line), and pressure p (thin black line) at $t = 6$ hr. Figure 52b is the same as Figure 52a, except that the shaded colors display temperature. The contours of B, θ, and streamlines are parallel before flows reach the turbulent areas on the lee side, indicating that the flow is steady, adiabatic, and inviscid, and the contours of B or θ can be interpreted as trajectories outside turbulence areas.

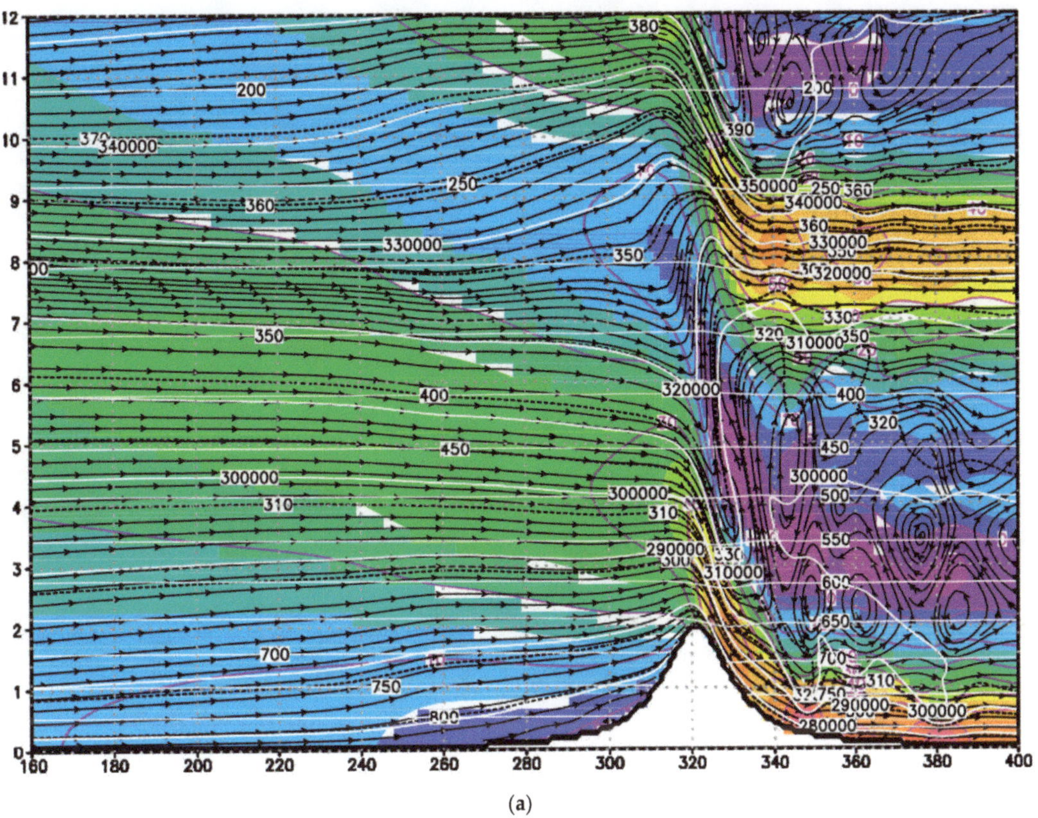

Figure 52. *Cont.*

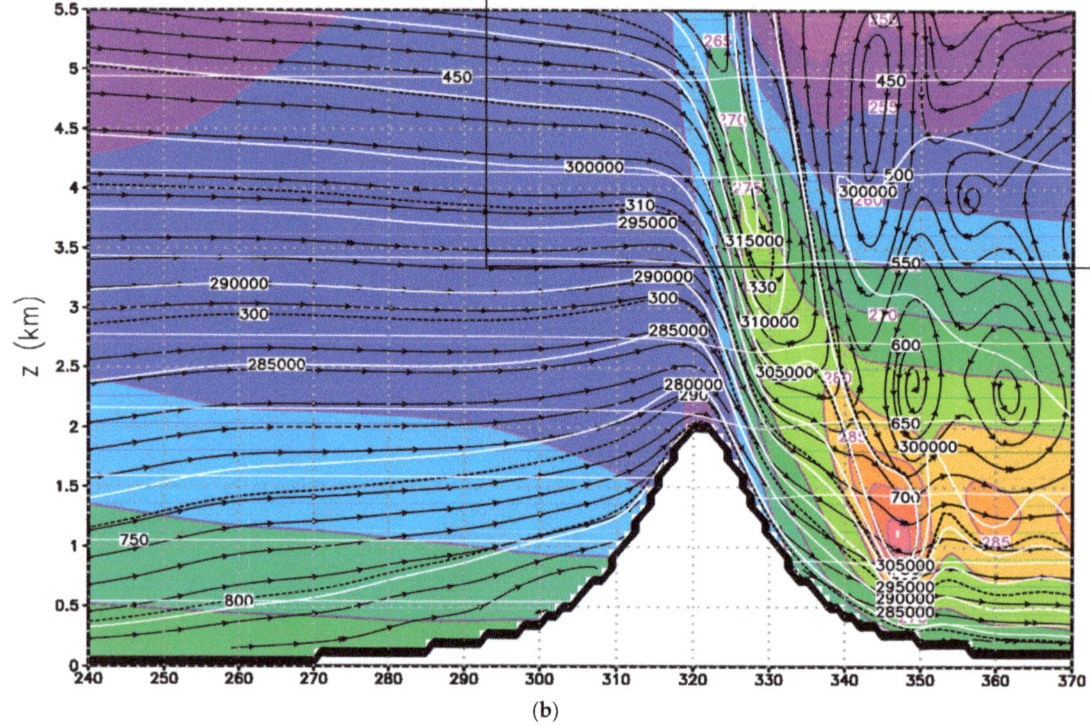

Figure 52. (a) Simulated wind U (shaded color), Bernoulli function B (thick white lines), potential temperature θ (dashed black lines), pressure p (thin white lines), and streamlines at t = 6 h integration for case; (b) Same as (a) except that shaded color represents temperature T with different scale in x and z. (Sun and Sun 2015) [60].

We can follow the trajectory of an air parcel along $B = 280{,}000$ m^2 s^{-2}, which passes 1A ($x = 160$ km, $z = 1099.14$ m) in the upstream region, where $\theta = 291.1$ K, $T = 268.01$ K, $U = 11.24$ m s^{-1}, and $\rho = 0.9733$ kg m^{-3}; 1B ($x = 320$ km, $z = 2391$ m) over the mountain peak where $\theta = 291.1$ K, $T = 254.91$ K, $U = 31.89$ m s^{-1}, and $\rho = 0.8588$ kg m^{-3}; and 1C ($x = 348$ km, $z = 378$ m) with strong downslope wind, $U = 74.44$ m s^{-1}, $\theta = 290.7$ K, $T = 272.30$ K, and $\rho = 1.0178$ kg m^{-3}, as shown in Table 1 of Sun and Sun (2015) [60]. Figure 52a,b also show that near the mountain top (d $PE > 0$), the wind becomes stronger (d $KE > 0$), but the temperature is lower than surrounding. Both PE and KE increase but EN decreases; and the parcel does not follow the hydrostatic adiabatic lapse rate β_{ha}.

6.7. Boulder Severe Downslope Windstorm

Severe downslope windstorm occurred during 11 and 12 January 1972. Here, the observed soundings at Grand Junction, CO are used at inflow. On a free-slip surface without the effect of the Earth's rotation, the 3 h simulations from 11-hydrostatci models (Doyle et al., 2000) [210] show that all models produce a significant leeside windstorm, but the wind maximum at the 3 h time varies from 86 m s^{-1} to 43 m s^{-1} and substantial variations exist in the movement of the jump. However, on non-slip surface without rotation, the simulated downslope wind decreases with time after 2–3 h integration. At $z\sim25$ m, u = 35 m s^{-1} at t = 2 h and u = 27 m s^{-1} after t = 6 h (Sun and Hsu 2005) [189]. Furthermore, the updraft moves farther downstream as time increases (Sun 2013) [211], and Figure 45 in Lin 2007 [29]). As Lilly (1978) [212] pointed out that the air flow in the upstream changed little during 11 and 12 January, and a surge of warmer air moved across the Northwestern US at 500 hPa, accompanied by rapid pressure falls and strong surface cyclogenesis in the lee of the Rocky Mountains, which triggered outbreak of the downslope windstorm. Hence, Sun (2013) [211] proposed that the northwesterly upper-level jet is unbalanced by the large-scale pressure gradient, the Coriolis force decreases the westerly wind and increases the northerly wind of the upper-level jet as it approaches mountain. Convergence forms in the upper layer and forces air to descend and form a severe, long-lasting downslope windstorm. The 2D equations of Sun (2013) [211] are:

$$\frac{\partial u}{\partial t} + Adv(u) = f(v - \delta_2 v_g) + Diff(u) \tag{177}$$

$$\frac{\partial v}{\partial t} + Adv(v) = -f(u - \delta_2 u_g) + Diff(v) \tag{178}$$

$$\frac{\partial w}{\partial t} + Adv(w) = -\frac{1}{\rho}\frac{\partial p}{\partial z} + bu - g + Diff(w) \tag{179}$$

Here $f = 2\Omega \sin \phi$; $b = 2\Omega \cos \phi$; $\phi = 36°$ N; Other equations are the same as in Section 6.1. It is assumed that $f u_g = -\left(\frac{1}{\rho}\frac{\partial p}{\partial y}\right)_{background}$ and $f v_g = \left(\frac{1}{\rho}\frac{\partial p}{\partial x}\right)_{background}$; and δ_2 is set to be 0, or 0.5, or 1.0. The background θ_{eg} and (u_g, v_g) satisfy the thermal wind equation initially. At t > 0 and away from inflow, the geostrophic adjustment occurs, if $\delta_2 \neq 1$. The detailed turbulent kinetic energy, parameterizations, and numerical schemes can be found in (Hsu and Sun, 2001 [190], Sun and Hsu, 2005 [189], Sun et al., 2012 [196]).

The Case D of Sun (2023) [211] is presented. The time-independent observed wind is used at inflow (x = 0, black line in Figure 53a) on non-slip surface with $f \neq 0, b \neq 0$. The red line D in Figure 53a shows the profiles of westerly winds at t = 4 h on the windward side at the mountain foot (x = 300 km). With $\delta_2 = 0$, at x = 300 km, the westerly wind of the upper-level northwesterly jet decreases with time as flow approaches the mountain, because $\frac{du}{dt} \approx f(v - \delta_2 v_g) = fv$, and $\frac{dv}{dt} \approx f(\delta_2 u_g - u) = -fu$ with u > 0 and v < 0. Convergence in the upper layer enhances the downward motion over the mountain. Consequently, a strong downslope wind develops on the lee side, as shown in Figure 53b,c, which reaches about 55 ms^{-1} at t = 2 h, and about 60 m s^{-1} at 6 h. The maximum downdraft reaches -18 m s^{-1} at t = 2 h, and -27 m s^{-1} at t = 6 h; and the peak updraft reaches 30 m s^{-1} at t = 2 h and 33 m s^{-1} at t = 6 h. They are comparable with the magnitude of 30 m s^{-1} of the vertical motion observed over the turbulent zone at z~6 km. The simulated westerly wind varies from -2 m s^{-1} to 52 m s^{-1} in 320 km < x < 360 km at z~6 km is also comparable with observations (-5 ms^{-1} < u < 55 ms^{-1}) presented in Figure 13 of Lilly (1978) [212]. A strong/weak westerly wind corresponding to a cold/warm potential temperature is consistent with observations as well. The vertical velocities and downslope wind are much stronger than the cases with $f = 0$ or geostrophic balance cases. The hydraulic jump locates at x~340 km. Waves, rotors, and turbulences appear in downstream regions. A zone of the reversed and weak winds develops above the strong downslope wind and on the left-hand side of a deep hydraulic jump. The existence of the reversed flow (u < 0) near z~6 km is consistent with observation, and also turbulences on the lee side and upper layers of the mountain. As the downslope wind intensifies with increasing time, the zone of the reversed/weak winds broadens and descends. On the lee side, the waves, rotors, and turbulences gradually develop over a deep layer from surface to 10 km, and in the upper layers (z > 16 km); meanwhile, strong wind zones exist near θ_v~310 K (z~4 km) and θ_v~400 K (z~14 km) at t = 6 h (Figure 53b). The westerly wind at z ~ 25 m is about 43 m s^{-1} at t = 2 h, and 48 ms^{-1} at t = 6 h. They are comparable with observations (~50 m s^{-1}). Furthermore, the intensive downslope wind and strong vertical motions last for the entire 9 h integration, and the hydraulic jump does not propagate downstream. They are different from the case with a free-slip surface.

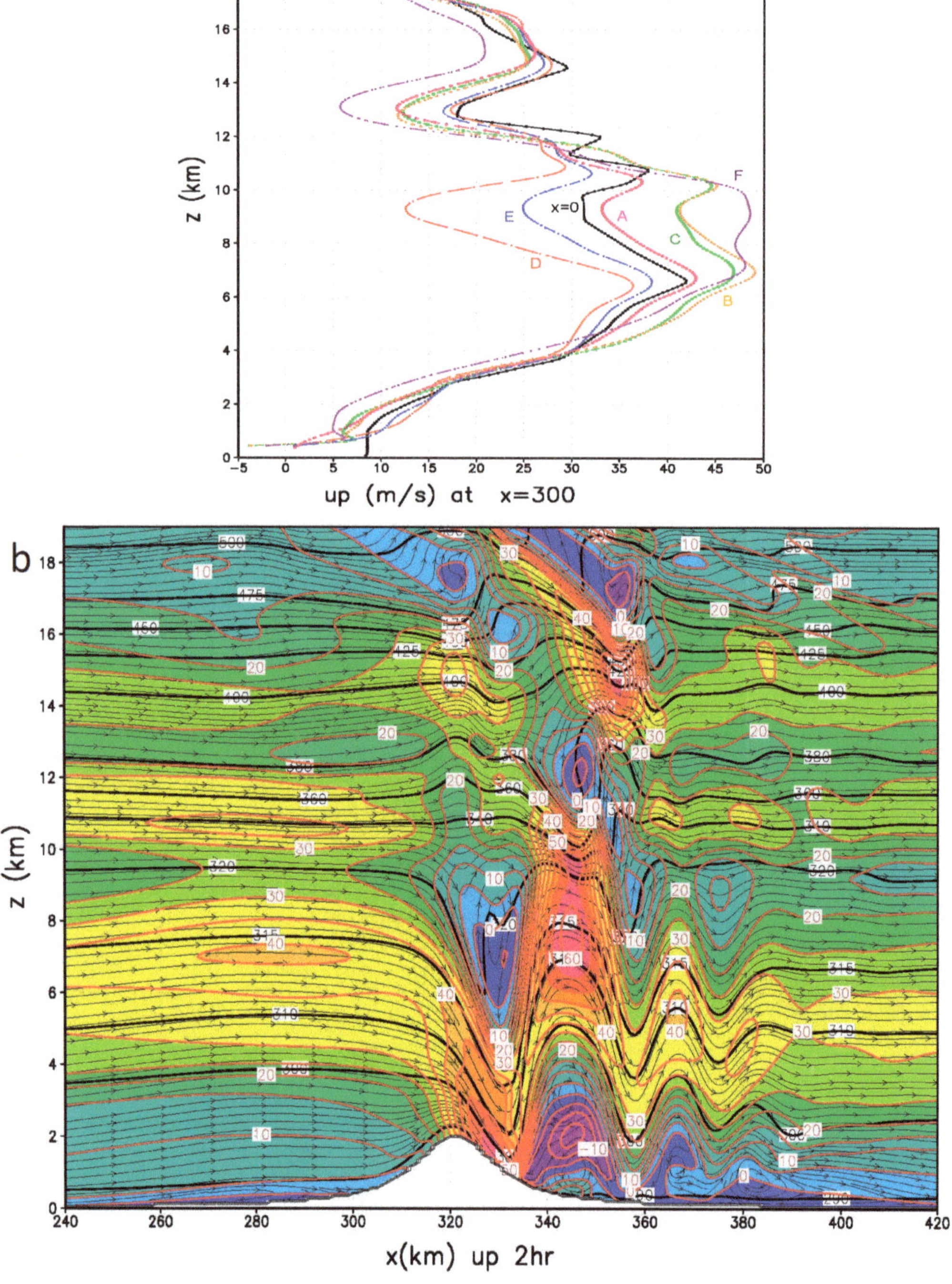

Figure 53. *Cont.*

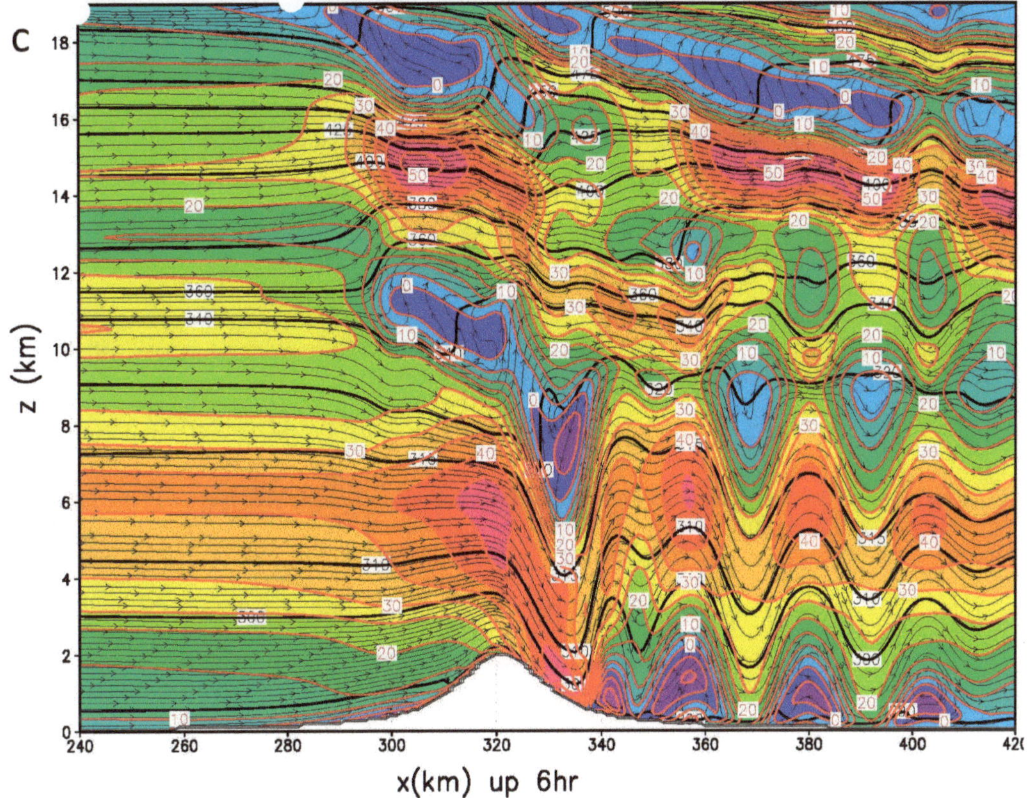

Figure 53. (**a**) Fix westerly wind profiles at X = 0 (black), and at X = 300 km (red line D) at t = 4 h for Case D: (**b**) Simulated u (shaded color with red contour and magnitude (m s^{-1})), streamlines (thin black), and θ$_v$ (thick black and in (K) at 2 h: (**c**) same as (**b**) except at t = 6 h for Case D (Sun 2013) [211].

6.8. Convective Available Potential Energy (CAPE) (Sun and Sun 2019) [202]

Sun and Sun (2015, 2019) [60,202] proposed that an increase in *KE* and *PE* can come from a decrease in *EN* according to conservation of Bernoulli function in dry atmospheres. Equations with moisture condensation are:

$$c_p \frac{dT}{dt} - \alpha \frac{dp}{dt} = -L \frac{dq_s}{dt} - D_h,$$ (180)

$$c_p \frac{d \ln \theta}{dt} = -\frac{1}{T}\left(L \frac{dq_s}{dt} + D_h\right), \text{ and } \theta_e = \theta + \frac{\theta}{T}\frac{q_s}{c_p},$$ (181)

where θ_e is the equivalent potential temperature.

$$\left(\frac{\partial \mathbf{V}}{\partial t}\right) + \mathbf{V} \cdot \nabla \mathbf{V} + 2\Omega \times \mathbf{V} = -\frac{\nabla p}{\rho} + \mathbf{g} - \mathbf{fr}$$ (182)

Integrating along the parcel trajectory, we obtain

$$\int_i \left(\frac{\partial \mathbf{V}}{\partial t}\right)_r \cdot d\mathbf{r} + \int_i d\left(\frac{\mathbf{V} \cdot \mathbf{V}}{2} + gz + c_p T + L q_s\right) = -\int_i [\mathbf{D}_h + \mathbf{fr}] \cdot d\mathbf{r}$$ (183)

The Bernoulli function with condensation in an inviscid, steady flow becomes:

$$\text{Bernoulli function } (B) = \frac{\mathbf{V} \cdot \mathbf{V}}{2} + gz + c_p T + L q_s = \text{constant,} \tag{184}$$

i.e., d $B = 0 = $ d $(KE + PE + EN) = 0$ for an inviscid, steady flow, where the moist *EN* (enthalpy) $= c_p T + L q_s$. Following (172)–(174) and putting (180) into the vertical momentum equation, we obtain

$$\begin{aligned} \frac{dw}{dt} &= -\frac{\partial p}{\rho \partial z} - g = -L \frac{\partial q_s}{\partial T} \frac{\partial T}{\partial z} - c_p \frac{\partial T}{\partial z} - g = -c_p \frac{\partial T}{\partial z}\left(1 + \frac{L}{c_p}\frac{\partial q_s}{\partial T}\right) - g \\ &= -c_p\left(1 + \frac{L}{c_p}\frac{\partial q_s}{\partial T}\right)\left(\frac{\partial T}{\partial z} + \beta_{hs}\right) = -\frac{\partial p''}{\rho \partial z} \end{aligned} \tag{185}$$

where p'' is nonhydrostatic pressure and the hydrostatic moisture adiabatic lapse rate

$$\beta_{hs} \equiv \frac{g}{c_p} \Big/ \left(1 + \frac{L}{c_p}\frac{\partial q_s}{\partial T}\right) \tag{186}$$

Both (184) and (185) are identical to the dry case, except the dry EN $= c_p T$ being replaced by the moist EN $= c_p T + L q_s$. It also confirms that vertical velocity does not change if it follows hydrostatic saturate adiabatic lapse rate. It is noted that (174) or (185) does not require a steady state condition (i.e., $\frac{\partial}{\partial t} = 0$). The change in momentum is driven by the nonhydrostatic pressure, which comes from the parcel departing from hydrostatic adiabatic lapse rate or hydrostatic moisture adiabatic lapse rate. More discussion can be found in Sun and Sun (2019) [202].

The convective available potential energy (CAPE), one of the most popular indicators in predicting severe weather, is the integration of the buoyancy ($Buoy = g\theta'/\bar{\theta}$) of a parcel from free convection (LFC) to the equilibrium level (EL): In steady Boussinesq fluid, it becomes:

$$CAPE = \int_{z_{LFC}}^{z_{EL}} \left(-g\frac{\rho'}{\bar{\rho}}\right) dz = \int_{z_{LFC}}^{z_{EL}} \left(g\frac{\theta'}{\bar{\theta}}\right) dz = \int_{z_{LFC}}^{z_{EL}} \left(g\frac{T'}{\bar{T}}\right) dz = \int_i d\left(\frac{w^2}{2}\right) \tag{187}$$

It shows that "the kinetic energy per unit mass of the outflow is equal to the energy per unit mass of the inflow plus the work conducted by buoyancy forcing during adiabatic ascent" (Moncrief and Green 1972) [213]. But CAPE of the hot towers observed in the Equator trough is negative (Riehl and Malkus 1958) [214]. Strong updrafts are also frequently observed in tropical cyclones without CAPE (Ebert and Holland 1992) [215]. Strong warm downdrafts without evaporation cooling but with positive buoyancy are also observed in the eyewall of hurricane Emily (1987) (Black et al., 1994 [216], etc.), and strong warm, downslope windstorms on the lee of mountains, as discussed previously.

The NTU/Purdue nonhydrostatic model (Hsu and Sun, 2001 [133], Sun and Hsu 2005 [189]), which includes warm cloud physics (Chen and Sun, 2002) [111] and turbulence parameterization (Sun and Chang 1986a, b [108,217]; Sun 1988, 1989 [218,219]), but no radiation with dx = 500 m, and dz $\sim$75 m, except dz = 25 m for the first layer above the surface was employed. A moist-stable (CAPE = 0) sounding of (98646 RPMT (Mactan Observations) at 00Z 18 June 2017 (Figure 54) was used as the initial temperature and moisture fields, unless the moisture was specified differently. We imposed a weak, vertical velocity w_{gr} at the surface:

$$w_{gr}(x,t) = \frac{w_{amp}\tanh(t/702s)}{\left(\frac{x-x_c}{m\cdot\Delta x}\right)^2 + 1}$$

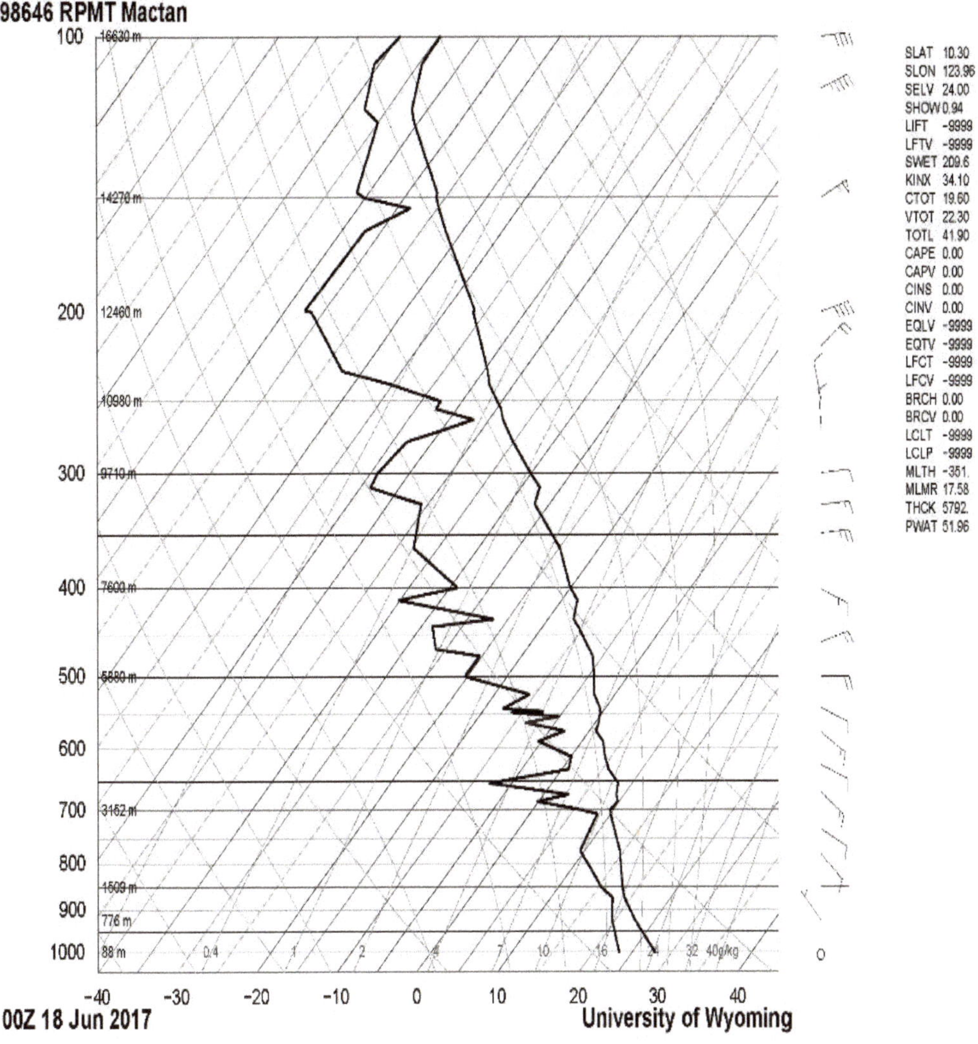

Figure 54. Sounding at RPMT (98646, at Mactan Observation Station) at 00Z18 Jun 2017 [202].

6.8.1. Dry Plume with w_{amp} = 3 m s^{-1} and m = 30, Case C of Sun and Sun (2019) [202]

Figure 55a shows the *Buoy* (shaded colors), w (purple line), θ (white), and streamlines (black) at t = 1.5 h. At x = 335 grid, *Buoy* = −0.04 m s^{-2} near the surface decreases to −0.62 m s^{-2} at 4.6 km, but w~3 m s^{-1} at surface increases to 12.4 m s^{-1} at 2 km, then deceases to −0.26 m s^{-1} at 4.9 km (Figure 55a,b). Figure 55c shows that θ (red) and B (green) remain constant from the surface to 4.4 km. Turbulence mixing decreases the values of θ and B at the top of the updraft. Part of EN is used to support the increase in KE and PE from surface to 2 km, as discussed in Sun and Sun (2015) [60]. The maximum adiabatic cooling (negative buoyancy) coincides with the maximum updraft along x = 335 grid in Figure 55a.

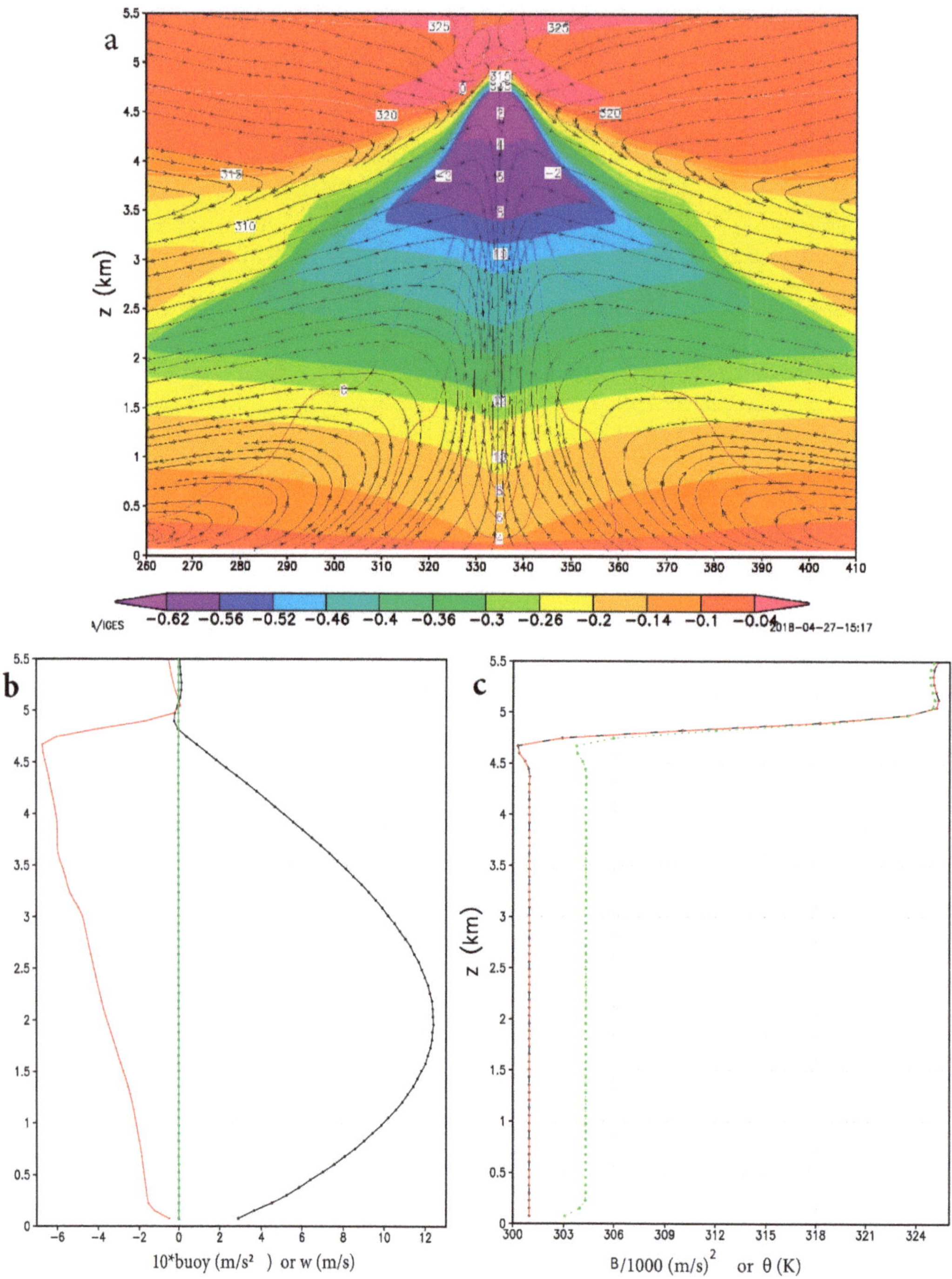

Figure 55. (**a**) Bernoulli function: B (shaded color), θ (white), w (purple), and streamline at 1.5 h. (**b**) Buoy (red) and w (black) at x = 335 grid at 1.5 h. (**c**) B (green) and θ (red) at x = 335 grid at 1.5 h (Sun and Sun 2019) [202].

6.8.2. Moist Plume with $w_{amp} = 1$ m s^{-1} and m = 30, Case D of Sun and Sun (2019) [202]

The observed temperature and humidity of the sounding in Figure 54 are used. Figure 56a shows the Buoy (shaded colors), which includes the loading of cloud water (ql; thick white) and rain (qr; thin white), θ_e (nearly horizontal white thin line), w (purple), and streamlines at t = 0.56 h. The cloud forms (thick white contour) between 0.4 and 3.7 km, with ql > 0.001. Strong adiabatic cooling occurs in the lower updraft around the center. Inside the cloud, adiabatic cooling is partially cancelled by the latent heat released. Strong negative buoyancy also occurs at cloud top and above. Although the buoyancy is negative, the vertical velocity increases from 1 m s^{-1} at surface to the maximum of 3.3 m s^{-1} at 0.8 km at x = 335 grid, where a stronger updraft releases more latent heat according to $-w(\partial q_s / \partial z)$, and the buoyancy becomes positive in 2.1 km $\leq$ z $\leq$ 3 km at t = 0.67 h (red area in Figure 56b). Meanwhile, loading and evaporative cooling from raindrops near and beneath cloud base strengthen negative buoyancy. Consequently, a red, pear-shaped positive buoyancy forms in the middle of cloud (Figure 56b). The updraft increases from 1 m s^{-1} at surface to 10.1 m s^{-1} at 1.5 km against negative buoyancy (Figure 56c). At x = 335 grid, B and θ_e are near constant from surface to 2.5 km (Figure 56c,d). They are not as straight as the dry case shown in Figure 55c due to stronger turbulent mixing as well as evaporation and loading of cloud water and rain.

Figure 56e shows that at t = 1.16 h, negative buoyancy still exists in the updraft below 2.6 km. Above it, positive buoyancy enhances the updraft. The vertical velocity reaches 18.3 m s^{-1} at 9.3 km (Figure 56e,f). A strong cooling occurs at the dome-shaped cloud top near the tropopause. The property of the updraft is modified above 6 km because of entrainment of dry air, as shown in streamline, θ_e and B at t = 1.16 h (Figure 56e,f). Figure 6e,f show adiabatic warming on downward motions outside the cloud at z~3 km at x~285 grid and 385 grid; and z~7 km at x~310 grid and x~360 grid, and gaps between clouds (at z~3 km and x~320 grid and x~350 grid. The downdraft is not due to evaporative cooling or loading of rain. More cases can be found in Sun and Sun (2019) [202]. Conservation of Bernoulli function shows that the enthalpy EN can be converted to KE and/or PE so the plume can ascend or descend vertically. Consequently, deep convection develops without CAPE, and downslope can form against positive buoyancy due to the dynamic pressure in (174) and (185).

Hot towers of tall cumulonimbus clouds in the equatorial trough have a horizontally span about 2–4 km across. They can reach altitudes as high as 12–18 km and exhibit high reflectivity. Hot towers are effectively undilute; as a result, the equivalent potential temperature within a hot tower remains nearly constant throughout their entire vertical extent. The release of latent heat caused by condensation of those hot towers supplied the energy necessary to maintain Hadley cells and the trade winds according to Riehl and Malkus (1958) [214]. They found that convections can rise against the stable environment with a temperature lower than that of the surroundings, where CAPE is negative according to Equation (187). The Bernoulli function provides plausible explanations that the cumulonimbus chimneys in the Equator, hurricanes and other severe storms can develop in an environment without CAPE, reach 12–18 km, and transport moisture and aerosols effectively into the stratosphere.

The observations show that weak thunderstorms may have vertical updraft speeds of 6–12 m s^{-1}. At such a speed, a thunderstorm will vertically develop by 10 km in about 15 min. A severe thunderstorm may have a vertical updraft going up at 30–35 m s^{-1}, and an overshooting top lasting for 10 min or longer. If the updraft at 20 m s^{-1}, the parcel takes about 10 min reaching the cloud top at 12 km from the ground, which is shorter than the life span of storms (about an hour or longer). The convection and moisture experiment in 1998 also showed that the equivalent potential temperature within hot towers was virtually constant across their entire vertical extent, confirming the lack of entrainment. Hence, the Bernoulli function, which assumed a steady state and inviscid fluid, should be applicable to storms as well as the flow passing over mountains.

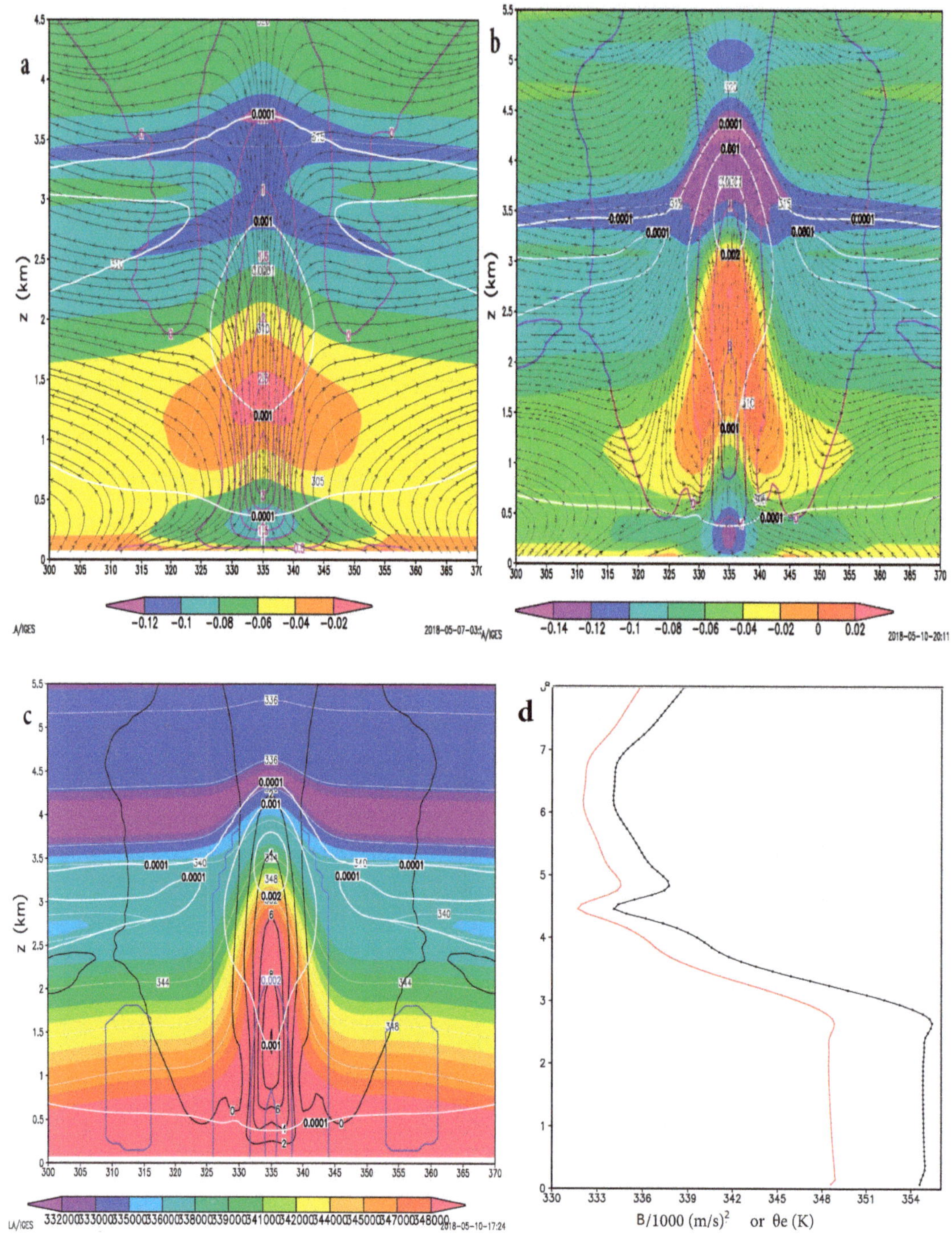

Figure 56. *Cont.*

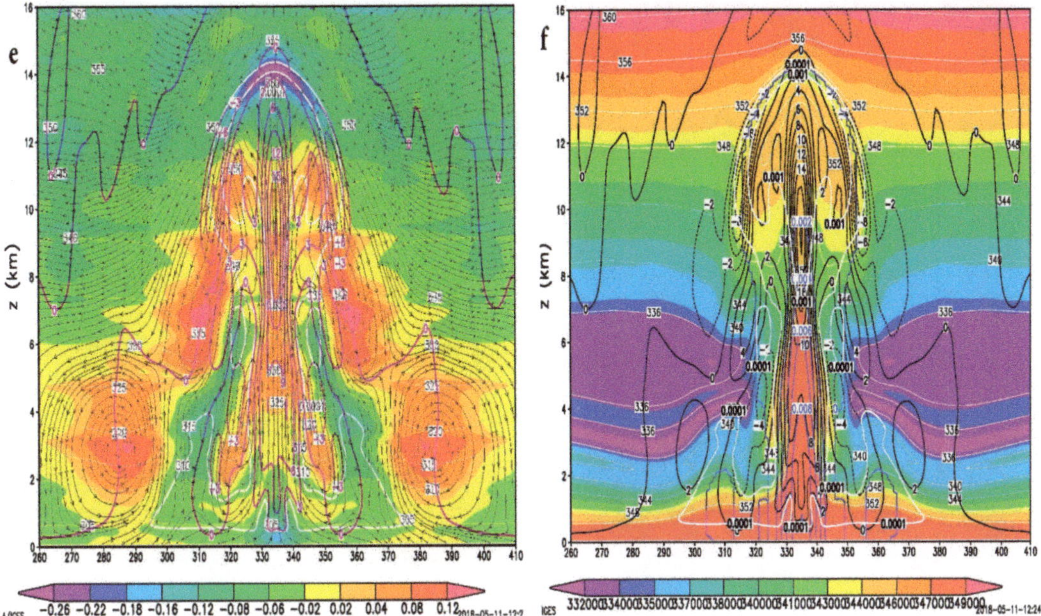

Figure 56. (**a**): Bernoulli function B (shaded color), ql (thick white), θe (thin white), w (purple), and streamline for Case D at 0.56 h, (**b**): buoy (shaded color), ql (thick white), w (purple) and streamline at 0.67 h [202], (**c**): same as (**a**) but at 0.67 h, (**d**) θe (black), B (red) at 0.67 h, (**e**): Buoyancy, buoy (shaded color), ql (thick white), θe (thin white), w (purple), and streamline for Case D. at 1.16 h. (**f**): B (shaded color), ql (thick white), qr (blue), θe (thin white), and w (black) at 1.16 h (Sun and Sun 2019) [202].

The hot towers in the Equator or deep convections of typhoons can interact with the large-scale systems and influence the large-scale systems in the atmosphere and the ocean. The individual cumulonimbus or storms are difficult to predict, but the development of typhoons or cluster of hot towers depends on the large-scale environment, including sea surface temperature, wind, and the supply of heat and moisture, etc., which are easier than the individual convection to predict.

6.8.3. Nonhydrostatic Pressure Inside a Cloud Model

The importance of nonhydrostatic pressure in an air parcel was discussed in Equations (174) and (185) for both dry and moist air parcels. In a one-dimensional time dependent cloud model including detailed cloud microphysics discussed in Section 4.1, entrainment, and nonhydrostatic (i.e., dynamic) pressure, Chen and Sun (2002) [111] found that the inclusion of the nonhydrostatic pressure can (1) affect vertical velocities, (2) help the cloud develop sooner, (3) help maintain a longer mature stage, (4) produce a stronger overshooting cooling, and (5) approximately double the precipitation amount as shown in Figure 57. It is noted that the dynamics pressure is solved from the continuity equation of an anelastic atmosphere. They also show that the one-dimensional cloud model can reproduce the time evolution of the area averaged of three-dimensional WRF simulations shown in Figure 58.

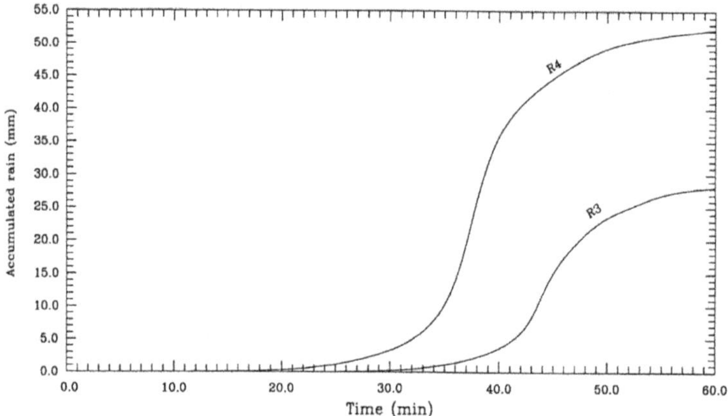

Figure 57. Accumulated rain (mm) without nonhydrostatic pressure (line R3) and with nonhydrostatic pressure (line R4) (Chen and Sun 2002) [111].

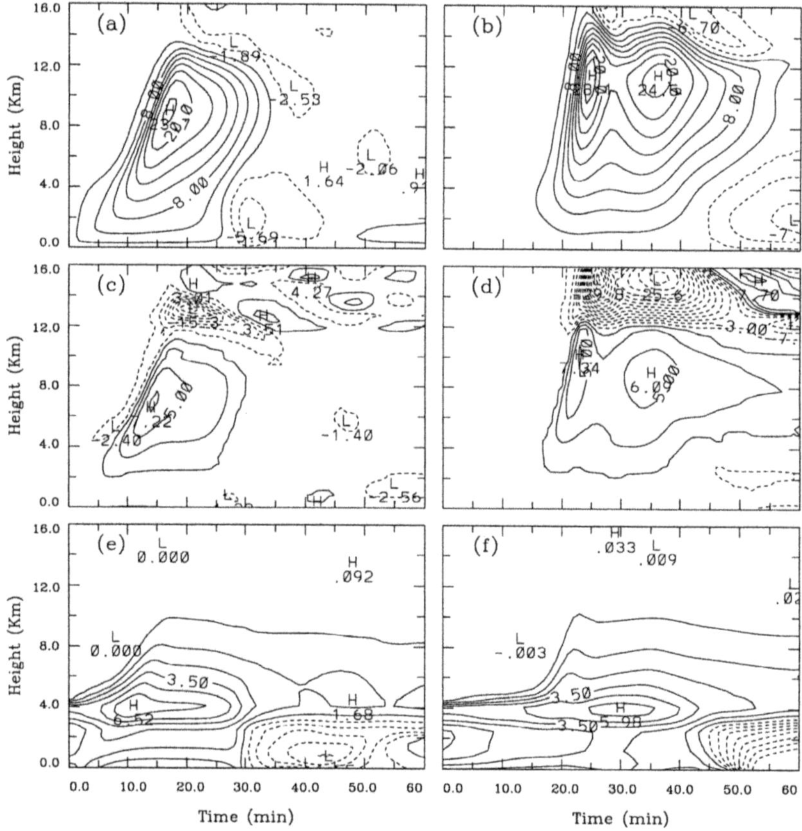

Figure 58. Time evolutions of (**a**) vertical velocity (m s^{-1}), (**c**) potential temperature anomaly (K), and (**e**) moisture anomaly (kg kg^{-1}) from the one-dimensional cloud model, and the averaged (**b**) vertical velocity (m s^{-1}), (**d**) potential temperature anomaly (K), and (**f**) moisture anomaly (kg kg^{-1}) within the radius of 5000 m cloud from the WRF model. The values in (**e**,**f**) are multiplied by 10^3 (Chen and Sun 2002), [111].

6.9. Lee Vortices and Hydraulic Jump in White Sand Missile Range (Haines et al., 2019) [220]

The National Taiwan University/Purdue University (NTU/PU) nonhydrostatic model using a modified forward–backward integration scheme (Sun et al., 2013c) [197] to retains internal gravity waves but suppresses unwanted sound waves was use to study the field experiment conducted at White Sand Missile Range (WSMR), New Mexico, USA on 25 January 2004, which was designed to study a downslope windstorm, lee vortices, and hydraulic jumps along the lee of the Organ and San Andres Mountains (Haines et al., 2006). It was set up with a vertical grid spacing of 300 m, and a horizontal resolution of 1 km on a domain of 201×201 km with the sounding at EL Paso, Texas at 1200 UTC on 25 January 2004 as the initial condition. After 4 h integration, the simulated flows including hydraulic jumps and vortex shedding in the lee of the Organ Mountains agree well with observations as show in Figure 59. The simulations from the WRF used the Gal-Chen and Somerville coordinate (1975) [221] are presented in Figure 60, which failed to reproduce the vortex shedding as observed (Haines et al., 2006) [222]. Figure 61 shows that the observed wind pattern on January 25 can be reproduced very well by an initial westerly wind of 5 m s^{-1} with $\Delta x = \Delta y = 2$ km and $\Delta z = 300$ after 4 h integration. Furthermore, the train of lee waves is also revealed clearly in this larger domain simulation, which has been reported frequently in this area (Haines et al., 2006) [222].

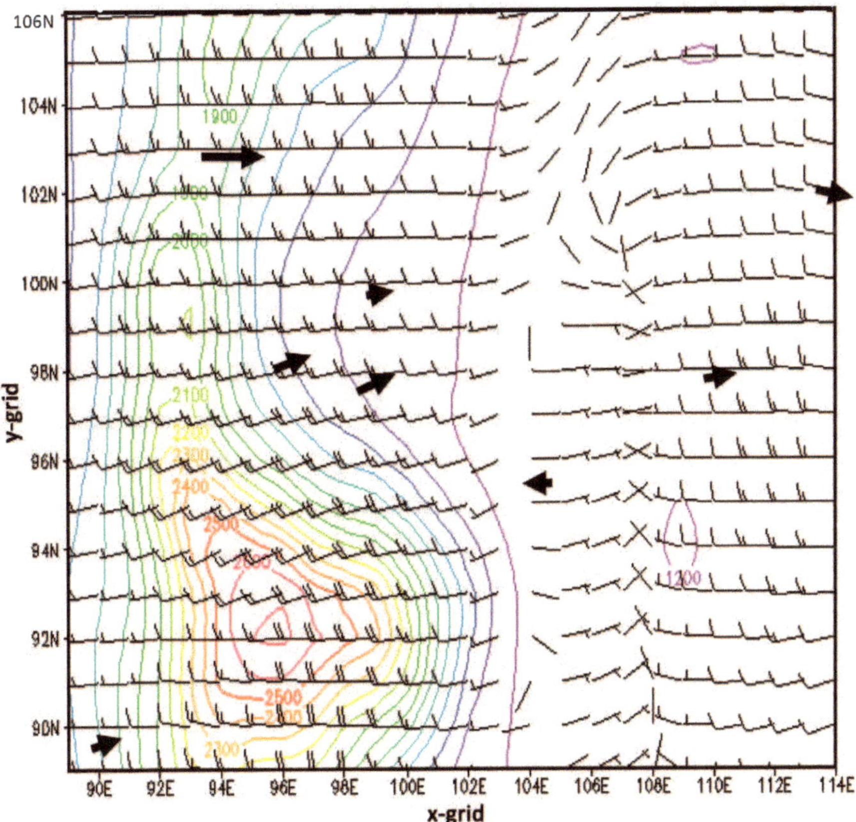

Figure 59. The NTU/PU simulation of January 25 downslope windstorm case at WSMR over a no-slip surface at 32 N after 4 h integration with a mountain height of 1.5 km. The vertical grid interval is 300 m and horizontal grid spacing is 1 km. The model's 10 m winds are shown by the wind barbs. Observed 10 m winds are shown by the thick arrows (Haines et al., 2019) [220].

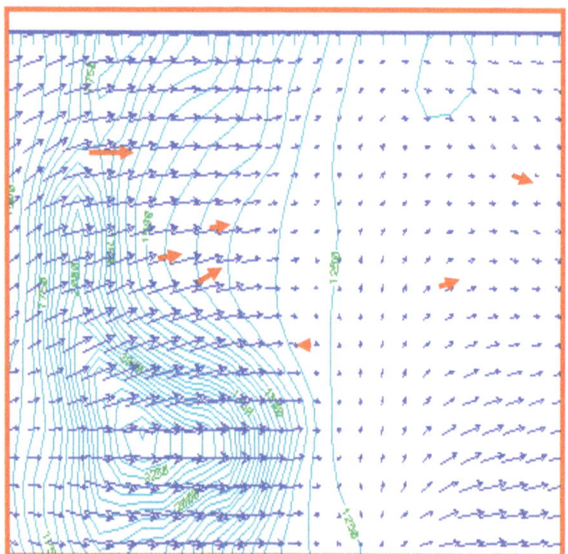

Figure 60. Same as Figure 59, except for WRF (Haines et al., 2006 [222]).

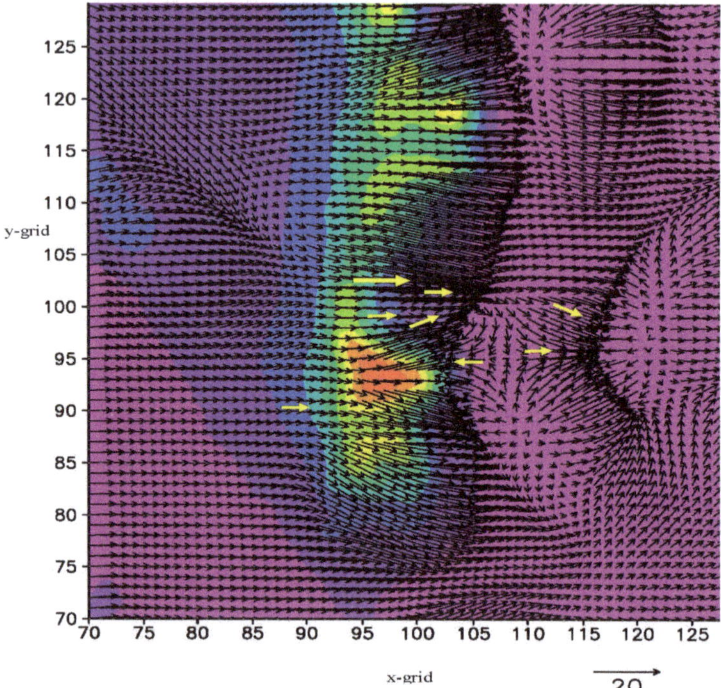

Figure 61. NTU/PU simulated wind vector at z = 10 m above ground after 4 h integration from a constant westerly wind of 5 m s^{-1} with Δx = Δy = 2 km and Δz = 300 m. The observed wind at 0800L25 25 January 2004, is indicated by yellow arrows. The arrow at right-lower corner indicates 20 m s^{-1}. The elevation is shaded by color. Yellow arrows are the observed wind at the time (Haines et al., 2019) [220].

7. Terrain Following Coordinate in Atmospheric Model (Sun 2021) [223]

Numerical atmospheric and oceanic models usually require terrain following coordinates to handle the complex terrain (Kasahara 1974 [198], Sun and Chern1993 [130], Pacanowski 1995 [224], Saito et al., 2001 [225], Staniforth et al., 2004 [226], Lin 2007 [29], Lin et al., 2018 [227], Sun and Sun 2019 [202], etc.) The coordinates proposed by Gal-Chen and Somerville (1975) [221] (will be referred as GS) have been applied to the Regional Atmospheric Modeling System (RAMS, Pielke et al., 2002) [228], Geesthacht Simulation Model of the Atmosphere (GESIMA, Kapitza and Eppel 1992) [229], Cloud Resolving Strom Simulator (CReSS, Tsuboki and Sakakibara 2007) [195], Japan Meteorological Research Institute (MRI-model, Saito et al., 2001) [225], Weather Research and Forecasting Model (WRF) (Skamarock et al., 2008) [230], and other models. The relationship between their coordinates $(\bar{x}, \bar{y}, \bar{z})$ and the Cartesian coordinates (x, y, z) satisfies: $\bar{x} = x$, $\bar{y} = y$, and $\bar{z} = \frac{z_t(z - z_b)}{z_t - z_b}$, where z_t and z_b are the height of the domain and the terrain elevation, respectively. This popular coordinate works well when it is applied to the gradient $\nabla\psi$ but fails to produce the accurate divergence or the curl over a sloped mountain. It cannot be applied to the Navier–Stokes equations either. We propose a new terrain-following coordinate, in which the spatial intervals are not constant.

7.1. Equations

In a curvilinear coordinate (Wikipedia), a position vector can be presented by

$$\mathbf{r} = \hat{x}^i \mathbf{g}_i \left(= \sum_1^3 \hat{x}^i \mathbf{g}_i\right) = \hat{x}_i \mathbf{g}^i \left(= \sum_1^3 \hat{x}_i \mathbf{g}^i\right) = x^i \mathbf{e}_i \tag{188}$$

where $x^i \mathbf{e}_i$ is in Cartesian coordinate, $\mathbf{g}_i$ is the tangent (contravariant) basis vector along the curvilinear coordinate, $\hat{x}^i$ (a contravariant component), $\hat{x}_i$ is covariant component, and covector (or dual basis vector) $\mathbf{g}^i$ are shown in Figure 62.

$$d\mathbf{r} = \frac{\partial \mathbf{r}}{\partial \hat{x}^i} d\hat{x}^i = \mathbf{g}_i d\hat{x}^i = \frac{\partial \mathbf{r}}{\partial \hat{x}_i} d\hat{x}_i = \mathbf{g}^i d\hat{x}_i = \mathbf{e}_i dx^i = \mathbf{e}_x dx + \mathbf{e}_y dy + \mathbf{e}_z dz \tag{189}$$

$$\mathbf{g}_i \equiv \frac{\partial \mathbf{r}}{\partial \hat{x}^i} = \mathbf{e}_j \frac{\partial x^j}{\partial \hat{x}^i} = \mathbf{e}_x \frac{\partial x}{\partial \hat{x}^i} + \mathbf{e}_y \frac{\partial y}{\partial \hat{x}^i} + \mathbf{e}_z \frac{\partial z}{\partial \hat{x}^i} \tag{190}$$

and

$$\frac{\partial \mathbf{r}}{\partial \hat{x}_i} = \mathbf{g}^i \tag{191}$$

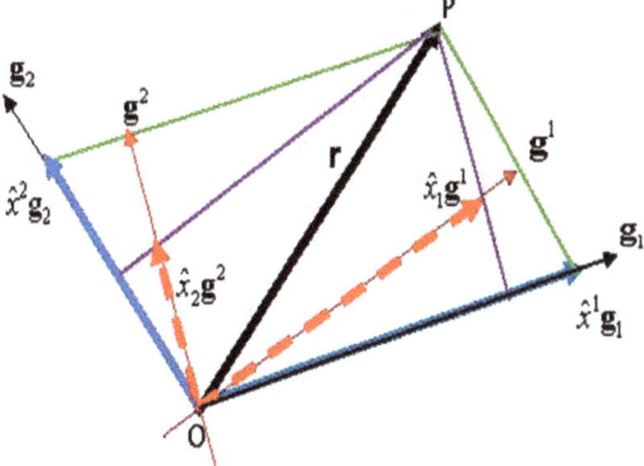

Figure 62. Contravariant (tangent) basis vectors $\mathbf{g}_i$, covector (dual) basis $\mathbf{g}^i$, and a 2D vector $\mathbf{r} = \hat{x}^i \mathbf{g}_i \left(= \sum_1^2 \hat{x}^i \mathbf{g}_i\right) = \hat{x}_i \mathbf{g}^i \left(= \sum_1^2 \hat{x}_i \mathbf{g}^i\right)$ (Sun 2021) [223].

They satisfy:

$$\mathbf{g}_i \cdot \mathbf{g}^j = \delta_i^j \tag{192}$$

$$\hat{x}_i = \mathbf{r} \cdot \mathbf{g}_i = \left(\hat{x}_j \mathbf{g}^j\right) \cdot \mathbf{g}_i; \; \hat{x}^i = \mathbf{r} \cdot \mathbf{g}^i = \left(\hat{x}^j \mathbf{g}_j\right) \cdot \mathbf{g}^i \tag{193}$$

and

$$\mathbf{g}^1 = \frac{\mathbf{g}_2 \times \mathbf{g}_3}{J}, \; \mathbf{g}^2 = \frac{\mathbf{g}_3 \times \mathbf{g}_1}{J}, \; \mathbf{g}^3 = \frac{\mathbf{g}_1 \times \mathbf{g}_2}{J}, \tag{194}$$

Jacobin is

$$J = \mathbf{g}_1 \cdot \mathbf{g}_2 \times \mathbf{g}_3 = \mathbf{g}_1 \times \mathbf{g}_2 \cdot \mathbf{g}_3 = \mathbf{g}_2 \times \mathbf{g}_3 \cdot \mathbf{g}_1 \tag{195}$$

$$J = [\mathbf{g}_1, \mathbf{g}_2, \mathbf{g}_3] = \det \begin{bmatrix} \dfrac{\partial \mathbf{r}}{\partial \hat{x}^1}, & \dfrac{\partial \mathbf{r}}{\partial \hat{x}^2}, & \dfrac{\partial \mathbf{r}}{\partial \hat{x}^3} \end{bmatrix} = \det \begin{bmatrix} \dfrac{\partial x}{\partial \hat{x}^1} & \dfrac{\partial x}{\partial \hat{x}^2} & \dfrac{\partial x}{\partial \hat{x}^3} \\ \dfrac{\partial y}{\partial \hat{x}^1} & \dfrac{\partial y}{\partial \hat{x}^2} & \dfrac{\partial y}{\partial \hat{x}^3} \\ \dfrac{\partial z}{\partial \hat{x}^1} & \dfrac{\partial z}{\partial \hat{x}^2} & \dfrac{\partial z}{\partial \hat{x}^3} \end{bmatrix} \tag{196}$$

and

$$df = \frac{\partial f}{\partial \hat{x}^i} d\hat{x}^i = \mathbf{g}^i \frac{\partial f}{\partial \hat{x}^i} \cdot \mathbf{g}_k d\hat{x}^k = \mathbf{g}^i \frac{\partial f}{\partial \hat{x}^i} \cdot d\mathbf{r} = \nabla f \cdot d\mathbf{r} \tag{197}$$

Hence, the gradient operator is:

$$\nabla \equiv \mathbf{g}^i \frac{\partial}{\partial \hat{x}^i} = \mathbf{e}_i \frac{\partial}{\partial x^i} \tag{198}$$

and

$$\nabla \hat{x}^i = \mathbf{g}^j \frac{\partial \hat{x}^i}{\partial \hat{x}^j} = \mathbf{g}^i = \mathbf{e}_j \frac{\partial \hat{x}^i}{\partial x^j} \tag{199}$$

According to the divergence theorem, $\int \nabla \cdot \mathbf{V} dv = \oint \mathbf{V} \cdot \mathbf{n} dA$, we can calculate the divergence in the volume of $dv = \mathbf{g}_1 d\hat{x}^1 \times \mathbf{g}_2 d\hat{x}^2 \cdot \mathbf{g}_3 d\hat{x}^3 = J d\hat{x}^1 d\hat{x}^2 d\hat{x}^3$ in Figure 63.

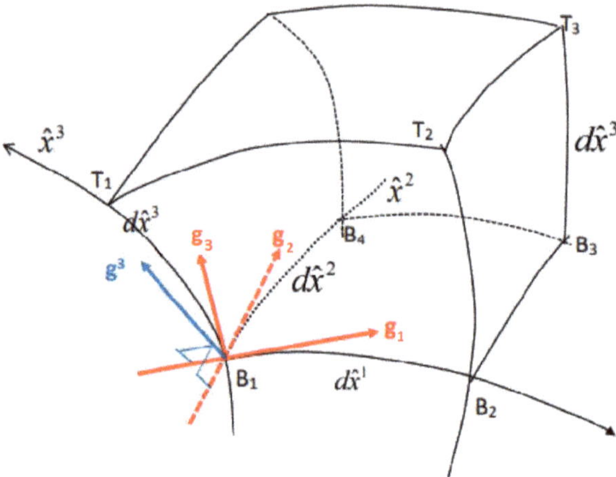

Figure 63. 3D curvilinear coordinates and volume dv = $\left(d\hat{x}^1 \mathbf{g}_1 \times d\hat{x}^2 \mathbf{g}_2 \cdot d\hat{x}^3 \mathbf{g}_3\right)$ (Sun 2021) [223].

The flux of a velocity **V** normal to the lower surface is

$$\begin{aligned} \mathbf{V} \cdot \mathbf{n} dA &= \mathbf{V} \cdot \left(\mathbf{g}_1 d\hat{x}^1 \times \mathbf{g}_2 d\hat{x}^2\right) = \mathbf{V} \cdot \mathbf{g}_1 \times \mathbf{g}_2 d\hat{x}^1 d\hat{x}^2 \\ &= J \mathbf{V} \cdot \mathbf{g}^3 d\hat{x}^1 d\hat{x}^2 = J \hat{v}^3 d\hat{x}^1 d\hat{x}^2 \end{aligned}$$

where $\hat{v}^3 = \mathbf{V} \cdot \mathbf{g}^3$, and the flux on the upper surfaces can be estimated by: $J\hat{v}^3 d\hat{x}^1 d\hat{x}^2 + \frac{\partial(J\hat{v}^3 d\hat{x}^1 d\hat{x}^2)}{\partial \hat{x}^3} d\hat{x}^3$. Hence, the net flux along $\mathbf{g}^3$ is $\sim \frac{\partial(J\hat{v}^3)}{\partial \hat{x}^3} d\hat{x}^1 d\hat{x}^2 d\hat{x}^3$. It is also applied to two other directions. The summation of the net fluxes becomes

$$\sum \mathbf{V} \cdot \mathbf{n} dA = \sum_{i=1}^{3} \frac{\partial(J\hat{v}^i)}{\partial \hat{x}^i} d\hat{x}^1 d\hat{x}^2 d\hat{x}^3 \tag{200}$$

Therefore:

$$\int \nabla \cdot \mathbf{V} dv = \sum \nabla \cdot \mathbf{V} \left(d\hat{x}^1 \mathbf{g}_1 \times d\hat{x}^2 \mathbf{g}_2 \cdot d\hat{x}^3 \mathbf{g}_3 \right) = \sum \nabla \cdot \mathbf{V}(\mathbf{g}_1 \times \mathbf{g}_2 \cdot \mathbf{g}_3) d\hat{x}^1 d\hat{x}^2 d\hat{x}^3$$
$$= \sum \nabla \cdot \mathbf{V} J d\hat{x}^1 d\hat{x}^2 d\hat{x}^3 = \sum \mathbf{V} \cdot \mathbf{n} dA = \sum_{i=1}^{3} \frac{\partial(J\hat{v}^i)}{\partial \hat{x}^i} d\hat{x}^1 d\hat{x}^2 d\hat{x}^3$$

and

$$\nabla \cdot \mathbf{V} = \frac{1}{J} \left(\frac{\partial(J\hat{v}^1)}{\partial \hat{x}^1} + \frac{\partial(J\hat{v}^2)}{\partial \hat{x}^2} + \frac{\partial(J\hat{v}^3)}{\partial \hat{x}^3} \right) \tag{201}$$

The curl of vector $\mathbf{V}$ becomes:

$$\nabla \times \mathbf{V} = \mathbf{g}^1 \times \frac{\partial}{\partial \hat{x}^1} \mathbf{V} + \mathbf{g}^2 \times \frac{\partial}{\partial \hat{x}^2} \mathbf{V} + \mathbf{g}^3 \times \frac{\partial}{\partial \hat{x}^3} \mathbf{V}$$
$$= J^{-1} \left((\mathbf{g}_2 \times \mathbf{g}_3) \times \frac{\partial}{\partial \hat{x}^1} \mathbf{V} + (\mathbf{g}_3 \times \mathbf{g}_1) \times \frac{\partial}{\partial \hat{x}^2} \mathbf{V} + (\mathbf{g}_1 \times \mathbf{g}_2) \times \frac{\partial}{\partial \hat{x}^3} \mathbf{V} \right)$$
$$= J^{-1} \begin{pmatrix} \mathbf{g}_3 \frac{\partial}{\partial \hat{x}^1}(\mathbf{g}_2 \cdot \mathbf{V}) - \mathbf{g}_3 \frac{\partial \mathbf{g}_2}{\partial \hat{x}^1} \cdot \mathbf{V} - \mathbf{g}_2 \frac{\partial}{\partial \hat{x}^1}(\mathbf{g}_3 \cdot \mathbf{V}) + \mathbf{g}_2 \frac{\partial \mathbf{g}_3}{\partial \hat{x}^1} \cdot \mathbf{V} + \\ \mathbf{g}_1 \frac{\partial}{\partial \hat{x}^2}(\mathbf{g}_3 \cdot \mathbf{V}) - \mathbf{g}_1 \frac{\partial \mathbf{g}_3}{\partial \hat{x}^2} \cdot \mathbf{V} - \mathbf{g}_3 \frac{\partial}{\partial \hat{x}^2}(\mathbf{g}_1 \cdot \mathbf{V}) + \mathbf{g}_3 \frac{\partial \mathbf{g}_1}{\partial \hat{x}^2} \cdot \mathbf{V} + \\ \mathbf{g}_2 \frac{\partial}{\partial \hat{x}^3}(\mathbf{g}_1 \cdot \mathbf{V}) - \mathbf{g}_2 \frac{\partial \mathbf{g}_1}{\partial \hat{x}^3} \cdot \mathbf{V} - \mathbf{g}_1 \frac{\partial}{\partial \hat{x}^3}(\mathbf{g}_2 \cdot \mathbf{V}) + \mathbf{g}_1 \frac{\partial \mathbf{g}_2}{\partial \hat{x}^3} \cdot \mathbf{V} \end{pmatrix} \tag{202}$$

But

$$\frac{\partial \mathbf{g}_1}{\partial \hat{x}^2} = \frac{\partial}{\partial \hat{x}^2} \frac{\partial \mathbf{r}}{\partial \hat{x}^1} = \frac{\partial}{\partial \hat{x}^1} \frac{\partial \mathbf{r}}{\partial \hat{x}^2} = \frac{\partial \mathbf{g}_2}{\partial \hat{x}^1}.$$

Therefore

$$\nabla \times \mathbf{V} = J^{-1} \begin{pmatrix} \mathbf{g}_3 \frac{\partial}{\partial \hat{x}^1}(\mathbf{g}_2 \cdot \mathbf{V}) - \mathbf{g}_2 \frac{\partial}{\partial \hat{x}^1}(\mathbf{g}_3 \cdot \mathbf{V}) + \\ \mathbf{g}_1 \frac{\partial}{\partial \hat{x}^2}(\mathbf{g}_3 \cdot \mathbf{V}) - \mathbf{g}_3 \frac{\partial}{\partial \hat{x}^2}(\mathbf{g}_1 \cdot \mathbf{V}) + \\ \mathbf{g}_2 \frac{\partial}{\partial \hat{x}^3}(\mathbf{g}_1 \cdot \mathbf{V}) - \mathbf{g}_1 \frac{\partial}{\partial \hat{x}^3}(\mathbf{g}_2 \cdot \mathbf{V}) \end{pmatrix}$$

Finally, we obtain:

$$\nabla \times \mathbf{V} = \frac{1}{J} \begin{bmatrix} \mathbf{g}_1 & \mathbf{g}_2 & \mathbf{g}_3 \\ \frac{\partial}{\partial \hat{x}^1} & \frac{\partial}{\partial \hat{x}^2} & \frac{\partial}{\partial \hat{x}^3} \\ \mathbf{g}_1 \cdot \mathbf{V} & \mathbf{g}_2 \cdot \mathbf{V} & \mathbf{g}_3 \cdot \mathbf{V} \end{bmatrix} = \frac{1}{J} \begin{bmatrix} \mathbf{g}_1 & \mathbf{g}_2 & \mathbf{g}_3 \\ \frac{\partial}{\partial \hat{x}^1} & \frac{\partial}{\partial \hat{x}^2} & \frac{\partial}{\partial \hat{x}^3} \\ \hat{v}_1 & \hat{v}_2 & \hat{v}_3 \end{bmatrix} \tag{203}$$

where $\hat{v}_i = \mathbf{V} \cdot \mathbf{g}_i = \hat{v}_j \mathbf{g}^j \cdot \mathbf{g}_i$.

7.2. New Terrain Following Coordinate

Figure 64 shows the 2D diagram of the curvilinear coordinate $(\hat{x}^1, \hat{x}^3)$, the inclination angle θ of the $\hat{z}$-surface. For convenience, we also define the vertical coordinate $\hat{z} = \hat{x}^3$ and:

$$\hat{z} = \hat{x}^3 = \frac{z - z_b}{z_t - z_b} \tag{204}$$

where z_b, is the terrain elevation and z_t is the domain height. From (204), we obtain:

$$\left(\frac{\partial \hat{z}}{\partial z} \right)_{x,y} = \left. \frac{\partial \left(\frac{z - z_b}{z_t - z_b} \right)}{\partial z} \right|_{x,y} = \frac{1}{(z_t - z_b)} \tag{205}$$

and

$$\left(\frac{\partial z}{\partial \hat{x}^3}\right)_{\hat{x}^1,\hat{x}^2} = \left(\frac{\partial z}{\partial \hat{z}}\right)_{\hat{x}^1,\hat{x}^2} = \left(\frac{\partial(z_b + \hat{z}(z_t - z_b))}{\partial \hat{z}}\right)_{\hat{x}^1,\hat{x}^2} = (z_t - z_b) \tag{206}$$

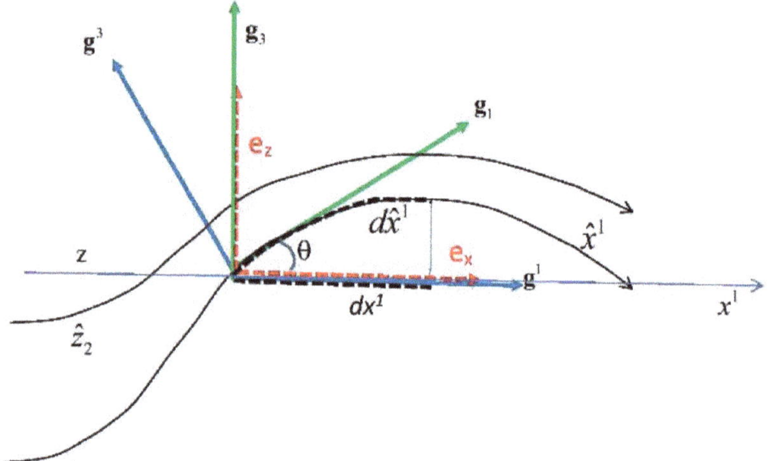

Figure 64. Basis vectors of 2D Cartesian coordinate and terrain following coordinates. (Sun 2021) [223].

The inclinations of $\hat{z}$ along x and y–directions are

$$\left(\frac{\partial z}{\partial x}\right)_{\hat{z}} = \left(\frac{\partial(z_b + \hat{z}(z_t - z_b))}{\partial x}\right)_{\hat{z}} = (1 - \hat{z})\frac{\partial z_b}{\partial x} \tag{207}$$

and

$$\left(\frac{\partial z}{\partial y}\right)_{\hat{z}} = (1 - \hat{z})\frac{\partial z_b}{\partial y} \tag{208}$$

The changes $\hat{z}$ with respect to x and y are:

$$\left(\frac{\partial \hat{z}}{\partial x}\right)_z = \frac{\partial\left(\frac{z-z_b}{z_t-z_b}\right)}{\partial x} = \frac{(z_t - z_b)\frac{\partial(z-z_b)}{\partial x} - (z - z_b)\frac{\partial(z_t-z_b)}{\partial x}}{(z_t - z_b)^2} = \frac{\frac{\partial z_b}{\partial x}(\hat{z} - 1)}{(z_t - z_b)} \tag{209}$$

$$\left(\frac{\partial \hat{z}}{\partial y}\right)_z = \frac{\partial\left(\frac{z-z_b}{z_t-z_b}\right)}{\partial y} = \frac{\frac{\partial z_b}{\partial y}(\hat{z} - 1)}{(z_t - z_b)} \tag{210}$$

The tangent (contravariant) basis vectors in the terrain coordinate system become:

$$\mathbf{g}_1 = \left(\frac{\partial \mathbf{r}}{\partial \hat{x}^1}\right)_{\hat{x}^3} = \left(\frac{\partial\left(x\mathbf{e}_x + y\mathbf{e}_y + z\mathbf{e}_z\right)}{\partial \hat{x}^1}\right)_{\hat{x}^3} = \frac{\partial x}{\partial \hat{x}^1}\left[\mathbf{e}_x + \mathbf{e}_z(1 - \hat{z})\frac{\partial z_b}{\partial x}\right] \tag{211}$$

$$\mathbf{g}_2 = \left(\frac{\partial \mathbf{r}}{\partial \hat{x}^2}\right)_{\hat{x}^3} = \left(\frac{\partial\left(x\mathbf{e}_x + y\mathbf{e}_y + z\mathbf{e}_z\right)}{\partial \hat{x}^2}\right)_{\hat{x}^3} = \frac{\partial y}{\partial \hat{x}^2}\left[\mathbf{e}_y + \mathbf{e}_z(1 - \hat{z})\frac{\partial z_b}{\partial y}\right] \tag{212}$$

and

$$\mathbf{g}_3 = \left(\frac{\partial \mathbf{r}}{\partial \hat{x}^3}\right)_{\hat{x}^1,\hat{x}^2} = \left(\frac{\partial\left(x\mathbf{e}_x + y\mathbf{e}_y + z\mathbf{e}_z\right)}{\partial \hat{x}^3}\right)_{\hat{x}^1,\hat{x}^2} = (z_t - z_b)\mathbf{e}_z \tag{213}$$

At surface $(\hat{z} = 0)$, $\mathbf{g}_1$ and $\mathbf{g}_2$ are tangent to the terrain surface, while $\mathbf{g}_3$ always points to the vertical direction. The Jacobian becomes:

$$J = \mathbf{g}_1 \cdot \mathbf{g}_2 \times \mathbf{g}_3 = \det \begin{bmatrix} \frac{\partial x}{\partial \hat{x}^1} & \frac{\partial x}{\partial \hat{x}^2} & \frac{\partial x}{\partial \hat{x}^3} \\ \frac{\partial y}{\partial \hat{x}^1} & \frac{\partial y}{\partial \hat{x}^2} & \frac{\partial y}{\partial \hat{x}^3} \\ \frac{\partial z}{\partial \hat{x}^1} & \frac{\partial z}{\partial \hat{x}^2} & \frac{\partial z}{\partial \hat{x}^3} \end{bmatrix}$$

$$= \det \begin{bmatrix} \frac{\partial x}{\partial \hat{x}^1} & 0 & 0 \\ 0 & \frac{\partial y}{\partial \hat{x}^2} & 0 \\ (1-\hat{z})\frac{\partial z_b}{\partial x}\frac{\partial x}{\partial \hat{x}^1} & (1-\hat{z})\frac{\partial z_b}{\partial y}\frac{\partial y}{\partial \hat{x}^2} & (z_t - z_b) \end{bmatrix} = \frac{\partial x}{\partial \hat{x}^1}\frac{\partial y}{\partial \hat{x}^2}(z_t - z_b) \tag{214}$$

From (210)–(214), the covector (dual) basis vectors become:

$$\mathbf{g}^1 = \frac{\mathbf{g}_2 \times \mathbf{g}_3}{J} = \mathbf{e}_x \frac{\partial y}{\partial \hat{x}^2}(z_t - z_b) / \left(\frac{\partial x}{\partial \hat{x}^1}\frac{\partial y}{\partial \hat{x}^2}(z_t - z_b) \right) = \mathbf{e}_x / \left(\frac{\partial x}{\partial \hat{x}^1} \right) = \left(\frac{\partial \hat{x}^1}{\partial x^j} \right)\mathbf{e}_j \tag{215}$$

$$\mathbf{g}^2 = \frac{\mathbf{g}_3 \times \mathbf{g}_1}{J} = \mathbf{e}_y \frac{\partial x}{\partial \hat{x}^1}(z_t - z_b) / \left(\frac{\partial x}{\partial \hat{x}^1}\frac{\partial y}{\partial \hat{x}^2}(z_t - z_b) \right) = \mathbf{e}_y / \frac{\partial y}{\partial \hat{x}^2} = \left(\frac{\partial \hat{x}^2}{\partial x^j} \right)\mathbf{e}_j \tag{216}$$

and

$$\mathbf{g}^3 = \frac{\mathbf{g}_1 \times \mathbf{g}_2}{J} = \frac{\frac{\partial z_b}{\partial x}(\hat{z}-1)\mathbf{e}_x}{(z_t - z_b)} + \frac{\frac{\partial z_b}{\partial y}(\hat{z}-1)\mathbf{e}_y}{(z_t - z_b)} + \frac{\mathbf{e}_z}{(z_t - z_b)} = \left(\frac{\partial \hat{x}^3}{\partial x^j} \right)\mathbf{e}_j \tag{217}$$

The transformation between the velocity (u, v, w) in the Cartesian coordinate (x, y, z) and the velocity $(\hat{u}, \hat{v}, \hat{w})$ in the new coordinate is

$$\begin{pmatrix} u \\ v \\ w \end{pmatrix} = \begin{pmatrix} \frac{dx}{dt} \\ \frac{dy}{dt} \\ \frac{dz}{dt} \end{pmatrix} = \begin{pmatrix} \frac{\partial x}{\partial \hat{x}^1}\frac{d\hat{x}^1}{dt} + \frac{\partial x}{\partial \hat{x}^2}\frac{d\hat{x}^2}{dt} + \frac{\partial x}{\partial \hat{x}^3}\frac{d\hat{x}^3}{dt} \\ \frac{\partial y}{\partial \hat{x}^1}\frac{d\hat{x}^1}{dt} + \frac{\partial y}{\partial \hat{x}^2}\frac{d\hat{x}^2}{dt} + \frac{\partial y}{\partial \hat{x}^3}\frac{d\hat{x}^3}{dt} \\ \frac{\partial z}{\partial \hat{x}^1}\frac{d\hat{x}^1}{dt} + \frac{\partial z}{\partial \hat{x}^2}\frac{d\hat{x}^2}{dt} + \frac{\partial z}{\partial \hat{x}^3}\frac{d\hat{x}^3}{dt} \end{pmatrix}$$

$$= \begin{pmatrix} \frac{\partial x}{\partial \hat{x}^1} & 0 & 0 \\ 0 & \frac{\partial y}{\partial \hat{x}^2} & 0 \\ (1-\hat{z})\frac{\partial z_b}{\partial x}\frac{\partial x}{\partial \hat{x}^1} & (1-\hat{z})\frac{\partial z_b}{\partial y}\frac{\partial y}{\partial \hat{y}^1} & (z_t - z_b) \end{pmatrix} \begin{pmatrix} \hat{u}^1 \\ \hat{u}^2 \\ \hat{u}^3 \end{pmatrix} = J \begin{pmatrix} \hat{u}^1 \\ \hat{u}^2 \\ \hat{u}^3 \end{pmatrix} \tag{218}$$

and

$$\begin{pmatrix} \hat{u}^1 \\ \hat{u}^2 \\ \hat{u}^3 \end{pmatrix} = \begin{pmatrix} \frac{d\hat{x}^1}{dt} \\ \frac{d\hat{x}^2}{dt} \\ \frac{d\hat{x}^3}{dt} \end{pmatrix} = \begin{pmatrix} \frac{\partial \hat{x}^1}{\partial x} & \frac{\partial \hat{x}^1}{\partial y} & \frac{\partial \hat{x}^1}{\partial z} \\ \frac{\partial \hat{x}^2}{\partial x} & \frac{\partial \hat{x}^2}{\partial y} & \frac{\partial \hat{x}^2}{\partial z} \\ \frac{\partial \hat{x}^3}{\partial x} & \frac{\partial \hat{x}^3}{\partial y} & \frac{\partial \hat{x}^3}{\partial z} \end{pmatrix} \begin{pmatrix} u \\ v \\ w \end{pmatrix}$$

$$= \begin{pmatrix} \frac{\partial \hat{x}^1}{\partial x} & 0 & 0 \\ 0 & \frac{\partial \hat{x}^2}{\partial y} & 0 \\ \frac{\frac{\partial z_b}{\partial x}(\hat{z}-1)}{(z_t-z_b)} & \frac{\frac{\partial z_b}{\partial x}(\hat{z}-1)}{(z_t-z_b)} & \frac{1}{(z_t-z_b)} \end{pmatrix} \begin{pmatrix} u \\ v \\ w \end{pmatrix} = J^{-1} \begin{pmatrix} u \\ v \\ w \end{pmatrix} \tag{219}$$

and

$$JJ^{-1} = \begin{pmatrix} \frac{\partial x}{\partial \hat{x}^1} & 0 & 0 \\ 0 & \frac{\partial y}{\partial \hat{x}^2} & 0 \\ (1-\hat{z})\frac{\partial z_b}{\partial x}\frac{\partial x}{\partial \hat{x}^1} & (1-\hat{z})\frac{\partial z_b}{\partial y}\frac{\partial y}{\partial \hat{x}^2} & (z_t - z_b) \end{pmatrix} \begin{pmatrix} \frac{\partial \hat{x}^1}{\partial x} & 0 & 0 \\ 0 & \frac{\partial \hat{x}^2}{\partial y} & 0 \\ \frac{\frac{\partial z_b}{\partial x}(\hat{z}-1)}{(z_t-z_b)} & \frac{\frac{\partial z_b}{\partial x}(\hat{z}-1)}{(z_t-z_b)} & \frac{1}{(z_t-z_b)} \end{pmatrix}$$

$$= \begin{pmatrix} 1 & 0 & 0 \\ 0 & 1 & 0 \\ 0 & 0 & 1 \end{pmatrix} = J^{-1}J \tag{220}$$

It shows that $\mathbf{g}_3$ parallels to $\mathbf{e}_z$, $\mathbf{g}^1$ parallels $\mathbf{e}_x$, and $\mathbf{g}^2$ parallels $\mathbf{e}_y$. In this new system, $\mathbf{g}_i$ is different from $\mathbf{g}^i$.

7.3. Gal-Chen and Somerville Terrain Following Coordinate

Gal-Chen and Somerville (1975) [221] assumed:

$$\overline{x} = x; \ \overline{y} = y; \ \text{and} \ \overline{z} = \frac{z_t(z - z_b)}{z_t - z_b} \tag{221}$$

And the inverse transformation:

$$x = \overline{x}; \ y = \overline{y}; \ z = \frac{\overline{z}(z_t - z_b)}{z_t} + z_b \tag{222}$$

The relationships of the basis vectors and other variables between the systems in Sections 7.2 and 7.3 are:

$$\hat{z} = (z - z_b)/(z_t - z_b) = \overline{z}/z_t \tag{223}$$

and

$$\frac{\partial x}{\partial \overline{x}^1} = 1 \text{ and } \frac{\partial y}{\partial \overline{x}^2} = 1, \text{ but } \frac{\partial x}{\partial \hat{x}^1} \text{ or } \frac{\partial y}{\partial \hat{x}^2} \text{ can be different from 1.} \tag{224}$$

Their basic vector $\overline{\mathbf{g}}_i$ and the relationship with $\mathbf{g}_i$ are:

$$\begin{aligned}
\overline{\mathbf{g}}_1 &= \frac{\partial \mathbf{r}}{\partial \overline{x}^1} = \left(\frac{\partial(x\mathbf{e}_x + y\mathbf{e}_y + z\mathbf{e}_z)}{\partial \overline{x}^1}\right)_{\overline{x}^3} = \left(\frac{\partial x}{\partial \overline{x}^1}\mathbf{e}_x + \mathbf{e}_z\frac{\partial}{\partial \overline{x}^1}\left(\frac{\overline{z}(z_t - z_b)}{z_t} + z_b\right)\right)_{\overline{x}^3} \\
&= \left(\frac{\partial x}{\partial \overline{x}^1}\mathbf{e}_x + \mathbf{e}_z\frac{\partial z_b}{\partial \overline{x}^1}\left(1 - \frac{\overline{z}}{z_t}\right)\right)_{\overline{x}^3} = \frac{\partial x}{\partial \overline{x}^1}\left(\mathbf{e}_x + \mathbf{e}_z(1 - \hat{z})\frac{\partial z_b}{\partial x}\right)_{\overline{x}^3} = \frac{\partial \hat{x}^1}{\partial \overline{x}^1}\mathbf{g}_1
\end{aligned} \tag{225}$$

$$\begin{aligned}
\overline{\mathbf{g}}_2 &= \frac{\partial \mathbf{r}}{\partial \overline{x}^2} = \left(\frac{\partial(x\mathbf{e}_x + y\mathbf{e}_y + z\mathbf{e}_z)}{\partial \overline{x}^2}\right)_{\overline{x}^3} = \frac{\partial y}{\partial \overline{x}^2}\left[\mathbf{e}_y + \mathbf{e}_z\frac{(z_t - \overline{z})}{z_t}\frac{\partial z_b}{\partial y}\right] \\
&= \frac{\partial y}{\partial \overline{x}^2}\left(\mathbf{e}_y + \mathbf{e}_z(1 - \hat{z})\frac{\partial z_b}{\partial y}\right) = \frac{\partial \mathbf{r}}{\partial \overline{x}^2} = \frac{\partial \hat{x}^2}{\partial \overline{x}^2}\mathbf{g}_2
\end{aligned} \tag{226}$$

$$\overline{\mathbf{g}}_3 = \left(\frac{\partial \mathbf{r}}{\partial \overline{x}^3}\right)_{\overline{x}^1,\overline{x}^2} = \left(\frac{\partial(x\mathbf{e}_x + y\mathbf{e}_y + z\mathbf{e}_z)}{\partial \overline{x}^3}\right)_{\overline{x}^1,\overline{x}^2} = \mathbf{e}_z\frac{(z_t - z_b)}{z_t} = \frac{\partial \mathbf{r}}{\partial \overline{x}^3} = \frac{\mathbf{g}_3}{z_t} \tag{227}$$

And the Jacobian

$$\begin{aligned}
\overline{J} &= \overline{\mathbf{g}}_1 \cdot \overline{\mathbf{g}}_2 \times \overline{\mathbf{g}}_3 = \det\left[\frac{\partial \mathbf{r}}{\partial \overline{x}^1} \cdots \frac{\partial \mathbf{r}}{\partial \overline{x}^3}\right] = \det\begin{bmatrix} \frac{\partial x}{\partial \overline{x}^1} & \frac{\partial x}{\partial \overline{x}^2} & \frac{\partial x}{\partial \overline{x}^3} \\ \frac{\partial y}{\partial \overline{x}^1} & \frac{\partial y}{\partial \overline{x}^2} & \frac{\partial y}{\partial \overline{x}^3} \\ \frac{\partial z}{\partial \overline{x}^1} & \frac{\partial z}{\partial \overline{x}^2} & \frac{\partial z}{\partial \overline{x}^3} \end{bmatrix} \\
&= \det\begin{bmatrix} \frac{\partial x}{\partial \overline{x}^1} & 0 & 0 \\ 0 & \frac{\partial y}{\partial \overline{x}^2} & 0 \\ \frac{\partial z_b}{\partial x}\frac{(z_t - \overline{z})}{z_t} & \frac{\partial z_b}{\partial y}\frac{(z_t - \overline{z})}{z_t} & \frac{(z_t - z_b)}{z_t} \end{bmatrix} = \frac{(z_t - z_b)}{z_t}
\end{aligned} \tag{228}$$

As well as

$$\overline{\mathbf{g}}^1 = \overline{\mathbf{g}}_2 \times \overline{\mathbf{g}}_3/\overline{J} = \frac{(z_t - z_b)}{z_t}\mathbf{e}_y \times \frac{z_t\mathbf{e}_z}{(z_t - z_b)} = \mathbf{e}_x = \mathbf{g}^1\frac{\partial \hat{x}^1}{\partial x} \tag{229}$$

$$\overline{\mathbf{g}}^2 = \overline{\mathbf{g}}_3 \times \overline{\mathbf{g}}_1/\overline{J} = \mathbf{e}_y = \mathbf{g}^2\frac{\partial \hat{x}^2}{\partial y} \tag{230}$$

$$\begin{aligned}
\overline{\mathbf{g}}^3 &= \overline{\mathbf{g}}_1 \times \overline{\mathbf{g}}_2/\overline{J} \\
&= \left(\frac{\partial x}{\partial \overline{x}^1}\mathbf{e}_x + \mathbf{e}_z\frac{\partial z_b}{\partial \overline{x}^1}\left(1 - \frac{\overline{z}}{z_t}\right)\right)_{\overline{x}^3} \times \left[\frac{\partial y}{\partial \overline{x}^2}\mathbf{e}_y + \mathbf{e}_z\frac{(z_t - \overline{z})}{z_t}\frac{\partial z_b}{\partial \overline{x}^2}\right] / \frac{(z_t - z_b)}{z_t} \\
&= \frac{z_t}{(z_t - z_b)}\mathbf{e}_z - \frac{(z_t - \overline{z})}{(z_t - z_b)}\frac{\partial z_b}{\partial y}\mathbf{e}_y - \frac{(z_t - \overline{z})}{(z_t - z_b)}\frac{\partial z_b}{\partial x}\mathbf{e}_x \\
&= z_t\left[\frac{(\hat{z} - 1)}{(z_t - z_b)}\frac{\partial z_b}{\partial x}\mathbf{e}_x + \frac{(\hat{z} - 1)}{(z_t - z_b)}\frac{\partial z_b}{\partial y}\mathbf{e}_y + \frac{1}{(z_t - z_b)}\mathbf{e}_z\right] = z_t\mathbf{g}^3
\end{aligned} \tag{231}$$

The velocity turns out:

$$\begin{pmatrix} \bar{u} \\ \bar{v} \\ \bar{w} \end{pmatrix} = \begin{pmatrix} \frac{d\bar{x}^1}{dt} \\ \frac{d\bar{x}^2}{dt} \\ \frac{d\bar{x}^3}{dt} \end{pmatrix} = \begin{pmatrix} \frac{\partial \bar{x}^1}{\partial x} & \frac{\partial \bar{x}^1}{\partial y} & \frac{\partial \bar{x}^1}{\partial z} \\ \frac{\partial \bar{x}^2}{\partial x} & \frac{\partial \bar{x}^2}{\partial y} & \frac{\partial \bar{x}^2}{\partial z} \\ \frac{(\bar{z}-z_t)\frac{\partial z_b}{\partial x}}{(z_t-z_b)} & \frac{(\bar{z}-z_t)\frac{\partial z_b}{\partial y}}{(z_t-z_b)} & \frac{z_t}{(z_t-z_b)} \end{pmatrix} \begin{pmatrix} u \\ v \\ w \end{pmatrix}$$
$$= \begin{pmatrix} 1 & 0 & 0 \\ 0 & 1 & 0 \\ \frac{(\bar{z}-z_t)\frac{\partial z_b}{\partial x}}{(z_t-z_b)} & \frac{(\bar{z}-z_t)\frac{\partial z_b}{\partial y}}{(z_t-z_b)} & \frac{z_t}{(z_t-z_b)} \end{pmatrix} \begin{pmatrix} u \\ v \\ w \end{pmatrix} \tag{232}$$

GS-derived Equation (232). It is noted that:

$$\bar{u} = \frac{d\bar{x}^1}{dt} = \frac{dx}{dt} = u = \hat{u}\frac{\partial x}{\partial \hat{x}^1}, \tag{233}$$

$$\bar{v} = \frac{d\bar{x}^2}{dt} = \frac{dy}{dt} = v = \hat{v}\frac{\partial y}{\partial \hat{x}^2}, \tag{234}$$

Hence, their system becomes quite simple, and

$$\bar{w} = \frac{d\bar{x}^3}{dt} = \left(\frac{(\bar{z}-z_t)\frac{\partial z_b}{\partial x}}{(z_t-z_b)}u + \frac{(\bar{z}-z_t)\frac{\partial z_b}{\partial y}}{(z_t-z_b)}v + \frac{z_t}{(z_t-z_b)}w \right)$$
$$= z_t\frac{d\hat{z}}{dt} = z_t \left(-\frac{(1-\hat{z})\frac{\partial z_b}{\partial x}}{(z_t-z_b)}u - \frac{(1-\hat{z})\frac{\partial z_b}{\partial y}}{(z_t-z_b)}v + \frac{1}{(z_t-z_b)}w \right) \tag{235}$$
$$= z_t\hat{u}^3$$

The corresponding gradient, the divergence, and curl are

$$\nabla S = \mathbf{g}^{-i}\frac{\partial S}{\partial \bar{x}^i}, \tag{236}$$

$$\nabla \cdot \mathbf{V} = \frac{1}{\bar{J}}\left(\frac{\partial(\bar{J}\bar{v}^1)}{\partial \bar{x}^1} + \frac{\partial(\bar{J}\bar{v}^2)}{\partial \bar{x}^2}\frac{\partial(\bar{J}\bar{v}^3)}{\partial \bar{x}^3} \right), \tag{237}$$

and

$$\nabla \times \mathbf{V} = \frac{1}{\bar{J}}\begin{bmatrix} \bar{\mathbf{g}}_1 & \bar{\mathbf{g}}_2 & \bar{\mathbf{g}}_3 \\ \frac{\partial}{\partial \bar{x}^1} & \frac{\partial}{\partial \bar{x}^2} & \frac{\partial}{\partial \bar{x}^3} \\ \bar{\mathbf{g}}_1 \cdot \mathbf{V} & \bar{\mathbf{g}}_2 \cdot \mathbf{V} & \bar{\mathbf{g}}_3 \cdot \mathbf{V} \end{bmatrix} = \frac{1}{\bar{J}}\begin{bmatrix} \bar{\mathbf{g}}_1 & \bar{\mathbf{g}}_2 & \bar{\mathbf{g}}_3 \\ \frac{\partial}{\partial \bar{x}^1} & \frac{\partial}{\partial \bar{x}^2} & \frac{\partial}{\partial \bar{x}^3} \\ \bar{v}_1 & \bar{v}_2 & \bar{v}_3 \end{bmatrix} \tag{238}$$

7.4. Numerical Simulations

The 2D model consists of two sets of grids: The first one is the C-grids in the Cartesian coordinate without mountain. The second set is derived from the current and GS coordinates. The Cartesian 2D model includes (20×20) uniform grids with $\Delta x = 1$ km, $z_t = 12$ km, and $\Delta z = 12$ km/20. The elevation of a bell-shaped mountain is given by:

$$z_b = z_{bm}/\left(\left(\frac{x-x_c}{wa} \right)^2 + 1.0 \right) \tag{239}$$

where $z_{bm} = 2000$ m, wa $= 4000$ m, and x_c is located at the mid-point in the x-axis. The coordinate $\hat{z}$ is given by (204) and $\bar{z}$ by (221).

7.4.1. Gradient

A scalar variable,

$$S = Cs(x-x_c)z \tag{240}$$

With Cs = 3.0. The second-order centered difference in the gradient in the Cartesian coordinate is:

$$\nabla S_a = \frac{\partial S}{\partial x}\mathbf{e}_x + \frac{\partial S}{\partial z}\mathbf{e}_z = \frac{S_{j+1/2,k} - S_{j-1/2,k}}{\Delta x}\mathbf{e}_x + \frac{S_{j,k+1/2} - S_{j,k-1/2}}{\Delta z}\mathbf{e}_z \tag{241}$$

The numerical scheme in the current terrain following coordinate is:

$$\nabla S_b = \mathbf{g}^j \frac{\partial S}{\partial \hat{x}^j} = \frac{\mathbf{e}_x}{\partial x/\partial \hat{x}^1}\frac{\partial S}{\partial \hat{x}^1} + \left(\frac{\frac{\partial z_b}{\partial x}(\hat{z}-1)\mathbf{e}_x}{(z_t - z_b)} + \frac{\mathbf{e}_z}{(z_t - z_b)} \right)\frac{\partial S}{\partial \hat{x}^3} \tag{242}$$

The scheme in GS is

$$\nabla S_c = \overset{-j}{\mathbf{g}} \frac{\partial S}{\partial \bar{x}^j}$$
$$= \frac{\mathbf{e}_x}{\partial x/\partial \bar{x}^1}\frac{\partial S}{\partial \bar{x}^1} + \left(\frac{(\bar{z}-z_t)}{(z_t-z_b)}\frac{\partial z_b}{\partial x}\mathbf{e}_x + \frac{z_t}{(z_t-z_b)}\mathbf{e}_z \right)\frac{\partial S}{\partial \bar{x}^3} \tag{243}$$

The continuous black lines in Figure 65 show the gradient in the Cartesian coordinate of ∇S_a along $\mathbf{e}_x$ and $\mathbf{e}_z$ directions, respectively. They are completely covered by the dashed red lines from ∇S_b and the blue dots of ∇S_c, i.e., $\nabla S_b = \nabla S_c$.

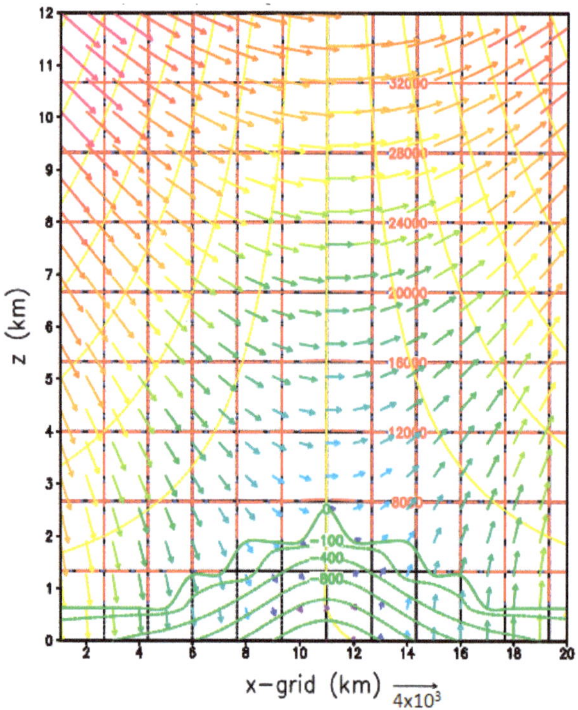

Figure 65. Function S in (240) is shown by yellow lines, ∇S_a by color vector; x and y components of gradient in Cartesian coordinate by black lines, which are covered by dashed red lines of ∇S_b and blue dots of ∇S_c in terrain following coordinates. Green lines show terrain and depth beneath (in meters.) (Sun 2021) [223].

7.4.2. Divergence

A vector

$$\mathbf{V} = Dc(x - x_c)z^2\mathbf{e}_x + Dc(x - x_c)^2 z\mathbf{e}_z \tag{244}$$

With Dc = 6×10^{-8}. We solve the divergences numerically:

$$Div_a = \nabla \cdot \mathbf{V} = \frac{\partial v}{\partial x} + \frac{\partial w}{\partial z}, \tag{245}$$

$$Div_b = \nabla \cdot \mathbf{V} = \mathbf{g}^j \frac{\partial}{\partial \hat{x}^j} \cdot (\hat{v}^i \mathbf{g}_i) = \frac{1}{\hat{J}} \frac{\partial}{\partial \hat{x}^j}(\hat{J}\hat{v}^j) \tag{246}$$

and

$$Div_c = \nabla \cdot \mathbf{V} = \overline{\mathbf{g}}^{-j} \frac{\partial}{\partial \overline{x}^j} \cdot (\overline{v}^i \overline{\mathbf{g}}_i) = \frac{1}{\overline{J}} \frac{\partial}{\partial \overline{x}^j}(\overline{J}\overline{v}^j) \tag{247}$$

Figure 66 shows that Div_b (dashed red lines) agrees with Div_a (black lines). But Div_c (blue dots) departs from Div_a (black lines) over the slope of the mountain, where the difference between $\delta\overline{x}^1$ and $\delta\hat{x}^1$ is large.

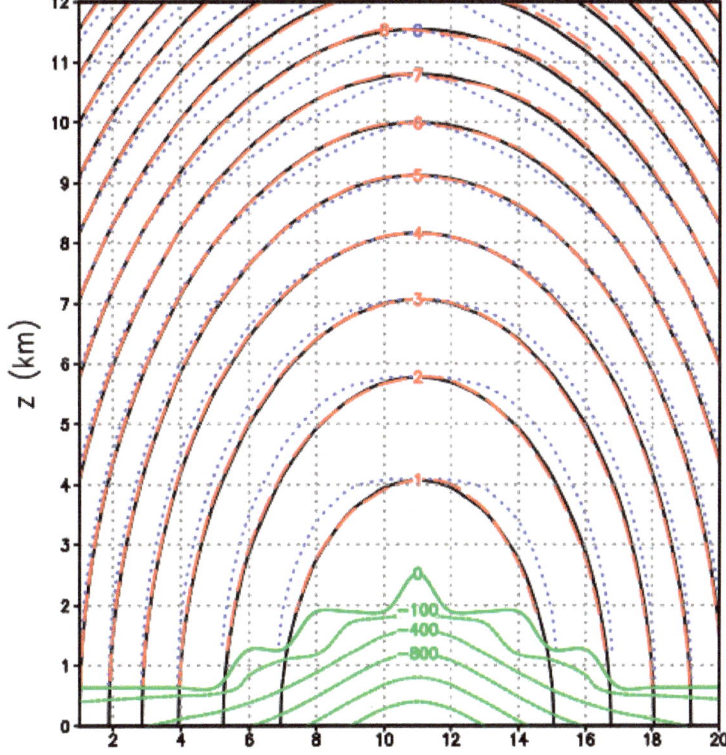

Figure 66. Black lines for divergence of **V** of (244) calculated from Div_a, dashed red lines from Div_b, and blue dots from Div_c. (Sun 2021) [223].

7.4.3. Curl along y-Direction

The vector **V** is given by

$$\mathbf{V} = 2.526 \times 10^6 (x - x_c)z\mathbf{e}_x + 2.526 \times 10^3 (x - x_c)(z - 10\Delta z)^2\mathbf{e}_z \tag{248}$$

The y-component of curls becomes:

$$Curl_a = \frac{\partial u}{\partial z} - \frac{\partial w}{\partial x}, \tag{249}$$

$$Curl_b = \frac{1}{J}\left(\frac{\partial \hat{v}_1}{\partial \hat{x}^3} - \frac{\partial \hat{v}_3}{\partial \hat{x}^1}\right), \tag{250}$$

and

$$Curl_c = \frac{1}{\bar{J}}\left(\frac{\partial \bar{v}_1}{\partial \bar{x}^3} - \frac{\partial \bar{v}_3}{\partial \bar{x}^1}\right) \tag{251}$$

The results of $Curl_a$ (black), $Curl_b$ (dashed red), and $Curl_c$ (blue dots) are shown in Figure 67.

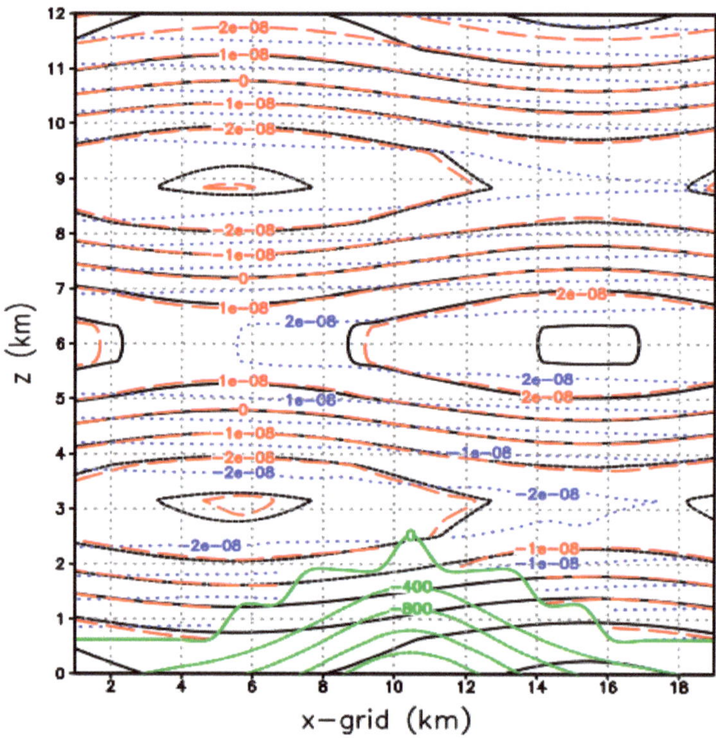

Figure 67. y-component curl: Black lines of **V** of (248) calculated from $Curl_a$, dashed red lines from $Curl_b$ and blue dots from $Curl_c$, and green lines for terrain. (Sun 2021) [223].

Overall, $Curl_b$ resembles $Curl_a$, except some errors in the shortwaves. But $Curl_c$ can be quite different from $Curl_a$.

7.4.4. The Navier–Stokes Equations in New Terrain-Following Coordinate

As discuss Sect 6, the advection form of nonlinear equations has been applied to the NTU/Purdue nonhydrostatic model (Hsu and Sun 2001 [190], MacCall et al., 2015 [231]). It is simple and can be solved numerically with high accuracy (Sun and Yeh 1997 [124], Sun and Sun 2004 [125]), but it does not conserve the total energy or momentum. Combining those equations, we obtain the flux forms:

$$\frac{\partial \rho}{\partial t} + \frac{\partial (\rho u_j)}{\partial x_j} = 0 \tag{252}$$

$$\frac{\partial u_i \rho}{\partial t} + \frac{\partial (u_i \rho u_j)}{\partial x_j} = -\frac{\partial p}{\partial x_j} + g_i + F_{ri} = F_i \tag{253}$$

and

$$\frac{\partial \rho \theta}{\partial t} + \frac{\partial \rho \theta u_j}{\partial x_j} = \frac{1}{c_p} \frac{\theta \rho}{T} \frac{dq}{dt} + D_\theta \rho = Q \tag{254}$$

Without forcing terms on the right-hand side, Equation (131) in the curvilinear coordinate becomes

$$\frac{\partial \hat{v}^k \mathbf{g}_k}{\partial t} = -\hat{v}^j \mathbf{g}_j \cdot \mathbf{g}^i \frac{\partial}{\partial \hat{x}^i}(\hat{v}^m \mathbf{g}_m) = -\hat{v}^i \frac{\partial}{\partial \hat{x}^i}(\hat{v}^m \mathbf{g}_m) = -\hat{v}^i \hat{v}^m \frac{\partial \mathbf{g}_m}{\partial \hat{x}^i} - \hat{v}^i \frac{\partial \hat{v}^m}{\partial \hat{x}^i} \mathbf{g}_m$$

or

$$\frac{\partial \hat{v}^k}{\partial t} = -\hat{v}^i \frac{\partial \hat{v}^k}{\partial \hat{x}^i} - \hat{v}^i \hat{v}^m \frac{\partial \mathbf{g}_m}{\partial \hat{x}^i} \cdot \mathbf{g}^k. \tag{255}$$

With terrain given by (239), and the velocity: $\mathbf{v} = 4\sin(\pi x / x_L)\sin(2\pi z / z_L)\mathbf{e}_x + \sin(2\pi x / x_L)\sin(\pi z / z_L)\mathbf{e}_z$, where $x_L = 20\,\Delta x$, is the length of the horizontal domain, z_L is the height of the domain, we calculate the local change in velocities $(\hat{u}, \hat{v}, \hat{w})$ and $(\overline{u}, \overline{v}, \overline{w})$ in the terrain following coordinates, and compare with (u, v, w) calculated from the Cartesian coordinate. The numerical results of time-derivative of x-component velocities in Figure 68 show that results in $(\hat{x}^1, \hat{x}^2, \hat{x}^3)$ coordinate is better than that in $(\overline{x}^1, \overline{x}^2, \overline{x}^3)$ coordinate as expected.

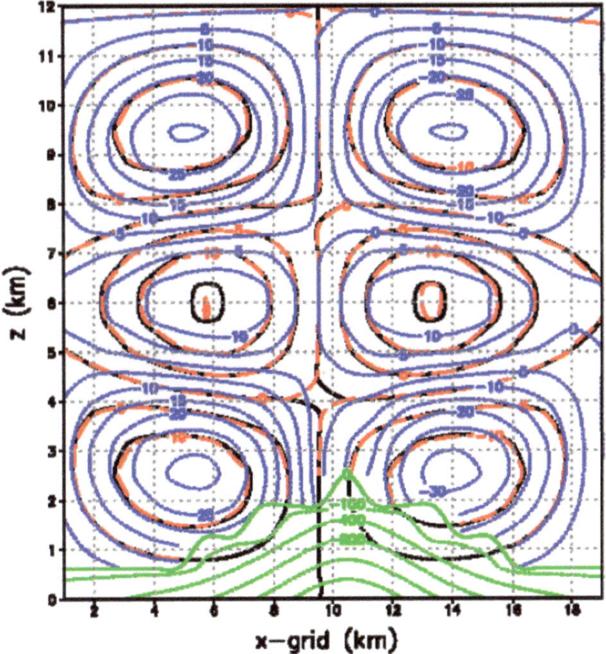

Figure 68. Local time derivative of u: black lines calculated from Cartesian coordinate, red from (253) based on new coordinate $(\hat{x}, \hat{y}, \hat{z})$, and blue dots based on Gal-Chen and Somerville coordinate $(\overline{x}, \overline{y}, \overline{z})$. (Sun 2021) [223].

The conventional flux form of momentum in the curvilinear is

$$\frac{\partial \rho \hat{v}^j}{\partial t} + \frac{1}{J}\frac{\partial}{\partial \hat{x}^i}(J\rho \hat{v}^i \hat{v}^j) - \mathbf{g}_k \hat{v}^i \hat{v}^k \cdot \frac{\partial \mathbf{g}^j}{\partial \hat{x}^i}$$
$$= -\mathbf{g}^j \cdot \mathbf{g}^i \frac{\partial p}{\partial \hat{x}^i} + \cdots \tag{256}$$

In the Cartesian coordinate, the component momentum over the entire domain $(\sum u_i \rho dv)$ in (253) is conserved with $F_i = 0$. $\sum \rho \theta dv$ in (254) is also conserved if $Q = 0$.

But (256) is not conserved if $F_i = 0$, it is also quite complicated. In the curvilinear systems, we may use the divergence operator to present the flux form of the equations, i.e.,

$$\frac{\partial \psi}{\partial t} = -\nabla \cdot (\psi \mathbf{v}) = -\frac{1}{J}\frac{\partial}{\partial \hat{x}^j}(J\psi \hat{v}^j) = -\frac{1}{J}\frac{\partial}{\partial \hat{x}^j}(J\tilde{v}^j) \tag{257}$$

where $\tilde{v}^j = \psi \hat{v}^j$, $\psi = \rho$ in (252), $\psi = \rho\theta$ in (254), and $\psi = \rho u_i$ in (253) where u_i is the i-component velocity in the Cartesian coordinate. Equation (257) is much simpler than (256) and can be solved easily. Equation (257) can also be written:

$$\frac{\partial(\rho J)}{\partial t} = -\frac{\partial(\rho J\hat{u}^i)}{\partial \hat{x}^i} \tag{258}$$

$$\frac{\partial(\rho\theta J)}{\partial t} = -\frac{\partial(\rho\theta J\hat{u}^i)}{\partial \hat{x}^i} + Q \tag{259}$$

and

$$\frac{\partial(\rho u_j J)}{\partial t} = -\frac{\partial(\rho u_j J\hat{u}^i)}{\partial \hat{x}^i} + F_j \tag{260}$$

The total mass is conserved, total energy is conserved for adiabatic flow, and total momentum along each component (u_j, in Cartesian coordinate) is also conserved when $F_j = 0$ in (260). The Navier–Stokes equations including forcing terms can be solved by the finite volume method (Sun 2011) [24]. It is noted that our new coordinate is consistent with Kasahara (1974) [198]. The latter is not in the flux form nor conserves the mass or momentum. The differences between simulations from the WFR based on GS coordinate and NTU/Purdue based on Kasahara are discussed in Section 6.8 and will be discussed more in the Summary.

8. Pollution Model

8.1. Pollution in Convective Boundary Layer (CBL)

Sun and Ogura (1980) [105] included the equation of the turbulent potential temperature–humidity covariance $(\overline{\theta q})$ to simulate the observed $(\overline{\theta q})$ in their turbulence parameterization. Sun and Chang (1986a, b) [108,217] and Sun (1988, 1989) [218,219] extended Sun and Ogura's work by including the temperature–concentration covariance $(\overline{c\theta})$ to simulate the dispersion of plumes in convective boundary layer (CBL). The 1D equations for wind, potential temperature Θ, and humidity are:

$$\frac{\partial U}{\partial t} = fV - fV_g - \frac{\partial \overline{uw}}{\partial z} \tag{261}$$

$$\frac{\partial V}{\partial t} = -fU + fU_g - \frac{\partial \overline{vw}}{\partial z} \tag{262}$$

$$\frac{\partial \Theta}{\partial t} = -\frac{\partial \overline{\theta w}}{\partial z} \tag{263}$$

$$\frac{\partial Q}{\partial t} = -\frac{\partial \overline{qw}}{\partial z} \tag{264}$$

The equations of subgrid-turbulences are shown in Section 4 of this article. The model does not include radiation or cloud physics, but they were included in Sun (1993a, b) [199,232] to study the nocturnal precipitation triggered by radiative cooling at stratocumulus cloud top. The equations for the crosswind integration concentration C_y, the concentration flux $\overline{wc_y}$ and the covariance of concentration and temperature $\overline{c\theta_y}$ are

$$\frac{\partial C_y}{\partial t} = -U\frac{\partial C_y}{\partial x} - \frac{\partial \overline{wc_y}}{\partial z} \tag{265}$$

$$\frac{\partial \overline{wc_y}}{\partial t} = -U\frac{\partial \overline{wc_y}}{\partial x} - \overline{w^2}\frac{\partial \overline{C_y}}{\partial z} + \frac{g}{\Theta}(1-\alpha_2)\overline{c\theta_y} + \frac{\partial}{\partial z}\left(\sqrt{2E}\lambda\frac{\partial \overline{wc_y}}{\partial z}\right) - \frac{\alpha_1\sqrt{2E}}{\lambda}\overline{wc_y} \quad (266)$$

$$\frac{\partial \overline{c\theta_y}}{\partial t} = -U\frac{\partial \overline{c\theta_y}}{\partial x} - \overline{w\theta}\frac{\partial \overline{C_y}}{\partial z} - \overline{wc_y}\frac{\partial \Theta}{\partial z} + \frac{\partial}{\partial z}\left(\sqrt{2E}\lambda\frac{\partial \overline{c\theta_y}}{\partial z}\right) - \frac{\alpha_3\sqrt{2E}}{\lambda}\overline{c\theta_y} \quad (267)$$

The results shown in Figure 69 are quite comparable with the laboratory experiments shown in Figure 70 (Deardorff and Willis 1975 [233], Willis and Deardorff 1978, 1981 [234,235]), in which the plume released from surface ascends to upper mixed layer, then gradually descends. If the sources are far away from ground, the plumes descend, touch the ground and bounce to the middle and upper mixed layer. The simulated surface concentration is also in good agreement with field experiments (Sun 1988, 1989) [218,219].

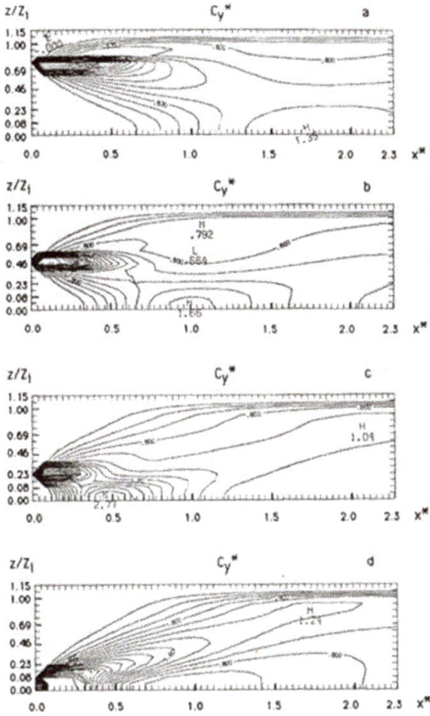

Figure 69. Simulated non-dimensional crosswind integrated concentration C_y^* as a function of non-dimensional distance, $x* = [w*/(U_m Zi)]$ $C_y^* = (U_m Zi/Q)C_y$ measured form point sources at (**a**) zs = 0.75 Zi; (**b**) zs = 0.49 Zi; (**c**) zs = 0.24 Zi; and (**d**) zs = 0.067 Zi (Zi is depth of mixed layer, U_m is mean wind, w* is convective vertical velocity at surface) (Sun 1988, 1989) [218,219].

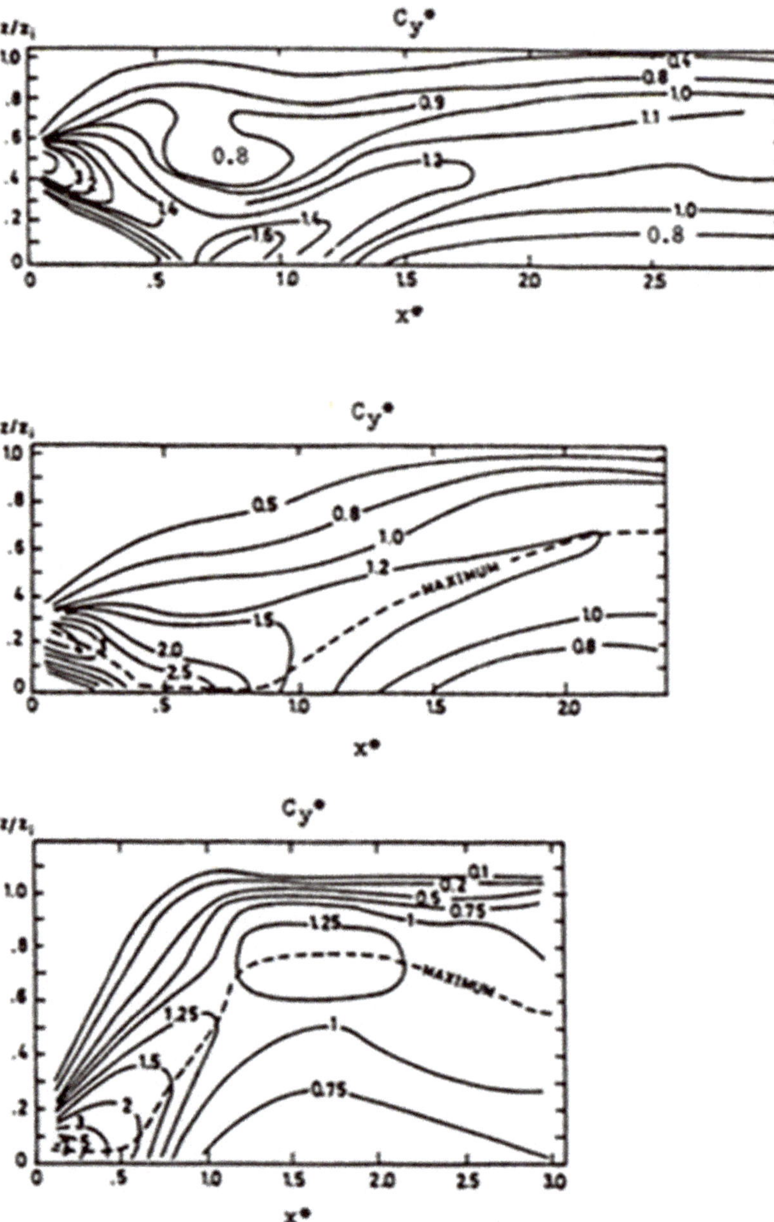

Figure 70. Crosswind integrated concentration C_y^* vs. x* from laboratory experiments for point sources of zs = 0.49, 0.24, and 0.067 Zi (Deardorff and Willis 1975, Willis and Deardorff 1978, 1981) [233–235].

8.2. Backward Integration of Diffusion Equation (Sun and Sun 2017) [236]

In geoscience, Lagrangian backward trajectory without mixing is widely used to estimate the property at the source. With mixing, the plume can mix with surrounding and become quite different form the source. Sun and Sun (2017) [236] proposed reverse-in-time integrating to assess the concentrations at the source. The model consists of five different sizes of concentric cylinders with different species. We integrate it forward in time, then use data collected at downwind to integrate the model backward in time. The Leibniz integral rule for density ρ_m^j for the m-th specie in the j-th cell with velocity $\mathbf{V}^j$ is:

$$\frac{D}{Dt}\int_{V_j}\rho_m^j dV = \int_{V_j}\frac{\partial \rho_m^j}{\partial t}dV + \oint \rho_m^j \mathbf{V}^j \cdot \mathbf{n} dA = -\int_{V_j}\rho_m^j S_m^j dV, \tag{268}$$

where S_m^j is sink, let us define $\mathbf{V}^j = \overline{\mathbf{V}^j}(\text{cell} - \text{sized mean}) + \mathbf{V}^{j\prime}$ (perturbation), $\rho_m^j = \overline{\rho_m^j} + \rho_m^{j\prime}$. It is also assumed $\nabla \cdot \mathbf{V}^j = 0$, and sizes of each cells remain constant. (268) becomes:

$$\int_{V_j}\frac{\partial \overline{\rho_m^j}}{\partial t}dV + \oint \overline{\left(\overline{\rho_m^j} + \rho_m^{j\prime}\right)\left(\overline{\mathbf{V}^j} + \mathbf{V}^{j\prime}\right)} \cdot \mathbf{n} dA = -\int_{V_j}\overline{\rho_m^j S_m^j}dV,$$

$$\int_{V_j}\frac{D\overline{\rho_m^j}}{Dt}dV = \frac{D}{Dt}\int_{V_j}\overline{\rho_m^j}dV = -\oint \overline{\rho_m^{j\prime}\mathbf{V}^{j\prime}} \cdot \mathbf{n} dA - \int_{V_j}\overline{\rho_m^j S_m^j}dV. \tag{269}$$

The bar of cell-sized mean will be dropped hereafter. Following the Lagrangian frame, the change in the density ρ_m^j for the *m-th* specie in the *j-th* cell with volume V_j is

$$\frac{D\rho_m^j}{Dt} = \frac{1}{V_j}\left\{-\rho_m^j\sum_{i\neq j}A_{i,j}u_{i,j}' + \sum_{i\neq j}\rho_m^i A_{i,j}u_{i,j}' - \rho_m^j V_j S_m^j\right\} \text{ for i} = 1,2.,\text{I; j} = 1,2,.,\text{J; and m} = 1,2,\ldots\text{M} \tag{270}$$

where $A_{i,j}$ is the surface area between the *i-th* and *j-th* cells, and $u_{i,j}'$ is the turbulent velocity on $A_{i,j}$. It is noted that (270) is equivalent to

$$\frac{\partial \overline{\rho}}{\partial t} = -\nabla\cdot\left[\overline{(\rho + \rho')(\mathbf{V} + \mathbf{V}')}\right] = -\nabla\cdot\left[\overline{\rho}\mathbf{V} + \overline{\rho'\mathbf{V}'}\right] = -\overline{\mathbf{V}}\cdot\nabla\overline{\rho} - \nabla\cdot\overline{\rho'\mathbf{V}'}; \ \frac{D\overline{\rho}}{Dt} = -\nabla\cdot\left[\overline{\rho'\mathbf{V}'}\right] \tag{271}$$

Traditionally, $\nabla\cdot\left[\overline{\rho'\mathbf{V}'}\right]$ is ignored in the continuity equation for air density, but the subgrid turbulence fluxes for moisture and aerosols are kept in the equations. For simplicity, V_j, $A_{i,j}$, and $u_{i,j}'$ are time independent. The finite difference form of Equation (270) becomes:

$$\frac{\left(\rho_m^j\right)^{n+1} - \left(\rho_m^j\right)^n}{\Delta t} =$$

$$\frac{1}{V_j}\left\{-\left[p\left(\rho_m^j\right)^n + q\left(\rho_m^j\right)^{n+1}\right]\sum_{i\neq j}A_{i,j}u_{i,j}' + \sum_{i\neq j}\left[p\left(\rho_m^i\right)^n + q\left(\rho_m^i\right)^{n+1}\right]A_{i,j}u_{i,j}' - \left[p\left(\rho_m^j\right)^n + q\left(\rho_m^j\right)^{n+1}\right]V_j S_m^j\right\} \tag{272}$$

for j $= 1,2,\ldots,$ J; i $= 1,2,\ldots,$ J; and m $= 1,2,\ldots,$ M.

An implicit time step is applied on the right hand, where q $= 1 - $ p, and $0 \leq p \leq 1$. (272) Becomes:

$$\left(\rho_m^j\right)^{n+1}\left\{1 + q\left[\sum_{i\neq j}\left(\frac{A_{i,j}}{V_j}\right)u_{i,j}'\Delta t + S_m^j\right]\right\} - q\sum_{i\neq j}\left(\frac{A_{i,j}}{V_j}\right)u_{i,j}'\Delta t\left(\rho_m^i\right)^{n+1} =$$

$$\left(\rho_m^j\right)^n\left\{1 - p\left[\sum_{i\neq j}\left(\frac{A_{i,j}}{V_j}\right)u_{i,j}'\Delta t + S_m^j\right]\right\} + p\sum_{i\neq j}\left(\frac{A_{i,j}}{V_j}\right)u_{i,j}'\Delta t\left(\rho_m^i\right)^n \tag{273}$$

$$\text{for j} = 1,..,\text{J; i} = 1,.,\text{I; and m} = 1,.,\text{M}.$$

We assume that $p = 0.5$ and $S_m^j = 0$. Equation (273) can be integrated forward-in-time from an initial condition or reverse-in-time from that collected at downwind region. The model consists of two layers of cylinders as shown in Figure 71. The depth of the lower layer is 2000 m, which is topped by the cylinder E (light yellow) with radius of 30,000 m and thickness of 2222.2 m. The lower layer includes four concentric cylinders (that will be referred as cylinder or cell hereafter): A (black), B (red), C (green), and D (blue), with radii of R_{AB} (between cylinders A and B) = 5000 m, R_{BC} (between cylinders B and C) = 10,000 m, R_{CD} (between cylinders C and D) = 20,000 m, and R_D (outside of D) = 30,000 m. The volume is $157,079 \times 10^6$ m^3 for cylinder A; $471,239 \times 10^6$ m^3 for cylinder B; $188,496 \times 10^7$ m^3 for cylinder C; $314,159 \times 10^7$ m^3 for cylinder D; and $628,256 \times 10^7$ m^3 for cylinder E (light blue).

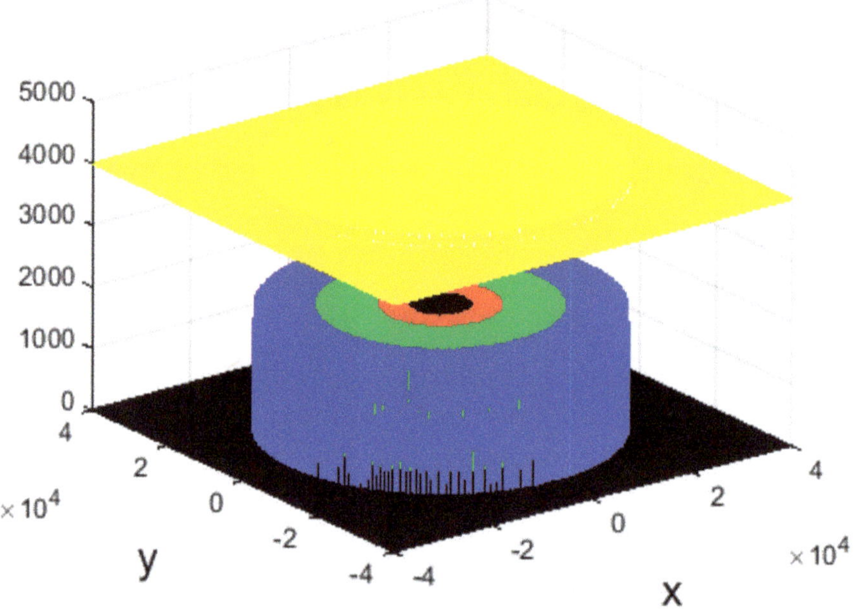

Figure 71. The size and arrangement of cylinders. (Sun and Sun 2017) [236].

Severe pollutions usually occur in a calm, stable condition. Hence, the value of horizontal u' is set equal to 2.5×10^{-2} m s^{-1}. The vertical mixing u' is set equal to 0.1 of the horizontal u' in the lower layer. The time interval $\Delta t = 50$ s is applied to both the forward and reverse integrations.

8.2.1. Forward-in-Time Integration

Each cylinder has its unique specie with one unit amount of density initially. The time evolutions of different species in each cylinder are shown in Figure 72a–e. Initially, ρ_1^A (specie 1 in cylinder A, black line) decreases with time exponentially, which is replaced by the material coming from cylinders B (ρ_2^A specie 2 in cylinder A, red line), C (ρ_3^A green), D (ρ_4^A blue), and E (ρ_5^A light blue). The density of those invaders starts from zero and increases with time. The decrease in ρ_1^A slows down as time increases, at the end of integration, tf = 4×10^5 s (~111.1 h), $\rho_1^A \cong 0.07$, while $\rho_2^A = 0.144$, $\rho_3^A = 0.26$, $\rho_4^A = 0.20$, and $\rho_5^A = 0.32$. Most of original material in cylinder A is replaced by the substances from outside, only 7% of the original material remains at t = 4×10^5 s.

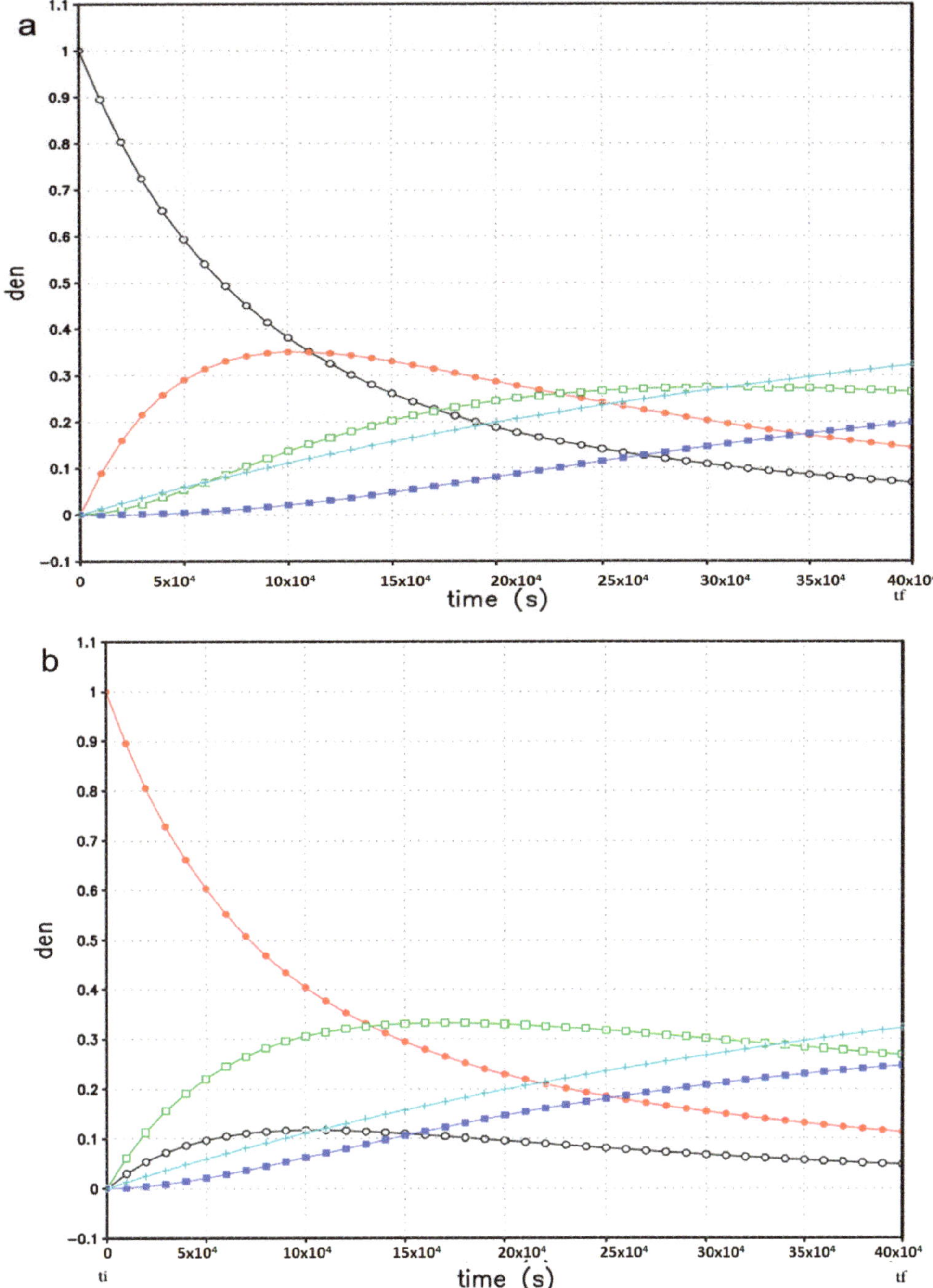

Figure 72. *Cont.*

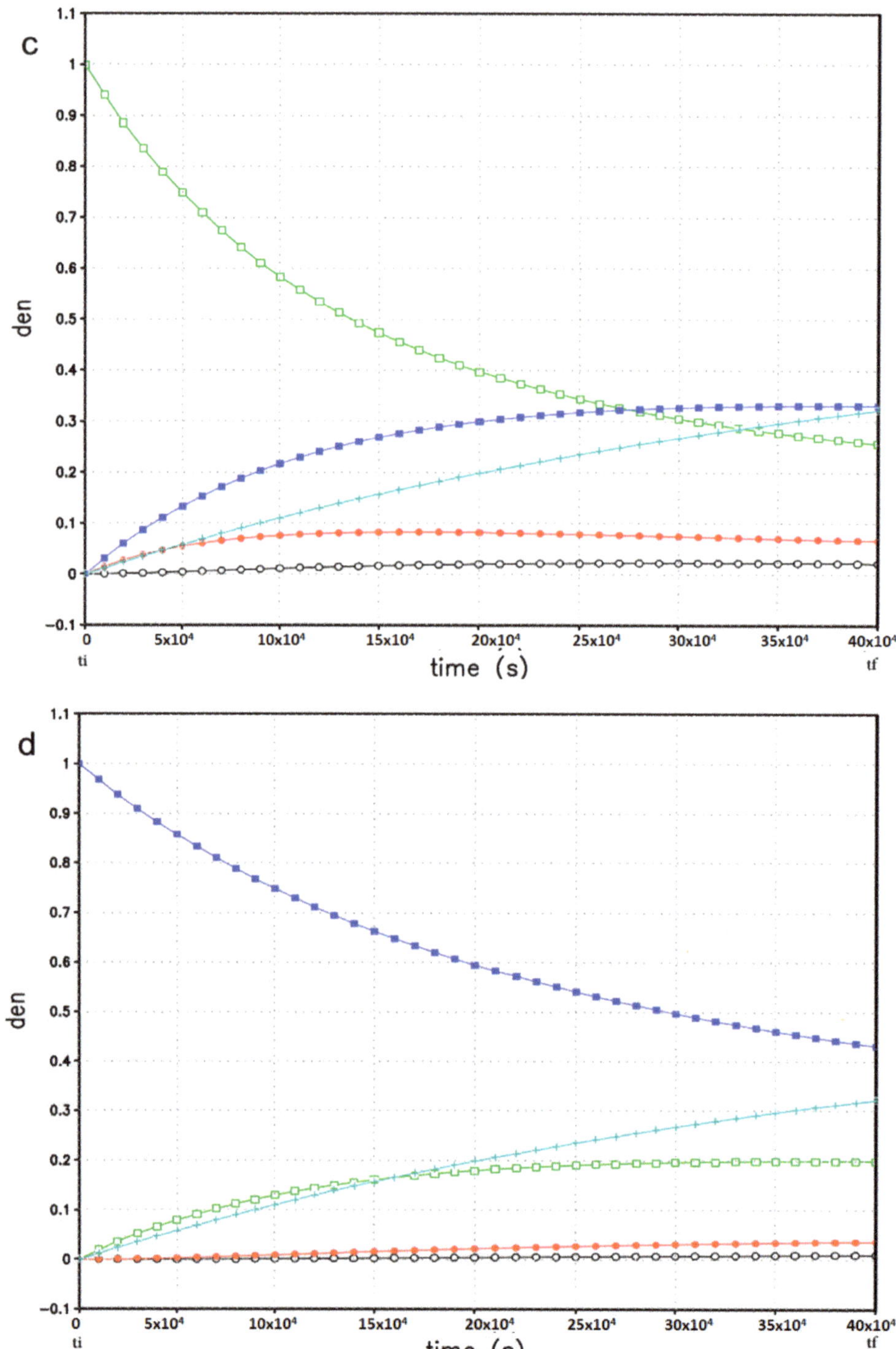

Figure 72. *Cont.*

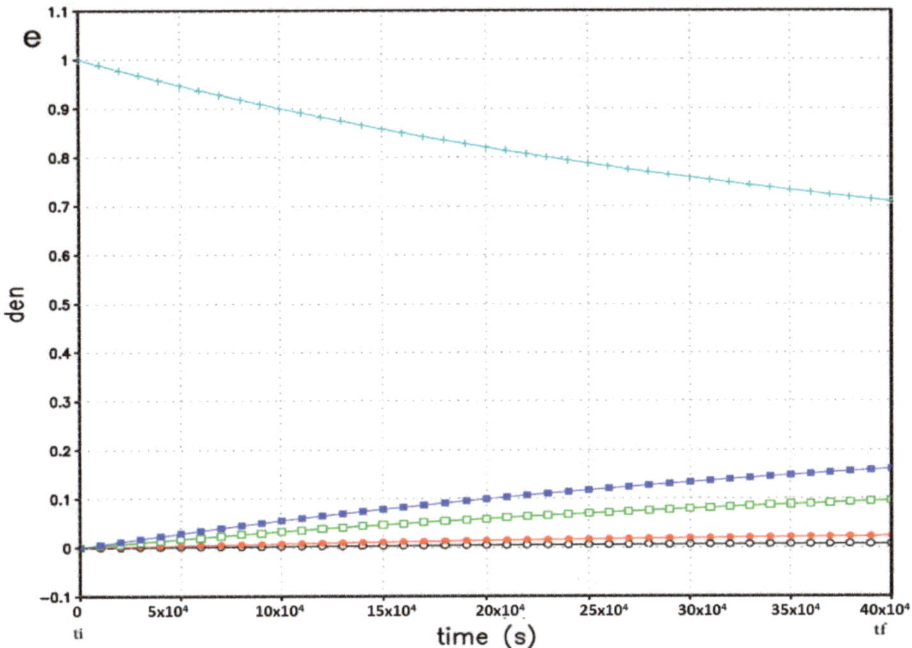

Figure 72. Different species (black, red, green, blue, and light blue for specie 1, 2, 3, 4, and 5, respectively) at each cell (A, B, C, D, E): (**a**) ρ_1^A, ρ_2^A, ρ_3^A, ρ_4^A, and ρ_5^A; (**b**) ρ_1^B, ρ_2^B, ρ_3^B, ρ_4^B, and ρ_5^B; (**c**) ρ_1^C, ρ_2^C, ρ_3^C, ρ_4^C, and ρ_5^C; (**d**) ρ_1^D, ρ_2^D, ρ_3^D, ρ_4^D, and ρ_5^D; and (**e**) ρ_1^E, ρ_2^E, ρ_3^E, ρ_4^E, and ρ_5^E (Sun and Sun 2017) [236].

The distribution of ρ_2^B (species 2 in cylinder B), shown in Figure 72b is quickly replaced by the material coming from other cylinders. Only about 11% remains at tf = 4 × 10⁵ s. Figure 72c–e show that the larger the volume, the more the original material can maintain. It is also noted that Cylinder E (Figure 72e) can keep up more than 70% of the original material because of large volume as well as a small u' between it and the cylinders in the lower layer. Figure 72e also shows that the contributions from the lower cylinders are proportionally to their contact areas with Cylinder E.

The forward-in-time integration (marketed with +) of each specie among different cylinders are indicated by black (in cylinder A), red (B), green (C), blue (D), and light blue (E) lines, and shown in Figure 73a–e.

8.2.2. Reverse-in-Time Integration

Using the density distribution of species at t = 4 × 10⁵ s from forward integration in Figure 73a–e as initial condition, we integrate Equation (273) backward-in-time, i.e., we calculate $\left(\rho_m^j\right)^n$ from a given $\left(\rho_m^j\right)^{n+1}$ by Equation (273). The integrated density from tf = 4 ×'10⁵ s to the end at ti = 0 are shown by black, red, green, blue, and light blue lines marked with "O" signs in Figure 73a–e for each species. They coincide with the forward-in-time integrations (with +). The results end up almost identical to the initial condition in the forward-in-time integration. ρ_5^A (Figure 73) has an error of 1.15% after 2 × 8000 times of forward and backward integrations. Figure 73 shows that ρ_2^A (=0.145) is larger than ρ_2^B (=0.113) at tf = 4 × 10⁵ s, but at the end of the reverse integration $\rho_2^A = 0$ and $\rho_2^B = 1$. The values of ρ_3^A, ρ_3^B, and ρ_3^C are around 0.26 at tf = 4 × 10⁵ s, after the reverse integration $\rho_3^A = \rho_3^B = 0$ and $\rho_3^C = 1$ at t = 0. It indicates that the model can handle the counter gradient flux during backward integration, a difficult task in diffusion modeling (Sun and Chang 1986a, b, Sun 1988, 1989).

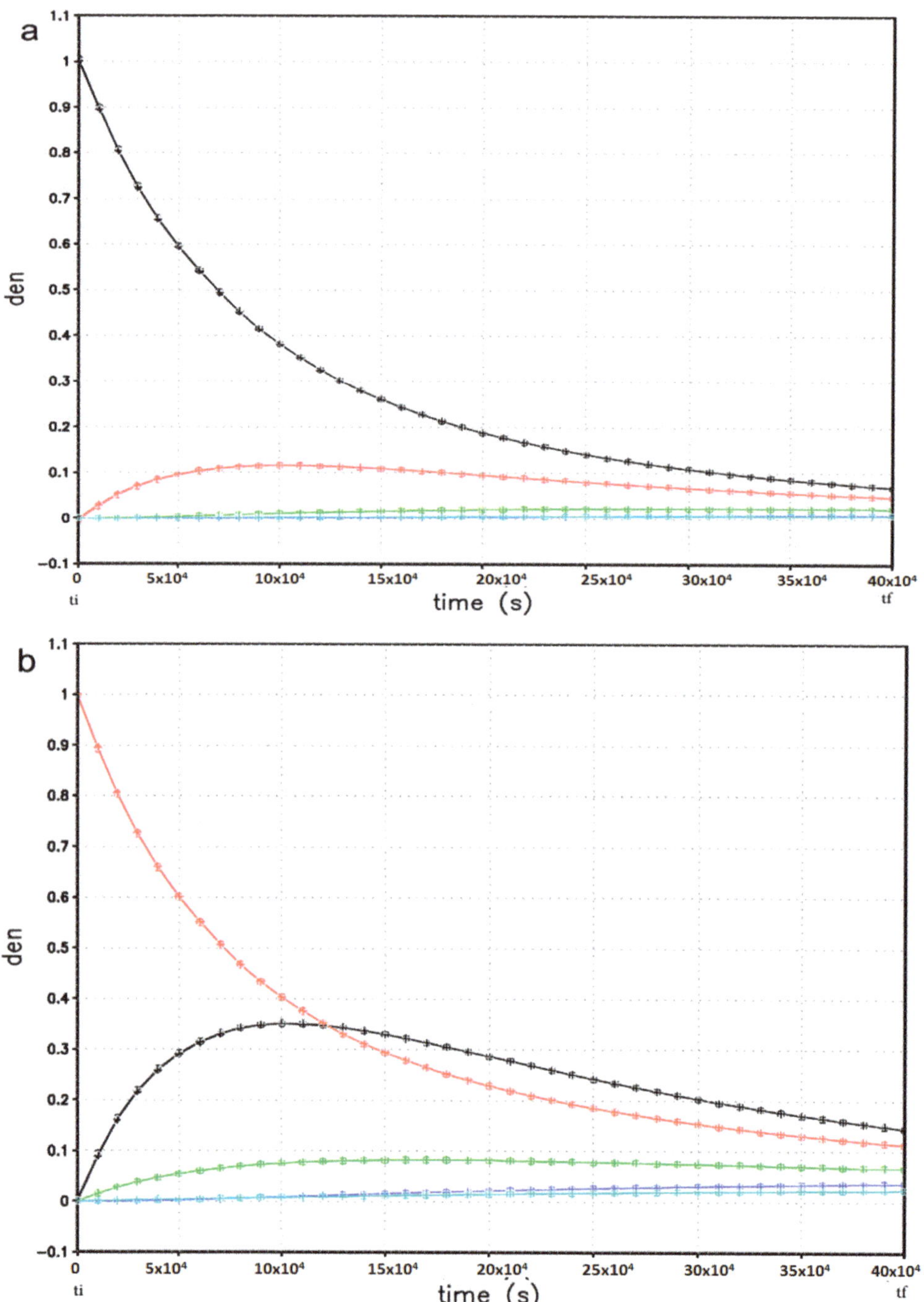

Figure 73. *Cont.*

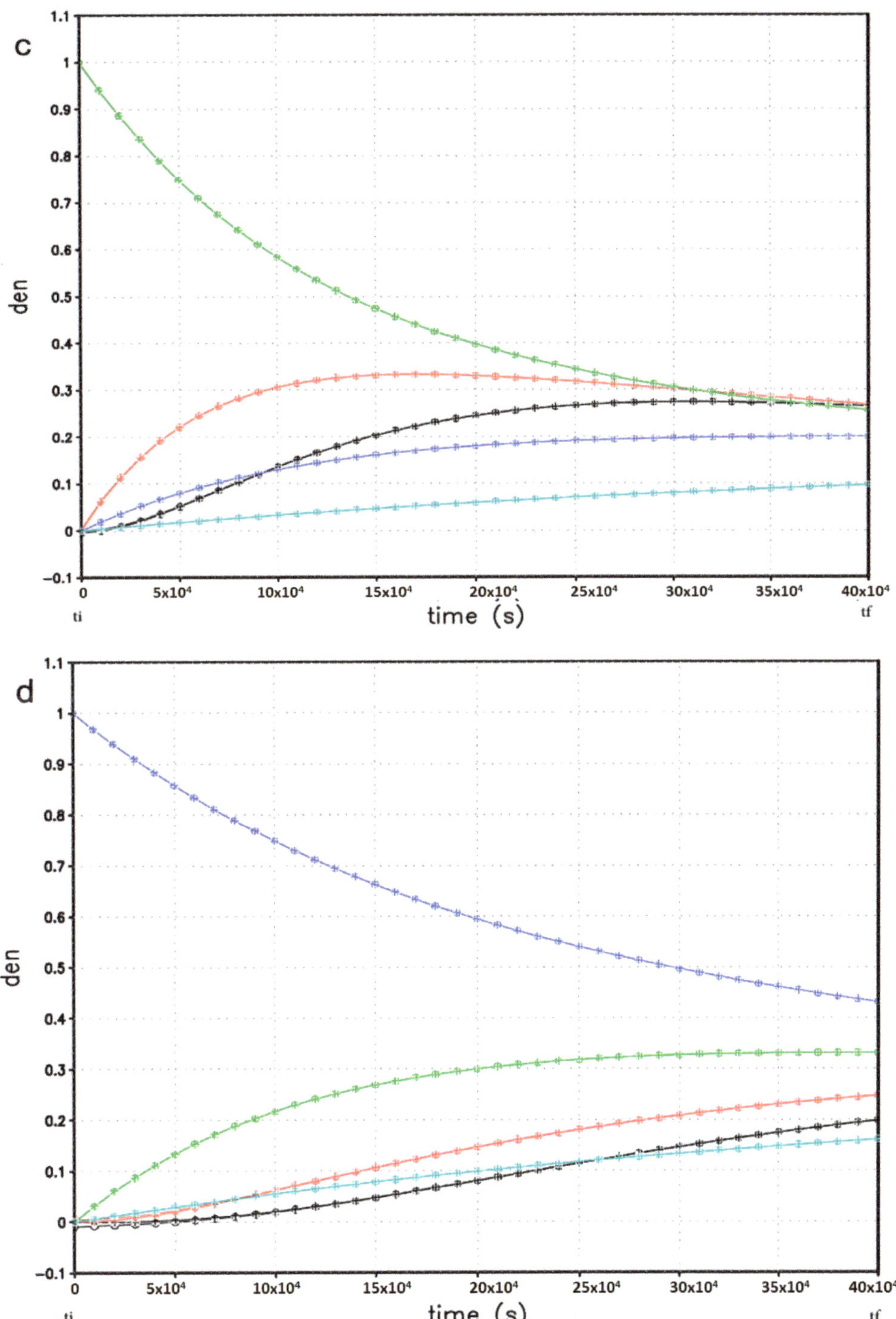

Figure 73. *Cont.*

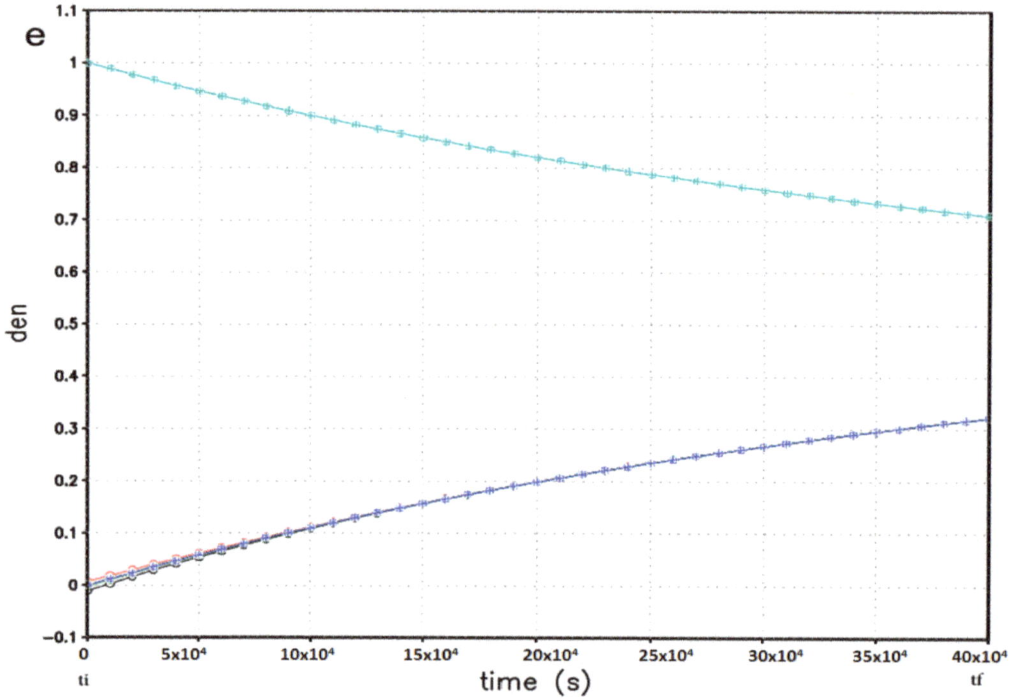

Figure 73. Forward-integrations (+) and backward integration $\otimes$ for species (1, 2, 3, 4, 5) among different cells (A, B, C, D. E), (a) ρ_1^A, ρ_1^B, ρ_1^C, ρ_1^D, and ρ_1^E; (b) ρ_2^A, ρ_2^B, ρ_2^C, ρ_2^D, and ρ_2^E; (c) ρ_3^A, ρ_3^B, ρ_3^C, ρ_3^D, and ρ_3^E; (d) ρ_4^A, ρ_4^B, ρ_4^C, ρ_4^D, and ρ_4^E; and (e) ρ_5^A, ρ_5^B, ρ_5^C, ρ_5^D, and ρ_5^E (Sun and Sun 2017) [236].

The simulations show the importance of mixing. Because of mixing, the material of each cell at the downwind region is significantly different from its beginning. Hence, the conventional forward or backward trajectory method without mixing can lead to a significant error to interpret the property at downwind regions or sources.

Equation (273) can be written as

$x^n = Ax^{n-1} = \ldots = A^n x^0$ for t = 0 to tf. in forward integration. Then $y^n = A^{-1}y^{n-1} = \ldots = A^{-n}y^0 = A^{-n}x^n$ in the backward integration starts from t = tf to t = 0 with $y^0 = x^n$ as initial condition. The period is 2 tf, and the system after a complete cycle can be determined by matrix: $y^n = A^{-n}x^n = A^{-n}A^n x^0 = Ix^0$ according to Floquet theorem (Hand and Finch 1998) [17]. The system starts from a simple system that the species mix by perturbation velocity and become quite different from the original situation in the forward integration. After backward integration, the species return to their initial position, which is like the Arnold cat map (Arnold 1967) [237]. It is also noted that in the forward integration, each species approaches a uniform distribution among different cells as time approaches infinity, then the system becomes a fixed point, f(x) = x, no matter forward or backward integration (Strogatz 2015) [238]. Because it is a first-order system, stability is much simpler than the nonlinear dynamics (Poincare, Lyapunnov [239], Lorenz [240], Hand and Finch 1998 [17], Greenwood 2003 [22], and others), but most inverse value problems are difficult to solve.

9. Summary

We discuss the equations of CGFD, numerical methods, applications, and comparisons with observations and earlier studies. In Section 2 of shallow water, in addition to stability analysis of the forward–backward and leapfrog schemes with or without diffusion, we successfully applied Sun's scheme (2011) [24] without smoothing or nudging to simulate the nonlinear dam break, solitary Rossby wave, hydraulic jump, and vortices merging, etc. Solving the linearized equations (in Section 3), we obtained the coexistence of the longitudinal, shallow cloud bands and transverse, deep cloud bands developed in the tropical region. We also reproduced the observed mesoscale convection system in Brazil due to the resonance of trapezoidal instability and the PBL diurnal oscillation. The coupled atmosphere–ocean Ekman layer shows the resonance instability due to the frequency of the PBL oscillation coinciding with the inertial frequency of the Coriolis force at 30°. It also proves that the weekly or monthly mean of Ekman profiles should not be obtained by solving the steady diffusion equation using different diffusion coefficients for phase angle and amplitude of the velocity.

In Section 4, the numerical schemes and the physical components of the PRCM, except radiation, are presented. The PRCM successfully captured the development of explosive severe winter storms in Western US from a fair weather. It reproduced the formation and advection of lee vortices in Taiwan, the winter storms, and a small observed vortex over the Rocky Mountains, the deformation of a weak cold front by the Central Mountain Range in Taiwan during TAMEX. The results also show that the vertical stretching is most important to the formation of lee vortex. The regional climate simulations in Section 5 show that snow–land module can affect the wind and geopotential at 200 mb. The detailed formulation of snow-soil-vegetation is crucial to the short-term weather forecasting or long-term climate simulations. The PRCM reproduced the 1993 flooding, and 1988 drought and hot dome in the US, it also well simulated the 1998 flooding in China and Korea, as well as the weather and northward movement of East Asian Monsoon. It predicted the onset and decay of the summer Monsoon. The model also replicated the potential vorticity at 850 mb in the South China Sea in the summers of 1991–1998. The PRCM can simulate the transition from a stable state to an unstable state and vice versa. It is also noted the PRCM has an excessive rainfall in the flooding areas. The model seems able to correct the small-scale errors if the synoptic systems are correct. The result confirms that the large-scale systems are more important, which can guard the small-scale disturbances. It implies that the predictability from the NWP model seems longer than the simplified dynamics models in chaos theory.

Despite the chaotic nature of the atmosphere, ECMWF also believes that long-term and seasonal (up to 13 months) forecasting are possible because ENSO shows variations on long time scales (seasons and years) and, to a certain extent, are predictable. The slow variations of snow and soil wetness also affect the long-term forecast. (https://effis.jrc.ec.europa.eu/about-effis/technical-background/seasonal-forecast-explained (accessed on 25 July 2023)). Lorenz (1982) [241] stated "Predictions at least ten days ahead as skillful as predictions now made seven days ahead appear to be possible. Additional improvements at extended range may be realized if the one-day forecast is capable of being improved significantly".

In Section 6, we discuss Boussinesq fluid versus compressible fluid and different approaches to cut down the computing time in the atmospheric nonhydrostatic models. We also discuss the mistakes of the parcel method, Froude number, and CAPE, because they ignored the dynamic pressure based on the hydrostatic system, in which the flow cannot move against buoyancy; and the KE remains constant if the parcel moves vertically following the hydrostatic–adiabatic lapse rate. On the other hand, the Bernoulli function conserved the summation of KE, PE, and EN based on the nonhydrostatic, compressible atmosphere. According to the Bernoulli function, the plumes can rise against buoyancy and reach 12–18 km as the hot towers of cumulonimbus clouds in the Equator, or the deep convections in hurricanes and severe storms. They can also effectively pump moisture and pollutants deeply into the stratosphere. The simulations also show an upper layer convergence is needed to sustain the downslope windstorm on the lee of mountain on a non-slip surface.

We present a new terrain-following coordinate and the comparisons with the Gal-Chen and Somerville (GS) model in Section 7. GS model introduces errors in divergence, curl, and the advection terms of the equations because they ignored the variation of the (horizontal) spatial interval over the mountain. Using the divergence formula in curvilinear coordinate, we also provide an efficient flux form of the Navier–Stokes equations which conserved the mass, energy, and momentum.

Section 8 shows the model and numerical results from a turbulence–pollution model, which replicates the ascending and descending plumes in a convective boundary layer. The backward integration scheme shows that the data collected from the downwind region can be used to assess the property of pollutants at the source. Because of including mixing, the results are more accurate than those based on the traditional Lagrangian backward trajectory method.

The numerical method is applied to integrate the Navier–Stokes equations from the initial condition (at time step n = 0) to the next time step, n = 1. Then, the results at n = 1 are used to calculate results at n = 2, 3, . . . The solutions at each time step depend upon not only the previous condition but also the equations and numerical method applied. With the same initial condition, Figure 74a,b show the simulated flows over an elliptic mountain at 10 and 20 h integration from NTU/Purdue and WRF. The vortices at t = 10 h are similar. The vortices developed earlier and moved to the farther downstream are still quite similar at t = 20 h in Figure 74b, but those developed later become quite different near the mountain. It is also noted that the stagnation exists on the windward side of the mountain in the NTU/Purdue simulation, and flow is also wider on the lee compared with the WRF simulation (Sun et al., 2010) [242]. The differences grow with time even for smooth, well resolved cases.

The solutions of the Navier–Stokes equations depend upon the initial and boundary conditions (Keiss and Lorenz 2004 [243]), they gradually deteriorate with time and eventually lose the memory of the initial conditions even without any strange attractor. Beyond that, the NWP cannot be applied (a predictability limit?). On the other hand, the numerical results are continuously influenced by the forcing and lateral boundary conditions, examples include the seasonable variations, land-sea contrast, terrain effect and ground/sea surface, etc. Advance in supercomputer does help the CGFD and NWP, and scientists will continue to face the challenge of formulating the physics, sub-grid fluxes, and handing the boundaries. Sometimes, a model may not be able to predict a well-resolved large system either, for example, it is difficulty to accurately replicate the strength and extent of the Western Pacific subtropical high, which affects the regional flooding in Southeast Asia (Hsu et al., 2004 [164], Sun et al., 2011 [165], Sun et al. [192]).

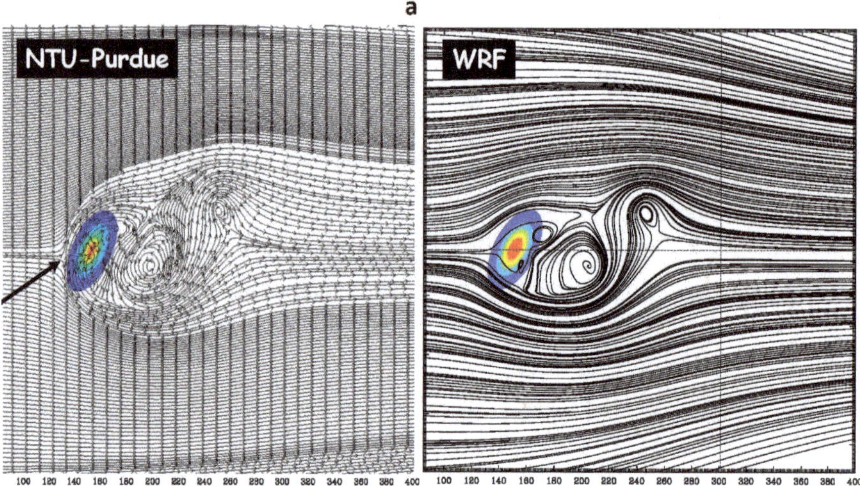

Figure 74. *Cont.*

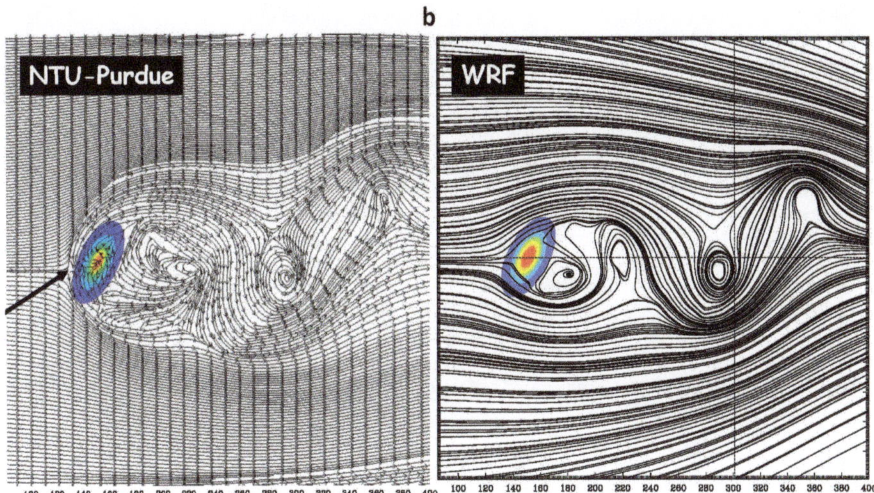

Figure 74. Numerical simulations from the NTU/Purdue model and WRF over an idealized elliptical mountain: They contain $600 \times 600 \times 50$ in x, y, and z directions with space interval $\Delta x = \Delta y = 1$ km; $\Delta z \sim 300$ m on a free-slip surface; initial wind $U = 4$ m s^{-1} and the Brunt–Vaisala frequency $N = 10^{-2}$ s^{-1}, mountain peak = 2 km; half width lengths: 5 and 10 km (tilted by 30 degrees), and the Froude number Fr = U/Nh = 0.2, (**a**) after 10 h integration, (**b**) after 20 h integration (Sun et al., 2010) [242].

Funding: This research received no external funding.

Institutional Review Board Statement: Not applicable.

Informed Consent Statement: Not applicable.

Data Availability Statement: Not applicable.

Acknowledgments: The author is indebted to H. L. Kuo and other teachers for their teaching and stimulating. He also appreciates his family for their support, and the opportunities working with the former students and colleagues, who have contributed significantly to the research and model development. He is grateful to N. H. Lin, H. H. Hsu, K. Tsuboki, B. Shen, Mandić and the reviewers for their helps, comments, and suggestions. He also thanks Purdue University and P. L. Lin at National Central University for providing computing facility. The author withdrew Sun 2023 [200] at the final stage of the gallery proof, because OJFD broke the agreement and refused including Taiwan in author's affiliation.

Conflicts of Interest: The author declares no conflict of interest.

References

1. Bjerknes, V. Das Problem der Wettervorhersage, Betrachtet vom Stanpunkt der Mechanik und der Physik. *Meteor. Zeits* **1904**, *21*, 1–7.
2. Richardson, L.F. *Weather Prediction by Numerical Processes*; Cambridge University Press: Boston, MA, USA, 1922; p. 66. ISBN 9780511618291.
3. Charney, J.; Fjørtoft, R.; von Neumann, J. Numerical Integration of the Barotropic Vorticity Equation. *Tellus* **1950**, *2*, 237–254. [CrossRef]
4. Phillips, N.A. The general circulation of the atmosphere: A numerical experiment. *Q. J. R. Meteorol. Soc.* **1956**, *82*, 123–154. [CrossRef]
5. Poincaré, H. Sur les Fonctions Fuchsiennes. *Comptes Rendus Hebd. L'académie Sci. Paris* **1882**, *94*, 1166–1167.
6. Poincaré, H. Sur le problème des trois corps et les équations de la dynamique. *Acta Math.* **1890**, *13*, 1–270. (In French)
7. Poincaré, H. *Solutions Periodiques, Non-Existence des Integrales Uniformes, Solutions Asymptotiques*; Gauthier-Villars: Paris, France, 1892; Volume 1. (In French)
8. Poincare, H. Memoire sur Jes courbes definies par une equation differentielle. *J. Math.* **1881**, *7*, 375–442.

9. Lorenz, E. Deterministic nonperiodic flow. *J. Atmos. Sci.* **1963**, *20*, 130–141. [CrossRef]
10. Lorenz, E. The predictability of a flow which possess many scales of motion. *Tellus* **1969**, *21*, 289–307. [CrossRef]
11. Smagorinsky, J. Problems and promises of deterministic extended range forecasting. *Bull. Am. Meteorol. Soc.* **1969**, *50*, 286–311. [CrossRef]
12. Arakawa, A. Computational design for long-term numerical integration of the equations of fluid motion: Two-dimensional incompressible flow. Part I. *J. Comput. Phys.* **1966**, *1*, 119–143. [CrossRef]
13. Paegle, P.; Yang, Q.; Wang, M. Predictability in limited area and global models. *Meteorol. Atmos. Phys.* **1997**, *63*, 53–69. [CrossRef]
14. Wiin-Nielsen, A. On the stable part of the Lorenz attractor. *Atmôsfera* **1998**, *11*, 61–73.
15. Charney, J.G.; Shukla, J. Predictability of monsoons. In *Monsoon Dynamics*; Lighthill, S.J., Pearce, R.P., Eds.; Cambridge University Press: Cambridge, UK, 1981; pp. 99–109.
16. Hoskins, B.J. Predictability Beyond the Deterministic Limit. *WMO Bull.* **2012**, *61*, 33–36.
17. Hand, L.N.; Finch, J.D. *Analytical Mechanics*; Cambridge University Press: Cambridge, UK, 2008; 575p.
18. Beltrami, E. *Mathematic for Dynamic Modeling*; Academic Press: Cambridge, MA, USA, 1987; 277p.
19. Shen, B.-W. Homoclinic Orbits and Solitary Waves within the Nondissipative Lorenz Model and KdV Equation. *Int. J. Bifurc. Chaos* **2020**, *30*, 15. [CrossRef]
20. Shen, B.-W.; Pielke, S.; Pielke, R.; Cui, J.; Faghih-Naini, S.; Paxson, W.; Kesarkar, A.; Atlas, R. The dual nature of chaos and order in the atmosphere. *Atmosphere* **2022**, *13*, 1892. [CrossRef]
21. Shen, B.-W.; Pielke, R.A.; Zeng, X.; Cui, J.; Faghih-Naini, S.; Paxson, W.; Atlas, R. Three kinds of butterfly effects within Lorenz Models. *Encyclopedia* **2022**, *2*, 1250–1259. [CrossRef]
22. Greenwood, D.T. *Advanced Dynamics*; Cambridge Press: New York, NY, USA, 2003; 425p.
23. Sun, W.Y. Instability in leapfrog and forward-backward Schemes. *Mon. Wea. Rev.* **2010**, *138*, 1497–1501. [CrossRef]
24. Sun, W.Y. Instability in Leapfrog and Forward-Backward Schemes: Part II: Numerical Simulation of Dam Break. *J. Comput. Fluids* **2011**, *45*, 70–76. [CrossRef]
25. Mesinger, F.; Arakawa, A. Numerical Methods Used in Atmospheric Models. *GARP Glob. Atmos. Res. Program* **1976**, *17*, 64.
26. Gill, A.E. *Atmospheric-Ocean Dynamics*; Academic Press: San Diego, CA, USA, 1982; 661p.
27. Haltiner, G.J.; Williams, R.T. *Numerical Prediction and Dynamic Meteorology*, 2nd ed.; Wiley: Hoboken, NJ, USA, 1980; 477p.
28. Pielke, S.R.A. *Mesoscale Meteorological Modeling*, 2nd ed.; Academic Press: Cambridge, MA, USA, 2002; 676p.
29. Lin, Y.L. *Mesoscale Dynamics*; Cambridge University Press: New York, NY, USA, 2007; 633p.
30. Holton, J.R.; Hakim, G.J. *An Introduction to Dynamic Meteorology*, 5th ed.; Academic Press: Cambridge, MA, USA, 2012; 531p.
31. Versteeg, H.K.; Malalasekera, W. *An Introduction to Computational Fluid Dynamics: The Finite Volume Method*; Longman Scientific & Technical: Harlow, UK, 1995; 257p.
32. Pielke, R.A. *Mesoscale Metorological Modeling*; Elsevier Inc.: Amsterdam, The Netherlands, 1984; 612p.
33. Pedlosky, J. *Geophysical Fluid Dynamics*; Springer: Berlin/Heidelberg, Germany, 1979; 626p.
34. Lorenz, E.N. Attractor sets and quasi-geostrophic equilibrium. *J. Atmos. Sci.* **1980**, *37*, 1685–1699. [CrossRef]
35. Nobel, B. *Applied Linear Algebra*; Prentice-Hall: Upper Saddle River, NJ, USA, 1969; 523p.
36. Sun, W.Y. A comparison of two explicit time integration schemes applied to the transient heat equation. *Mon. Wea. Rev.* **1982**, *110*, 1645–1652. [CrossRef]
37. Sun, W.Y.; Sun, O.M.T. A modified leapfrog scheme for shallow water equations. *Comput. Fluids* **2011**, *52*, 69–72. [CrossRef]
38. Sun, W.Y. Numerical analysis for hydrostatic and nonhydrostatic equations of inertial-internal gravity waves. *Mon. Wea. Rev.* **1984**, *112*, 259–268. [CrossRef]
39. Roe, P.L. Approximate Riemann Solvers, Parameter Vectors and Difference Schemes. *J. Compu. Phys.* **1981**, *43*, 357–372. [CrossRef]
40. Toro, E.F. *Riemann Solvers and Numerical Methods for Fluid Dynamics*; Springer: Berlin/Heidelberg, Germany, 1999; 309p.
41. Wang, H.; Yeh, G.T. A characteristic-based semi-Lagrangian method for hyperbolic systems of conservation laws. *Chin. J. Atmos. Sci.* **2005**, *29*, 21–42.
42. Oh, T.J. Development and Testing of Characteristic-Based Semi-Lagrangian Two-Dimensional Shallow Water Equations Development and Testing of Characteristic-Based Semi-Lagrangian Two-Dimensional Shallow Water Equations Model. Ph.D. Thesis, Purdue University, W. Lafayette, IN, USA, 2007; 148p. Available online: https://docs.lib.purdue.edu/dissertations/AAI3278686/ (accessed on 1 January 2008).
43. MacCormack, R.W. *The Effect of Viscosity in Hyper-Velocity Scattering*; AIAA: Washington, DC, USA, 1969; pp. 69–354.
44. Fujihara, M.; Borthwick, A.G.L. Godunov-type solution of curvilinear shallow-water equations. *J. Hydraul. Eng. ASCE* **2000**, *126*, 827–836. [CrossRef]
45. Anastasiou, K.; Chan, C.T. Solution of the 2D shallow water equations using the finite volume method on unstructured triangular meshes. *Int. J. Numer. Methods Fluids* **1997**, *24*, 1225–1245. [CrossRef]
46. Tseng, M.H.; Chu, C.R. Two-dimensional shallow water flows simulation using TVD-MacCormack scheme. *J. Hydraul. Res.* **2000**, *38*, 123–131.
47. Liang, Q.; Borthwick, A.G.L.; Stelling, G. Simulation of dam- and dyke-break hydrodynamics on dynamically adaptive quadtree grids. *Int. J. Numer. Methods Fluids* **2004**, *46*, 127–162. [CrossRef]
48. Sun, W.Y.; Sun, O.M.T. Numerical Simulation of Rossby Wave in Shallow Water. *Comput. Fluids* **2013**, *76*, 116–127. [CrossRef]

49. Chu, P.C.; Fan, C. Space-Time Transformation in Flux-form Semi-Lagrangian Schemes. *Terr. Atmos. Ocean. Sci.* **2010**, *21*, 17–26. [CrossRef]
50. Boyd, J.P. Equatorial solitary waves. Part-1: Rossby solitons. *J. Phys. Oceanogr.* **1980**, *10*, 1699–1717. [CrossRef]
51. Boyd, J.P. Equatorial solitary waves. Part 3: Westward-traveling Motions. *J. Phys. Oceanogr.* **1985**, *15*, 46–54. [CrossRef]
52. Korteweg, D.J.; de Vries, G. On the change of long waves advancing in a rectangular canal, and on a new type of long stationary waves. *Phil. Mag.* **1895**, *39*, 422–433. [CrossRef]
53. Houghton, D.D.; Kasahara, A. Nonlinear shallow fluid flow over an isolated ridge. *Commun. Pure Appl. Math* **1968**, *11*, 1–23. [CrossRef]
54. Sun, W.Y. Hydraulic jump and Bernoulli equation in nonlinear shallow water model. *Dyn. Atmos. Ocean.* **2018**, *82*, 37–53. [CrossRef]
55. Sun, W.Y. The Vortex Moving toward Taiwan and the influence of the Central Mountain Range. *Geosci. Lett.* **2016**, *3*, 21. [CrossRef]
56. Lin, Y.-L.; Chen, S.-Y.; Hill, C.M.; Huang, C.-Y. Control parameters for tropical cyclones passing over mesoscale mountains. *J. Atmos. Sci.* **2005**, *62*, 1849–1866. [CrossRef]
57. Huang, C.-Y.; Lin, Y.-L. The influence of mesoscale mountains on vortex tracks: Shallow-water modeling study. *Meteor. Atmos. Phys.* **2008**, *101*, 1–20. [CrossRef]
58. Mastsuno, T. Quasi-geosprophic motions in the equator area. *J. Meteor. Soc. Jpn.* **1966**, *44*, 25–43. [CrossRef]
59. Sun, W.Y.; Chern, J.D. Numerical experiments of vortices in the wake of idealized large mountains. *J. Atmos. Sci.* **1994**, *51*, 191–209. [CrossRef]
60. Sun, W.Y.; Sun, O.M.T. Bernoulli equation and flow over a mountain. *Geosci. Lett.* **2015**, *2*, 7. [CrossRef]
61. Wu, C.C.; Li, T.-H.; Huang, Y.-H. Influence of mesoscale topography on tropical cyclone tracks: Further examination of the channeling effect. *J. Atmos. Sci.* **2015**, *72*, 3032–3050. [CrossRef]
62. Sun, W.Y.; Oh, T.J. Vortex Merger in Shallow Water Model. *Asia-Pac. J. Atmos. Sci.* **2022**, *58*, 533–547. [CrossRef]
63. Waugh, D.W. The efficiency of symmetric vortex merger. *Phys. Fluid Dyn.* **1992**, *4*, 1745–1758. [CrossRef]
64. Kimura, Y.; Herring, J.R. Gradient enhancement and filament ejection for a non-uniform elliptic vortex in two-dimensional turbulence. *J. Fluid Mech.* **2001**, *439*, 43–56. [CrossRef]
65. Cerretelli, C.; Williamson, C.H.K. The physical mechanism for vortex merging. *J. Fluid Mech.* **2003**, *475*, 41–77. [CrossRef]
66. Melander, M.V.; McWilliams, J.C.; Zabusky, N.J. Axisymmetrization and vorticity-gradient intensification of an isolated two-dimensional vortex through filamentation. *J. Fluid Mech.* **1987**, *178*, 137–159. [CrossRef]
67. Melander, M.V.; Zabusky, N.J.; McWilliams, J.C. Symmetric vortex merger in two dimensions: Causes and conditions. *J. Fluid Mech.* **1988**, *195*, 305–340. [CrossRef]
68. Huang, M.J. The physical mechanism of symmetric vortex merger: A new viewpoint. *Phys. Fluids* **2005**, *17*, 1–7. [CrossRef]
69. Moore, D.W.; Saffman, P.G. The instability of a straight vortex filament in a strain field. *Proc. R. Soc. Lond. A* **1975**, *346*, 413–425.
70. Brandt, L.K.; Nomura, K.K. The physics of vortex merger: Further insight. *Phys. Fluids* **2006**, *18*, 051701. [CrossRef]
71. Sun, W.Y.; Chern, J.D. Numerical study of influence of mountain ranges on Mei-yu front. *J. Meteorol. Soc. Jpn.* **2006**, *84*, 27–46. [CrossRef]
72. Fjortoft, R. On the changes in the spectral distribution of kinetic energy for two dimensional nondivergent flow. *Tellus* **1953**, *5*, 225–230. [CrossRef]
73. Kraichnan, R.H. Tnertial renges in two-dimensional turbulence. *Phys. Fluid* **1967**, *10*, 1417–1423. [CrossRef]
74. Kolmogorov, A.N. Local structure of turbulence in an incompressible fluid at very high Reynolds numbers. *Dokl. Akad. Nauk. SSSR* **1941**, *31*, 99–101. [CrossRef]
75. Obukhov, A.M. Spectral energy distribution in a turbulent flow. *Dokl. Akad. Nauk. SSSR* **1941**, *32*, 22–24.
76. Eady, E.T. Long wave and cyclone waves. *Tellus* **1949**, *1*, 33–52. [CrossRef]
77. Sun, W.Y. Stability of deep cloud streets. *J. Atmos. Sci.* **1978**, *35*, 466–483. [CrossRef]
78. Malkus, J.S.; Riehl, H. *Cloud Structure and Distribuions over the Tropical Pacific Ocean*; University of California Press: Berkeley, CA, USA; Los Angeles, CA, USA, 1964; 229p.
79. Kuo, H.L. Perturbations of plane Couette flow in a stratified fluid and origin of cloud streets. *Phys. Fluid* **1963**, *6*, 195–211. [CrossRef]
80. Sun, W.Y. Rainbands and symmetric instability. *J. Atmos. Sci.* **1984**, *41*, 3412–3426. [CrossRef]
81. Fjortoft, R. Application of integral theorems in deriving criteria of stability for laminar flows and for the baroclinic circular vortex. *Geofys. Pub.* **1950**, *17*, 1–52.
82. Ooyama, K. On the stability of baroclinic circular vortex: A sufficient criterion for instability. *J. Atmos. Sci.* **1966**, *23*, 43–53. [CrossRef]
83. Hoskins, B.J. *Baroclinic Instability and Frontogensis in Rotating Fluids in Geophysics*; Roberts, P.H., Ed.; Academic Press: Cambridge, MA, USA, 1978; pp. 171–204.
84. Sun, W.Y. Convection in the Surface Layer of the Atmosphere. Ph.D. Thesis, The University of Chicago, Chicago, IL, USA, 1975; 130p.
85. Kuo, H.L.; Sun, W.Y. Convection in the lower atmosphere and its effects. *J. Atmos. Sci.* **1976**, *33*, 21–40. [CrossRef]
86. Agee, E.; Church, C.; Morris, C.; Snow, J. Some synoptic aspects and dynamic features of vortices associated with the tornado outbreak of 3 April 1974. *Mon. Wea. Rev.* **1975**, *103*, 318–353. [CrossRef]

87. Sun, W.Y.; Orlanski, I. Large mesoscale convection and sea breeze circulation. Part I: Linear stability analysis. *J. Atmos. Sci.* **1981**, *38*, 1675–1693. [CrossRef]

88. Sun, W.Y.; Orlanski, I. Large mesoscale convection and sea breeze circulation. Part 2: Non-Linear numerical model. *J. Atmos. Sci.* **1981**, *38*, 1694–1706. [CrossRef]

89. Orlanski, I.; Polinsky, L.J. Spectral distribution of cloud cover over Africa. *J. Meteor. Soc.* **1977**, *55*, 483–494. [CrossRef]

90. Sun, W.Y. Unsymmetrical symmetric instability. *Q. J. R. Meteorol. Soc.* **1995**, *121*, 419–431. [CrossRef]

91. Sun, W.Y.; Sun, O.M. Inertia and diurnal oscillations of Ekman layers in atmosphere and ocean. *Dyn. Atmos. Ocean.* **2020**, *90*, 289. [CrossRef]

92. Blackadar, A.K. Boundary layer wind maxima and their significance for the growth of nocturnal inversion. *Bull. Am. Meteorol. Soc.* **1957**, *38*, 283–290. [CrossRef]

93. Palmer, E.; Newton, C.W. *Atmospheric Circulation Systems*; Academic Press: London, UK, 1969; 606p.

94. Price, J.F.; Weller, R.A.; Pinkel, R. Diurnal cycling: Observations and models of the upper ocean response to heating, cooling, and wind mixing. *J. Geophys. Res.* **1986**, *91*, 8411–8427. [CrossRef]

95. Kalashnik, M.V. Resonant excitation of baroclinic waves in the presence of Ekman friction. *Izvestiya. Atmos. Ocean. Phys.* **2018**, *54*, 109–113. [CrossRef]

96. Brown, R.A. A secondary flow model for the planetary boundary layer. *J. Atmos. Sci.* **1970**, *27*, 742–757. [CrossRef]

97. Polton, J.A.; Lenn, Y.D.; Elipot, S.; Chereskin, T.K.; Sprintall, J. Can Drake Passage observations match Ekman's classic theory? *J. Phys. Oceanogr.* **2013**, *43*, 1733–1740. [CrossRef]

98. Chereskin, T.K. Direct evidence for an Ekman balance in the California Current. *J. Geo. Res.* **1995**, *100*, 261–269. [CrossRef]

99. Price, J.F.; Sundermeyer, M.A. Stratified Ekman layers. *J. Geoph. Res.* **1999**, *104*, 20467–20494. [CrossRef]

100. Lenn, Y.-D.; Chereskin, T.K. Observations of Ekman currents in the Southern Ocean. *J. Phys. Ocean.* **2009**, *39*, 768–799. [CrossRef]

101. Fujita, T.T. Structure and movement of a dry front. *Bull. Am. Meteor. Soc.* **1958**, *39*, 574–582. [CrossRef]

102. Wexler, H. A boundary layer interpretation of low-level jet. *Tellus* **1961**, *12*, 268–378.

103. Sun, W.Y.; Ogura, Y. Boundary layer forcing as a possible trigger to a squall-line formation. *J. Atmos. Sci.* **1979**, *36*, 235–254. [CrossRef]

104. Sun, W.Y.; Wu, C.C. Formation and diurnal oscillation of dryline. *J. Atmos. Sci.* **1992**, *49*, 1606–1619. [CrossRef]

105. Sun, W.Y.; Ogura, Y. Modeling the evolution of the convective planetary boundary layer. *J. Atmos. Sci.* **1980**, *37*, 1558–1572. [CrossRef]

106. Deardorff, J.W. Stratocumulus-capped mixed layers derived from a three-dimensional model. *Bound-Layer Meteor.* **1980**, *18*, 495–527. [CrossRef]

107. Caughey, S.J.; Palmer, S.G. Some aspects of turbulence structure through the depth of the convective layer. *Quart. J. Roy. Meteor. Soc.* **1979**, *105*, 811–827. [CrossRef]

108. Sun, W.Y.; Chang, C.Z. Diffusion model for a convective layer: Part 1. Numerical simulation of a convective boundary layer. *J. Clim. Appl. Meteor.* **1986**, *25*, 1445–1453. [CrossRef]

109. Sun, W.Y. Air pollution in a convective boundary. *Atmos. Environ.* **1986**, *20*, 1877–1886. [CrossRef]

110. Chern, J.D. Numerical Simulation of Cyclogenesis over the Western United States. Ph.D. Thesis, Purdue University, W. Lafayette, IN, USA, 1994; 178p.

111. Chen, S.H.; Sun, W.Y. A one-dimensional time dependent cloud model. *J. Meteor. Soc. Jpn.* **2002**, *80*, 99–118. [CrossRef]

112. Bosilovich, M.G.; Sun, W.Y. Formation and Verification of a land surface parameterization for atmospheric models. *Bound.-Layer Meteorol.* **1995**, *73*, 321–341. [CrossRef]

113. Chern, J.D.; Sun, W.Y. Formulation and validation of a snow model. In Proceedings of the A11A-06, 1998 Fall Meeting, AGU, Supplement to EOS, AGU, 79, 45, F96, Washingto, DC, USA, 10 November 1998.

114. Sun, W.Y.; Chern, J.D. One dimensional snow-land surface model applied to Sleepers experiment. *Bound.-Layer Meteorol.* **2005**, *116*, 95–115. [CrossRef]

115. Chou, M.-D.; Suarez, M.J. A solar radiation parameterization for atmospheric studies. *NASA Tech. Memo* **1994**, *15*, 40.

116. Chou, M.-D.; Suarez, M.J.; Liang, X.Z.; Yan, M.H.H. A thermal infrared radiation parameterization for atmospheric sciences. *NASA Tech. Memo* **2001**, *19*, 68.

117. Kuo, H.L. On formation and intensification of tropical cyclones through latent heat release by cumulus convection. *J. Atmos. Sci.* **1965**, *22*, 40–63. [CrossRef]

118. Kuo, H.L. Further studies of the parameterization of the influence of cumulus convection on large-scale flow. *J. Atmos. Sci.* **1974**, *31*, 1232–1240. [CrossRef]

119. Anthes, R.A. A cumulus parameterization scheme utilizing a one-dimensional cloud model. *Mon. Wea. Rev.* **1977**, *105*, 270–286. [CrossRef]

120. Molinari, J. A method for calculating the effects of deep cumulus convection in numerical models. *Mon. Wea. Rev.* **1982**, *110*, 1527–1534. [CrossRef]

121. Sun, W.Y. A forward-backward time integration scheme to treat internal gravity waves. *Mon. Wea. Rev.* **1980**, *108*, 402–407. [CrossRef]

122. Sun, W.Y. Numerical experiments for advection equation. *J. Comput. Phys.* **1993**, *108*, 264–271. [CrossRef]

123. Sun, W.Y.; Yeh, K.S.; Sun, R.Y. A simple Semi-Lagrangian Scheme for advection equation. *Q. J. R. Meteorol. Soc.* **1996**, *122*, 1211–1226. [CrossRef]
124. Sun, W.Y.; Yeh, K.S. A general semi-Lagrangian advection scheme employing forward trajectories. *Quart. J. Roy. Meteor. Soc.* **1997**, *123*, 2463–2476. [CrossRef]
125. Sun, W.Y.; Sun, O.M.T. Mass Correction Applied to Semi-Lagrangian Advection Scheme. *Mon. Wea. Rev.* **2004**, *132*, 975–984. [CrossRef]
126. Sun, W.Y. Pressure gradient in a sigma coordinate. *Terr. Atmos. Ocean. Sci.* **1995**, *4*, 579–590. [CrossRef] [PubMed]
127. Yang, K.J.-S. Numerical Simulations of Asian Aerosols and Their Regional Climatic Impacts: Dust Storms and Bio-Mass Burning. Master's Thesis, Purdue University, W. Lafayette, IN, USA, 2004; 141p.
128. Yang, K.J.-S. Regional Climate-Chemistry Model Simulations of Ozone in the Lower Troposphere and Its Climatic Impacts. Ph.D. Thesis, Purdue University, W. Lafayette, IN, USA, 2004; 157p.
129. Schaefer, J.T. The life cycle of the dryline. *J. Appl. Meteorol. Climatol.* **1974**, *13*, 444–449. [CrossRef]
130. Sun, W.Y.; Chern, J.D. Diurnal variation of lee-vortexes in Taiwan and surrounding area. *J. Atmos. Sci.* **1993**, *50*, 3404–3430. [CrossRef]
131. Sun, W.Y.; Chern, J.D.; Wu, C.C.; Hsu, W.R. Numerical simulation of mesoscale circulation in Taiwan and surrounding area. *Mon. Wea. Rev.* **1991**, *119*, 2558–2573. [CrossRef]
132. Kuo, Y.; Chen, G. The Taiwan Area Mesoscale Experiment (TAMEX): An overview. *Bull. Am. Meteor. Sci.* **1990**, *71*, 488–503. [CrossRef]
133. Hsu, W.R.; Sun, W.Y. A numerical study of a low-level jet and its accompanying secondary circulation in a Mei-Yu system. *Mon. Wea. Rev.* **1994**, *122*, 324–340. [CrossRef]
134. Haines, P.A.; Chern, J.D.; Sun, W.Y. Numerical simulation of the valentine's Day storm during WISP 1990. *Tellus* **1997**, *49*, 595–612. [CrossRef]
135. Yildirim, A. Numerical Study of an Idealized Cyclone Evolution and Its Subsynoptic Features. Ph.D. Thesis, Purdue University, W. Lafayette, IN, USA, 1994; 145p.
136. Bosilovich, M.G.; Sun, W.-Y. Numerical simulation of the 1993 Midwestern Flood: Local and remote sources of water. *J. Geophys. Res.* **1999**, *104*, 415–423. [CrossRef]
137. Bosilovich, M.G.; Sun, W.-Y. Numerical simulation of the 1993 Midwestern flood: Land-Atmosphere Interactions. *J. Clim.* **1999**, *12*, 1490–1505. [CrossRef]
138. Oglesby, R.J.; Erikson, D. Soil moisture and persistence of North American drought. *J. Clim.* **1989**, *2*, 1362–1380. [CrossRef]
139. Oglesby, R.J.; Leap, D.I.; Sun, W.Y.; Agee, E.M. Modeling the effects of climatic changes on availability and quality of water. *Comput. Mech.* **1994**, *2*, 9–17.
140. Sun, W.Y.; Bosilovich, M.G.; Chern, J.D. CCM1 regional response to anomalous surface properties. *Terr. Atmos. Ocean. Sci.* **1997**, *8*, 271–288. [CrossRef]
141. Sun, W.Y.; Chern, J.D.; Bosilovich, M. Numerical Study of the 1988 Drought in the United States. *J. Meteorol. Soc. Jpn.* **2004**, *82*, 1667–1678. [CrossRef]
142. Bosilovich, M.B.; Sun, W.Y. Simulation of soil moisture and temperature. *J. Atmos. Sci.* **1998**, *55*, 1170–1183. [CrossRef]
143. Sun, W.Y.; Bosilovich, M.G. Planetary boundary layer and surface layer sensitivity to land surface parameters. *Bound.-Layer Meteorol.* **1996**, *77*, 353–378. [CrossRef]
144. Sun, W.Y. Chapter 6. Interactions among atmosphere, soil, and vegetation. In *Applications and Development of Agrometeorology*; Yang, L., Ed.; Taiwan Agricultural Research Institute and Chinese Society of Agrometeorology: Taipei, Taiwan, 2001; pp. 69–92.
145. Jordan, R. *A One-Dimensional Temperature Model for a Snow Cover*; Special Report 91-16; U.S. Army Corps of Engineers, Cold Regions Research and Engineering Laboratory: Hanover, NH, USA, 1991; 49p.
146. Johansen, O. Thermal Conductivity of Soils. Ph.D. Thesis, Trondheim University, Trondheim, Norway, 1975.
147. Farouki, O. *Evaluation of Methods for Calculating Soil Thermal Conductivity*; CRREL Rep. 82-8; CRREL: Hanover, NH, USA, 1982; 90p.
148. Clapp, R.B.; Hornberg, G.M. Empirical equations for some soil hydraulic properties. *Water Resour. Res.* **1978**, *13*, 601–604. [CrossRef]
149. Dingman, S.L. *Physical Hydrology*, 2nd ed.; Prentice Hall: Upper Saddle River, NJ, USA, 2002; 646p.
150. Verseghy, D.L. CLASS- A Canadian land surface scheme for GCMS. I: Soil model. *Int. J. Climatol.* **1991**, *11*, 111–133. [CrossRef]
151. Oke, T.R. *Boundary Layer Climates*; Routledge: London, UK, 1988; 464p.
152. Bohren, C.; Barkstrom, B. Theory of the optical properties of snow. *J. Geophys. Res.* **1974**, *79*, 4527–4535. [CrossRef]
153. Anderson, E.A.; Greenan, H.J.; Whipkey, R.Z.; Machell, C.T. 1977: NOAA-ARS cooperative snow research project—Watershed hydro-climatology and data for water year 1960–1974. *NOAA Tech. Rep.* **1977**, *316*. Available online: https://onlinebooks.library.upenn.edu/webbin/book/lookupid?key=olbp53867 (accessed on 25 July 2023).
154. Yen, Y.C. *Heat Transfer Characteristics of Naturally Compacted Snow*; Research Report 166; U.S. Army Cold Regions Research and Engineering Laboratory: Hanover, NH, USA, 1965; 9p.
155. Anderson, E.A. *A Point Energy Balance Model for a Snow Cover*; NOAA Tech. Rep. NWS 19; Office of Hydrology, National Weather Service: Chanhassen, MN, USA, 1976.
156. LaChapelle, E.R. *Field Guide to Snow Crystals*; University of Washington Press: Seattle, WA, USA; London, UK, 1969; 101p.
157. Lynch-Stieglitz, M. The development and validation of a simple snow model for GISS GCM. *J. Clim.* **1994**, *7*, 1842–1855. [CrossRef]

158. Pitman, A.J.; Yang, Z.-L.; Cogley, J.G.; Henderson-Sellers, A. *Description of Bare Essentials of Surface Transfer for the Bureau of Meteorological Research Centre AGCM*; BMRC Research Report No. 32; BRMC: Ballarat, Australia, 1991; 117p.
159. Sun, W.Y.; Chern, J.D. One-Dimensional Sea Ice-Ocean Model Applied to SHEBA Experiment in 1997–1998 Winter: Terr. *Atmos. Ocean. Sci.* **2010**, *21*, 1–15. [CrossRef]
160. Min, K.H. Coupled Cryosphere Model Development for Regional Climate Study and the Initialization of Purdue Regional Model with the Land Data Assimilation System. Ph.D. Thesis, Purdue University, W. Lafayette, IN, USA, 2005; 149p.
161. Min, K.-H.; Sun, W.Y. Atmosphere-Cryosphere Coupled Model Development and Its Application for Regional Climate Studies. *Adv. Meteorol.* **2015**, *2015*, 764970. [CrossRef]
162. Yu, Y.C.; Hsu, H.H.; Kau, W.S.; Tsou, T.H.; Hsu, W.R.; Sun, W.Y. An evaluation of the East Asian summer monsoon simulation by the Purdue Regional Model. *J. Environ. Prot. Soc. Taiwan* **2004**, *27*, 41–56.
163. Yu, Y.C.; Hsu, H.-H.; Kau, W.S.; Tsou, T.H.; Hsu, W.R.; Sun, W.Y.; Yu, W.N. An evaluation of the east Asian summer monsoon simulation using the Purdue Regional Model. In Proceedings of the Third Workshop on Regional Climate Modeling for Monsoon System, Honolulu, HI, USA, 17 February 2004. 23p.
164. Hsu, W.-R.; Hou, J.P.; Wu, C.C.; Sun, W.Y.; Tcheng, S.C.; Chang, H.Y. Large-eddy simulation of cloud streets over the East China Sea during cold-air outbreak events, pm.2.4. In Proceedings of the 16th Symposium on Boundary Layers and Turbulence, Portland, ME, USA, 9–13 August 2004; pp. 1–7.
165. Sun, W.Y.; Min, K.H.; Chern, J.-D. Numerical study of 1998 flooding in East Asia. *Asia-Pac. J. Atmos. Sci.* **2011**, *47*, 123–135. [CrossRef]
166. Molinari, J.; Dudek, M. Parameterization of Convective Precipitation in Mesoscale Numerical Models: A Critical Review. *Mon. Wea. Rev.* **1992**, *120*, 326–344. [CrossRef]
167. Shen, B.-W. Aggregated negative feedback in a generalized Lorenz model. *Int. J. Bifurc. Chaos* **2019**, *29*, 1950037–1950091. [CrossRef]
168. Ginoux, P.; Chin, M.; Tegen, I.; Prospero, J.M.; Holben, B.; Dubovik, O.; Lin, S.-J. Sources and distributions of dust aerosols simulated with the GOCART model. *J. Geophys. Res.* **2001**, *106*, 255–273. [CrossRef]
169. DeFries, R.S.; Townshend, J.R.G. NDVI-derived land cover classification a global scale. *Int. J. Remote Sens.* **1994**, *15*, 3567–3586. [CrossRef]
170. Tegen, I.; Fung, I. Contribution to the atmospheric mineral aerosol load from land surface modification. *J. Geophys. Res.* **1995**, *100*, 707–726. [CrossRef]
171. Marticorena, B.; Bergametti, G. Modeling the atmospheric dust cycle, I, Design of a soil-derived dust emission scheme. *J. Geophys. Res.* **1995**, *100*, 16415–16430. [CrossRef]
172. Sun, W.Y.; Yang, K.J.-S.; Lin, N.-H. Numerical Simulations of Asian Dust-Aerosols and Regional Impacts on Weather and Climate-Part I: Control Case-PRCM Simulation without Dust-Aerosols. *Aerosol Air Qual. Res.* **2013**, *13*, 1630–1640. [CrossRef]
173. Sun, W.Y.; Yang, K.J.-S.; Lin, N.-H. Numerical Simulations of Asian Dust-Aerosols and Regional Impacts on Weather and Climate-Part II: PRCM-Dust Model Simulation. *Aerosol Air Qual. Res.* **2013**, *13*, 1641–1654. [CrossRef]
174. Koepke, P.; Hess, M.; Schult, I.; Shettle, E.P. *Global Aerosol Data Set*; Tech. Rcp. 243; Max Planck Inst. for Meteorol: Hamburg, Germany, 1997; 44p.
175. Myhre, G.; Stordal, F. Global Sensitivity Experiments of the Radiative Forcing due to Mineral Aerosols. *J. Geophys. Res.* **2001**, *106*, 18193–18204. [CrossRef]
176. Husar, R.B.; Tratt, D.M.; Schichtel, B.A.; Falke, S.R.; Li, F.; Jaffe, D.; Gasso, S.; Gill, T.; Laulainen, N.S.; Lu, F.; et al. Asian Dust Events of April 1998. *J. Geophys. Res.* **2001**, *106*, 18317–18330. [CrossRef]
177. In, H.J.; Park, S.U. A Simulation of Long-range Transport of Yellow Sand Observed in April 1998 in Korea. *Atmos. Environ.* **2002**, *36*, 4173–4187. [CrossRef]
178. Wu, C.C.; Yu, Y.C.; Hsu, W.R.; Hsu, K.J.; Sun, W.Y. Numerical Study on the Wind Fields and Atmospheric Transports of a Typical Winter Case in Taiwan and Surrounding Area. *Atmos. Sci.* **2003**, *31*, 29–54. (In Chinese)
179. Sun, W.Y. *Uncertainties in Climate Models*, 10th ed.; McGraw-Hill Encyclopedia of Sciences & Technology: New York, NY, USA, 2006.
180. Hsu, W.R.; Sun, W.Y. Numerical study of mesoscale cellular convection. *Bound.-Layer Meteor.* **1991**, *57*, 167–186. [CrossRef]
181. Hsu, W.-R.; Sun, W.Y. Reply to "Comment on the numerical study of mesoscale cellular convection". *Bound.-Layer Meteorol.* **1992**, *60*, 405–406. [CrossRef]
182. Laprise, R. The Euler equations of motion with hydrostatic pressure as an independent variable. *Mon. Wea. Rev.* **1992**, *120*, 197–207. [CrossRef]
183. Yeh, K.S.; Cote, J.; Gravel, S.; Methot, A.; Patoine, A.; Roch, M.; Staniforth, A. The CMC–MRB Global Environmental Multiscale (GEM) Model. Part III: Nonhydrostatic Formulation. *Mon. Wea. Rev.* **2002**, *130*, 339–356. [CrossRef]
184. Janjic, Z.I. Nonhydrostatic model based on a new approach. *Meteorol. Atmos. Phys.* **2003**, *82*, 271–285. [CrossRef]
185. Chen, S.H.; Sun, W.Y. The applications of the multigrid method and a flexible hybrid coordinate in a nonhydrostatic model. *Mon. Wea. Rev.* **2001**, *129*, 2660–2676. [CrossRef]
186. Ogura, Y.; Phillips, N.A. Scale analysis of deep and shallow convection in the atmosphere. *J. Atmos. Sci.* **1962**, *19*, 173–179. [CrossRef]

187. MacDonald, A.E.; Lee, J.L.; Sun, S. QNH: Design and test of a quasi-nonhydrostatic model for mesoscale weather prediction. *Mon. Wea. Rev.* **2000**, *128*, 1016–1036. [CrossRef]
188. Klemp, J.; Wiehelmson, R. The Simulation of Three-Dimensional Convective Storm Dynamics. *J. Atmos. Sci.* **1978**, *35*, 1070–1096. [CrossRef]
189. Sun, W.Y.; Hsu, W.R. Effect of surface friction on downslope wind and mountain waves. *Terr. Atmos. Ocean. Sci.* **2005**, *16*, 393–418. [CrossRef]
190. Hsu, W.R.; Sun, W.Y. A time-split, forward-backward numerical model for solving a nonhydrostatic and compressible system of equations. *Tellus* **2001**, *53*, 279–299. [CrossRef]
191. Hsu, H.H.; Yu, Y.C.; Kau, W.S.; Hsu, W.R.; Sun, W.Y. Simulation of the 1998 East Asian summer monsoon using Purdue regional model. *J. Meteor. Soc. Jpn.* **2004**, *82*, 1715–1733. [CrossRef]
192. Sun, W.Y.; Hsu, W.R.; Chern, J.-D.; Chen, S.H.; Yang, K.J.-S.; Yeh, K.; Bosilovich, M.G.; Haines, P.A.; Min, K.-H.; Oh, T.-J.; et al. Purdue atmospheric models and applications. In *Recent Progress in Atmospheric Sciences: Applications to the Asia-Pacific Region*; Liao, K.N., Chou, M.D., Eds.; World Scientific Publishing: Singapore, 2009; 496p, pp. 200–230.
193. Hsieh, M.N. A Monotone Semi-Lagrangian Advection Scheme in a 3D Nonhydrostatic Model and Applications in Squall Line Simulations. Ph.D. Thesis, NTU, Taipei, Taiwan, 2006; 146p. (In Chinese).
194. Hsieh, M.-E.; Hsu, W.R.; Sun, W.Y. Applications of a three-dimensional nonhydrostatic atmospheric model on uniform flows over an idealized mountain. In Proceedings of the 2010 Computational Fluid Dynamics Taiwan, Chung-Li, Taiwan, 29–31 July 2010; pp. 1–12.
195. Tsuboki, K.; Sakakibara, A. Numerical Prediction of High-Impact Weather Systems. In *The Seventeenth IHP Training Course (International Hydrological Program)*; Nagoya University and UNESCO HYARC: Nagoya, Japan, 2007; 273p.
196. Sun, W.Y.; Sun, O.M.T.; Tsuboki, K. A Modified Atmospheric Nonhydrostatic Model on Low Aspect Ratio Grids. *Tellus A* **2012**, *64*, 17516. [CrossRef]
197. Sun, W.Y.; OSun, M.; Tsuboki, K. A Modified Atmospheric Nonhydrostatic Model on Low Aspect Ratio Grids-Part II. *Tellus A* **2013**, *65*, 19681. [CrossRef]
198. Kasahara, A. Various vertical coordinate systems used for numerical weather prediction. *Mon. Wea. Rev.* **1974**, *102*, 509–522. [CrossRef]
199. Sun, W.Y. Numerical simulation of a planetary boundary layer: Part I. Cloud free case. *Beitr. Phys. Atmos.* **1993**, *66*, 3–16.
200. Sun, W.Y. An Efficient Forward Semi-Lagrangian Model. *Open J. Fluid Dyn. (OJFD)* **2023**. accepted.
201. Doswell, C.A., III. A kinematic analysis associated with a nondivergent vortex. *J. Atmos. Sci.* **1984**, *41*, 1242–1248. [CrossRef]
202. Sun, W.Y.; Sun, O.M.T. Revisiting the parcel method and CAPE. *Dyn. Atmos. Ocean.* **2019**, *86*, 134–152. [CrossRef]
203. Queney, P. The problem of airflow over mountains: A summary of theoretical studies. *Bull. Am. Meteor. Soc.* **1948**, *29*, 16–26. [CrossRef]
204. Queney, P.; Corby, G.A.; Gerbier, N.; Koschmieder, H.; Zierep, J. The airflow over mountains. *WMO Tech.* **1960**, *34*, 135.
205. Long, R.R. Some aspects of the flow of stratified fluids. Part I. A theoretical investigation. *Tellus* **1953**, *5*, 42–58. [CrossRef]
206. Long, R.R. A laboratory model resembling the Bishop-waves phenomena. *Bull. Am. Meteorol. Soc.* **1953**, *34*, 205–211. [CrossRef]
207. Smith, R.B. On severe downslope winds. *J. Atmos. Sci.* **1985**, *42*, 2597–2603. [CrossRef]
208. Clark, T.L.; Peltier, W.R. Critical level reflection and the resonant growth of nonlinear mountain waves. *J. Atmos. Sci.* **1984**, *41*, 3122–3134. [CrossRef]
209. Peltier, W.R.; Clark, T.L. The evolution and stability of finite-amplitude mountain waves. Part II: Surface drag and severe downslope windstorms. *J. Atmos. Sci.* **1979**, *36*, 1498–1529. [CrossRef]
210. Doyle, J.D.; Durran, D.R.; Chen, C.; Colle, B.A.; Georgelin, M.; Grubisic, V.; Hsu, W.R.; Huang, C.Y.; Landau, D.; Lin, Y.L.; et al. An intercomparison of model predicted wave breaking for the 11 January 1972 Boulder windstorm. *Mon. Wea. Rev.* **2000**, *128*, 901–914. [CrossRef]
211. Sun, W.Y. Numerical Study of Severe Downslope Storm. *Weather Clim. Extrem.* **2013**, *2*, 22–30. [CrossRef]
212. Lilly, O.K. A severe downslope windstorm and aircraft turbulence induced by a mountain wave. *J. Atmos. Sci.* **1978**, *35*, 59–77. [CrossRef]
213. Moncrieff, M.W.; Green, J.S.A. The propagation and transfer properties of steady convective overturning in shear. *Q. J. R. Meteorol. Soc.* **1972**, *98*, 336–352. [CrossRef]
214. Riehl, H.; Malkus, J.S. On the heat balance in the equatorial trough zone. *Gephysica* **1958**, *6*, 503–538.
215. Ebert, E.; Holland, G.J. Observations of record cold cloud-top temperatures in tropical cyclone Hilda (1990). *Mon. Wea. Rev.* **1992**, *120*, 2240–2251. [CrossRef]
216. Black, R.A.; Bluestein, H.B.; Black, M.L. Unusually strong vertical motions in a Caribbean hurricane. *Mon. Wea. Rev.* **1994**, *122*, 2722–3739. [CrossRef]
217. Sun, W.Y.; Chang, C.Z. Diffusion model for a convective layer: Part II: Plume released from a continuous point source. *J. Clim. Appl. Meteor.* **1986**, *25*, 1454–1463. [CrossRef]
218. Sun, W.Y. Air pollution in a convective atmosphere. In *Library of Environmental Control Technology*; Cheremisinoff, Ed.; Gulf Publishing: Houston, TX, USA, 1988; pp. 515–546.
219. Sun, W.Y. Diffusion modeling in a convective boundary layer. *Atmos. Environ.* **1989**, *23*, 1205–1217.

220. Haines, P.A.; Sun, W.Y.; Chen, S.H.; Hsu, W.R.; Hsieh, M.E. NTU/PU model simulations and observed flow over mountain. *Terr. Atmos. Ocean. Sci.* **2019**, *30*, 171–184. [CrossRef]
221. Gal-Chen, T.; Somerville, R.C.J. Numerical solution of the Navier-Stokes equations with topography. *J. Comput. Phys.* **1975**, *17*, 276–309. [CrossRef]
222. Haines, P.A.; Grove, D.J.; Sun, W.Y.; Hsu, W.R. Observations and Numerical Modeling of Sub and Super-critical Flow at White Sands Missile Range. In Proceedings of the 12th Conference on Mountain Meteorology, the American Meteorological Society, Santa Fe, NM, USA, 28 August–2 September 2006.
223. Sun, W.Y. Coordinates over complex terrain in atmospheric model. *J. Atmos. Sci. Res.* **2021**, *4*, 3–49. [CrossRef]
224. Pacanowski, R.C. *MOM 2 Documentation User's Guide and Reference Manual Version 1.0*; GFDL Ocean Technical Report #3; GFDL, NOAA: Princeton, NJ, USA, 1995; 233p.
225. Saito, K.; Kato, K.T.; Eito, H.; Muroi, C. *Numerical Prediction Division Unified Nonhydrostatic Model*; MRI Report 42; Documentation of the Meteorological Research Institute: Singapore, 2001; 133p.
226. Staniforth, A.; White, W.; Wood, N.; Thuburn, J.; Zerroukat, M.; Corderro, E. *Unified Model Documentation Paper. No. 15. Joy of U.M. 6.0-Model Formulation. Dynamics Research Numerical Weather Prediction*; Met Office: Devan, UK, 2004.
227. Lin, M.-Y.; Sun, W.Y.; Chiou, M.-D.; Chen, C.-Y.; Cheng, H.-Y.; Chen, C.-H. Development and evaluation of a storm surge warning system in Taiwan. *Ocean. Dyn.* **2018**, *68*, 1025–1049. [CrossRef]
228. Pielke, R.A.; Cotton, W.R.; Walko, R.L.; Tremback, C.J.; Lyons, W.A.; Grasso, L.D.; Nicholls, M.E.; Moran, M.D.; Wesley, D.A.; Lee, T.J.; et al. A comprehensive meteorological modeling system-RAMS. *Meteor. Atmos. Phys.* **1992**, *49*, 69–91. [CrossRef]
229. Kapitza, H.; Eppel, D.P. The non-hydrostatic mesoscale GESIMA. Part I: Dynamical equations and tests. *Beitr. Phys. Atmos.* **1992**, *65*, 129–145.
230. Skamarock, W.C.; Klemp, J.B.; Dudhia, J.; Gill, D.O.; Barker, D.M.; Duda, M.G.; Huang, X.-Y.; Wang, W.; Powers, J.G. *A Description of Advanced Research WRF Version 3*; NCAR/TN–475+STR; NCAR: Boulder, CO, USA, 2008; 113p.
231. MacCall, B.-T.; Wang, Y.; Sun, W.Y. *A New Semi-Implicit Time Integration Scheme for the Time-Dependent Atmospheric Boundary Layer Environment (ABLE) Model*; ARL-TR-7428; ARL: Aberdeen Proving Ground, MD, USA, 2015.
232. Sun, W.Y. Numerical simulation of a planetary boundary layer: Part II. Cloudy case. *Beitr. Phys. Atmos.* **1993**, *66*, 17–30.
233. Deardorff, J.W.; Willis, G.E. A parameterization of diffusion into the mixed layer. *J. Appl. Met.* **1975**, *14*, 1451–1458. [CrossRef]
234. Willis, G.E.; Deardorff, J.W. A laboratory study of dispersion from an elevated source within a modeled convective planetary boundary layer. *Atmos. Environ.* **1978**, *12*, 1305–1311. [CrossRef]
235. Willis, G.E.; Deardorff, J.W. A laboratory study of dispersion from a source in the middle of the convective mixed layer. *Atmos. Environ.* **1981**, *15*, 109–117. [CrossRef]
236. Sun, W.Y.; Sun, O.M.T. Backward integration of diffusion equation. *Aerosol Air Qual. Res.* **2017**, *17*, 278–289. [CrossRef]
237. Arnold, V.I.; Avez, A. *Problèmes Ergodiques de la Mécanique Classique*; Gauthier-Villars: Paris, France, 1967. (In French)
238. Strogatz, S.H. *Nonlinear Dynamics and Chaos. With Applications to Physics, Biology, Chemistry, and Engineering*; Westpress View: Boulder, CO, USA, 2015; 513p.
239. Lyapunov, A.M. Problème général de la stabilité du mouvement. *Ann. Fac. Sci. Toulouse Mathématiques Série 2* **1947**, *9*, 203–474.
240. Lorenz, E.N. Predictability: Does the flap of a butterfly's wings in Brazil set off a tornado in Texas? In Proceedings of the 139th Meeting of AAAS Section on Environmental Sciences, New Approaches to Global Weather: GARP, AAAS, Cambridge, MA, USA, 29 December 1972. 5p.
241. Lorenz, E.N. Atmospheric predictability experiments with a large numerical model. *Tellus* **1982**, *34*, 505–513. [CrossRef]
242. Sun, W.Y.; Hsu, W.R.; Haines, P.K.; Hsieh, M.E.; Chen, S.H.; Grove, D. Numerical simulations over mountains. In Proceedings of the Third International Workshop on Next- Generation NWP Models: Bridging Parameterization, Explicit Clouds and Large Eddies, Jeju Island, Republic of Korea, 29 August–1 September 2010.
243. Kreiss, H.-O.; Lorenz, J. *Initial-Boundary Value Problems and the Navier-Stokes Equations*; SIAM, Publications Library: Philadelphia, PA, USA, 2004; 488p.

MDPI

St. Alban-Anlage 66

4052 Basel

Switzerland

www.mdpi.com

Atmosphere Editorial Office

E-mail: atmosphere@mdpi.com

www.mdpi.com/journal/atmosphere